Electrical
Measuring Instruments
and
Measurements

Electrical Measuring Instruments and Measurements

DR. S.C. BHARGAVA

Retired Specialist (Electromagnetic Phenomena),
BHEL, Corporate R & D, Hyderabad

and

formerly

Asst. Professor, Dept. of EE, BITS, Pilani,
Professor & Head, Dept. of EEE, Mahatma Gandhi Institute of Technology,
and
Professor, Dept. of EEE, Sreenidhi Institute of Science and Technology,
Hyderabad

BSP BS Publications

 CRC Press
Taylor & Francis Group

A BALKEMA BOOK

Distributed in India, Pakistan, Nepal, Myanmar (Burma), Bhutan, Bangladesh and Sri Lanka by **BS Publications**

Distributed in the rest of the world by
CRC Press/Balkema, Taylor & Francis Group, an **informa** business
Schipholweg 107C
2300 AK Leiden, The Netherlands
www.crcpress.com
www.taylorandfrancis.com

First published in India by
BS Publications, 2013

ISBN: 978-0-415-62151-9

Printed in India by
Sanat Printers,
Kundli, Haryana, India

Published by
BS Publications
A Unit of BSP Books Pvt. Ltd.
4-4-309, Giriraj Lane, Sultan Bazar, Hyderabad - 500 095 (A.P.)
India.

"It is a capital mistake, to theorize before one has data. Insensibly one begins to twist facts to suit theories, instead of theories to suit facts".

Sir Arthur Conan Doyle
(1859-1930)

Dedicated to

the memory of my parents

Preface

The subject of electrical measurements, with the help of a variety of general-purpose and special instruments, devices and equipment forms an important part of electrical engineering education, perhaps more important than other subjects, for obvious reasons. This must therefore be studied in detail with full comprehension by students of electrical engineering and must be followed with the utmost care and precision by practicing engineers associated with manufacturing and testing of electric machinery and with maintenance of power systems and utility.

However, over the past few decades, the course of electrical measurements has acquired the dubious status of just one of several courses hurriedly taught during the undergraduate study (B.Tech. or B.E.), usually in the third year, to be "covered" in just one semester in contrast to what was earlier taught during the entire session or year. Even in the laboratory classes which invariably follow the theory course in the preceding semester, there is little scope for improvement since the experiments are generally conducted as a routine without any enthusiasm, usually in ill-equipped labs staffed by disinterested technicians.

For the teaching of the course, there have been books in the past on electrical measurements written mainly by authors from abroad, starting with the well-known classic on the subject by E. W. Golding, first published in 1933, as well as numerous others by authors like Harris, Baldwin and Buckingham and Price et al. Over the preceding years, quite a few books have been published by Indian authors, too, mostly aimed at 'covering the syllabi' on the subject as prescribed by the various universities across the country. Quite a few of these books contain, by and large, what constitutes the typical "textbook" material meant to be a quick guide for the students to prepare for their exams. To do full justice to the subject, therefore, it is the *quantitative* knowledge and experience acquired through good reading and sound study, followed by properly organised and conducted experiments, that forms the basis of all electrical measurements, later to help in the design, development, manufacture and usage of various electrical equipment. It is ironic that the subject of measurement, signified by two important aspects, viz., accuracy and errors, is treated rather casually and with a lack of seriousness even by practicing engineers who might regard the job of testing of machinery as routine to the detriment of, for example, the utility and industrial consumers.

Hence, the need for a *comprehensive* book on electrical measurements for **students as well as practicing engineers** had long been felt by this author. The present book, based on vast experience of the author in teaching the subject at different colleges and decades of practical and industrial research, whilst covering the syllabi of most universities, is characterised by the following special features:

(a) Inclusion of several important topics such as "Units, Dimensions and Standards", "Magnetism, Electricity and Electromagnetism", "Electrical Circuit Analysis" and "Visual Display and Analyses" which may not usually be included in a 'standard' textbook with a view to 'covering' the syllabi, and yet are significant for the grasp and application of the subject.

(b) Much enlarged chapters on "Magnetic Measurements" – a subject often neglected and not studied with due seriousness, but vital for the design, development and analyses of all electrical machines and equipment – and "Measurement of Non-electrical Quantities", particularly involving electrical means for medical scanning and imaging. The former is enriched with input from exclusive research experience of the author obtained first during his doctoral work at the University of Aston in Birmingham (the thesis being primarily based on experimental research involving magnetic fields) and later during his 25 years spent at Corporate R & D of BHEL, Hyderabad.

(c) A unique style of presentation of material in each chapter in a "reader-friendly" language, beginning with a "recall" to link the chapter subject with the basic science or phenomena, explanatory footnotes where necessary and comments, in addition to clear figures, diagrams and illustrations, in colour where required.

(d) A number of worked examples, exercises (with answers) and quiz questions at the end of each chapter to enhance the understanding of the subject matter.

(e) Various appendices at the end of the book dealing with many associated aspects of the topics dealt within the chapters, particularly those on organisation of experiments in the lab, so important for the conduct of successful experiments, and instrumentation based on experimental research experience of the author.

(f) Where possible, inclusion of material related to development and usage of solid-state instruments, a typical case being that of energy meters.

(g) An elaborate subject and authors' index to help the reader to locate the matter easily in the book.

(h) References (quire a few being research publications by the author) dealing with various aspects of measurement, and bibliography.

(i) Chapter-by-chapter list of symbols and notation for easy reference.

The book necessarily involves appropriate mathematical treatment in several chapters. However, this has been kept as simple as possible so as to be followed easily by the readers, the emphasis being on elucidating the various concepts of measurements.

It is hoped that the readers of the book – especially the students, faculty and practicing engineers – will derive full benefit from this unique publication, describing a number of measurement aspects not easily available elsewhere.

- The Author

Acknowledgements

The author would like to express his sincere thanks to

- his supervisor, Dr. M.J. Jevons, for the doctoral work at the University of Aston, UK, for initiating the author in experimental research incorporating magnetic fields, the staff of the department of electrical engineering and (the late) Prof. E.J. Davies for freely providing all research facilities and constant encouragement during his research;

- his several colleagues and staff in industry, particularly the BHEL at the Corporate Research and Development, Hyderabad, and other units (at Bhopal, Hardwar, R C Purarm and Bangalore), for their help and support for carrying out the extensive experimental work and analyses, and in general for providing all facilities – some of them *very* special – for decades of experimental research;

- my students whom I taught over the years and whose feedback by seeking clarification made me revise many aspects of the subject in a way that contributed to the improved presentation of material in the book;

- all others, including my family members, whose help and support in more ways than one made this book possible.

About the Author

S. C. Bhargava, a gold medalist of University of Roorkee in 1966, obtained his Ph.D. from the University of Aston in Birmingham, UK, in 1972. His research for the doctorate was based heavily on experimental studies and analyses involving electromagnetic fields. He has had a chequered professional career spanning nearly 50 years, mainly in teaching at various engineering institutions and universities and industrial research.

Dr. Bhargava worked as Specialist (electromagnetic phenomena) at Corporate R & D of Bharat Heavy Eelectricals Limited (BHEL) – a Govt. of India Undertaking - since 1974 from where he retired in 1999. His span of activities at BHEL involved dealing with the analyses of various electrical machines, in particular large turboalternators to analyse their operation under unbalanced conditions. Whilst at the R & D, he was also instrumental in establishing a unique "electromagnetic phenomena lab", the only one of its kind in India, for experimental research involving electromagnetic fields. His work at the R & D dealt extensively with the experimental studies of actual machines and, later, with a specially designed and fabricated model turbogenerator.

He has published nearly 50 research/technical papers on varied topics in national and international journals, many of these having been presented at conferences first. His current field of research is "Biomedical Effects of Power Frequency Electromagnetic Fields on Living Beings" on which he has also published and co-authored several papers.

Dr. Bhargava is a Fellow of Institution of Engineers (India), Instituion of Electrical Engineers (UK) and Institution of Electronics and Telecommunications Engineers, and a Senior Member of Computer Society of India. He is also Life Senior of the Institute of Electrical and Electronics Engineers (USA) with which he has been very actively associated since 1980, having been the Hyderabad Section Chair in 1991 and 1992, and having organised various national and international conventions and conferences.

His latest interest is designing and developing solar electric modules and distribution systems.

Contents

List of Symbols and Notation

[General, other than as specified or defined at the point of use]

Chapter I

A	ampere, area
a	acceleration
B	flux density
C	capacitance, coulomb
c	velocity of light
cm	centimetre
D	distance, electric flux density
d	distance
E, e	electromotive force (EMF), energy
F	force
g	acceleration due to gravity
H	magnetising field or force
I, i	current
J	joule
J	imaginary factor ($= \sqrt{-1}$)
K, k	constants (in general)
kWh	energy (kilowatt hour)
L	distance/length
L	self inductance
M	mass, mutual inductance
m	mass, metre, (magnetic) pole strength
mm	millimetre
N	newton, number of turns
P	power
PD	potential difference
ppm	parts per million
Q	charge

R	resistance
r	distance, radius
s	second
t	time
T	time, torque
v	velocity
V	EMF, volt, volume
W	watt, work
w	work
Z	impedance
$\dfrac{d}{dt}$	time derivative

M	metre
K	kilogramme
S	second
T	time

in MKS system of units

α	acceleration
ε	permittivity
ε_o	absolute permittivity/permittivity of free space (= 8.854 x 10^{-12})
ε_r	relative permittivity
$\in$	electric field strength
$\mathcal{F}$	magnetomotive force
Ω	ohm
π	constant (= 3.1415927 . . .)
ω	angular frequency/velocity (=2 π r)
μ	permeability
μ_o	absolute permeability/permeability of free space (4 π x 10^{-7})
μ_r	relative permeability
Ø	flux
Ψ	electric flux
σ	surface charge density

θ angle (as defined/specified)

$\dfrac{d}{d\theta}$ "angle" derivative

Chapter II

f   frequency

G   (potential) gradient

H   intensity of magnetic field

J   intensity of magnetisation, current density

L, l, l length

m   pole strength

M   moment, mutual inductance

Q, q   charge

S   reluctance

α, β, γ temperature coefficients

x   (variable) distance

Δ increment

$\in$ base of natural logarithm

Chapter III

a   operator : $-0.5 + j\,0.866$

a^2 operator : $-0.5 - j\,0.866$

a,b   sides of a coil

B   susceptance

D   detector

G   conductance

n   angular speed (rpm or rps)

v   peripheral velocity, instantaneous value of voltage

X   reactance

Y   admittance

α	angle
℧	mho
φ	phase angle, phase displacement
θ	angular 'distance', rotational angle

Chapter IV

| e | charge on an electron |
| T | time period |

Chapter V

b	breadth
d	distance
G	galvanometer constant
J	moment of inertia
w	width
μ	micro [$= 10^{-6}$]
ρ	resistivity

Chapter VI

M	mutual inductance
n	turns ratio
α, δ, θ	phase angles
Δ	area, phase angle, small increment, triangle

Chapter VII

CC	current coil
°E	degree(s) electrical
H	heat energy
K	constant
PC	potential (or pressure) coil
p	instantaneous power

Q	reactive power
r	(internal) resistance
T	torque
Y	"star"-connected (load/network)
β	phase angle
δ	increment
Δ	"delta"-connected (load/network)
φ	angle of impedance/pf angle
π	constant

Chapter VIII

e	induced EMF
i	induced (eddy) current
F	force
K, K'	constants
N	rotational speed (rpm or rps)
p	instantaneous power
T	torque
R, Y, B	phases of 3-phase supply, reckoned anti-clockwise
α, β	phase angles (of lag)
Δ	angle
Ø	flux
φ	phase angle
θ	angular deflection ($=\omega t$)

Chapter IX

G	galvanometer
T	transformer
VG	vibration galvanometer

Chapter XI

Det	detector
δ	"loss" angle (of a capacitor), skin depth [tan δ : loss factor]

Chapter XII

A	magnetic vector potential
BG	ballistic galvanometer
J	induced current density
RS	reversing switch
σ	electrical conductivity
τ	time, time period
^	symbol for amplitude

Chapter XIII

c	specific heat
cp	candle power (unit of intensity of light)
db	decibel (unit of sound)
Q	generated heat
T	temperature
κ	thermal conductivity
ω	solid angle
ρ	density

Appendices

I

cd	candela (unit of luminous intensity)
Hz	hertz (unit of frequency)
lm	lumen (unit of luminous flux)
lx	lux (unit of illumination)
K	degree Kelvin (absolute temperature)
Pa	pascal (unit of pressure)
sr	steradian (unit of solid angle)

II

dl	incremental length
R	radius
s	distance
⊗	current "in"
⊙	current "out"

VI

A	magnetic vector potential
D	displacement current density
E	electric field intensity
J	current density
k	constant
q	heat input
$\dfrac{\partial}{\partial t}$	partial (time) derivative
$\dfrac{\partial}{\partial x}$	partial (displacement) derivative
∇	"del" operator
σ	electrical conductivity

VII

$\oint$	"closed" integral

I : Units, Dimensions and Standards

I

UNITS, DIMENSIONS AND STANDARDS

RECALL

Units have been associated with physical quantities from the ancient times, inadvertently or with full awareness by way of trade or exchange. They have had special significance in engineering, increasingly so with the continual advances and developments of technology. Thus, whilst in the beginning units generally pertained to simple entities such as mass, distance and assorted articles, as well as time, there are now more important physical quantities like force, pressure, current, potential and so on which involve a variety of units associated with them.

The process of measurement of a quantity – electrical or non-electrical – in particular cannot be considered complete, however elaborately and accurately the measurement might have been carried out, unless associated with the *proper* unit. Accordingly, the result of measurement of a physical quantity must be expressed *both* in terms of quantum and a unit.

For example:

10 metres

15 kilograms

25 seconds

230 volts

5 amperes

100 ohms

etc.

Here, the second part of each expression refers to a particular unit, *in a given system*, the first being the magnitude, pertaining to a physical quantity; that is, distance, mass, time, potential difference, current and resistance, respectively.

FUNDAMENTAL AND DERIVED UNITS

Fundamental or absolute units may be defined as those in which the various other units may be expressed, either in whole or small number or fraction of the fundamental units. The word "absolute" in this sense does not necessarily imply supremacy and extreme accuracy; rather it is used as opposed to "relative."

The Committee of the British Association of Electrical Units and Standards which met in 1863 came with the decision that even electrical units should be defined by some *natural* law that expresses the relation between the quantity concerned and the fundamental quantities of length, mass and time.

The units that relate to these fundamental quantities are known as fundamental units[1].

Each one of these units itself may be indentified in terms of mile, kilometre, metre; gramme, pound, tonne, ton; and second, minute, hour, respectively, depending on the *system* of units used or the 'ease' of expression.

Units in Electrical Engineering

In electrical engineering, and measurements related to electricity and magnetism, two more aspects related to units in use, in addition to the fundamental units (of length, mass and time) must be considered depending on the properties oi the media in which the electrical and/or magnetic actions take place. These are known as *specific electric constant*, also called specific inductive capacitance or "permittivity of free space" [ε_0], and *magnetic space constant* or "permeability of free space" [μ_0].

DERIVED UNITS

CGS esu and emu Systems of Unit

The earliest system of units used to express nearly all quantities of electricity and magnetism was based on the use of centimetre[c], gramme[g] and second[s] as the fundamental units *and* either one or both of the above constants. Owing to the process involved in *deriving* the required units in terms of length, mass, time and/or ε_0 and μ_0, these units are called derived units and are generally related mathematically, usually in the form of ratios, to the fundamental units.

---

[1] As seen, these (fundamental) quantities represent the basic existence and motion of beings in the universe.

The CGS esu system

This system involves only the permittivity ε of the medium[1] as well as units of length, mass and time. In the basic system, evolved in the very beginning, the permittivity *as a whole* is taken as unity, or $\varepsilon = 1$. The acronym esu stands for "electrostatic units".

The CGS emu system

This system is based on the use of permeability, μ, as well as units of length, mass and time and is known as CGS "electromagnetic unit" system. The emu system is found to be more convenient from the point of view of most electrical measurements and hence has been more generally used than the esu system.

The MKS (or Giorgi) System of Units

The CGS systems of unit are reduced to only historic interest, being of not much practical use in the modern engineering practices due to the units being too "small" or too "large", or otherwise. For example, the magnetic flux used to be expressed as maxwells (or the number of lines of force) and flux density by gauss (or number of lines of force per square centimetre)[2]. Over the decades, therefore, systems of practical applications of units have been evolved, consistent with engineering practices.

The system of practical units in vogue is called the MKS system, or rather the "rationalised MKS system" of units.

The system was originally suggested by Prof. G. Giorgi in 1901. The characteristic feature of the system is that the units of length, mass and time are expressed in terms of metre [M], kilogramme [K] and second [S], respectively, and hence the acronym MKS. The system was finally adopted by the International Electrotechnical Commission (the IEC) at its meeting in 1938 at Torquay.

The MKS system is an "absolute" system of units and has the advantage that a single set of units covers all electrical and magnetic quantities, and is applicable both to electromagnetic and electrostatic effects. It differs from the historic CGS system in the expression of values of the "permeability and permittivity of free space". Thus, μ_0 the absolute permeability is assigned the value 10^{-7} and, using the relation

---

[1] ε being the product of ε_0 and ε_r, the *relative* permittivity, depending on the medium, or $\varepsilon = \varepsilon_0 \, \varepsilon_r$.

[2] The modern, practical units of these quantities being weber ($= 10^8$ maxwell) and tesla ($= 10^4$ gauss), respectively.

$$\frac{1}{\sqrt{\mu_0 \varepsilon_0}} = c = 3 \times 10^8 \, \text{m/s (the velocity of light)}$$

the absolute permittivity ε_0 is obtained to be $\varepsilon_0 = 1.113 \times 10^{-10}$.

As an example, some of the common practical mechanical and electrical units in the MKS system and their symbols are

Area, A	:	square metre, m^2
		[*note that the unit for area may best be written sq.m.*]
Volume, V	:	cubic metre, m^3
Velocity, v	:	metre per second, m/s
Acceleration, a	:	metre per second per second, m/s^2
Force, F	:	newton, N
Work, W	:	newton-metre, Nm
Energy, E	:	joule, J
Current, I	:	ampere, A
Potential difference, PD	:	volt, V
Electric power, P	:	watt, W

Rationalised MKS System

Further considerations of the MKS system of units to be more applicable, esp. to *electrical* engineering practice, gave rise to "rationalised" MKS system of units, also known as SI system as the abbreviation for "System International d' Units" in French, and is now being followed universally. The main difference between the MKS and the rationalised system of units is in the values assigned to the permeability and permittivity (of free space) as in the case of the difference between CGS and MKS systems. This is because the basis of the rationalised system is the conception of unit magnetic flux issuing from a unit magnetic pole, or unit electric flux issuing from a unit charge, instead of a flux of 4π as was assumed earlier. Thus, in the rationalised system of units, the permeability of free space, μ_0, is assigned the value $4\pi \times 10^{-7}$ (instead of simply equal to 10^{-7} in the un-rationalised MKS system). Correspondingly, by still using the relation

$$c = 1/\sqrt{\mu_0 \varepsilon_0}$$

the permittivity of free space is found to be

$$\varepsilon_0 = 8.854 \times 10^{-12}$$

(as against 1.113×10^{-10} in the un-rationalised system).

Definitions of Some Units in Rationalised MKS System

In the definitions given below, the standard symbol for the quantity under steady-state (DC or rms AC) and the name of each unit is followed by the symbol of the unit as indicated in parenthesis.

Unit *force* (F, *newton*, N) is the force which produces an acceleration of one metre per second per second to a mass of one kilogram.

Unit *work* or *energy* (W, *newton-metre* or *joule*, Nm or J) is the work or energy associated with a force of one newton when it acts through a distance of one metre.

Unit *power* (P, *watt*, W) is the rate of work done at one joule per second. (accordingly, the unit of energy, joule, can be specified as

1 joule = 1 watt-second).

Unit *current* (I, *ampere*, A) is the current which, flowing in a long, straight conductor at one metre distance (in vacuum) from a similar conductor carrying an equal current, experiences a force of 2×10^{-7} N per metre length.

Unit *charge* (Q, *coulomb*, C) is the charge, or quantity of electricity, which passes in one second through any cross-section of a conductor through which a current of one ampere is flowing.

Unit *potential difference* (PD or V, *volt*, V) exists between two points if one joule of work is done in transferring one coulomb from one point to the other.

[The unit of *electromotive force* (E) is also the volt].

Unit *resistance* (R, *ohm*, Ω) is a resistance such that a potential difference of one volt exists across it when a current of one ampere is flowing.

Unit *inductance* (L or M, *henry*, H) is an inductance such that a rate of change of one ampere per second induces an EMF of one volt[1].

Unit *capacitance* (C, *farad*, F) is a capacitance such that a charge of one coulomb results in a potential difference of one volt.

Unit *magnetic flux* (Ø, *weber*, Wb) is the flux such that when its linkage with a single turn is removed in one second, an EMF of one volt is induced in the turn.

Unit *magnetic flux density* (B, *tesla*, T) at a point exists when the flux per unit area over a small surface perpendicular to the direction of flux, surrounding the point, is one weber per square metre.

---

[1]The definition applies to *self inductance* (L) if the changing current and induced EMF are in the *same* circuit, or to *mutual inductance* (M) if the current is in one circuit and the induced EMF is in the other.

Unit *magneto motive force* ($\mathcal{F}$, *ampere*, I) is that associated with a current of one ampere flowing in a single turn.

Unit *magnetising force* (H, *ampere per metre*, A/m) is an MMF gradient of one ampere per metre of (magnetic) flux-path length.

Unit *electric flux* (ψ, *coulomb*, C) is the flux emanating from a charge of one coulomb.

Unit *electric flux density* (D, *coulomb per square metre*, C/m^2) at a point exists when the electric flux per unit area over a small surface perpendicular to the direction of flux surrounding the point is unity.

Unit *electric field strength* ($\mathcal{E}$, *volt per metre*, V/m) is the electric potential gradient of one volt per metre of (electric) flux-path length.

Multiple and sub-multiple units

For convenience in practice, multiples and sub-multiples of the above units are frequently used. The multiplying factor in such cases is usually a *power of ten* with the corresponding standard prefixes being milli, Mega (or simply Meg) etc. In addition, for certain purposes, other multiples may be convenient; for example, watt-hour, kilowatt-hour, ampere-hour for units of energy and quantity of electricity, respectively.

The various MKS units in common use are summarised in Table 1.1.

Table 1.1 : Some practical units and symbols : Rationalized MKS system[1]

Quantity	Symbol	Unit	Remarks
Length	L (or l, *l*)	metre, m	Fractional units: millimetre(mm); centimetre(cm) etc. Multiples : kilometre(km) etc.
Mass	M (or m)	kilogram, kg	Fractional units: milligram(mg); gram(gm) or simply (g) etc.
Time	s	second, s	Fractional units: micro-second(μs); millisecond(ms) Multiples : minute, hour etc.
Force	F	newton, N	Also, m × g when m is mass in kg and g acceleration due to gravity in m/s/s (= 9.81)
Distance	d (or D)	metre, m	See "length"
Acceleration	α (or a)	metre/s/s, m/s^2	
Work	w (or W)	newton-metre, Nm	
Torque	T	newton-metre, Nm	Same unit as work as both quantities are dimensionally identical

---

[1]See Appendix I for more details about various units and their conversions.

Table 1.1 *Contd*.....

Quantity	Symbol	Unit	Remarks
Power	W	watt, W	Fractional units : milliwatt (mW) etc. Multiples: kilowatt (kW); Megawatt (MW) etc. Can also be expressed in terms of (the old unit) hp where 1 hp = 746 W
Energy	E	joule, J	Also, W-s
Current	I	ampere, A	Fractional units : milliamp(mA) etc. Multiples : kiloamp(kA) etc.
Charge	Q	coulomb, C	Also, electric flux (symbol : ψ)
Charge surface density	σ	coulomb per metre^2, C/m^2	
Potential difference	V	joule per coulomb, J/C; volt, V	Also, electromotive force (EMF)
Electric field strength	$\in$	volt per metre, V/m	
Electric flux density	D	coulomb per metre^2, C/m^2	
Capacitance	C	coulomb per volt, C/V or farad, F	A very large unit; common practical unit : micro-farad (μF) Very small fraction: pico-farad (pF)
Permittivity	ε	farad per metre, F/m	Actually refers to unit of permittivity of free space ε_0. Relative permittivity of media, ε_r is a constant.
Resistance	R	volt per ampere,	Fractional units : milliohm (mΩ), micro ohm $\left(\mu\Omega\right)$ etc.
		V/A or ohm (Ω)	Multiples : kilo ohm (kΩ), Megaohm (MΩ) etc.
Magnetic flux	$\emptyset$	weber, Wb or volt-sec	Equal to 10^8 maxwell
Magnetic flux density or magnetic induction	B	tesla, T	Also, weber per metre^2; equal to 10^4 gauss Fractional unit millitesla (mT) etc.
Magneto-motive force	$\mathcal{F}$	ampere, A	In an N turns coil, equal to NI, I the current in each turn
Magnetic field strength	H	ampere per metre, A/m	Also called magnetising force
Inductance	L	weber per ampere, Wb/A or henry, H	Fractional units : millihenry (mH); microhenry (μH) etc.
Permeability	μ	henry per metre, H/m	Actually refers to unit of permeability of free space μ_0. Relative permeability of media, μ_r, is a constant

International Units

With the universal adoption of the rationalised MKS system of units, the following four units were defined as international or absolute units.

The international ohm

The ohm was chosen as the "primary" reference quantity. This is the resistance offered to the passage of an unvarying electric current (or DC) by a column of mercury at the temperature of melting ice, of mass equal to 14.4521 gm of uniform cross sectional area and of length equal to 106.30 cm.

The cross sectional area would work out to be very nearly 1 mm^2.

The international ampere

This is defined as the steady (or direct) electric current which when made to pass through a solution of silver nitrate in water would deposit the quantity of silver at the rate of 0.0011180 gm per second[1].

The international volt

This is the steady electric potential difference which, when applied across a conductor of resistance of one international ohm, would produce a current of one international ampere. (As would be expected, this is straightforward, derived from the definitions of the international resistance and current).

The international watt

Once again, following the definitions of the resistance and current, the international watt, as the basic reference of power, is defined as the electrical energy per second expanded when an unvarying (direct) current of one international ampere flows under a potential difference of one international volt.

In accordance with the above, the international units of charge (one coulomb), capacitance (one farad) and inductance (one henry) were later obtained as "derived" units.

---

[1]It is interesting that definitions of both, the international ohm and ampere, are based on natural phenomena : the former on physical properties of mercury and the latter on the process of electrolysis involving silver.

> **Comment**
>
> *Note that the names of all units are written in small-case letters: for example, metre, kilogram, second, ampere, volt, watt, hertz, newton, weber, tesla etc.*
>
> *The symbol of units are generally denoted by upper case letters in most cases, esp. when the name of the unit relates to a scientist: for example Hz (for hertz), N (for newton), A (for ampere) and so on. Symbols for others may be written in small-case: for example, m (for metre), s (for second), cd (for candela), lm (for lumen), lx (for lux) etc.*

DIMEMSIONS OF UNITS

Dimensions of units – of mechanical, electrical or other quantities – form an important aspect, especially when dealing with the derived units and expressions incorporating the same.

Units Pertaining to Mechanical Quantities

All mechanical quantities can be expressed in terms of the three (fundamental) quantities, viz., mass (denoted as [M]), length (denoted as [L]) and time (denoted as [T]) for "dimensional" representation – the basics to motion of bodies and in mechanics. Thus, dimensional "equations" are used to relate *any* chosen quantity to any two or all of the above three quantities.

For example, velocity of a body in motion is defined as the distance, or length, traversed per unit time. Or

$$\text{velocity} = \frac{\text{length}}{\text{time}}$$

This relation can be expressed in the dimensional notation as

$$[v] = \frac{[L]}{[T]}$$

where [v], [L] and [T] represent the dimensions of velocity, length and time, respectively. The square brackets indicate that the equality is dimensional only and *does not refer to numerical values.*

Writing the above expression as

$$[v] = [LT^{-1}]$$

gives the dimensions of velocity in "standard" form. Thus, velocity has the "dimensions" $[LT^{-1}]$, *independent of any system of units.*

Similarly,

$$\text{acceleration} = \frac{\text{velocity}}{\text{time}} = \frac{\text{distance or length}}{\text{time} \times \text{time}}$$

and, therefore, dimensionally

$$[a] = \frac{[L]}{[T^2]} = [LT^{-2}]$$

Likewise

$$force = mass \times acceleration$$

thus, representing the dimensions of mass by [M]

$$[F] = [M] \times [LT^{-2}] = [MLT^{-2}]$$

Accordingly, the dimensions of other *derived* units in mechanics can be obtained by first writing the expression of the quantity in terms of combination of length, mass and time as applicable and working out the final dimensions as above.

Dimensions in Electrostatic and Electromagnetic Systems

In electrostatics

To obtain the dimensions of *charge*, the Coulomb's Inverse Square Law can be applied, that is,

$$F = \frac{Q_1 \times Q_2}{\varepsilon \, r^2}$$

where Q_1 and Q_2 are the two charges, distance r apart, ε is the permittivity of the medium, and F is the resulting force.

Dimensionally,

$$F = \frac{[Q]^2}{\varepsilon \, [L^2]}$$

in which, for the purpose of obtaining the dimensions of charge, the product $Q_1 \times Q_2$ is replaced by the term (quantity of charge)^2. Now the dimensions of force are $[MLT^{-2}]$ as derived above.

Therefore

$$[MLT^{-2}] = \frac{[Q]^2}{\varepsilon \, [L^2]}$$

from which

$$[Q] = [M^{1/2} \; L^{3/2} \; T^{-1} \; \varepsilon^{1/2}]^{[1]}$$

---

[1]Thus, in addition to the dimensions of the fundamental quantities, L, M, and T, another fundamental quantity, the permittivity of medium, ε, has to be introduced in the (CGS) electrostatic system, as an example of deriving dimensions of Q. Similarly, the constant μ will have to be introduced in the (CGS) electromagnetic system.

In electromagnetics

The corresponding inverse square law in electromagnetics involves the pole strengths m_1 and m_2 of a magnet to express the force between the two magnetic poles, given by

$$F = \frac{m_1 \times m_2}{\mu \, r^2}$$

where r is the distance between the poles and μ the permeability of the medium.

That is,

$$\text{force} = \frac{\text{pole strength} \times \text{pole strength}}{\mu \times (\text{length})^2}$$

Therefore, dimensionally

$$[MLT^{-2}] = \frac{[m]^2}{[\mu \, L^2]}$$

from which

$$[m] = [M^{1/2} \, L^{3/2} \, T^{-1} \, \mu^{1/2}]$$

Dimensions of ε and μ

If ε and μ are assumed to be "fundamental" quantities, their dimensions cannot be expressed in terms of length, mass and time. However, a relationship *between them* can still be deduced as follows:

The dimensions of the charge in electrostatic system is expressed as

$$[Q] = [\varepsilon^{1/2} \, L^{3/2} \, M^{1/2} T^{-1}]$$

Now, in electromagnetism, the force exerted upon a magnetic pole of strength m units, placed at the centre of a circular wire of radians r due to a current i flowing in an arc of the circle of length *l* is given by

$$F = \frac{m \, i \, l}{r^2}$$

or

$$i = \frac{F \, r^2}{m \, l}$$

and the quantity of electricity flowing in time t, or the charge, is

$$Q = i \times t = \frac{F \, r^2 \, t}{m \, l}$$

Dimensionally,

$$[Q] = \frac{[MLT^{-2}][L^2][T]}{[\mu^{1/2}L^{3/2}M^{1/2}T^{-1}][L]}$$

by substituting the dimensions of F, r^2, t and l, and that of m derived previously.

Therefore

$$[Q] = [M^{1/2}L^{1/2}\mu^{-1/2}]$$

giving the dimensions of Q in the electromagnetic system, in an alternative form.

Since Q must have the same dimensions in *either* system,

$$[\varepsilon^{1/2} L^{3/2} M^{1/2} T^{-1}] = [M^{1/2}L^{1/2}\mu^{-1/2}]$$

or $\qquad [\varepsilon^{1/2} LT^{-1}] = [\mu^{-1/2}]$

or $\qquad [LT^{-1}] = [\mu^{-1/2} \varepsilon^{-1/2}]$

Now $[LT^{-1}]$ are the dimensions of a velocity;

$\therefore \qquad \dfrac{1}{\sqrt{\mu\varepsilon}}$ = a velocity

In any system of units, the "permeability of free space", μ_0, and the "permittivity of free space", ε_0, are related by the equation

$$\mu_0\varepsilon_0 = \frac{1}{c^2}$$

where c is the velocity of light *in the system of units considered.*

From this relationship, the dimensions of any electrical quantity can be converted from those of the electrostatic system to those of the electromagnetic system and *vice versa.*[1]

Dimensions of Electrical and Magnetic Quantities

The dimensions of the various electrical and magnetic quantities can be derived from their known relationships between them and using one of the fundamental quantities μ or ε in addition, as illustrated below.

---

[1]Instead of having to use either μ or ε as the necessary fourth fundamental quantity, any of the electrical or magnetic quantity could be used; for example, the quantity of electricity Q. Thus, using the product QV which represents work done, dimensionally

$$[QV] = [work] = [force \times distance]$$
$$= [MLT^{-2}] \times [L] = [ML^2 T^{-2}]$$

from which $\qquad [V] = [ML^2 T^{-2} Q^{-1}]$

being independent of either μ or ε.

Electric current

The electric current can be defined by

$$\text{current} = \frac{\text{quantity of electricity or charge}}{\text{time}}$$

$\therefore \qquad [I] = \dfrac{[\varepsilon^{1/2} L^{3/2} M^{1/2} T^{-1}]}{[T]} \qquad$ using the dimensions of electric charge derived earlier

$$= [\varepsilon^{1/2} \, L^{3/2} \, M^{1/2} T^{-2}] \quad \text{in the ES system.}$$

To convert these dimensions to the ones of the EM system, involving μ instead of ε, substitute $[\mu^{-1/2} L^{-1} T]$ for $\varepsilon^{1/2}$ as derived previously. Then, in the EM system,

$$[I] = [\mu^{-1/2} L^{-1} T L^{3/2} M^{1/2} T^{-2}]$$

$$= [\mu^{-1/2} M^{1/2} L^{1/2} T^{-1}] .$$

Electric potential[1]

The electric potential is defined by

$$\text{potential} = \frac{\text{``work''}}{\text{quantity of electricity or charge}}$$

Then, expressing work and charge in their dimensional forms

$$[V] = \frac{[ML^2 T^{-2}]}{[\varepsilon^{1/2} L^{3/2} M^{1/2} T^{-1}]}$$

$$= [\varepsilon^{-1/2} M^{1/2} L^{1/2} T^{-1}] \quad \text{in the ES system.}$$

And by the same reasoning as before, the potential in the EM system will be given by

$$[V] = [\mu^{1/2} LT^{-1} L^{1/2} M^{1/2} T^{-1}]$$

$$= [\mu^{1/2} L^{3/2} M^{1/2} T^{-2}]$$

Resistance

The dimensions of resistance can be best derived as the ratio of electric potential to electric current

or $\qquad [R] = \dfrac{[\varepsilon^{-1/2} L^{1/2} M^{1/2} T^{-1}]}{[\varepsilon^{1/2} L^{3/2} M^{1/2} T^{-2}]}$

$$= [\varepsilon^{-1} L^{-1} T] \qquad \text{in the ES system}$$

---

[1]Also, electromotive force or EMF.

Also,
$$[R] = \frac{[\mu^{1/2}L^{3/2}M^{1/2}T^{-2}]}{[\mu^{-1/2}M^{1/2}L^{1/2}T^{-1}]}$$

or
$$[R] = [\mu LT^{-1}] \qquad \text{in the EM system}$$

Since the two dimensions must 'equal'; by doing so

$$\frac{1}{\sqrt{\mu\varepsilon}} = [LT^{-1}]$$

or (dimensionally) velocity as was obtained earlier.

Magnetic flux

To obtain the dimensions of flux, use can be made of Faraday's law which states that

EMF (induced in any circuit) = rate of change of flux

or
$$\text{the EMF} = \frac{\text{flux}}{\text{time}}$$

or
$$\text{flux} = \text{EMF} \times \text{time}$$

$\therefore$ dimensionally

$$[\varnothing] = [\varepsilon^{-1/2}L^{1/2}M^{1/2}T^{-1}] \times [T]$$
$$= [\varepsilon^{-1/2}L^{1/2}M^{1/2}], \qquad \text{in the ES system}$$

Likewise
$$[\varnothing] = [\mu^{1/2}L^{3/2}M^{1/2}T^{-1}], \qquad \text{in the EM system}$$

Flux density

Since by definition flux density is flux per unit area

$$[B] = \frac{[\varnothing]}{[L^2]} = [\varepsilon^{-1/2}L^{-3/2}M^{1/2}], \qquad \text{in the ES system}$$

or
$$[B] = [\mu^{1/2}L^{-1/2}M^{1/2}T^{-1}], \qquad \text{in the EM system}$$

and similarly for the other derived units.

The dimensions of some of the important electrical and magnetic quantities, other than the above, in both ES and EM systems, together with the relationships from which they are derived are given in Table 1.2.

Table 1.2 : Dimensions of electrical and magnetic quantities

Quantity	Symbol	Relevant equation	Dimensions electrostatic system : ES	Dimensions electromagnetic system : EM	Practical (MKS) unit
Quantity of electricity, charge	Q	$F = \dfrac{Q_1 Q_2}{\varepsilon\, r^2}$	$L^{3/2}M^{1/2}T^{-1}\varepsilon^{1/2}$	$L^{1/2}M^{1/2}\mu^{-1/2}$	coulomb/ ampere-hour
Electric power	P	$P = V\,I$	$L^2\,M\,T^{-2}$	$L^2\,M\,T^{-2}$	watt
Electric energy	E	$E = V\,I\,t$	$L^2\,M\,T^{-1}$	$L^2\,M\,T^{-1}$	watt-hour/ joule
Inductance	L	$e = L\,\dfrac{di}{dt}$	$L^{-1}\,T^2\,\varepsilon^{-1}$	$L\,\mu$	henry
Capacitance	C	$C = \dfrac{Q}{V}$	$L\,\varepsilon$	$L^{-1}\,T^2\,\mu^{-1}$	farad
Impedance	Z	$Z = \dfrac{V}{I}$	$L^{-1}\,T\,\varepsilon^{-1}$	$L^{3/2}M^{1/2}T^{-1}\mu^{1/2}$	ohm
Magnetic field intensity	H	$H = \dfrac{B}{\mu}$	$L^{1/2}M^{1/2}T^{-2}\varepsilon^{1/2}$	$L^{-1/2}M^{1/2}T^{-1}\mu^{-1/2}$	ampere per metre
Magneto-motive force	$\mathcal{F}$	$\mathcal{F} = H\,l$	$L^{3/2}M^{1/2}T^{-2}\varepsilon^{1/2}$	$L^{1/2}M^{1/2}T^{-1}\mu^{-1/2}$	ampere
Permeability	μ	$\mu = \dfrac{B}{H}$	$L^{-2}T^2\varepsilon^{-1}$	μ	

Observe that in all the above cases, one of the two fundamental quantities, μ or ε, in addition to the other three, appears invariably in the dimensions. This is the basic characteristic of ES or EM system.

Dimensions of Electrical and Magnetic Quantities in Rationalised MKS System

The dimensions of the various quantities in the Rationalised System are based on FOUR fundamental quantities and their dimensions. These are: metre [M], kilogramme [K], second [S] (or time [T]), and *ampere* (or current) [I]. The use of ampere as the fourth fundamental unit was recommended by the IEC in July, 1950 and was subsequently adopted universally. The main feature of the system is that it does away with the use of either permittivity or permeability which do not appear in any of the dimensions.

The current itself being the rate of electric charge (charge per unit time), can be dimensionally expressed as

$$[I] = \frac{[Q]}{[T]} \qquad \text{or} \qquad [Q] = [I]\,[T]$$

It follows that when viewed in *electrical* context, therefore, the fourth fundamental quantity will appear to be charge instead of the ampere or current.

The dimensions of the various quantities can now be derived as illustrated below.

Electromotive force or potential difference

Since potential difference = work per unit charge

$$[PD] = [V] = \frac{[W]}{[Q]}$$

$$= \frac{[ML^2T^{-2}]}{[Q]}$$

$$= [M\,L^2T^{-3}\,I^{-1}]$$

substituting for the dimensions of [Q].

Electric flux

This is dimensionally equal to the charge producing it,

or $\qquad [\psi] = [Q] = [I]\,[T] = [I\,T]$

Similarly worked out, the dimensions of other quantities are listed in Table 1.3.

Table 1.3 : Dimensions of quantities in rationalised MKS system

Quantity	Symbol	Relevant equation	Dimensions
Electric flux density	D	$D = \dfrac{\psi}{\text{area}}$	$[I\,T\,L^{-2}]$
Electric field strength	$\mathcal{E}$	$\mathcal{E} = \dfrac{V}{L}$	$[M\,L\,T^{-3}I^{-1}]$
Capacitance	C	$C = \dfrac{Q}{V}$	$[M^{-1}L^{-2}T^4I^2]$
Magnetomotive force	$\mathcal{F}$	$\mathcal{F} = (N)\,I$	$[I]$
Magnetising force	H	$H = \dfrac{\mathcal{F}}{L}$	$[I\,L^{-1}]$
Magnetic flux	$\varnothing$	$\varnothing = V \times t$	$[ML^2T^{-2}I^{-1}]$
Magnetic flux density	B	$B = \dfrac{\varnothing}{\text{area}}$	$[MT^{-2}I^{-1}]$
Inductance	L	$L = \dfrac{V \times t}{I}$	$[ML^2T^{-2}I^{-2}]$
Resistance	R	$R = \dfrac{V}{I}$	$[ML^2\,T^{-3}I^{-2}]$
Power	P	$P = V \times I$	$[ML^2T^{-3}]$
Energy	E	$E = P \times t$	$[ML^2T^{-2}]$

Then, the permeability and permittivity are associated with the field quantities as follows:

$$B = \mu \, H, \qquad \text{where} \quad \mu = \mu_0 \, \mu_r; \quad \mu_0 = 4\pi \times 10^{-7}$$

and

$$D = \varepsilon \, \mathcal{C}, \qquad \text{where} \quad \varepsilon = \varepsilon_0 \, \varepsilon_r; \quad \varepsilon_0 = 8.854 \times 10^{-12}$$

Comment

What is the significance associated with units and dimensions of quantities in engineering practice?

In the course of measurement, quantities are generally 'measured' using a given supply as an input and a current or PD as a manifestation in parts of a circuit. For example, in the simple case of measurement of a "medium" resistance by the voltmeter-ammeter method, the input is a given DC voltage, resulting in a current that is measured by an ammeter; whilst the PD across the unknown resistance is in volts. Dividing the PD by the current gives the unknown resistance directly in ohms. However, if the resistance is of the order of a few kilo ohms, the current may be measured in mA. And the unit associated with the measured quantity is to be carefully recorded.

As to the importance of dimensions, consider the measurement of inductance of a coil using Hay's bridge (to be discussed in a later chapter), the expression for the inductance being obtained as

$$L = \frac{R_2 R_3 C}{1 + \omega^2 R_4^2 C^2}$$

where R_2, R_3, R_4 and C denote the various branch components of the bridge and ω the angular frequency of the supply.

Here, if R_2, R_3 and R_4 are in ohm, ω in rad/s, and C is in farad, the inductance must be obtained in henry if the derivation of the above expression is correct. However, whilst the *magnitude* of L will be obtained in terms of the values of resistances etc. in the expression, it is essential that the expression must be *dimensionally balanced* to yield the true value of the inductance. Thus, in the above expression, after some simplification,

$$[L] = \left[\frac{1}{[\omega^2][C]} \right]$$

or

$$= \left[\frac{1}{T^{-2} L^{-1} T^2 \mu^{-1}} \right]$$

by substituting the dimensions of $[\omega^2]$ and $[C]$ in EM system.

or $[L] = [L\,\mu] = [L]$, in the EM system.

Hence, the expression provides the correct value of the inductance.

STANDARDS

All measurements, whether of electrical or non-electrical quantities, are characterised by a numeric value or a number representing the magnitude of the quantity followed by the unit. For example, the *voltage* of the single-phase domestic supply may be expressed as 230 volt (written 230 V) ; or the supply *frequency* as 50 Hz. Here, the numbers 230 and 50 are the magnitudes of the supply voltage and frequency, respectively, whilst volt and hertz are the corresponding units. How accurately the voltage and frequency are known? Could the voltage be 235, 220 or 215 V? And the frequency 52 or 48 Hz?

When measuring these quantities, a simple voltmeter and frequency meter may be employed to measure the voltage and frequency, respectively. How accurate are these instruments?

Obviously, the accuracy of measurement of the quantities of interest – voltage and frequency in the present case – would depend on the *accuracy* (or the error, in other words) of the measuring instruments. How would one ensure the accuracy of these instruments? A logical answer would be: to compare the instruments with those having a better or higher accuracy; for example, to compare a ±2.0% accuracy instrument with the one having an accuracy of ±0.5%, and the latter to be compared with the instruments that are ±0.1% accurate and so on[1]. The ultimate thus would be to approach the condition of comparing instruments with the ones having zero error, at least theoretically.

If the physical quantity being measured is just 1 volt, the device having a value of 'exactly' one volt can be considered as a "standard" of the unit of voltage. The term standard is thus applied to a piece of equipment or instrument, having a known measure of a given physical quantity, which can be used, usually by a 'comparison-method' to obtain the values of the physical properties of other equipment.

---

[1]This aspect acquires a particular significance when extremely low or high magnitudes of the quantity of interest are to be measured; for example, mV, μV, mA or μΩ or MV and MΩ. In such cases, a slight deviation in the accuracy of the instrument(s) based on a specified percentage (±1.0 or ±0.5% etc.) can lead to considerable deviation in the *actual* value of the quantity.

Thus, care must be exercised in the choice or selection of the measuring instruments for a given application, the accuracy of the instruments being compatible with the accuracy with which the measurement is being carried out. Note, however, that measuring instruments of superior accuracy are, in general, quite expensive and may not be justified for routine measurements.

Classification of Standards

The standards can be broadly classified as

(a) Absolute standards

(b) Primary standards

(c) Laboratory or practical standards.

In addition, there may also be "legal standards" which 'enforce' a legal binding on the users of a class of instruments, or their manufacturers.

Absolute Standards

An absolute standard is one whose value can be determined directly from physical dimensions of the device; for example, the self–inductance of a suitably shaped coil whose value can be deduced directly from its size and number of turns etc. Similarly, the mutual inductance between a long solenoid and a short co-axial search coil situated at mid-length can be calculated from physical dimensions of the two coils and their number of turns. In such cases, there is no reference that may be necessary to any other physical quantity or constant.

Primary Standards

Primary standards which can be devised by means not simply dependent on physical dimensions are standards of such high accuracy that they can be used as the *ultimate* reference standards, especially for all electrical instruments and equipment. It is imperative that in addition to highest possible accuracy, such standards must possess extremely good stability, that is, their values may vary only negligibly over long periods of time, at times stretching to many years, despite changes in atmospheric and other conditions (typically the ambient temperature). Thus, the construction of such standards must include consideration of

(i)   long-term stability of materials;

(ii)   very small temperature coefficient responsible for expansion, or change of resistance etc.;

(iii) avoidance of deterioration of materials caused by moisture or other atmospheric conditions;

(iv)   precision of machining and fabrication;

(v)   accuracy of measurement of dimensions;

(vi)   rigidity of construction.

By refinements perfected over past many decades, the accuracy of such standards has been raised to extremely high levels. For example, the accuracy of a few of the quantities in use is now ensured as follows:

unit of inductance : 5-10 ppm

unit of current   : 10-20 ppm

unit of resistance : 10-20 ppm

or better.

Standards of fundamental quantities

As previously specified, these pertain to the (fundamental) units of length, mass and time.

Length

Metre

The *working standard* of metre now universally adopted as unit of length has been defined for many decades as the distance between two precision blocks of steel having two parallel surfaces. With this arrangement, an accuracy of 1 ppm is claimed.

The International Standard (or unit) of Metre

This is specified as 1,650,763.73 times the wave-length in vacuum of the orange-red light radiation of Krypton-86 atom under controlled atmospheric conditions.

The latest (since 1983) definition of metre is defined as the distance travelled by light in 1/299, 792, 458 second; that is, the inverse of the speed of flight in m/s in 'vacuum'.

Mass

Kilogramme

The international prototype of a kilogramme since 1889 is the mass of a cylinder of platinum-iridium about 39 mm high and 39 mm in diameter maintained under controlled conditions[1]. The latest proposal is to define $(1/1000)^{th}$ of a kilogramme (that is, one gramme) as the mass of exactly 18×140744813 carbon-12 atoms. Also, it is being attempted to link the kilogramme to a fundamental unit of measurement based on quantum physics – the Planck's constant.

Time

Second

The earliest standard or unit of time emerged from the time the earth took to rotate on its axis and the duration of its revolutions around the Sun.

The modern and latest standard of time, and definition of a second, is related to frequency of oscillations of crystals and quartz under controlled environmental conditions; that is, on their atomic properties and thus form

---

[1]However, not being completely independent of time, the "standard" kilogramme is known to have 'lost' around 50 micro-gramme (almost the weight of a grain of sand!) over the past more than 100 years.

the basis of "atomic clocks" regarded to be accurate to within ±1 second in 300, 000 years![1]

Practical or Laboratory Standards

These standards are of direct relevance to the measurement practices in general and are dealt with in detail; here and in later chapters. The main purpose of these standards is either to check the accuracy of general laboratory instruments or to act as a "standard" variable in precision methods, such as in bridge measurements discussed later [see Chapter XI], or both. The design and construction of such standards may vary.

For example

(i) standards of resistance for general laboratory use take the form of specially fabricated resistance coils whose values are determined by comparison, the ultimate reference being a resistance derived by the absolute method;

(ii) laboratory standards of inductance may again be obtained by comparison methods or in terms of physical dimensions as in the case of primary standards, but with less emphasis on accuracy;

(iii) a standard of EMF may be in the form of a "standard cell", or the voltage drop across a known resistor when a known current is flowing through it.

The accuracy and stability of laboratory standards will range from values approaching those of the primary standards to relatively (somewhat) low accuracies which are adequate for most purposes. The term "standard" may be applied to fixed or variable resistors, inductors and capacitors, and to "high-quality" indicating instruments as well which are meant to accurately measure voltage or PD, current and power[2].

The precautions mentioned in connection with the design and construction of primary standards are also relevant to practical standards, even if to a lesser extent in many cases. Other stringent requirements may apply for some variable standards. For example, in continuously variable types, the bearings of the moving part must be of the highest quality whilst in switched types, the switches must be extremely reliable having low contact resistance and lasting for thousands of operations.

---

[1]The first such atomic clock built decades ago is housed in the British Science Museum, London.

[2]There are a class of instruments which are inherently more accurate than others owing to their principle of operation, design and construction as dealt with in a later chapter. [See Chapter V : Measurement of Power].

Laboratory standards of resistance

Resistance standards range in values from $0.001\ \Omega$ to $10^6\ \Omega$ or more and usually are required for use at frequencies ranging from DC to very high values. No single design or method of fabrication may be suitable over such wide range, but the construction and resistance material must comply with the following properties :

- permanence
- an extremely low temperature coefficient
- very low thermoelectric EMF, usually with copper
- adequate heat dissipating capacity
- low values of residual inductance and capacitance
- robustness

The most suitable material for constructing coils that satisfies nearly all requirements generally is **manganin**, an alloy of manganese, copper and nickel having almost all desirable properties; its main disadvantage being its high cost.

In general, a resistor in practice, comprising 'large' number of turns, is bound to have some inductance due to the magnetic field of the current through it (which may be quite large in some measurements), and capacitance due to appreciable electrostatic field in high-voltage applications. Both these effects can be significant at higher frequencies.

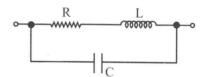

Fig.1.1 : Equivalent circuit of a practical resistor

The resistor may therefore be represented by the equivalent circuit of Fig.1.1 in which the resultant *impedance* of the two parallel branches can be given by

$$\dot{Z} = \frac{(R + j\omega L)\ (-j/\omega C)}{R + j\omega L - j/\omega C}$$

at the angular frequency, ω rad/s.

or

$$\dot{Z} = \frac{R(1 + j\omega L/R)}{1 + j\omega CR - \omega^2 LC}$$

which simplifies to

$$\dot{Z} \simeq R\ [1 + \omega^2\ LC + j\omega\ (L/R - CR)]$$

neglecting "squares" and higher powers of the terms containing L and C as both are too small. The resulting impedance will be non-reactive if

$$L/R = CR$$

(that is when the imaginary term is equal to zero)

or $\qquad L = CR^2$

Hence, in the design of resistors for *AC measurements* care is taken to fulfill the above condition as much as possible. Then, the component behaves as a pure resistance of the value $R(1 + \omega^2 LC)$. Clearly, at DC when $\omega = 0$, there is no error due to inductance or capacitance. Over the commonly used range of frequency, too, the term $\omega^2 LC$ may be quite small.

Construction of standard resistors

Standard resistors constitute the most frequently used component in electrical measurements. The form of a laboratory standard resistor depends greatly on its value, purpose and the frequency range over which it is to be used.

Low-resistance Standards

Typical values of these standards may lie between 0.001 Ω to 0.1 or 0.5 Ω. These are usually constructed using strips of manganin (typically about 15 cm long, 3 cm wide and 1 mm or less thick, depending on the value and design of the resistance), arranged in parallel leaving about 5 mm gap between adjacent strips to allow for cooling, and rigidly joined at the two ends to blocks of copper or brass. A typical feature of these resistors is the "4-terminal"

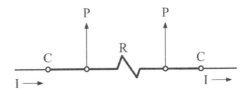

Fig.1.2 : A 4-terminal low, standard resistance

construction, with two current terminals to carry the load current in and out and two parallel or potential terminals for external connections as shown in Fig.1.2. This is further discussed in later chapters.

Medium-resistance Standards

These range in values from about 1 Ω or tens of ohm to several hundred or thousands of ohms. The typical construction of resistances in this class comprises coils that utilise "bifilar" loops as shown in Fig.1.3. The wire is first doubled back on itself and then wound helically on a non-metallic, non-magnetic cylindrical former. With the neighbouring portions

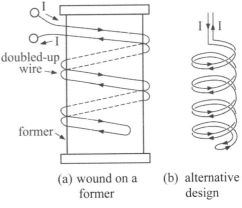

(a) wound on a former

(b) alternative design

Fig.1.3 : Bifilar construction of resistance

of wire carrying currents in opposite direction at any instant, the magnetic field due to the current and hence the inductance effect is cancelled. However, the frequency range is somewhat limited because of capacitive effect.

The Ayrton-Perry Winding

This is often employed being better than the bifilar construction in terms of compensation' for inductive and capacitive effects. Two wires of the winding are connected in parallel and are wound with opposite sense of winding on a flat insulator strip or card as shown in Fig.1.4.

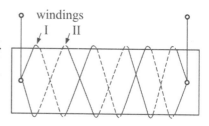

Fig.1.4 : Ayrton-Perry winding

Here, too, the magnetising effect of the two wires is in opposition and so the resulting inductance is extremely small. Since the portions of wire do not run in parallel the capacitance effect would also be very small.

For higher values of resistance (> 10,000 Ω), it is impossible to completely eliminate the capacitive effect. It is therefore usual to wind the coil allowing a small amount of inductance in order to compensate for the capacitance. A simple, single winding is used on a flat card for the purpose for resistance values up to about 100,000 Ω.

Resistance Boxes

These boxes, also known as decade resistance boxes, find extensive use in bridge methods; for example, the Wheatstone bridge discussed in detail later. Typically, the boxes comprise a number of sets of precision resistance coils (for example, the bifilar type) housed in a box, the appropriate coils being connected into circuit by means of rotary switches. Accordingly, a ten ohm decade box consists of ten 1 Ω coils in series and the switch can connect any number of these from 0 to 10 Ω into the measuring circuit. Thus, four decades, 10 Ω, 100 Ω, 1000 Ω and 10,000 Ω each, will provide an adjustable resistor having any value from 1 Ω to 11,110 Ω in steps. However, the accuracy of a resistance value at a given setting, especially at lower range may be affected by various factors from individual coil due to the contact error(s) of the switches; the switch contacts may at times be gold plated to minimise such errors.

Laboratory standards of inductance

Fixed standards of inductance are simple in principle and are constructed in the form of a coil or coils based on considerations of stability and high-frequency performance. A single coil can act as a self-inductance standard whilst two coils wound in close proximity on rigidly connected formers can act as a standard of mutual inductance. For low values, it is customary to adopt air-core construction; for higher values, use of high permeability soft iron core may be imperative, *operating in non-saturating state of iron.*

It is important to adopt suitable choice of the size of the former and number of turns of a particular wire size. Use of a relatively thin wire and a very large number of turns, to obtain high value of inductance, may result in appreciable resistance introduced in series with the inductance, with corresponding heating and temperature rise for long periods of testing. For improved accuracy, elaborate methods of compensation for the resistance and means for cooling and ventilation may be employed.

Variable Standards

Whilst the self-inductance of a coil may be varied by altering number of turns, the mutual inductance of two coils can be adjusted by altering the relative position of the coils. One of the commonly used device consists of a fixed coil inside which a second coil can be rotated about a diametre, along horizontal or vertical axis. Clearly, the maximum inductance is obtained when the two coils are co-axial and very nearly zero when the axes of the coils are perpendicular to each other.

Variable standards of inductance can also be used as standards of self-inductance by connecting the fixed and moving coils in series. The resulting net inductance L is then given by

$$L = L_1 + L_2 + 2 M$$

where L_1 and L_2 are the self inductances of the two individual coils and M the mutual inductance between them. If the coupling between the coils is fairly close and the variation of M can be effected to be both positive and negative, a wide range of variation of L can be achieved.

Laboratory standards of capacitance

Fixed Standards

For the very low capacitances, in the range of pico-farad, parallel-plate or concentric-cylinder configuration, with *air* as dielectric, is usually employed which provides fairly good accuracy. For higher values, up to about 0.01 μF, multiple capacitors form very dependable standards. These may use some dielectric other than air and may entail some dielectric loss, especially at higher frequencies. By careful design and use of high-grade materials, the loss may be minimised, with the power factors (or values of tan δ) brought down to as low as 0.00001.

For high-voltage work, in special bridge methods, capacitors which are practically loss-free, yet having adequate dielectric strength, are constructed using compressed nitrogen or carbon-dioxide as the dielectric in the concentric-cylinders form.

Variable Standards

Variable standards of capacitance covering a range of about $0.001\mu F$ to $1.0 \ \mu F$ are available in the form of decade boxes. Capacitor units with mica as the dielectric are selected by rotary switches, the arrangements being similar to a decade resistance box. A usual 4-decade box may provide capacitance values accurate to 0.1% or better, with tan δ values being 0.0001 to 0.0005.

Laboratory standards of EMF

Standards of EMF, or what are known as "standard cells", find extensive use in nearly all potentiometer methods of measurement. In this context, the EMF of certain primary cells is found to be remarkably constant with respect to time and environmental conditions. A few of these have thus been developed into EMF standards of high accuracy and precision. The most satisfactory and commonly employed of these

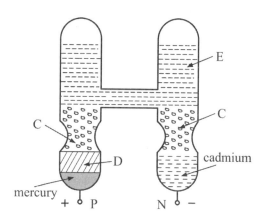

Fig.1.5 : Schematic of the Weston cell

is the Weston cell developed and patented by Dr. Weston in 1892. It has a "standard" EMF of 1.01859 V (usually rounded off to 1.0186 V) at 20°C, decreasing by about 40 µV per 1°C temperature rise. The construction of a typical cell is shown schematically in Fig.1.5. A sealed H-shaped vessel contains the electrolyte E, being a saturated solution of cadmium sulphate. The two lower ends constitute the positive and negative poles, P and N; these being mercury and cadmium amalgam, respectively. A layer of mercurous sulphate D is the depolariser (to maintain the two polarities), and the solution is maintained saturated by crystals of cadmium sulphate, C^1.

---

[1]Characteristically, the value of the EMF is specified and accurate to *four* decimal places. This has special significance in the standardisation process of various potentiometers as discussed later in Chapter IX.

Laboratory standards of current

It is not possible, nor practicable, to set up a standard of current in the same manner as the standard of resistance, or indeed that of the EMF. In practice, therefore, recourse is made to the combination of a standard EMF and a resistance to achieve a "standard" of current. The voltage drop across a suitable standard resistance (for example a 4-terminal resistance of $1\,\Omega$) is measured using a precision potentiometer by reference to a standard cell in the first instance. The PD divided by the resistance then provides the (standard) current in the circuit, or otherwise.

Alternatively, as a simple, direct means, the current can be measured by using a high precision ammeter, the reading being assumed to be of sufficient accuracy.

WORKED EXAMPLES

On Units

1. A force of 500 N pulls a sledge of mass 100 kg and overcomes a constant frictional force to motion of 100 N. What would be the acceleration of the sledge in motion?

    Resultant force on the sledge for acceleration

    $$F = 500 - 100 = 400 \text{ N}$$

    Using the expression : force = mass × acceleration,

    the acceleration of the sledge

    $$\alpha = \frac{400}{100} = 4 \text{ m/s/s} \quad (4 \text{ m/s}^2)$$

2. A body of 10.0 kg mass is attached to the hook of a huge spring balance, the same being suspended vertically below the top of a lift cabin as shown in the adjoining figure.

    What would be the reading on the spring balance when the lift is

    (a) ascending with an acceleration of 0.5 m/s² ;

    (b) descending with an 'acceleration' of 1.0 m/s² ?

    Assume g = 10 m/s².

    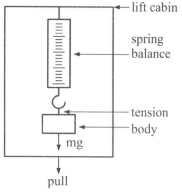

    As seen, the suspended body is acted upon by *two* forces :

    (i) the *tension*, T, (in N) in the spring balance, acting *upwards*;

    (ii) the *weight* of the body, resulting in a gravitational pull acting *downwards*.

    The weight is given by m × g or 10.0 × 10 or 100 kg-force or 100 N.

    (a) when the lift is ascending, the body is experiencing an acceleration which is upwards.

        This means   T > 100 N

        The net force, F, acting on the body will then be

        $$F = (T - 100) \text{ N}$$

Also $F = m\,\alpha$

where α is the acceleration by which the body is "moving" upward.

or $F = 10 \times 0.5$ or   5 N

$\therefore$ $T - 100 = 5$

whence $T = 105$ N

(b) when the lift is descending, the body weight would appear to be greater than the tension in the spring balance.

or $100 - T = F = m\,\alpha$

$$= 10 \times 1.0$$

$$= 10$$

$\therefore$ $T = 90$ N

3. A body of mass 800 kg is made to move upward on an inclined plane, the slope of the plane being given by "1 in 100". Down the slope, the frictional force acting on the body is 500 N. Calculate the force required to move the body when it is

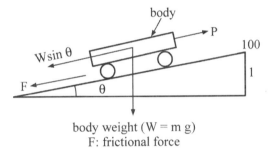

body weight $(W = m\,g)$
F: frictional force

(a) to accelerate at 5 m/s^2;

(b) to move with a constant velocity of 10 m/s.

Assume $g = 10$ m/s^2.

The configuration of the body movement is shown in the above figure.

As seen, the body is acted upon by *three* forces :

(i) the upward (applied) force, P, to pull the body up;

(ii) the component of body's weight along the inclined plane, acting downward 'opposite' to P;

(iii) the frictional force, F, acting downward, again in opposition to P.

Now the weight of the body acting *vertically* downward is

$$W = m \times g$$

$$= 800 \times 10 \quad \text{or} \quad 8000 \text{ N}$$

and its component downhill is

$$W' = W \sin \theta$$

$$= 8000 \times (1/100) \quad \text{or} \quad 80 \text{ N}$$

With $F = 500$ N, the net force required *uphill* is given by

$$F' = P - 80 - 500 \; ; \; \text{also, } F' = 800 \times 5 \; \text{ or } 4000 \text{ N}$$

$$= P - 580$$

whence $P = 4000 + 580$ or 4580 N

(b)   when the body is supposed to be moving with a constant velocity, there is no acceleration.

Hence, $\alpha = 0$

and $F' = 0$

[the velocity does not figure in the calculation]

Then

$$0 = P - 580$$

or $P = 580$ N

[Note that if the incline was steeper, say 1 in 50, the pulling force in the second case would be 660 N].

4.   A train weighing 300 tonne is hauled up an incline of 1 in 80 at a steady speed of 40 km per hour by means of an electric locomotive fed from an overhead traction system at 750 V. The overall efficiency of the locomotive is 75% in terms of the energy input. The rotational inertia of the train is 10% whilst the resistance to motion is 20 N per tonne.

Calculate

(a)   the energy expended in the traction in kWh for pulling the train for a distance of 1 km;

(b)   the current fed to the motors of the locomotive;

(c)   specific energy consumption of the train in kWh per tonne.

When the train is being hauled up the incline, the component of its weight to oppose the motion will be

$$W \sin \theta$$

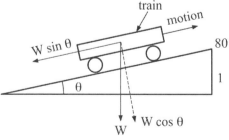

where W is the weight of the train and θ the inclination of the track as shown in the adjoining figure.

Here   W = 300 tonne

$$\sin \theta = \frac{1}{80} \text{ [the incline being 1 in 80]}$$

'Effective' weight of the train

$$W_e = 1.1 \times 300 = 330 \text{ tonne,}$$
$$\text{considering rotational inertia @ 10\%}$$

Tractive effort required to pull the train up the incline

$$T_1 = W_e \times g \times \sin \theta$$
$$= 330 \times 9.81 \times (1/80) \times 1000,$$
$$\text{assuming } g = 9.81 \text{ m/s}^2$$
$$\text{[since the weight of the train is in tonne]}$$
$$= 40466 \text{ N}$$

Tractive effort required to overcome the resistance to train motion

$$T_2 = W_e \times 20$$
$$= 330 \times 20$$
$$= 6600 \text{ N}$$

Total tractive effort required

$$T = T_1 + T_2 = 47066 \text{ N}$$

When the train is hauled for a distance of 1 km or 1000 m, the work done

$$W = 47066 \times 1000$$
$$= 47.066 \times 10^6 \text{ Nm or J or W-s}$$
$$= \frac{47.066 \times 10^6}{1000 \times 3600} \text{ kWh}$$
$$= 13 \text{ kWh}$$

And the energy *input* at the overall efficiency of 75%

(a) $$W_{in} = \frac{13}{0.75} = 17.34 \text{ kWh}$$

At a steady speed of 40 km per hour, time taken to travel 1 km

$$t = \frac{1}{40} \text{ hr}$$

Therefore power input

$$P_{in} = \frac{17.34 \times 1000}{(1/40)}$$

$$= 693.6 \times 10^3 \text{ W or } 693.6 \text{ kW}$$

(b)   At the supply voltage of 750 V, the current input to the motors

$$I_{in} = \frac{693.6 \times 1000}{750}$$

$$= 925 \text{ A}$$

(c)   Specific energy consumption

$$SEC = \frac{17.34}{300}$$

$$= 0.0578 \text{ kWh/tonne (or 57.8 Wh/tonne)}$$

On Dimensional Analysis

5.   Obtain suitable units for μ_0 and ε_0.

$$[\mu_0] = \frac{[B]}{[H]}, \text{ by definition and with } \mu_r = 1$$

$$= \frac{[MT^{-1}Q^{-1}]}{[L^{-1}T^{-1}Q]}$$

in practical units, with $Q = I T$ as the fourth "fundamental" quantity.

or $[\mu_0] = [M L Q^{-2}]$

Comparison of this result and the list of dimensional expressions of various units [see Table 1.3] shows that $[\mu_0]$ may conveniently be expressed as

$$[\mu_0] = [M L Q^{-2}] = \frac{[ML^2Q^{-2}]}{[L]}$$

$$= \frac{[\text{inductance}]}{[L]}$$

Hence the unit of μ_0 may be called *henry per metre*.

Now $[\varepsilon_0] = \frac{[D]}{[\mathcal{E}]} = \frac{[L^{-2}Q]}{[MLT^{-2}Q^{-1}]}$

$$= [M^{-1}L^{-3}T^2Q^2]$$

Again, by comparison, it is seen that this may conveniently be written

$$[\varepsilon_0] = \frac{[M^{-1}L^{-2}T^2Q^2]}{[L]}$$

$$= \frac{[\text{farad}]}{[L]}$$

Therefore, the unit of ε_0 may be called *farad per metre*.

6. In a derivation, the resistance portion of a capacitive component is given by

$$R' = \frac{1 + \omega^2 C^2 R^2}{\omega^2 CR}$$

where R and R′ are resistances, C a capacitance and ω the angular frequency.

Check whether the expression is dimensionally correct.

It is necessary to first check the dimensional consistency of the numerator of the right-hand side of the expression, remembering that the dimensions of $[\omega]$ is simply $[T^{-1}]$.

Then

$$[\omega^2 C^2 R^2] = [T^{-1}]^2 [M^{-1}L^{-2}T^2Q^2]^2 [ML^2T^{-1}Q^{-2}]^2$$

$$= [M^0 L^0 T^0 Q^0]$$

that is, $[\omega^2 C^2 R^2]$ is dimensionless which is consistent with the fact that the numeric 1 is added to it in the numerator of the given expression.

Thus, the dimensions of the whole of right-hand side are

$$\left[\frac{1}{\omega^2 CR}\right] = [\omega^{-2} C^{-1} R^{-1}]$$

$$= [T^{-1}]^{-2} [M^{-1}L^{-2}T^2Q^2]^{-1} [ML^2T^{-1}Q^{-2}]^{-1}$$

$$= [T], \text{ after simplification}$$

But the dimensions of the left-hand side, that is of the resistance R′, are

$$[R'] = \left[ML^2T^{-1}Q^{-2}\right]$$

which can be balanced only if a "C" is added to the denominator of the expression, making it

$$\frac{1}{\omega^2 C^2 R}$$

and R' then given by

$$R' = \frac{1 + \omega^2 C^2 R^2}{\omega^2 C^2 R}$$

[Clearly, 'disregarding' 1 from the numerator, R' simplifies to

$$R' = \frac{\omega^2 C^2 R^2}{\omega^2 C^2 R} = \text{``R''} \qquad \text{- a resistance]}$$

7. In an electric circuit fed from a supply voltage V and comprising a 'load' resistance, the power 'consumed' in the load is known to be a function of applied voltage and the resistance, R. Determine the expression of proportionality in terms of exponents of V and R.

Assume that the electric power, P, can be expressed as

$$P = K\, V^x\, R^y$$

where K is a constant. It is required to deduce the values of x and y.

For this, substitute the dimensions of P, V and R, respectively, in the above expression, say in the EM system.

Then

$$[L^2 M T^{-2}] = K\{[L^{3/2} M^{1/2} T^{-2} \mu^{1/2}]^x \times [LT^{-1}\mu]^y\}$$

To balance this equation *dimensionally*, equate the indices of L, M, T and μ on both sides.

Thus, for L

$$2 = (3/2)\, x + y$$

and for M

$$1 = (1/2)\, x$$

whence $x = 2$ and $y = -1$

Also, for T

$$-3 = -2x - y$$

which itself is satisfied when $x = 2$ and $y = -1$.

Further, for μ

$$0 = (1/2)\, x + y$$

which is also satisfied for $x = 2$ and $y = -1$.

Therefore, substituting the results in the assumed expression for power,

$$P = K V^2 R^{-1}$$

or
$$P \propto \frac{V^2}{R}$$

8. The expression for the mean torque T of an electrodynamic wattmeter may be written

$$T \propto M^p E^q Z^t$$

where   M = mutual inductance between fixed and moving coils

E = applied voltage

Z = impedance of the load circuit.

Determine the values of p, q, and t from the dimensions of the quantities involved.

Assuming a dimensions-less constant, the torque expression can be written

$$T = K\ M^p E^q Z^t$$

Now the dimensions of the various quantities in the EM system are

$$[M] \rightarrow [L\mu]$$
$$[E] \rightarrow [L^{3/2} M^{1/2} T^{-2} \mu^{1/2}]$$
$$[Z] \rightarrow [LT^{-1}\mu]$$

and
$$[T] \rightarrow [ML^2 T^{-2}]$$

Substituting these dimensions in the above expression,

$$[ML^2 T^{-2}] = K\{[L\mu]^p\ [L^{3/2} M^{1/2} T^{-2} \mu^{1/2}]^q\ [LT^{-1}\mu]^t\}$$

$$= K\{[L^{(p+3q/2+t)}]\ [M^{q/2}]\ [T^{(-2q-t)}]\ [\mu^{(p+q/2+t)}]\}$$

Equating corresponding indices on both sides

$$p + (3/2)\, q + t = 2$$
$$q/2 = 1$$
$$-2q - t = -2$$

and $p + (q/2) + t = 0$ (for μ)

Solving the first three equations gives

$$p = 1,\ q = 2,\ t = -2$$

and these values satisfy the fourth equation.

Hence, the torque expression is given by

$$T = K E^2 M Z^{-2}$$

or
$$T = K M I^2$$

where I is the load current, equal to E/Z[1].

---

[1]As dealt with later, the expression for torque of a dynamometer wattmeter is obtained to be

$$T = K I_1 \times I_2 \times \frac{dM}{d\theta}$$

in which I_1 represents the current through the fixed coil (or load current), I_2 the current through the moving coil (or current proportional to the supply voltage) and $dM/d\theta$ is the mutual inductance variation with deflection angle, θ.

II : Magnetism, Electricity and Electromagnetism

II

MAGNETISM, ELECTRICITY AND ELECTROMAGNETISM

RECALL

One of the most revolutionary discoveries in the history of science that has been the key to almost all technological developments to date was the one related to magnetic properties of various materials – notably the "iron" – and "magnetism" which preceded the other important phenomenon of electricity in its various forms.

Aristotle attributes the first of what might be called a reference to magnetism to Thales, circa 600 BC. Around the same time in ancient India, it is believed that the Indian surgeon, Sushruta, was the first to make use of a magnet for surgical purposes.

The earliest reference to the use of a "magnet" in some form was a "loadstone", actually a magnetic entity that could "attract iron", and which found extensive use in navigation as "one part of the loadstone always pointed towards geographical north of the earth[1]". In the eleventh century (AD), the Chinese actually invented the magnetic compass which consisted of a small magnet, or rather a magnetic needle, floating on a buoyant support in a dish of water. The basic properties of a magnet, or any material behaving like a magnet, were that it invariably had two poles, each pole attracting an iron piece. In the thirteenth century, a scientist called Peter Peregrinus showed that like poles of magnets repel each other whilst unlike poles attract.

[1]It was only in the year 1600 that William Gilbert proved 'conclusively' that the earth actually possessed magnetism in north-south direction and that this was the reason why the loadstone pointed north, proving wrong the belief that it was the pole star (Polaris) or a large "magnetic island" on the north pole that attracted the loadstone.

About the time of discovery of magnetism came another important discovery – that of *static electricity* or *electrostatics*. Once again, it was Thales who noted that rubbed amber could attract silk[1].

In terms of its uses and significance, the form of electricity that really changed the world was *current* electricity, relating also to the phenomenon of magnetism. The genesis of current electricity can be traced to the invention of the "voltaic cell" by Alexander Volta around 1800 AD who observed that a plate of copper and another of zinc when separated by a piece of cloth soaked in brine could develop a potential difference across them, and when connected through a wire or conductor would result in flow of a current. A number of such pairs of plates when arranged in succession could result in a pile of voltaic cells, representing what is now called a battery, increasing the net potential difference across the bottom and top most plate considerably, as also the current from such a pile to be many-fold.

Such a current, always flowing from one of the plates, copper or the "positive" terminal to the other, zinc or the "negative" terminal came to be identified as "direct current", also being the primary input for the process(es) of electrolysis, then an important industrial requirement.

---

[1]The Greek word for amber is *electron*, and a body made attractive by rubbing is said to be 'electrified' or *charged*. Accordingly, this branch of electricity, the earliest discovered, came to be known as *electrostatics*.

MAGNETISM

The property of a body whereby it *attracts* a magnetic material, say a piece of iron or steel, can be defined as magnetism, and the body itself may be identified as a magnet. A "magnetic field" is said to be surrounding the magnet which can be imagined to be built of "magnetic lines (of force)" and, theoretically, extend to infinity. Thus, "a magnetic field can be defined as the space in which a magnetic effect can be detected". The most familiar example is that of earth's magnetic field whose presence is easily indicated by a compass in that the compass needle aligns itself in the direction of the earth's field (in north-south direction).

In practice, the simplest example is that of a bar (permanent) magnet, exhibiting its magnetic field somewhat as shown in Fig.2.1.

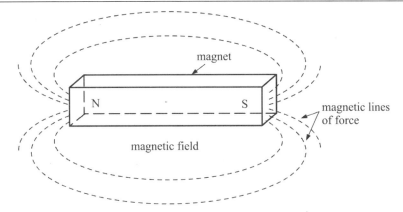

Fig.2.1: Magnetic field of a bar magnet[1]

Clearly, the stronger the pole or higher the "pole strength", stronger will be the magnetic field as also the force (of attraction) exerted on a magnetic material placed within the field; the closer to the pole, the more the force.

Coulomb's Law

For quantitative analysis, a law relating pole strengths of a magnet to the force was provided by Coulomb, expressed as

$$F \propto \frac{m_1 m_2}{r^2}$$

where F is the force between the magnetic poles of pole strengths m_1 and m_2 (units), respectively, and r the distance between them[2].

Lines of Force and Intensity of Magnetic Field

As seen, a magnetic field encompasses a 'large' number of lines of force, being infinite in theory. An indication of how strong the field is at a given point in the field can be obtained from *"intensity of magnetic field"* which can be expressed by the "force that would be exerted on a north pole of unit strength placed at the point." Thus, the intensity at a point, distant r from a pole of strength m (units) in air, will be given by

$$H = \frac{m}{r^2}$$

following Coulomb's law.

---

[1]The figure shows a very simplified, rather hypothetical picture of the magnetic field where the magnet itself is shown by the plan of its flat surface. In actual case, the magnetic field which is truly 3-dimensional will be quite complicated, as can be visualised by dipping a pole of the magnet into a vessel containing iron filings and observing the 'orientation' of the filings.

[2]Strictly, poles of a magnet always occur in pairs and cannot be separated. Still, the law in theory expresses the important aspects of a magnet.

Expressed alternatively, if the intensity at a point were H, then H lines of magnetic intensity would cross unit area in a direction perpendicular to the direction of the field at the point.

Magnetic Flux and Flux density

In a given magnetic field, the number of lines of magnetic intensity are defined, collectively, as "magnetic flux", denoted by Ø. The number of lines, or the flux per unit area of cross section, *perpendicular* to the direction of flux lines, is defined as the "flux density" and is denoted by the letter B. Thus,

$$\text{Ø} = B \times A$$

or
$$B = \frac{\text{Ø}}{A}$$

where A represents the area crossed by Ø lines of force.

Clearly, both B and H being directional quantities in a magnetic field and dependent on lines of force, the two are related as

$$B = \mu H$$

where μ is the permeability of the medium in which B or H exist. In MKS units,

$$B = \mu_o \mu_r H$$

where μ_o is the permeability of free space (= $4\pi \times 10^{-7}$) and μ_r the relative permeability (= 1 for air).

Magnetic Moment and Intensity of Magnetisation

The magnetic moment of a (bar) magnet, having poles of strength m units each, distant *l* apart, is defined as m*l* units. This can be explained in term of the fact that if the magnet were placed in a magnetic field of unit intensity in a direction perpendicular to the line joining the two poles, the magnet would be acted on by two forces, each of m units forming a couple of turning moment m *l*. The symbol of the moment is M; hence

$$M = m\ l$$

The intensity of magnetisation, expressed by letter J, is defined by the ratio

$$\frac{\text{pole strength}}{\text{cross-sectional area}}$$

or
$$J = \frac{m}{A}$$

"Soft" and "Hard" Magnetic Materials

From the point of magnetic behaviour, the various magnetic materials – largely known as ferro-magnetic materials – can be classified as "soft" or "hard" magnetic materials. In simple terms, the materials which when placed in a magnetic field of intensity H are magnetised to the flux density B, such that B = μ H (μ being the permeability of the material), *but lose their magnetism on removal of H,* are known to be *soft* magnetic materials, a classic example being ordinary iron or mild steel. In contrast, the materials which retain their magnetism, even if not at the same level, when the field is removed are categorised as *hard* magnetic materials[1]. A number of (metal) alloys belong to this class; for example, ALNICO – an alloy of aluminium, nickel and cobalt.

At the molecular level, the difference between soft and hard magnetic materials lies in the phenomenon related to fast, reversible v/s slow and irreversible rotation of dipoles, as a function of applied field of magnetisation[2].

A Simple (Permanent) Bar Magnet

In the sense of permanent magnetism, a permanent bar magnet of round or rectangular cross-section can be considered as a 'storage' of magnetic flux, exhibiting essential magnetic properties as depicted in Fig.2.2.

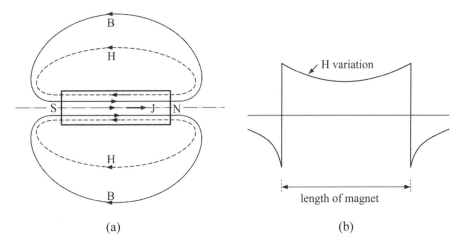

(a)   (b)

Fig.2.2 : Variation of H and B in a permanent magnet

---

[1]In usage in practice, the soft materials find their application in electromagnets (as discussed later) and a variety of electric machinery and transformers, whereas the hard magnetic materials are specifically employed in the form of permanent magnets.

[2]See, for example, the classic "Ferromagnetism" by Bozorth.

In the figure, J represents the intensity of magnetisation, H the magnetising field and B the corresponding flux density.

Note the direction of flux lines v/s the variation of H: in a permanent magnet the lines of flux density (B) coincide with the field (H) *outside* the magnet, but are oppositely directed *inside* as shown in Fig.2.2(a).

The flux density is governed by the expression

$$B = \mu (H - J)$$

The graphical variation of H inside and outside the magnet is shown in Fig.2.2(b), with the amplitude at mid-length being somewhat lower on account of leakage (phenomenon).

STATIC ELECTRICITY

ELECTROSTATICS

General

If a rod of ebonite is rubbed with fur, or a fountain pen with a coat-sleeve, the rod or the pen acquires the power to attract, in turn, light bodies such as small pieces of paper or tin-foil, or a piece of cork. The body being rubbed is said to have been charged – a process of accumulation of "free" electrons – and then inducing a charge of opposite polarity on the body being attracted.

As mentioned, the phenomenon was first discovered by ancient Greek philosopher Thales of Miletus about 600 BC and paved the way for a number of important discoveries related to electric charges; the corresponding 'science' being called *electrostatics*.

Basics of Electrostatics

It was observed that

- unlike magnetic poles, electric charges could exist in space singularly, that is, a body holding a positive charge, the other holding negative charge independently;
- similar to magnets, charged bodies would produce an *electric* field, the strength of which would depend on how strongly the bodies were charged [see Fig.2.3];
- quantitatively, the behavior of charged bodies could be related to various laws;
- the other fundamental property of charges was that unlike charges (of opposite polarities) would attract each other; the charges of similar polarities would repel.

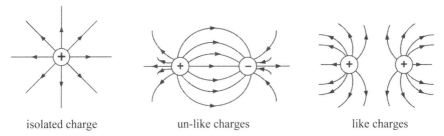

| isolated charge | un-like charges | like charges |

Fig.2.3 : Electric field patterns due to charges

Fundamental Law of Force

The charged bodies always exhibit a force, of attraction or repulsion, which can be related to the quantum of charge on the bodies and distance of separation between them. The law, similar to that related to magnetic pole strengths, enunciated by Coulomb in 1875, is known as the "inverse–square" law, given by

$$F \propto \frac{Q_1 Q_2}{d^2}$$

where Q_1 and Q_2 are the charges on the bodies and d the distance between them. In MKS system of units and in "free space",

$$F = \frac{1}{4\pi \, \varepsilon_0} \frac{Q_1 Q_2}{d^2} \, N$$

when Q_1 and Q_2 are in coulomb (the unit of charge), d in metre and ε_0 is the permittivity of free space ($= 8.854 \times 10^{-12}$). The force is then obtained in newton. Or, approximately,

$$F = 9 \times 10^9 \frac{Q_1 Q_2}{d^2} \, N$$

Electric Potential

A charged body, having a charge Q, is said to be at a *potential* or 'level' of energy, similar to the potential energy of a body held at a height and acted upon by the gravitational force. The "classic" definition of electric potential at any point is given by "the work that would have to be done in bringing a unit positive charge from an infinite distance to that point", or from a place of zero potential.

Potential due to a "point" charge

Let there be an 'isolated' "point" charge of +Q coulomb situated at a point O (the datum) as depicted in Fig.2.4. Now assume a positive charge of 1 coulomb being brought from an infinite distance along the straight line as shown. Naturally, there would be a force of repulsion between the charges and work will have to be done against that force to bring the unit charge close to the charge Q.

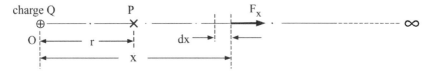

Fig.2.4 : Potential due to a "point" charge

From Coulomb's law, the force on the unit positive charge at any distance, x (metre) from O, along the line will be

$$F_x = \frac{Q \times 1}{4\pi \, \varepsilon_0 \, x^2} \quad N$$

Since, by definition, the potential at any point P, distant r from O, is equal to the work done in bringing the unit positive charge of 1 coulomb from infinity to P, the potential will be given by

$$V_p = \frac{1}{4\pi \, \varepsilon_0} \int_r^\infty \frac{Q}{x^2} \, dx$$

$$= \frac{Q}{4\pi \, \varepsilon_0 r} \quad V$$

Potential difference: PD

If, as above, potentials at two points P_1 and P_2 were derived to be V_{P_1} and V_{P_2}, the *potential difference* between points P_1 and P_2 will be given by

$$PD = V_{P_1} \sim V_{P_2}^1 \quad V$$

Potential gradient

This is defined as the rate of change of potential measured in the direction of the electric force and is a measure of how the electric field may vary around the charged body. Consider a charged sphere of radius r(m) as shown in Fig.2.5.

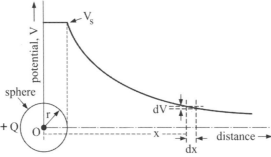

Fig.2.5 : Potential of a sphere at a point away from the charge

---

[1]This has an important bearing on the production of electric current and associated phenomena in external circuit(s) when connected across a PD as discussed later.

From the centre of the sphere to its surface, inside and outside, the potential is given by

$$V_s = \frac{Q}{4\pi \, \varepsilon_0 r} \quad V$$

and is constant as shown. At a distance x from O the potential is $\left(\frac{Q}{4\pi \, \varepsilon_0 x}\right)$ V, indicating that as x increases V decreases, meaning that a positive increment of x (+dx) is associated with a negative increment (or decrement) of V (–dV). Hence, for the gradient at distance x,

$$G_x = -\frac{dV}{dx} \quad V/m \qquad \text{(if V is in volt and x in metre)}$$

Equipotentials

A concept of importance in electric field analyses is that of equipotentials, or rather equipotential surfaces. An equipotential surface is a surface such that all points on it are at the same potential. Clearly, the potential gradient, dV/dx, for such a surface is zero. The concept is analogous to contours of equal heights encountered in land surveys or navigation.

Equipotential surfaces can be drawn throughout any space where there is an electric field. For example, consider an isolated positive *point* charge Q shown in Fig.2.6(a). At a distance r from the charge, the potential is $\left(\frac{Q}{4\pi \, \varepsilon_0 r}\right)$; a sphere of radius r and centre at Q is therefore an equipotential surface, of potential $\left(\frac{Q}{4\pi \, \varepsilon_0 r}\right)$. In fact, all spheres centred on the charge would thus be equipotential surfaces, their potentials being inversely proportional to their radii. Note that for a sphere "equipotential" is a 3-dimensional phenomenon.

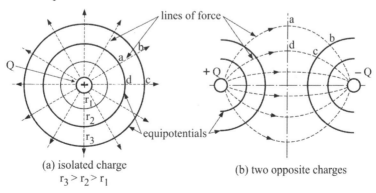

(a) isolated charge
$r_3 > r_2 > r_1$

(b) two opposite charges

Fig.2.6 : Electric field and equipotentials of charges

An equipotential surface has the property that, along any direction lying on the surface, there is no electric field. As illustrated in the above figure, *equipotentials*, or equipoential surfaces, *are always at right angles to the lines of force*[1]. This is also borne out by the lines of force and equipotentials formed by two opposite charges, placed a distance apart, as shown in Fig.2.6(b).

Capacitance

A significant application of electrostatic concepts is in the device called **capacitor**, invariably used in electrical and electronic circuits; they are essential, for example, in radio and television receivers, in transmitters, and now in all types of digital equipment. Primarily, a capacitor is a device for storing charge, at the given electric potential. Capacitance can be explained as electrostatic property of a capacitor, or the capacity of holding charge at the given potential.

Since charge, Q, is proportional to electric potential V, or potential difference, that is

$$Q \propto V$$

then $\dfrac{Q}{V}$ would be a constant. This is known as *capacitance*, C

or
$$C = \frac{Q}{V}$$

$$[Q = CV; \quad V = Q/C]$$

Thus, capacitance of a capacitor can be defined as the ratio of stored charge to the potential at which the charge is held[2].

Parallel-plate capacitor

The commonest form of a capacitor consists of two parallel plates, each having an 'active' surface area of, say, A m^2 and separated by a gap of height d as shown in Fig.2.7. A PD of V volt is maintained across the plates,

---

[1]The 'squares' formed by the intersection of adjacent lines of force and equipotentials, such as a, b, c, d, (Fig. 2.6) are known as "curvilinear" squares and form the basis of field mapping, by graphical means or otherwise.

[2]In an electric/electronic circuit a number of capacitors may be connected in series or parallel. In the first case, the individual potential drops across each of the capacitors would be added, the resultant capacitance being given by

$$\frac{1}{C} = \frac{1}{C_1} + \frac{1}{C_2} + \frac{1}{C_3} + \cdots ;$$

in the latter, capacitors would share a common PD, V, the resultant capacitance in this case will be C = C_1 + C_2 + C_3 . . ., derived from the Q v/s V relationship as above where C_1, C_2, C_3, ... denote individual capacitances.

each of which having a charge numerically equal to Q. The gap between the plates may simply be filled with air (the capacitor thus being called an air-capacitor) or any other dielectric or insulator such as mica, ebonite or even paper.

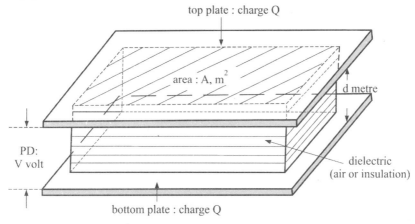

Fig.2.7 : A parallel-plate capacitor

The electric field strength between the plates is given by

$$E = \frac{Q}{\varepsilon A}$$

while the potential gradient is V/d V/m.

Since both are numerically equal,

$$\frac{V}{d} = \frac{Q}{\varepsilon A}$$

or $C = \dfrac{Q}{V} = \varepsilon \dfrac{A}{d}$ F , or farad, the unit of capacitance in MKS system.

For an "air-capacitor"

$$C = \varepsilon_0 \frac{A}{d}, \quad \text{since } \varepsilon_r = 1.0$$

Energy stored in a charged capacitor

A capacitor may simply be charged by connecting its plates to a source of electricity; for example, a storage battery when the capacitor itself becomes a store of electrical energy, with a potential difference across its plates and a charge on each plate corresponding to the PD and capacitance of the capacitor.

During the process of charging, let the potential difference at any instant across the plates be v when a corresponding quantity of charge dq is transferred to the plates, and the work done will be v dq. Since q is also equal to C v

$$dq = C \, dv$$

and the small work done

$$dw = C\ v\ dv$$

Hence, the work done in establishing the full PD, V

$$W = \int_0^V C\ v\ dv$$

or the energy stored,

$$W = \frac{1}{2}\ CV^2$$

this work done being stored as potential energy in the medium between the plates, the unit of stored energy being expressed in terms of the units of C and V. If C is in farad and V in volt, the stored energy will be in joule (J).

Charging of a capacitor

Let an uncharged (or fully discharged) capacitor of capacitance C be connected across a battery of potential difference V and let a suitable resistance R be included in the charging circuit[1]. Starting from zero, the voltage slowly builds up across the plates as the current flows into the circuit. At any instant when the voltage is v and corresponding charge dq in time dt, the current will be given by

$$i = \frac{dq}{dt}$$

Since

$$q = C\ v, \quad i = C\frac{dv}{dt}.$$

Also, in the series circuit the current in terms of R is given by

$$i = \frac{V - v}{R}$$

$$\therefore \qquad \frac{V - v}{R} = C\frac{dv}{dt}$$

or

$$v = V - CR\left(\frac{dv}{dt}\right)$$

The solution of this equation is

$$v = V\left(1 - \epsilon^{-t/CR}\right)$$

showing that the PD across the capacitor builds up exponentially with time, with a time constant of CR, the corresponding charge at any instant being

$$q = C\ v = C\ V\left(1 - \epsilon^{-t/CR}\right)$$

---

[1]The series resistance R is invariably included in the circuit to control or limit the charging *rate* (or flow of current) of the capacitor with time.

Similarly, if the capacitor is initially charged to a PD of V and charge Q, the expression for discharge with time through a series resistance R can be obtained to be

$$v = V \in^{-t/CR}$$

and $$q = C V \in^{-t/CR}$$

The variations of PD across a capacitor with time during charging/ discharging are shown in Fig.2.8 for two arbitrarily different values of time constant. Note that the time constant being a product of C and R, the same can be controlled by a proper selection of C and R[1].

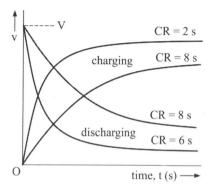

Fig.2.8 : Charging and discharging of a capacitor

CURRENT ELECTRICITY

Current electricity or now commonly known simply as electricity differs from static electricity by the flow of electric current in a circuit.

Comment

Historically, the concept of electric current in practice originated from the discovery of voltaic pile by Volta as the easily produced source of electric potential. When connected to a 'circuit', it resulted in the flow of a current. The pile worked like the modern-day battery.

In this context, the next historic development was the enunciation of a law by Alexander Ohm in 1826 (known as the Ohm's law) who stated that the current in a circuit for a given PD was proportional to a property of the circuit, later came to be known as "resistance", such that

$$I \propto V$$

and $$I = \frac{V}{R}$$

---

[1]An interesting aspect of the relationship between Q, V and I is observed in a simple dry-cell battery used to light a (torch) bulb. The battery is known to hold a certain charge at a PD or voltage of (say) 3 V and supplies current to the bulb for a period of time, itself getting discharged in the process. In partially discharged condition whilst the terminal voltage may still be very close to 3 V and the battery may light up an LED, it may not provide even a faint glow in the torch bulb. This is because an LED can light up to full brightness taking only about 15 mA whilst the torch bulb may require up to 250 mA and the *charge* in the battery may not be sufficient for the purpose!

being the circuit resistance, largely dependent on physical properties of the conductor making up the circuit, viz., the material and size or length and cross-sectional area of the conductor; the above relationship being commonly known as the **Ohm's law**. The current so obtained was constant in magnitude with time of flow, given by the above expression, and called the "direct current", (DC); its one of the main usage being in the process of electrolysis, then a flourishing 'industry'.

In fact, till the advent and development of alternating current (abbreviated AC) and generators (largely known as ALTERNATORS) ånd transformers towards the close of the 19th century, everything electric, including motors and street lighting, was in terms of direct current usage.

The Direct-current Electricity

From the historic voltaic pile to development of batteries (lead-acid and other types) as means of producing and storing DC electricity, to huge dynamos later generating electricity based on electromagnetic phenomenon, the DC electricity and its working is essentially based on Ohm's Law. An important usage of DC is in various measurements as a source of supply, as discussed in later chapters.

Conductors in DC circuits

The conductors or wires in DC circuits are invariably made of copper, esp. in measurement experiments, and may exhibit one or all of the following effects:

(i) *resistance-related effect* : the resistance of a wire is temperature dependent and the variation of resistance at any temperature can be expressed by

$$R_t = R_0 [1 + \alpha (t_2 - t_1) + \beta (t_2 - t_1)^2 + \gamma (t_2 - t_1)^3 + ...]$$

where R_t is the resistance of the wire at (the elevated) temperature t_2; R_0 the initial resistance at temperature t_1; and α, β, γ, . . . are the "temperature coefficients" of copper;

(ii) *thermo-emfs effects* : resulting from joints/contact of the wire with metal parts/terminals other than copper; for example brass or steel;

(iii) *effects in humid conditions* : the possibility of electrolysis and corrosion at dissimilar joints, and formation of oxide at bare ends.

DC supply from rectifiers

Increasingly, in most modern usage, DC supply at desired voltage(s) is obtained from rectifier units of which a variety are now commercially available. A rectifier supply essentially consists of a transformer, a (bridge) rectifier, a filter and a stabiliser. The waveform may, however, deviate from smooth DC and harmonics or ripples in the output may adversely affect the operation of some applications.

Alternating-current (or AC) Electricity

The fundamental feature of AC electricity, now used universally, and indeed in many measurements, is the *time dependence* of the voltage at the given frequency – largely the power frequency (50 or 60 Hz) – and the waveform which should ideally be sinusoidal (or a sine wave) under all circumstances.

The relationship between the supply voltage, V, and corresponding current, I, in an AC circuit is given by

$$V = I \times Z$$

a modified form of Ohm's law where Z is called IMPEDANCE (from the physical property of impeding the flow of current) of the circuit and depends on various combinations of circuit elements such as resistance, inductance and capacitance[1].

The other important aspects of AC electricity are:

Single phase or three phase

A single-phase supply involves the use of two conductors in the circuit: the LIVE and NEUTRAL[2]. It is important to maintain a few simple requirements: for example, the switch and fuse (if used) must be placed in the live wire, the latter following the former and wires of different colours to be used to distinguish live from the neutral wire.

In the three-phase AC supply, some more considerations are:

(a) *balanced or unbalanced*

A balanced supply is characterised by equal magnitudes of voltage in all the three phases, displaced by 120°(E) from each other; whilst an unbalanced supply may differ in both aspects, even if having the same phase sequence, as shown in Fig.2.9[3].

---

[1]See Chapter III for a detailed discussion.

[2]In some cases, there may be a third wire: the EARTH or ground wire.

[3]Fed from a balanced supply, if the (three-phase) load is also balanced, that is, it has identical impedances in the three phases, the load currents in the phases will also be balanced.

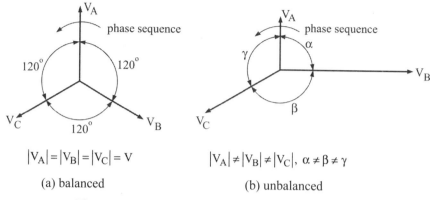

$$|V_A| = |V_B| = |V_C| = V$$

(a) balanced

$$|V_A| \neq |V_B| \neq |V_C|, \ \alpha \neq \beta \neq \gamma$$

(b) unbalanced

Fig.2.9 : Balanced and unbalanced 3-phase supply

(b) *phase sequence*

This depicts the arbitrary 'rotation' of the three phases (taken as anticlockwise to be normal, 'derived' from the orientation of the three co-ordinate axes X, Y, Z) in the time reference frame; for example V_A followed by V_B, followed by V_C as in Fig.2.9. Proper phase sequence is an important requirement in several applications. For example, the direction of rotation in a 3-phase induction motor depends on the phase sequence of the supply to the stator and may lead to excessive heating of the rotor if the phase sequence is suddenly reversed whilst the machine is running.

In most 3-phase usage, the supply is essentially balanced at normal phase sequence. However, the load may or may not be balanced.

"Colour code" for conductors in AC usage

For ease of working (wiring, connections and maintenance), a universally adopted colour code is generally adhered to as follows[1]:

Single Phase :   live – red, yellow, blue or brown
                           neutral – black
                           ground – green
                           of appropriate gauge or size

Three Phase :   phase conductors/connections – red[R], yellow[Y] and
                                               blue[B] (phase sequence : R-Y-B)
                           neutral – black
                           ground/earth – green

---

[1]A strict adherence to the colour code helps in making quick, un-ambiguous measurement in the laboratory. It may also result in 'balanced' distribution of connected loads in big residential and commercial building complexes, and easy fault repairs.

ELECTRICITY AND MAGNETISM

Magnetic Effect of Electric Current

Oersted's discovery

The magnetic effect of the electric current was discovered by Oersted in 1820, who like many others suspected a relationship between electricity and magnetism. In his classic experiment he held a current-carrying conductor near a compass and observed the compass needle to deflect. A reversal of current resulted in a deflection in the opposite direction.

Specifically, the magnetic lines circle around the conductor in a plane perpendicular to the axis of the conductor as shown in Fig.2.10. The flux lines flow in counter-clockwise direction for the current flowing upward; clockwise for current in the downward direction *when viewed from the top*[1]. More explicitly, the directional relationship of the magnetic field with the current is also given by the Maxwell's cork screw rule in which the direction of current is represented by the

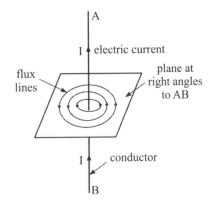

Fig.2.10 : Magnetic flux due to a current-carrying conductor

forward movement of the screw whilst the magnetic field direction is given by the clockwise circular rotation of the screw, moving away from oneself.

Force on a Current-carrying Conductor in a Magnetic Field

Oersted's discovery was followed by the classic experiments by Ampere about the same time, showing that if a current carrying conductor is held at right-angles to flux lines of a magnetic field, it would experience a mechanical force as illustrated on Fig.2.11. In the process, the original magnetic field is distorted in orientation as shown. Clearly, the force produced on the conductor is a consequence of the interaction between the original field and the field produced by the current in the conductor.

---

[1]A visual indication of the flux lines can be obtained by performing a simple experiment using a wire carrying a direct current, a paper held in a plane at right angles to the wire, passing through the paper as in Fig.2.10, and sprinkling iron filings on the paper which immediately assume the shape of flux lines, but without showing the direction.

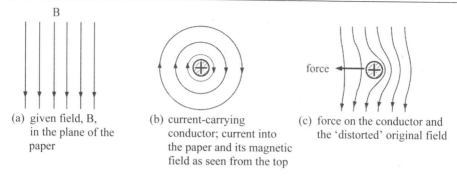

(a) given field, B, in the plane of the paper

(b) current-carrying conductor; current into the paper and its magnetic field as seen from the top

(c) force on the conductor and the 'distorted' original field

Fig.2.11 : Current-carrying conductor in a magnetic field and force on it

Note that the direction of the given field, the current through the conductor and the force acting on the conductor are *mutually perpendicular*, making the phenomenon to be three-dimensional in space, a fact established by Ampere after a great deal of experimentation.

This important phenomenon, forming the basis of working of all electric motors, can be pictorially depicted by the (Alexander) Fleming's "left-hand rule" as shown in Fig.2.12.

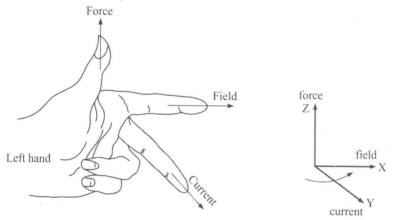

(a) the thumb, fore- and middle-finger held mutually at right angles

(b) force, magnetic field and current along three mutually perpendicular axes

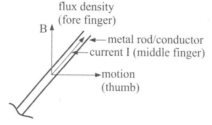

(c) Fleming's rule applied to a current-carrying conductor

Fig.2.12 : Fleming's left-hand rule

A reference to Fig.2.11(c) shows that the resultant lines of force are *stressed*, a simple example of Maxwell's stress. Such lines of force, being in a state of tension act like a stretched elastic thread, and so the system would resemble a kind of magnetic catapult resulting in exertion of the mechanical force on the conductor.

Numerically, the magnitude of the force in the case of a conductor of length *l* metre, arranged at right-angles to the magnetic field B tesla and carrying a current I ampere is given by

$$F = B \, l \, I \quad N$$

The magnetic force due to an electric current

The law of Biot and Savart

To derive the magnetic field, B, due to the current flowing in a conductor, Biot and Savart provided with a law which can be expressed as

$$\Delta B \propto \frac{I \times \Delta l \, \sin\alpha}{r^2}$$

at a point, P, as illustrated in Fig.2.13.

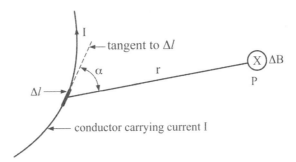

Fig.2.13 : Biot and Savart law

The law cannot be proved directly in its basic form as shown; yet it forms the basis of calculating field due to various 'regular' forms and combination of conductors as detailed out in Appendix II. The constant of proportionality in the law depends on the medium in which the conductor is situated. In air (or, more precisely, in vacuum), the law can be expressed as

$$\Delta B = \frac{\mu_0}{4\pi} \frac{I \times \Delta l}{r^2} \sin\alpha$$

and is also known as Ampere's formula.

The corresponding magnetising *field* in air will be given by

$$\Delta H = \frac{\Delta B}{\mu_0}$$

or

$$\Delta H = \frac{1}{4\pi} \frac{I \times \Delta l}{r^2} \sin \alpha$$

at the same point.

Magnetomotive force: MMF

Assume a magnetic force H at a given point that remains constant over a length of *magnetic path l*. Then the product $H \times l$ is defined as the "magneto-motive force", MMF, impressed on that path. Over the same length, H from the Biot-Savart law can be written as

$$H \times l = K I, \text{ where K is a constant}$$

so that

$$H = \frac{K I}{l}$$

Drawing an analogy to the (DC) electric circuit where the current is given by

$$I = E/R, \qquad R = \rho \left(l/a \right)$$

where E is the EMF and R the circuit resistance, in a magnetic circuit, the total flux $\varnothing$ can be considered analogous to current and related to the MMF.

Now

$$\varnothing = B\, a$$

$$= \mu\, H\, a$$

where $\mu = \mu_0\, \mu_r$, μ_r being the relative permeability of the magnetic material.

Substituting for H

$$\varnothing = \mu\, a \times \frac{KI}{l}$$

$$= \frac{KI}{\left(l/\mu\, a \right)}$$

Comparing the expression for $\varnothing$ with that for I, it is seen that

$$K I \text{ is analogous to E}$$

and

$$\left(l/\mu\, a \right) \text{ is analogous to R}$$

Hence, if the product KI is defined as MMF and $\left(l/\mu\, a \right)$ be named as "reluctance", there would be a *magnetic*-circuit 'law' similar to Ohm's law, viz.,

$$\text{Flux} = \frac{\text{MMF}}{\text{reluctance}}$$

The unit of MMF in MKS system will be ampere and that of reluctance would be A/Wb.

Magnetisation of Iron or Steel

In the above, it is seen that MMF is simply the current that produces the magnetising field. Also, in the expression a single conductor has been considered. Clearly, if there were N conductors, each carrying the current I, the MMF will increase to N I ampere.

A particular case is that of a very long solenoid of length l metre, wound with N turns placed close to each other for which the MMF is N I and the magnetising field is given by

$$H = N\ I/l \quad A/m$$

[See Appendix II].

An Electromagnet

So far, the discussion about the magnetic field has assumed air as the medium. However, the effect of current can be increased many-fold if the field is produced in a ferromagnetic material with a high relative permeability; for example, mild steel. A body made of such ferromagnetic material and excited by current – DC or AC – is called an "electromagnet". The important thing about an electromagnet is that it would exhibit magnetic properties only when excited by the current and would revert to non-magnetic state when the current is switched off. A simple example of a DC electromagnet is the historic telephone relay (not in much use now) and, on the other extreme, the field system of an alternator comprising windings wound around pole sections and carrying huge currents.

A Toroid as an Electromagnet

Imagine a *closed* ring of iron (or mild steel) of mean length l and uniform cross section a as shown in Fig.2.14. Let it be wound with a coil of N turns, carrying a current I. Such an electromagnet is called a "toroid", an important and much useful device in many magnetic measurements, and indeed the basic constructional requirement for a bar-primary current transformer [see Chapter VI]. For analysis, the toroid may be considered as a long solenoid, closed at its ends, esp. if the mean length is very large compared to its sectional dimensions, such that the flux inside the solenoid is 'continuous' and H is uniform.

Then

$$H = \frac{N\ I}{l}$$

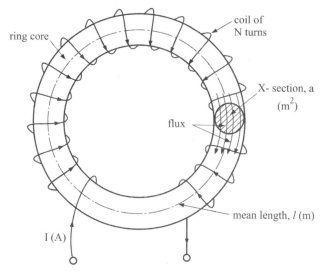

Fig.2.14 : A toroid wound with N turns

If the relative permeability of the iron (for this particular value of H^1) is μ_r, the flux density will be

$$B = \mu_0\,\mu_r\,H$$

$$= \mu_0\,\mu_r\,\frac{N\,I}{l}$$

and flux

$$\varnothing = B \times a$$

$$= \mu_0\,\mu_r\,\frac{N\,I\,a}{l}$$

or

$$\varnothing = \frac{N\,I}{\left(\dfrac{l}{\mu_0\mu_r a}\right)}$$

Comment

Although the expression for H and B in the toroid appears to be simple and, correspondingly, the design to be straightforward, in practice several factors may have to be taken into account to construct the toroids, known as "ring specimens" and frequently used in magnetic measurements as described in Chapter XI. Also, note that the process of winding large to very large number of turns manually and uniformly around the closed ring using thin wires can be tedious and time consuming. However, advanced machines are now available to do the job automatically in a short time.

---

[1] As discussed later, the relationship between H and μ_r is, in general, not proportional and is dependent on excitation.

Magnetic Circuits

A magnetic circuit may consist of several sections of iron (or any other magnetic material) joined to form various "series"-"parallel" branches, each characterised by its own length and area of cross-section such that the flux in each branch or section may be different. An analysis of such a circuit may be carried out analogous to that of an electric circuit. Just as the resistances in an electric circuit can be combined in series, parallel or series-parallel to yield a resultant resistance, in a magnetic circuit *reluctances* of the various branches can be treated similarly. For example, in the case of a magnetic circuit with three branches in series, the total reluctance may be given by

$$S = S_1 + S_2 + S_3$$

where S_1, S_2, S_3 represent individual reluctances, so that

$$S = \frac{l_1}{a_1\mu_1} + \frac{l_2}{a_2\mu_2} + \frac{l_3}{a_3\mu_3},$$

μ_1, μ_2, μ_3 themselves being given by $\mu_1 = \mu_0\,\mu_{r_1}$, $\mu_2 = \mu_0\,\mu_{r_2}$ and $\mu_3 = \mu_0\,\mu_{r_3}$, respectively.

Magnetic circuit and flux distribution of a typical salient pole DC machine

A part of the field-system of a multi-pole DC machine is shown in Fig.2.15.

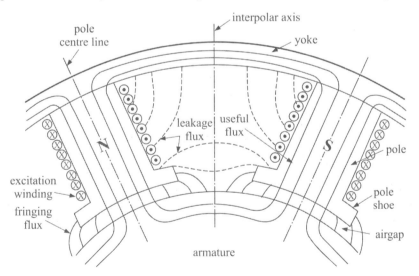

Fig.2.15 : Flux distribution of a pole-pair of a DC machine field-system

Specifically, the figure illustrates that all the flux produced by the excitation windings on the pole(s) *does not* reach the armature: there is the "useful" flux (including the "fringing" flux) that enters the armature through the airgap; additionally, there may be appreciable "leakage" flux as shown.

This phenomenon, in varying degree, is common to nearly all magnetic systems/circuits.

The ratio of *total* flux to useful flux is called the "leakage factor" which is a measure of utilisation of flux in a given system[1].

Tractive Effort of an Electromagnet

This forms another example of a magnetic circuit/phenomenon.

The basis of a "tractive" force, or the force of attraction, in a magnet can be explained by reference to the pair of poles shown in Fig.2.16. In the first figure, a north pole is in close contact with a south pole such that there will be certain force of attraction, F, between them. The second figure shows the two poles separated from each other. Assuming the flux distribution to be unaltered, the same force will now attract, or pull, the lower magnet so as to 'close' the gap.

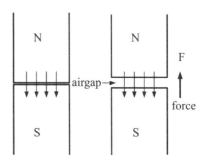

Fig.2.16 : Force of attraction in a magnet

Neglecting the effect of gravity the force of attraction between the poles is given by

$$F = \frac{B^2 a}{2\,\mu_0}\ N$$

where   F is the force in newton,

B the airgap flux density, T, and

a the "common" area in m^2.

A much useful application of the tractive effort of an electromagnet in industry (in machine shops etc.) is to lift ferrous materials or large steel products, as shown in Fig.2.17, and move them around with the help of an overhead crane. The electromagnet is moved close to the product to be moved and energised by switching on the excitation whereby the product is lifted and stuck to the magnet. It is then carried to its destination by the crane and placed at the desired location, and the excitation switched off. Heavy to very heavy steel products can thus be easily moved around.

---

[1]Although the field pattern in Fig.2.15 is shown to be qualitative, such field plots for a given field system can be obtained in quantitative term by using appropriate analytical methods; for example, a suitably applied finite-element technique defining Laplacian field in the region (and solving the Laplace equation with appropriate boundary conditions). The field distribution so obtained forms an important asset in the design of electric machines.

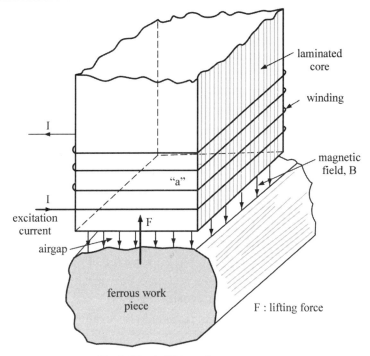

Fig.2.17 : A lifting electromagnet

Non-linear and Multi-valued Relationships Between B and H

Nearly all ferrous/magnetic materials exhibit non-linear relationship between the applied magnetising field and resulting flux density (a 'point' phenomenon); additionally the value of B may not be the same for a given H for ascending or descending values of H.

'Virgin' magnetisation curve, or B/H characteristic

When a specimen of magnetic material is magnetised for the first time or after having been completely demagnetised, the variation of flux density in it with the applied magnetising field will be as shown in Fig.2.18 and is inherently non-linear (curve A).

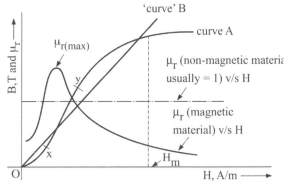

Fig.2.18 : Mangetisation curve or B/H characteristic

The corresponding variation of the material's relative permeability, μ_r, plotted against H is also shown and is highly non-linear with the maximum value of μ_r occurring somewhere when the B/H curve is linear (portion x y). By contrast the behavior of a non-magnetic material, for example copper or the air medium, will be a straight line, 'curve' B, and the μ_r variation will be a horizontal line with the μ_r value being simply unity. A particular feature of the B/H curve is to reflect the behaviour of the magnetic material for progressively increasing values of H; for example about H_m when the curve begins to flatten showing the setting up of magnetic saturation in the material, and the μ_r curve may tend to level off.

Multi-valued Relationship Between B and H

The B-H or hysteresis loop

If, after the beginning of saturation (B/H curve beyond $H=H_m$), the magnetising force H is gradually reduced, it will be observed that the original B/H curve is *not* traversed in the reverse direction, but follows a variation as shown by portion PQ in Fig.2.19, such that when H is reduced to zero the sample under test still has a non-zero flux density, marked B_r in the figure and known as "residual flux density" or *remanance*.

If now the magnetising force is reversed (for example by reversing the direction of excitation current in DC tests), some value of H in the reversed direction would lead to flux density becoming zero (point R on the – H axis). This value of H (= H_c) is known as coercive force or *coercivity*. Note that both B_r and H_c may be quite high for hard magnetic materials and play a significant role in the design and application of permanent magnets.

If H is continued to be varied *cyclically* as shown by the arrows, the magnetisation of the sample will result in a "loop" (or a closed curve) depicted by P, Q, R, S, T, U, V (or P), called the "hysteresis loop[1]".

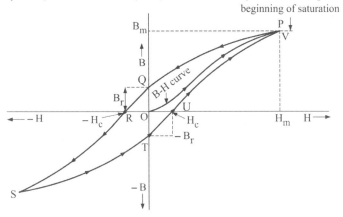

Fig.2.19 : Hysteresis loop of a magnetic material

---

[1]Hysteresis from the Greek word ὑστερέω (pronounced este'rio), meaning "to lag behind".

Major and minor hysteresis loops

The shape and size of the hysteresis loop would depend on the magnetic properties of the material at the molecular level and the process of magnetisation. When the sample is magnetised first well into saturation and the magnetising field is then gradually reduced, reversed and increased in a cyclic manner as indicated by arrows in Fig.2.19, the loop thus obtained is called the "major" B-H loop with the saturation condition being denoted by the point H_m, B_m. However, if a 'symmetrical' magnetising force is applied in full cyclic form, but of decreasing amplitudes H_{m_1}, H_{m_2}, H_{m_3}, etc., . . .

such that $H_{m_3} < H_{m_2} < H_{m_1}$, etc., B-H loops of nearly the same shape will result, fitting one inside the other within the major loop as shown in Fig.2.20. Such loops which can be countless in number corresponding to chosen values of H_{m_1}, H_{m_2}, (and B_{m_1}, B_{m_2} etc.) are called "minor" B-H

loops.

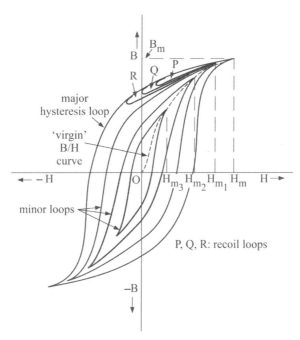

Fig.2.20 : Major, minor and recoil loops

If carried out appropriately, the apexes of the minor loops would lie on the virgin B-H curve of the material as shown in the figure.

Recoil loops

If, for some reasons, the magnetising force is reduced from its maximum value, traversing the B-H loop, and is increased again in the original sense, from some intermediate point, then small closed loops are produced, known as "recoil" loops. Three such loops, P, Q, R, depicting different conditions of decreasing and again increasing the magnetising field are shown in Fig.2.20.

The Area of the Hysteresis Loop and Hysteresis Loss

The hysteresis loop of a magnetic material can be visualised as a kind of "indicator diagram", similar to that of a heat engine; the area of the loop then being a measure of the work that has to be done so as to take the material through the complete cycle of magnetisation, that is, the B-H loop.

Referring to Fig.2.21, if the excitation or the magnetising force H is obtained by varying a current I (for example in a toroid) for a value of H between $+H_m$ and $-H_m$ and the corresponding flux density is B in tesla, the flux in the material cross section will be

$$\emptyset = B \, a \quad Wb$$

where a is the area of cross-section in m^2.

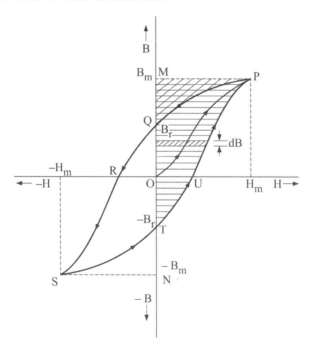

Fig.2.21: Hysteresis loop and hysteresis loss

Suppose the mean length of the toroid core is *l* metre and it is wound with N turns. Then the EMF induced in it at any instant will be

$$e = N \frac{d\emptyset}{dt} = N \frac{d(B \, a)}{dt} \quad V$$

and the power *supplied* at the instant to overcome this back EMF (in other words to build up the magnetic field in the core) will be

$$p = e\,i = i\,N\frac{d(B\,a)}{dt}\ \text{W}$$

and the energy supplied for the same in time t sec will be

$$\int_0^t e\,i\,dt = \int_0^t a\,i\,N\frac{dB}{dt} = a\int_{-B_r}^{+B_m} i\,N\,dB$$

since, proceeding from point T (H = 0) along T,U,P when t = 0, B = − B_r, equal to OT. [See Fig.2.21].

Also, the magnetising force acting upon the ring at any instant is

$$H = \frac{N\,i}{l}\ \text{A/m} \quad \text{or} \quad N\,i = l\,H$$

Hence, the energy supplied is

$$l\,a\int_{-B_r}^{+B_m} H\,dB\ \text{J}$$

Since (*l* a) is the volume of the toroid core, it follows that the energy supplied *per unit volume* to the system is

$$\int_{-B_r}^{B_m} H\,dB\ \text{J}$$

This energy is *stored* in the magnetic field in the toroid core, represented in the figure by T,U,P,M,Q,O,T (horizontal hatching).

Upon reducing the excitation (and hence the flux or flux density from B_m, traversing the curve PQ), the induced EMF is in the *same direction* as the applied EMF so that energy is now *returned* to the excitation circuit as the flux is reduced. From the same reasoning as before, the energy returned during the reduction of H from H_m to zero (or B from B_m to B_r) is

$$l\,a\int_{B_r}^{B_m} H\,dB\ \text{J}$$

or

$$\int_{B_r}^{B_m} H\,dB\ \text{J} \quad \text{per unit volume}$$

represented in the figure by (double hatched) area P,M,Q.

Thus, the energy *absorbed by the core material due to hysteresis*, or the net work in the process, is the *difference* between the energy supplied and the energy returned, or

$$\left[\int_{-B_r}^{B_m} H\,dB - \int_{B_r}^{B_m} H\,dB\right] \quad \text{per unit volume,}$$

represented by the area T,U,P,Q,O,T.

If the current be now reversed in the excitation winding, H being varied along O,R,S,T traversing the other half of the loop, the energy absorbed will be represented by the area Q,R,S,T,O,Q. Hence, the *total* energy absorbed, or work done in carrying the specimen through the entire cycle, will be the area T,U,P,Q,R,S,T, or the *area of the hysteresis loop*.

The quantum of work so supplied is lost during the cycle of magnetisation and demagnetisation and therefore the area of the hysteresis loop represents the loss in the material per unit volume and during each cycle[1].

If the loop is plotted to scale with H in A/m and B in tesla, the loss per unit volume per cycle will be the measured area of the loop, given in joule. It follows that larger the area of the loop pertaining to a magnetic material sample, greater will be the loss[2].

Steinmetz's (Hysteresis) Law

Based on prolonged practical studies, (Charles P.) Steinmetz, a German scientist, came out with an empirical formula for hysteresis loss in a magnetic material, given by

$$W_h = k\ B_{max}^{1.6}\quad \text{J/unit volume/cycle}$$

and is found to be of sufficient accuracy for most practical purposes, provided that the maximum flux density, B_{max}, encountered in the application lies between 0.1 to 1.2 T approximately.

The constant k, called the Steinmetz's hysteresis coefficient, depends on, and is constant for, a given magnetic material. Typical values for k may lie between 250 and 500 for annealed steels and mild steel, and about 200 for silicon steel.

Hysteresis with Alternating Current Excitation

When a magnetic material, for example the core of the toroid referred to above, is subjected to excitation by alternating current (say of the power frequency of 50 Hz), it is clear that this automatically results in cyclic

---

[1] It is reckoned as "loss" since it would manifest as heat and temperature rise in the material, if unchecked.

[2] This points to some important aspects of design of magnetic devices and electric equipment. For example, in devices where it is desirable to 'store' as much magnetic energy as possible, such as permanent magnets, materials having a "fat" hysteresis loop are preferred. In contrast, in electromagnets, such as relays, as well as AC excited equipment (in which the hysteresis loss is magnified by number of cycles for a given volume), it is essential to choose magnetic materials having the smallest hysteresis loop.

magnetisation of the material, each cycle of AC corresponding to the traverse of one hysteresis loop. In general, if the frequency of the AC supply is f Hz, the number of B/H loops traversed *per second* will be f and so the hysteresis loss per unit volume, per second will be f times that for one loop.

According to the Steinmetz's law, the total loss per second per unit volume will be

$$W_h = \left(k\, B_m^{1.6} \right) \times f \quad J$$

ELECTROMAGNETIC INDUCTION

Faraday's Discovery

Whilst the 19th-century scientists, following Oersted in 1820, established the link between electric current (or rather the direct current obtained from a battery or later the dynamo) and magnetic field due to it, it was left to the genius of Michael Faraday, a British scientist, to produce a *current* by means of a *varying magnetic field* linking an electric circuit or system. He began his 'research' around 1825, but succeeded only in 1831 to discover what are now universally known as "laws of electromagnetic induction" – the basis of all (AC) generation and utilisation all through the centuries.

For his experiments, Faraday[1] used a ring-shaped magnetic core of about 25 cm mean diameter, wound with two windings overlapping each other, but *not connected* in any way. A picture of the (original) ring used by Faraday is shown in Fig.2.22.

---

[1] A blacksmith's son, Michael Faraday was born on 22nd September, 1791, at Newington, Surrey, England. He began his 'career' as a book-binder, later to be engaged as a laboratory assistant at the Royal Institution, London. He was inspired and initiated into 'science' by the noted scientist Humphry Davy, and became his assistant.

In 1821, Faraday discovered the principle of (DC) electric motor (and built one); two years later learnt to liquify chlorine, and discovered the revolutionary laws of electromagnetic induction on 29 August, 1831 – the laws which gave electricity to the world and changed the destiny of mankind!

Among his other accomplishments, Faraday produced the first dynamo, stated the basic laws of electrolysis, discovered that a magnetic field would rotate the plane of polarisation of light, and discovered Benzene.

He died on 25th August, 1867, at Hampton Court, Surrey.

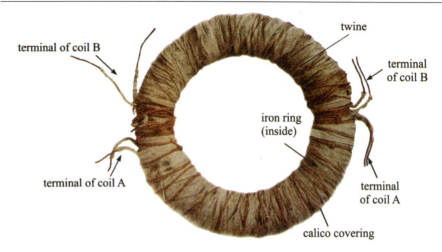

Fig.2.22 : The Faraday's ring

A schematic representation of the Faraday's ring[1] and experiment is shown in Fig.2.23. One of the two coils is connected to a battery in series with a key whilst the other is connected to a "centre-zero" DC galvanometer. The current in the winding can be switched on or off by pressing the key or opening it.

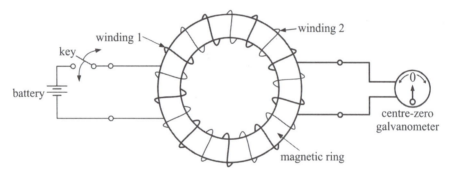

Fig.2.23 : Schematic of Faraday's experiment

The Experiment

Faraday observed that whilst the key was pressed making the current to grow in winding 1, a throw was noticed in the galvanometer in a certain direction (clockwise or anticlockwise). If the key was then opened to switch off current in the coil, the galvanometer needle again showed a deflection, *but in the opposite direction*. More importantly, the galvanometer showed no deflection for *steady* current in the coil (winding 1). This cycle is shown graphically in Fig.2.24.

---

[1]The original ring is housed in the Science Museum, London.

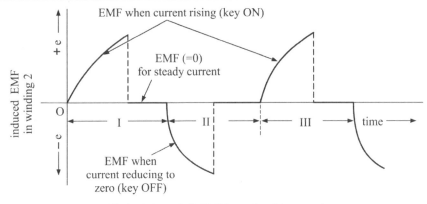

Note: intervals I, II, III need not be equal

Fig.2.24 : Induced EMFs in winding 2 during current ON and OFF cycles

Clearly, a small EMF was produced in winding 2 resulting in the deflection of the galvanometer whilst the current in winding 1 was *rising*, or establishing, causing a flux to be set up in the core, but there being no EMF for steady current or flux, showing that it was the *time variation* of current, and hence the flux linking the second winding through the core, which resulted in the induction of EMF in winding 2. Also, a *decay* of current in winding 1 caused collapse of flux in the core and induction of EMF in winding 2, *but of reverse sign*.

Faraday then repeated his experiment using an air-cored, simple coil with its terminals connected to a galvanometer as before and using a movable permanent magnet as depicted in Fig.2.25.

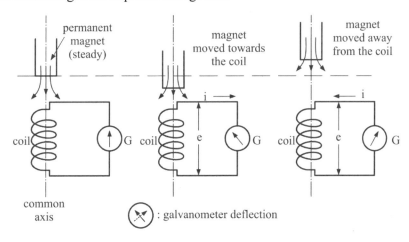

Fig.2.25 : Alternative experiment to show electromagnetic induction

The 'lateral' movement of the magnet relative to the coil resulted in momentary deflection of the galvanometer: in one direction if the magnet was brought closer to the coil; in opposite direction if the magnet was moved away. Faraday observed that the deflection occurred only when there was *relative* motion between the coil and the magnet. Also, that the deflection became stronger if the air-cored coil was replaced with one having an iron core. Another simple model to illustrate the phenomenon of electromagnetic induction is shown in Fig.2.26. By pulling the magnet in and out, a self EMF is induced in the coil, only

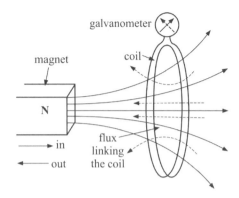

Fig.2.26 : Change of flux linking the coil and induction of EMF

during the action (or relative motion), resulting in the galvanometer deflection.

The Laws of Electromagnetic Induction

From his years of sustained experiments and observations, Faraday finally enunciated the laws of electromagnetic induction as follows:

1. A changing magnetic field *induces* an electromotive force (EMF) in a conductor.

2. The electromotive force is proportional to the *rate of change* of the field.

3. The direction of the induced electromotive force depends on the *orientation* of the field.

Here, the change of field is implied to be relative: a magnet moving relative to the coil or vice-versa. Also, the strength of the field and speed of relative movement both play a key role.

Mathematical Expressions of Faraday's Laws

A. The "flux-cutting" rule or dynamically-induced EMF in a moving conductor

Imagine a conductor of length L (m), moving with a velocity v (m/s) in a direction so as to '*cut*' magnetic field of flux density B (T) at right angles at all times. According to the concepts of Faraday's laws, an EMF will be induced in the conductor, given by

$$e = B\,L\,v \quad V$$

This can be demonstrated as follows:

Consider the circuit shown in Fig.2.27. PQ is a straight conductor of length *l* metre, in electrical contacts at its ends with two parallel wires QR and PS and free to slide over them while maintaining an electric contact. All the circuit elements are situated in a "uniform" *vertical* magnetic field of flux density B tesla, perpendicular to the horizontal plane PQRS containing the conductor and wires as shown.

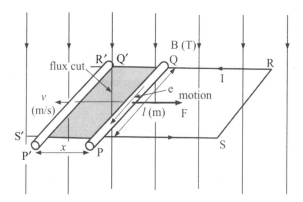

Fig.2.27 : EMF induced in a conductor by the flux-cutting rule

Suppose PQ moves to the left by a distance x metre with a uniform velocity v m/s to the position $P'Q'$. Let the time taken for the movement be t sec.

Then the flux cut

$$\emptyset = B \times area\ P\ Q\ Q'\ P'$$

$$= B\ l\ x$$

So, numerically, the induced EMF would be

$$e\ =\ \frac{flux\ cut}{time} = \frac{B\ l\ x}{t}$$

$$= B\ l\ v\ V^{1}, \qquad since\ \ v = x/t$$

Observe that the three quantities B, l and v are mutually at right angles with respect to each other at any instant. If a conductor cuts through magnetic flux whilst moving in a direction making an angle θ with that of the magnetic field, then the component of its velocity in a direction perpendicular to the field is v sin θ and the induced EMF is then given by

$$e = B\ l\ v\ sin\ \theta$$

The above relationship is also explained by the well-known Fleming's right-hand rule depicted in Fig.2.28, wherein the thumb, fore-finger and

---

[1]It follows that if the conductor PQ were carrying a *current* in the direction indicated by e, a force F will be produced on the conductor in the direction shown in the figure, as given by Fleming's left-hand rule, resulting in the motion of the conductor.

second finger are held mutually perpendicular to each other to indicate induced emf v/s magnetic field and motion of the conductor, respectively.

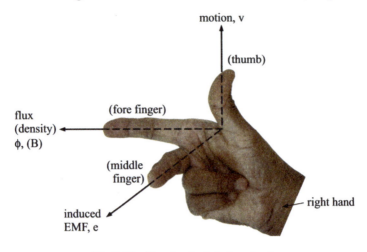

Fig.2.28 : Fleming's right-hand rule

Comment

The flux cutting rule forms the fundamental basis for generation of EMF in nearly all electric generators – from historic times to date – which comprise conductors moving relative to a magnetic field. In DC dynamos, the field system consisting of salient poles carrying the excitation windings [see Fig.2.15] is stationary whilst the conductors housed in the slots of a rotating system called the armature cut the radial flux lines in the airgap, the armature being rotated by means of a prime-mover. In giant-size alternators, designed to produce AC electricity, reverse is the case.

An important aspect is that the rating of a given generator would depend on the *electromagnetic loading*, proportional to D^2L where D is the machine diameter at the airgap and L the *active* axial length of the conductors. Thus, in the case of a 2-pole turbo-alternator having a 'rotor' diameter of about 1.5 m, the peripheral speed may be about 240 m/s corresponding to a rotational speed of 3000 rpm (to match the supply frequency of 50 Hz), and the axial length of conductors or the 'active' part of the machine may run into several metres. In contrast, a 20-pole hydro-generator having a diameter of about 6 metre on the airgap, the axial length may be only about a metre or so, the peripheral velocity of conductors being about 90 m/s. In both cases the electromagnetic loading or power output of either machine would be nearly the same.

B. EMF due to transformer action[1] or statically-induced EMF

This is the alternative form of EMF, based on the direct application of Faraday's laws where the induction of EMF does not involve movement or motion of conductor(s) relative to a given magnetic field.

The EMF is essentially induced due to the effect of a *time-varying* magnetic field in a circuit that is always *linked* with the field, Thus, the principle involved is the "changing of time-varying *flux linkages* "[2] rather the flux cutting.

Then, according to Faraday's (also Neumann's) law, if N Ø represents the flux linkages of a circuit of N turns and Ø is time-varying (or a function of time), the electromotive force induced in the circuit will be given by

$$e = N\frac{d\emptyset}{dt} \text{ volt}$$

when Ø is in weber.

Clearly, e in the above expression is the magnitude of EMF *at any instant*, varying with time, identical to Ø.

If, instead of Ø it is the flux density of the field that is defined, also time varying and linking a circuit of area A m[2], for example core of a magnetic circuit, the induced EMF will be

$$e = NA\frac{dB}{dt} \text{ volt}$$

where B is in tesla.

Lenz's Law

A concept that is sometimes neglected, or poorly understood, is the basic "cause-effect" relationship in electromagnetism. For example, in the case of the EMF induced in a stationary coil due to a movement of a magnet to change the flux linkages what would happen if the coil terminals are closed so as to result in the flow of a current? The current would set up its own magnetic field. What should be the 'direction' of this field? Very logically, this must *oppose* the field due to the magnet to 'balance' the flux linkages with the coil which were zero in the first place[3].

---

[1]So called because it forms the fundamental principle of design and operation of all transformers.

[2]*Flux linkages*: Usually, the term flux linking a circuit implies a coil of any shape through which the flux Ø 'threads'. If the coil has more than one turn, then the flux through the coil is the sum of the fluxes through the individual turns. This is known as flux linkages. Mathematically, it is expressed as N Ø Wb, or Wb-turns, where N is the number of turns and Ø (in weber) the flux linking *each* turn.

[3]An effect closely associated with the "constant flux linkages theorem", as also an example of the principle of the "conservation of energy".

These concepts were generalised into an important law by Lenz in 1835 which states that "the induced EMF in a (closed) circuit is such that the flux produced (by the current) due to the EMF will oppose the flux responsible for producing the EMF." Alternatively, "the direction of the indused EMF is such as to oppose the time-varying inducing flux".

The mathematical expression of this law, in conjunction with Faraday's law of induction, is a modification of the induced EMF expression. Thus

$$e = -N\frac{d\emptyset}{dt} \text{ V}$$

The "minus" sign in the expression accounts for the Lenz's law. Note, however, that an EMF (alone) cannot tend to prevent the change of flux in the inducing circuit.

Self and Mutual Induction

A phenomenon that directly derives from the statically-induced EMF together with the Lenz's law is known as "self induction", first discovered by Joseph Henry in 1832, and forms the basis of the important physical property *inductance* of an "inductor" - a component widely used in electronic circuits and devices. The term inductance is also associated with a winding (of single- or multi-turn) used variously in electrical equipment.

It is seen that an EMF is induced in a coil (or circuit) whenever its flux linkages change. This can result from the

(a) movement of a magnet relative to the stationary coil;

(b) change of current in the coil itself : a change in current resulting in a change of flux linking the coil and thus inducing an EMF;

(c) change of current in a *neighbouring* coil when the EMF induced in the circuit is said to be *mutually* induced.

Self inductance

Considering the second case if the change in flux is due to a change in the current (that is, a time-varying current) in the circuit, the induced EMF will be given by

$$e = -N\frac{d\emptyset}{dt}$$

and is called a "self-induced" EMF.

In the expression, $d\emptyset/dt$ is the "equivalent" rate of change of flux as the current is changed. Assuming that magnetic properties of the circuit remain unchanged, the magnetic flux can be related to the current as

$$\emptyset = k\,i$$

at any instant where k is a constant and i the current.

Thus, substituting for Ø,

$$e = -N\, k \frac{di}{dt} \text{ volt}$$

or

$$e = -\left(N \frac{\text{Ø}}{i}\right) \frac{di}{dt} \quad \text{since } k = \text{Ø}/i \text{ as defined}$$

or

$$\boxed{e = -L \frac{di}{dt}}$$

in which the quantity L to be given by

$$\boxed{L = \frac{N\text{Ø}}{i}}$$

or *flux linkages per unit current* is defined as "co-efficient of self induction" or simply the "self inductance" of the circuit.

If Ø is expressed in weber and i in ampere, L will be in Wb-turns/ampere, and the unit of self inductance is called "henry" (or H), named after Joseph Henry.

Also, in a circuit of cross-sectional area A m^2 and flux density B tesla, the flux will be given by

$$\text{Ø} = B \times A \text{ Wb}$$

With H as the magnetising field (proportional to I) producing the flux density B,

$$\text{Ø} = \mu_0\, \mu_r\, H\, A \text{ Wb} \qquad\qquad \mu_r > 1$$

and

$$L = \frac{N(\mu_0\, \mu_r\, H\, A)}{I}$$

$$= (N\, \mu_0\, \mu_r\, A) \times k' \text{ where } H = k'I$$

This shows that inductance of a circuit will be constant if relative permeability (μ_r) of the medium would be constant; for example, air ($\mu_r = 1$). Other media, esp. magnetic materials with non-linear μ_r variation possess variable inductance.

In a magnetic circuit of which the reluctance is known from the knowledge of 'magnetic' length, area of cross-section and relative permeability, viz.,

$$S = \frac{l}{a\, \mu_0\, \mu_r}$$

the flux Ø in the circuit will be given by

$$\emptyset = \frac{Ni}{S}$$

where N denotes the number of turns in the winding and i the current flowing through it. The self inductance of the circuit will then be given by

$$L = \frac{N}{i} \times \frac{Ni}{S}$$

$$= \frac{N^2}{S} H$$

in MKS system of units.

This expression is particularly useful in calculating self inductance of a solenoid when the number of turns on the winding and dimensions of the solenoid are known.

Mutual inductance

If two coils (or circuits) are held in close vicinity with their planes oriented in appropriate directions, for example parallel to each other, and a *unit* current flows in one of them, then the number of *flux linkages* with the other coil, of the magnetic flux due to this current, is called "mutual inductance" between the two coils or circuits.

Mathematically, if the current in coil 1 is i_1 and varies such that the rate of change is di_1/dt, then the EMF, e_2, induced in coil 2 will be given by

$$e_2 = -M \frac{di_1}{dt} \, V$$

where M is the mutual inductance of the two coils.

Likewise, if a time-varying current i_2 flows in coil 2 instead of in coil 1, then the EMF induced in coil 1 when the rate of change of current in coil 2 is di_2/dt, is given by

$$e_1 = -M \frac{di_2}{dt} \, V$$

with the mutual inductance M being still the same.

The unit of mutual inductance is also henry. In quantitative terms, a mutual inductance of 1 henry exists between two circuits when a rate of change of current of 1 ampere per second in one circuit induces an EMF of 1 volt in the other circuit.

Comment

1. Clearly, the relative orientation and disposition of the two coils with respect to each other would play the key role in the phenomenon of mutual induction and induction of EMF. For example, in order to maximise it, or to enhance the magnetic coupling, the two coils should be

   (a) placed as close to each other as possible with their planes parallel;

   (b) one of the coils may be completely embedded inside the other on a suitable support, as will be seen in one of the requirements of magnetic measurements.

Fig.2.29 : A bifilar coil

2. In many measurements, it is required to use a coil wound with desired number of turns but non-inductive in effect, that is, whose net self-inductance is very nearly zero. A practical method to achieve this, based on the property of mutual inductance, is to wind the coil as parallel-wire winding or the wire doubled-back on itself before being coiled up as shown in Fig.2.29, known as the *bifilar* principle. Every part of the coil is thus traversed by the same current in *opposite* direction such that its resultant magnetic field and hence the linkages are negligible. [See also Fig.1.3, Chapter I].

A common application of the above design is in the fabrication of precision, *non-inductive* resistance(s) used in bridge methods, or the resistances used in decade resistance boxes, discussed later.

Relation between self- and mutual-inductance

Suppose that two coils, having N_1 and N_2 turns, respectively, are so close together that the whole of the flux produced by a current in one coil links completely with the other. Let this flux be $\emptyset$(Wb) when the current in coil 1 is i_1(A).

Then the *self-inductance* of coil 1 is

$$L_1 = N_1 \frac{\emptyset}{i_1} \quad H$$

and the mutual-inductance is

$$M = N_2 \frac{\emptyset}{i_1} = \frac{N_2}{N_1} L_1 \quad H$$

Similarly, if a current i_2 flows in coil 2, its self-inductance is

$$L_2 = N_2 \frac{\varnothing}{i_2} \text{ H}$$

and

$$M = N_1 \frac{\varnothing}{i_2} = \frac{N_1}{N_2} L_2 \text{ H}$$

Therefore

$$\frac{N_2}{N_1} L_1 = \frac{N_1}{N_2} L_2 = M$$

or

$$M^2 = L_1 L_2$$

and

$$\boxed{M = \sqrt{L_1 L_2}}$$

This relationship holds only when the whole of the flux from one coil links with the other. In practice, this condition may not be fulfilled and the mutual inductance may be given by

$$M = K \sqrt{L_1 L_2}$$

where $K = \dfrac{M}{\sqrt{L_1 L_2}}$ is called the "coefficient of coupling", to be ideally equal to 1.

Stored Energy in an Inductor

A circuit possesses self induction by virtue of the magnetic flux set up by the current. However, a magnetic flux is the seat of stored energy and consequently an inductor carrying a current and associated with a flux will have energy stored in it. Consider a closed iron circuit for simplicity, for example a toroid, of mean length l (m), cross-section a (m^2) and wound with a coil of N turns. If the coil is excited with a current i (A), the flux in the core will be

$$\varnothing = \frac{Ni}{\left[l/(\mu_0\mu_r a)\right]} \text{ Wb}$$

where μ_r is the relative permeability *corresponding to current i*.

Let the current be increased by a small amount di and correspondingly the flux to increase by $d\varnothing$. Then

$$d\varnothing = \frac{N \, di}{\left[l/(\mu_0\mu_r a)\right]} \text{ Wb}$$

Now when the flux through a coil changes, the work done is the product of current and the flux change, so that for a coil of N turns the work done is

$$dw = Ni \, d\varnothing = \frac{N^2}{\left[l/(\mu_0\mu_r a)\right]} i \, di \text{ J}$$

If the current in the coil increases from zero to some final value I, the work done is thus

$$W = \frac{N^2}{[l/(\mu_0\mu_r a)]} \int_0^I i \, di$$

$$= \frac{1}{2}\frac{N^2}{[l/(\mu_0\mu_r a)]} I^2 \text{ J}$$

Now for the given magnetic circuit, the flux per unit of current is

$$\frac{\varnothing}{i} = \frac{N}{[l/(\mu_0\mu_r a)]} \text{ Wb/A}$$

∴ (Flux per ampere) × (No. of turns)

$$= \frac{N^2}{[l/(\mu_0\mu_r a)]}$$

and this is clearly the self-inductance of the circuit in henry.

Therefore, substituting in the expression for the work done

$$\boxed{W = \frac{1}{2}LI^2 \text{ J}}$$

which also represents the stored energy in the circuit (or the inductor).

Statically Induced EMF and Transformer

A static transformer is a practical realisation of the principle of mutual induction. The simplest possible transformer would thus consist of two coils in close vicinity, P and S, as shown in Fig.2.30, the former of which called the "primary" being connected to, or excited by, a suitable AC supply. The time-varying alternating current which flows in P sets up a corresponding time-varying magnetic flux, most of which links with the coil S called the "secondary". An EMF is induced in coil S due to the mutual-induction principle. Note the orientation of the two coils with respect to each other.

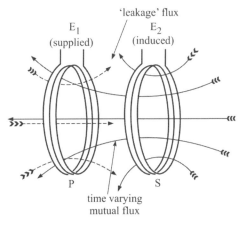

Fig.2.30 : Schematic of the "basic" transformer

Basic theory

Let N_1 = number of turns in coil P

N_2 = number of turns in coil S

$\emptyset_m$ = the maximum value of the mutual flux linking P and S

Assuming that the flux due to the applied EMF and current flowing in P to be varying sinusoidally, the flux at any instant will be

$$\emptyset = \emptyset_m \sin \omega t$$

where ω ($= 2\pi f$) is the *angular* frequency of variation of $\emptyset$ (and the applied EMF E_1).

Hence, e_2, the EMF induced in S at any instant will be

$$e_2 = - \text{(rate of change of flux)} \times N_2 \quad V$$

$$= (-\emptyset_m \, \omega \cos \omega t) \times N_2 \quad V$$

The maximum *magnitude* of e_2 is thus

$$\left| E_{2(max)} \right| = \emptyset_m \, \omega \, N_2 \quad V$$

$$= 2 \pi N_2 \, \emptyset_m \, f \quad V$$

and for the rms value of e_2 under the assumptions,

$$E_2 = E_{2(max)} / \sqrt{2} \quad V$$

$$= \sqrt{2} \, \pi \, \emptyset_m \, N_2 f \quad V$$

$$\text{or} \quad \boxed{E_2 = 4.44 \, \emptyset_m N_2 f \quad V}$$

On the primary side, this should correspond to an rms *applied* voltage given by

$$\boxed{E_1 = 4.44 \, \emptyset_m \, N_1 \, f \quad V}$$

from which it follows that $\dfrac{E_1}{E_2} = \dfrac{N_1}{N_2}$.

Comment

Just as the motionally-induced EMF by the flux-cutting rule forms the basis of induction of EMF in all rotating machines – dynamos and alternators – the statically induced EMF is basic to the design and operation of *all* transformers. From the elementary developments in the closing years of the 19th century, transformers now comprise the *most important* link in the transmission, distribution and utilisation of electrical energy. So much so that before the electricity reaches the door-steps of consumers – industrial, commercial or domestic – there would have been nearly half-a-dozen stages of transformation of voltage using a variety of transformers. For example, efficient transmission of electric power over long to very long distances (up to thousands of km) would be impossible without the use of step-up (and step-down) transformers. Like-wise the end-use distribution to users, single or three-phase, entails a typical 11kV/400V "distribution" transformer.

[Note that each distribution transformer also renders a kind of "free service" to LV-supply consumers in that the cost of 24-hour iron loss is borne out by the utility; hence the prediction of iron loss at design stage and its accurate measurement during manufacture acquires a special significance.]

A special application of transformers is in providing high voltage input for some special measurements (discussed later) where they also act as an isolating device.

Then there are special-purpose transformers known as instrumentation transformers, the CTs and PTs, used extensively for measurements, protection and control in all modern power systems.

A fact little known to, and appreciated by consumers of electronic gadgets and devices – from DVD players to cell-phone chargers – using very low voltage DC output is that they end up paying practically nothing for the use of the supply modules! This is because the current drawn on the AC side of the input transformer in such appliances may be only a few mA, being not recordable by the energy meters.

In yet another important field of usage, specially designed and commissioned convertor transformers form the basis of HVDC transmission and DC traction.

A special, "open-core" transformer

This is a uniquely designed and fabricated *electromagnetic* device, *similar* to a (1-phase) transformer, with an "open-ended" core meant to demonstrate a variety of aspects of electromagnetic induction following Faraday's laws.

A photographic view of the "transformer" is given in Fig.2.31 whilst its basic constructional details are shown in Fig.2.32. The device consists of a soft iron core in strip or wire form, tightly packed inside a PVC tube of about 20 mm diameter. The projected vertical length of the core is about 30 cm. The tube/core is inserted rigidly inside a wooden or PVC bobbin at the bottom. The bobbin carries a winding wound using (24 SWG) enameled copper wire, typically of 2300 turns[1]. When excited by an alternating current, the core is magnetised, the magnetic field strength gradually

---

[1]So that when excited by a 1-phase, 230 V AC supply, the turns/volt would be 10.

reducing towards the 'open' end. A variety of experiments can be organised around the core, using specially fabricated accessories.

[See Appendix III for details of experiments].

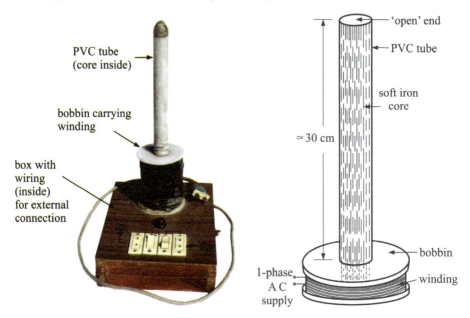

Fig.2.31 : Photographic view of the "open-ended" transformer

Fig.2.32 : Constructional detail of the transformer

WORKED EXAMPLES

1. The figure below shows three small charges A, B and P in a line. The charge at A is positive, that at B is negative and   the one at P is positive. The values are as shown.

    (a) Calculate the force on the charge at P due to A and B

    (b) At what point X on the line AB could there be no force on the charge P due to A and B if P were placed there?

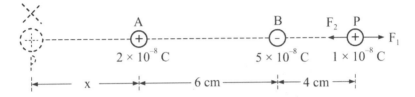

    (a) The distance from A to P is 10 cm or 0.1 m. So charge at A *repels* charge at P with a force given by

$$F_1 = 9 \times 10^9 \frac{Q_1 Q_2}{r^2}$$

$$= \frac{9 \times 10^9 \times 2 \times 10^{-8} \times 1 \times 10^{-8}}{(0.1)^2}$$

$$= 1.8 \times 10^{-4} \text{ N}$$

The distance from B to P is 4 cm or 4×10^{-2} m. So the charge at B *attracts* the charge at P with a force F_2 given by

$$F_2 = 9 \times 10^9 \frac{Q_1 Q_2}{r^2}$$

$$= \frac{9 \times 10^9 \times 5 \times 10^{-8} \times 1 \times 10^{-8}}{\left(4 \times 10^{-2}\right)^2}$$

$$= 2.8 \times 10^{-3} \text{ N}$$

∴   the resultant force *towards* B

$$= F_2 - F_1 = 2.8 \times 10^{-3} - 1.8 \times 10^{-4}$$

$$= 2.62 \times 10^{-3} \text{ N}$$

    (b) If the charge at P were taken to a point X to the *left* of A on the line AB, let the distance from A at which the net force on P is zero be x metre.

Then the force of repulsion on P at X due to A will be

$$F_1 = 9 \times 10^9 \frac{2 \times 10^{-8} \times 1 \times 10^{-8}}{x^2}$$

and the force of attraction between P and B will be

$$F_2 = 9 \times 10^9 \frac{5 \times 10^{-8} \times 1 \times 10^{-8}}{(0.06 + x)^2}$$

For the net force $F_1 \sim F_2$ to be zero,

$$9 \times 10^9 \frac{2 \times 10^{-8} \times 1 \times 10^{-8}}{x^2} = 9 \times 10^9 \frac{5 \times 10^{-8} \times 1 \times 10^{-8}}{(0.06 + x)^2}$$

or $2 \times (0.06 + x)^2 = 5x^2$

or $2x^2 + 0.24x + 0.0072 = 5x^2$

or $x^2 - 0.08x - 0.0024 = 0$

or

$$x = \frac{0.08 \pm \sqrt{0.0064 + 0.0096}}{2}$$

$$= 0.1032 \text{ m} \quad \text{or} \quad 10.32 \text{ cm}$$

2. Calculate the force of repulsion between two alpha particles separated by 10^{-10} cm, given that the charge on each particle is $+3.2 \times 10^{-19}$ C.

$$F = 9 \times 10^9 \frac{q_1 \times q_2}{d^2}$$

$$= 9 \times 10^9 \frac{(3.2) \times 10^{-19} \times (3.2) \times 10^{-19}}{(10^{-12})^2}$$

$$= 9.2 \times 10^{-4} \text{ N}$$

3. Two positive point charges of 12 and 8 micro-coulomb, respectively, are 10 cm apart at A and B. Find the work done in bringing them 4 cm closer. (Assume $1/4\pi\varepsilon_0 = 9 \times 10^9$ m/F).

Suppose the 12 μC charge is fixed in space/position. Since 6 cm or (10 − 4) cm is 0.06 m and 10 cm is 0.1 m, then the potential difference between points 6 and 10 cm from it is given by

$$V_{AB} = \frac{Q}{4\pi\,\varepsilon_0}\left(\frac{1}{a} - \frac{1}{b}\right)$$

Here, $a = 0.06$ m; $b = 0.1$ m

$$\therefore \qquad V_{AB} = \frac{12\times10^{-6}}{4\pi\,\varepsilon_0}\left(\frac{1}{0.06} - \frac{1}{0.1}\right)$$

$$= 12\times10^{-6}\times 9\times10^{9}\times\left(16\frac{2}{3} - 10\right)$$

$$= 720{,}000 \ \ V$$

(Note the extremely high PD due to quite small charges)

The *work done* in moving the 8 μC charge from 10 cm to 6 cm away from the 12 μC charge is given by the expression

$$W = Q\,V$$

or $\qquad W = 8\times10^{-6}\times720{,}000$

$$= 5.8 \ \ J$$

4. An oil drop of mass 2×10^{-14} kg carries a charge Q. The drop is stationary between two parallel plates, one above the other, 20 mm apart with a PD of 500 V between them. Calculate Q. Take g = 10 m/s^2.

The configuration is as shown in the figure.

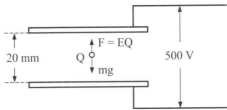

Since the drop is stationary,

upward force on the charge, F

$\qquad\qquad\qquad$ = downward force due to its weight, mg

Potential gradient across the plates,

$$E = \frac{PD}{d} = \frac{500}{20\times10^{-3}} \ \ V/m$$

Therefore $\qquad 2\times10^{-14}\times10 = Q\times\dfrac{500}{20\times10^{-3}}$

or

$$Q = \frac{2 \times 10^{-14} \times 10 \times 20 \times 10^{-3}}{500}$$

$$= 8 \times 10^{-18} \ C$$

5. Two 'infinite' parallel plates 1 cm apart are maintained at a potential difference of 100 V. Calculate the acceleration of an electron between them if the charge on the electron is 1.603×10^{-19} C and its mass is 9.1×10^{-31} kg.

PD across the plates, $\quad\quad V = 100 \ V$

$\therefore$ electric gradient $\quad\quad V/d = 100/10^{-2} \quad$ or $\quad 10^4 \ V/m$

This is also numerically equal to the electric field strength in N/C

$\therefore$ force on the electron, $\quad\quad F = 10^4 \times 1.603 \times 10^{-19}$

$$= 1.603 \times 10^{-15} \ N$$

and the acceleration, $\quad\quad \alpha = \dfrac{F}{m}$

$$= \frac{1.603 \times 10^{-15}}{9.1 \times 10^{-31}}$$

$$= 1.76 \times 10^{15} \ m/s/s$$

6. A capacitor of capacitance 0.01 µF in series with a non-inductive resistance of 50 kΩ is connected to a DC supply of 460 V. Determine the voltage to which the capacitor has been charged when the charging current has decreased to 90 percent of its initial value. Calculate also the time elapsed since the commencement of charge.

Time constant, $\quad\quad T = C \, R$

Voltage on the capacitor at any instant

$$v = V \, (1 - \in^{-t/T})$$

Charging current

$$i = \frac{V - v}{R}$$

When $\quad\quad t = 0, \quad v = 0$

$\therefore \quad\quad i = V/R$

When current has fallen to 90%,

$$0.9 \left(V/R \right) = \left(V - v \right)/R$$

∴ $$v/R = 0.1 \left(V/R \right)$$

or $$v = 0.1 \times 460$$

$$= 46 \ V$$

To find the time elapsed

$$1 - \epsilon^{-t/T} = v/V = 0.1$$

∴ $$\epsilon^{-t/T} = 0.9$$

or $$t = T \ \log_\epsilon \left(1/0.9 \right)$$

$$= CR \ \log_\epsilon \left(10/9 \right)$$

$$= \left(0.01/10^6 \right) \times 50{,}000 \times 2.3 \ (1 - 0.954)$$

$$= 52.9 \times 10^{-6} \ s \quad \text{(or 52.9 μs)}$$

7. An iron ring made up in the form shown in the figure below has $l_1 = 10$ cm and $a_1 = 5$ cm^2; $l_2 = 8$ cm and $a_2 = 3$ cm^2; $l_3 = 6$ cm and $a_3 = 2.5$ cm^2. It is wound with a coil of 250 turns. Assuming that the whole flux passes through the ring (no leakage), calculate the current required to produce a total flux, Ø, of 0.0004 Wb (4×10^4 maxwell). The magnetic properties of the material are given in the following table:

H (A/m)	199	239	398	724	1178	1934
B (T)	0.6	0.8	1.0	1.2	1.4	1.6

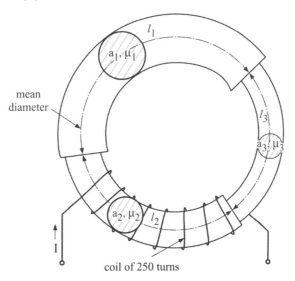

coil of 250 turns

From the relation $B = \mu H$, the corresponding values of relative permeability, μ_r, are

$$\mu_r \quad 2400 \quad 2666 \quad 2000 \quad 1320 \quad 946 \quad 65$$

In the various parts of the magnetic ring the flux densities are, for $\varnothing = 0.0004$ Wb

$$B_1 = \frac{\varnothing}{a_1} = \frac{0.0004}{5 \times 10^{-4}} = 0.8 \text{ T}$$

$$B_2 = \frac{\varnothing}{a_2} = \frac{0.0004}{3 \times 10^{-4}} = 1.333 \text{ T}$$

$$B_3 = \frac{\varnothing}{a_3} = \frac{0.0004}{2.5 \times 10^{-4}} = 1.6 \text{ T}$$

To obtain corresponding values of relative permeabilities, the available data are plotted in the graph shown below.[1]

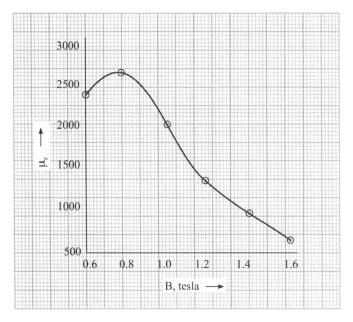

From this, the μ_r values are found to be

$$\mu_{r_1} = 2670$$

$$\mu_{r_2} = 1050$$

$$\mu_{r_3} = 650$$

---

[1]Note the progressive reduction in μ_r values, especially when the saturation sets in - a function of cross-sectional area for the same flux.

Calculating the reluctances of the various parts, and expressing as summation for the reluctances to be *in series* for the ring, the total reluctance

$$S = \frac{1}{\mu_0}\left(\frac{l_1}{a_1\mu_{r_1}} + \frac{l_2}{a_2\mu_{r_2}} + \frac{l_3}{a_3\mu_{r_3}}\right)$$

Substituting the respective values of l, a, μ_r etc.

$$S = \frac{1}{4\pi\times10^{-7}}$$

$$\times\left(\frac{1\times10^{-1}}{2670\times5\times10^{-4}} + \frac{0.8\times10^{-1}}{1050\times3\times10^{-4}} + \frac{0.6\times10^{-1}}{650\times2.5\times10^{-4}}\right)$$

$$= \frac{10^{-1}}{4\pi\times10^{-7}}(0.7491+2.539+3.69)$$

$$= 5.55\times10^5$$

$\therefore$ the MMF $= 0.0004\times5.55\times10^5 = 222$ A

and the current required in the winding/coil

$$I = \frac{220}{250} = 0.88 \text{ A}$$

8. A wrought-iron ring 50 cm in mean circumference and 5 cm^2 cross section has an airgap of 2 mm cut in it, the iron on either side of the gap being chamfered so as to give a gap area of 3 cm^2. If the ring has a magnetising coil of 1000 turns and the leakage factor is 1.5, calculate the current required to produce a flux across the gap of 0.0003 Wb.

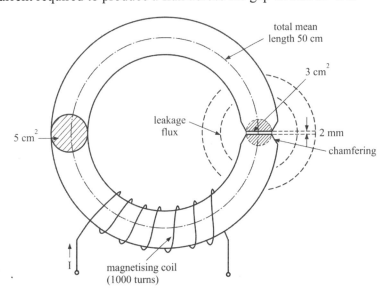

total mean length 50 cm

3 cm^2

leakage flux

5 cm^2

2 mm

chamfering

magnetising coil (1000 turns)

I

The magnetic circuit is as shown in the figure. The airgap or "useful" flux is 0.0003 Wb. When the ring is excited, some flux leaks across the airgap as shown.

"Total" flux in the iron ring $= 1.5 \times 0.0003$

or $\qquad\qquad\qquad \varnothing_i = 0.00045$ Wb

Flux density in the iron

$$B_i = \frac{\varnothing_i}{a_i} = \frac{4.5 \times 10^{-4}}{5 \times 10^{-4}} = 0.9 \text{ T}$$

From the graph of the previous example, assuming it to apply in this case, too, the relative permeability of iron corresponding to this flux density is 2350.

$\therefore\qquad$ reluctance of the iron part of the ring

$$S_i = \frac{l_i}{a_i \mu_0 \mu_r} = \frac{0.498}{5 \times 10^{-4} \times 4\pi \times 10^{-7} \times 2350}$$

and the MMF to produce the required flux in the *iron*

$$\text{MMF}_i = S_i \times \varnothing_i$$

$$= \frac{0.498}{5 \times 10^{-4} \times 4\pi \times 10^{-7} \times 2350} \times 4.5 \times 10^{-4}$$

$$\text{(since } l_i = 500 - 2 = 498 \text{ mm or } 0.498 \text{ m)}$$

$$= 152 \text{ A}$$

For the airgap,

the reluctance, $\qquad S_a = \dfrac{l_a}{a_a \times \mu_0}\quad$ since $\quad \mu_r = 1.0$

$$= \frac{2 \times 10^{-3}}{3 \times 10^{-4} \times 4\pi \times 10^{-7}} = \frac{10^8}{6\pi}$$

and the MMF$_a \qquad = S_a \times \varnothing_a$

$$= \frac{10^8}{6\pi} \times 3 \times 10^{-4}$$

$$= 1592 \text{ A}[1]$$

Therefore total MMF $= 152 + 1592 = 1744$ A

and the excitation current $= \dfrac{1744}{1000} = 1.744$ A

---

[1]Typically, even though the airgap length is only 2 mm, compared to nearly 50 cm of the iron, the MMF required for the airgap is almost TEN times that for the iron.

9. A smooth-core armature working in a 4-pole field magnet has a gap (from iron-to-iron) of 12.7 mm. The area of the surface of each pole is 929 cm². The flux from each pole is 7×10^{-2} Wb.

Calculate

(a) the mechanical force exerted by each pole on the armature;

(b) the energy in joules stored in the *four* airgaps.

With the data given, it is convenient to transform the equation for tractive effort or force as follows:

$$F = \frac{B^2 a}{2\,\mu_0} = \frac{(Ba)^2}{2\,\mu_0 a} = \frac{\emptyset^2}{2\,\mu_0 a}\ N$$

Now $\emptyset$ in the gap $= 7 \times 10^{-2}$ Wb

and area of each pole, $a = 929 \times 10^{-4}\ m^2$

Assuming no leakage of flux, this would also be the airgap flux.

Therefore

$$F = \frac{\left(7 \times 10^{-2}\right)^2}{2 \times 4\pi \times 10^{-7} \times 9.29 \times 10^{-2}}$$

$$= 2.098 \times 10^4\ N$$

Total stored energy

$$W = \left(\frac{B^2 a}{2\,\mu_0}\right) \times \left(\text{volume per gap}\right) \times \left(\text{no. of gaps}\right)$$

$$\uparrow$$

energy density

The volume per gap will be the area of the pole (surface) × the gap length or $a \times r$ where r is the radial gap length.

$$W = \frac{(Ba)^2}{2\,a^2\,\mu_0} \times a \times r \times \left(\text{no. of gaps}\right)$$

$$= \frac{\emptyset^2 \times r}{2\,a\,\mu_0} \times \left(\text{no. of gaps}\right)$$

$$= \frac{\left(7 \times 10^{-2}\right)^2 \times 1.27 \times 10^{-2}}{2 \times 9.29 \times 10^{-2} \times 4\pi \times 10^{-7}} \times 4$$

$$= 1.066 \times 10^3\ J$$

10. The hysteresis loop for an iron specimen weighing 12 kg is equivalent to 3000 ergs per cc. Find the loss of energy per hour at 50 Hz. Specific gravity of iron 7.5 gm/cc. Assume 1 erg = 10^{-7} J.

$$\text{The volume of the specimen} \; = \frac{12 \times 10^3}{7.5} = 1600 \text{ cc}$$

∴ loss in the specimen = 3000 × 1600 ergs

or 3000 × 1600 × 10^{-7} J

The frequency of magnetisation being 50 Hz,

the hysteresis loss per sec

= 3000 × 1600 × 10^{-7} × 50 W

= 24 W or 0.024 kW

∴ energy loss/hr = 0.024 × 1 or 0.024 kWh

11. A horizontal metal frame PQST moves with a uniform velocity v of 0.2 m/s into a uniform field B of 10^{-2} T acting vertically downwards. PT is 0.1 m and PQ = 0.2 m. The resistance of the frame is 5 Ω. The sides QS and PT enter the field in a direction normal to the field boundary as shown.

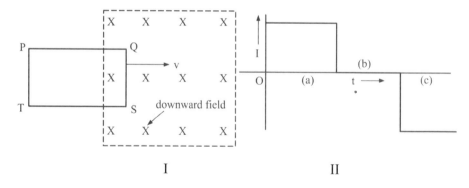

I   II

What current flows in the metal frame when

(a) QS just enters the field, starting to cut it;

(b) the entire frame is within the field and moving through it;

(c) QS just moves out of the field on the other side.

The configuration is shown in figure I above.

(a) An EMF given by

$$E = B\,l\,v = 10^{-2} \times 0.1 \times 0.2 = 2 \times 10^{-4} \text{ V}$$

is induced in only the conductor QS and none in others.

The current through the frame will be

$$I = V/R = 2 \times \left(10^{-4}\right)/5 = 4 \times 10^{-5} \text{ A}$$

(b) With the whole of the frame moving through the field, EMFs are induced in conductors QS and PT, *but in opposite directions;* there being thus no net EMF in the frame and hence no current.

(c) When the conductor QS has just moved out, EMF is induced in conductor PT, equal to that was induced in QS as in case (a), but of opposite sign.

or $\qquad\qquad\qquad\qquad I = -I$

Therefore, the variation of current in the frame with time will be as shown in figure II.

12. In a 2–pole DC machine, the armature has a diametre of 8.5 cm and the 'active' length of each conductor is 8 cm. If the average strength of the airgap field is 0.45 T and the armature rotates at 2000 rpm, calculate the EMF induced in each conductor.

$$\text{Circumference of the armature } = \pi D$$

$$= \pi \times 8.5 \times 10^{-2}$$

$$= 0.267 \text{ m}$$

$$\text{Speed of the armature } = 2000 \text{ rpm or } 2000/60 \text{ rps}$$

∴ peripheral speed of conductors,

$$v = 0.267 \times \frac{2000}{60} = 8.9 \text{ m/s}$$

Therefore EMF induced in each conductor

$$E = B\,l\,v = 0.45 \times 8 \times 10^{-2} \times 8.9$$

$$= 0.32 \text{ V}$$

13. A 100 kW, 125 V DC generator has 4 poles. There are 650 turns per field coil (on the poles) with a field current of 12 A. The flux per pole with this excitation is 8.5×10^{-2} Wb per pole. Calculate the self-inductance of the field. If the field winding is suddenly switched on to 125 V supply, calculate the time taken for the current to attain one half of its final value.

$$\text{Flux per ampere} = \frac{8.5 \times 10^{-2}}{12} = 7.08 \times 10^{-3} \text{ Wb/A}$$

$\therefore$ inductance of each coil $=$ (flux/ampere)$\times$ (number of turns)

$$= 7.08 \times 10^{-3} \times 6.5 \times 10^{2}$$

$$= 4.6 \text{ H}$$

and total inductance of the entire field

$$= 4 \times 4.6 = 18.4 \text{ H}$$

Resistance of the field system,

$$R = \frac{V}{I} = \frac{125}{12} = 10.4 \ \Omega$$

When the field supply is suddenly switched on, the current at any time t is given by

$$i = I[1 - \epsilon^{-(R/L)t}]$$

where I is the final steady current

or $i/I = 1 - \epsilon^{-0.565t}$ since $R/L = 0.565$

When $i/I = 0.5$

$$0.5 = 1 - \epsilon^{-0.565t}$$

whence $t = 1.21$ s

14. A solenoid 1 m long has 1000 turns, wound on a long iron bar of cross-section 5 cm^2. A second coil of 200 turns is wound on the same bar some distance from the end of the first coil; this distance being such that it is linked by one-half of the flux produced by the first coil. Find the coefficient of mutual induction of the second coil with respect to the first. Assume relative permeability of iron to be 1200 for the purpose.

The arrangement of the coils is as shown in the figure.

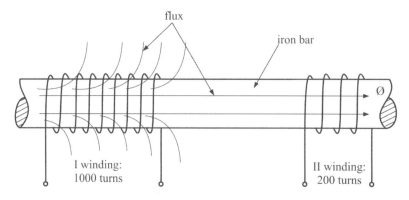

flux

iron bar

I winding:
1000 turns

II winding:
200 turns

Denoting the two coils/windings as P and Q, respectively, reluctance

of P, $S_P = \dfrac{l}{a\,\mu_0\,\mu_r}$

$= \dfrac{1}{5\times 10^{-4}\times 1200\ \mu_0} = 1.667/\mu_0$

$\therefore$ flux in coil P, $\varnothing_P = \dfrac{N_P I_P}{S_P} = \dfrac{1000}{1.667}\times \mu_0\ I_P = 600\ \mu_0\ I_P$

and   flux/A $= 600\ \mu_0$

$\therefore$ flux per ampere in coil Q

$= $ one-half of that in coil P $= 300\ \mu_0$

and the coefficient of mutual induction *of the second coil with respect to first,*

$$M = 300\ \mu_0 \times 200$$

$$= 6 \times 10^4 \times 4\pi \times 10^{-7} = 7.54 \times 10^{-2}\ \text{H}$$

15. Five turns of fine wire are wound closely about the centre of a long solenoid of radius 20 mm. If there are 500 turns stretched to 1 m length of the solenoid, calculate the mutual inductance of the two coils. State your reasoning.

The arrangement of the two coils on the solenoid is as shown below.

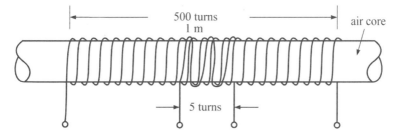

Area of X-section of the solenoid

$$a = \dfrac{\pi}{4}\left(\dfrac{40}{1000}\right)^2 = 4\pi \times 10^{-4}\ \text{m}^2$$

Assume this to be the cross-sectional area of each coil

The self inductance of the coil with 500 turns,

$$L_1 = \frac{N_1^2}{S_1} \text{ where } S_1 \text{ is the reluctance of this coil}$$

$$S_1 = \frac{l_1}{a\,\mu_0} = \frac{1.0}{4\pi \times 10^{-4}\,\mu_0}$$

$$\therefore \qquad L_1 = \frac{(500)^2}{1.0/(4\pi \times 10^{-4}\,\mu_0)} = (500)^2 \times 4\pi \times 10^{-4} \times \mu_0$$

$$= 100\,\pi\,\mu_0$$

Self inductance of the second coil, assuming it to link the whole flux produced in the solenoid,

$$L_2 = \frac{N_2^2}{S_2}$$

where $S_2 = \dfrac{l_2}{a\,\mu_0}$

Assuming lengths of coils to be proportional to number of turns and having the same area of cross section,

$$l_2 = 10 \text{ mm or } 10 \times 10^{-3}\,\text{m}$$

$$\therefore \qquad S_2 = \frac{10 \times 10^{-3}}{4\pi \times 10^{-4}\,\mu_0}$$

and $\qquad L_2 = \dfrac{5^2}{10 \times 10^{-3}} \times 4\pi \times 10^{-4}\,\mu_0 = \pi\,\mu_0$

$\therefore$ mutual inductance,

$$M = \sqrt{L_1\,L_2} = 10\,\pi\,\mu_0$$

$$= 39.4\ \mu H$$

EXERCISES

1. In the figure below, two small equal charges of 2×10^{-8} C are placed at A and B, one positive and the other negative. The distance AB is 6 cm. Determine the force on a charge of $+1 \times 10^{-8}$ C placed at P, where P is 4 cm from the line AB along the perpendicular bisector XP.

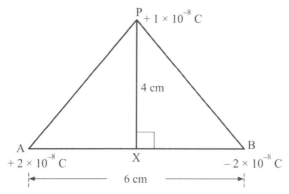

$$[8.64 \times 10^{-4} \text{ N}$$

2. An oil drop of mass 4.0×10^{-15} kg is held stationary between two horizontal plates when an electric field is applied between the plates. If the drop carries 6 electric charges, each of value 1.6×10^{-19} C, calculate the electric field strength across the plates.

$$[41.6 \text{ kV/m}$$

3. The figure below shows an arrangement of two point charges in air, Q being 0.30 μC.

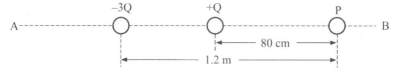

Find

(i) the electric field strength and the electric potential at P

(ii) the point on AB between the two charges at which the electric potential is zero.

$$[(i) \ 1.41 \times 10^{3} \text{ N/C}, -3.38 \text{ kV, (ii) } 0.3 \text{ m from } -3Q$$

4. A 'point' charge is held at $q = 100 \times 4 \pi \varepsilon_0$. Calculate the radii of the equipotential surfaces for a progressive difference of 10 V, from 100 to 60.

$$[1 \text{ m}, 1.11 \text{ m}, 1.25 \text{ m}, 1.429 \text{ m}, 1.667 \text{ m}$$

5. Two horizontal parallel plates, each of area 500 cm^2, are mounted 2 mm apart in vacuum. The lower plate is earthed and the upper plate is given a positive charge of 0.05 μC. Neglecting edge effects, find the electric

field strength between the plates. Deduce also (a) the potential of the upper plate, (b) the capacitance between the plates, (c) the electric energy stored in the system.

$$[1.13 \times 10^5 \text{ V/m, (a) } 227 \text{ V, (b) } 2.2 \times 10^{-10}\text{F, (c) } 5.67 \times 10^{-6} \text{ J}$$

6. A parallel plate capacitor is made up of a number of separate plates, each of 1 m square, the separation between adjacent plates being 1.3×10^{-3} m of air. If the *total* capacitance is to be not less than 0.4 μF and if the external plates are of the same polarity, how many plates will be required.

    [60

7. An iron ring of cross-sectional area 3 cm^2 and *mean* diametre 12 cm, has an airgap of 2 mm width, cut at one point at right angles to the circumference. Calculate the mmf required to produce a flux of 3.3×10^{-4} Wb in the gap. Neglect leakage.

    The iron has the following magnetic properties:

H, A/m	0.201	0.239	0.289	0.364
B, T	0.9	1.0	1.1	1.2

    [2450 A

8. An iron ring has a cross-section of 3×10^{-4} m^2 and a mean diametre of 0.25 m. An airgap of 4×10^{-3} m has been made by a cut across the section of the ring. The ring is wound with a coil of 200 turns through which a current of 2 A is passed. If the total magnetic flux is 2.1×10^{-4} Wb, determine the relative permeability of the iron assuming no magnetic leakage.

    [2470

9. Calculate the force in newtons required to separate two magnets, each of area of 100 cm^2 when the flux between them is 10^{-2} Wb.

    [3981 N

10. Using Steinmetz's hysteresis law, $P_h = \eta \text{ v } B_m^{1.6} f$ W, calculate the *total* hysteresis loss in a sample of sheet steel weighing 23 kg. The specific gravity of steel is 8000 kg/m^3, the hysteretic coefficient η is 500 and the sample is subjected to a maximum flux density of 1.2 T at 50 Hz.

    [96 W

11. A 750 kW DC generator has 16 poles, its armature diameter is 3 m and the active length of each conductor is 0.51 m. If the average strength of the airgap field is 1.2 T and the rotational speed is 100 rpm, calculate the emf per conductor.

    [0.96 V

12. A 300 kW, 500 V DC generator has 8 poles each with a flux of 0.73 mWb. There are 1300 turns on each pole and the field current is 6.95 A. If the field winding is suddenly switched on to a 500 V supply, calculate the time taken for the field current to attain (a) $1/10$, (b) $1/2$ of its final value.

Calculate also the stored energy in the field system when the current has attained its final value.

[(a) 0.163 s, (b) 1.055 s; 26,300 J

13. Calculate the mutual inductance between two coils A and B when an EMF of 0.02 V is induced in coil B whilst a current changes in coil A at the rate of 2 A/s. Also, if the flux linking coil B is 0.04 Wb when the current in neighbouring coil is 2 A what is the mutual inductance?

[0.01 H; 0.02 H

III : Electrical Circuit Analysis

III

ELECTRICAL CIRCUIT ANALYSIS

RECALL

Until the advent and 'working' of alternating current electricity towards the end of 1880s, the electric circuit and analysis revolved around the use of direct current. The prime circuit element was a resistance, largely single-valued or linear in electric behaviour, and followed the well-known Ohm's law when it was part of an active electric circuit. There was some non-linear behaviour, especially due to heating and temperature–rise effects, but this was studied more from physical point of view rather in the circuit analysis which assumed constant resistance(s). The electric power loss was given by the simple expression I^2R or V^2/R W and was a time-invariant or constant quantity so long as the applied voltage and resistance were constant

The present-day electric circuits in practice essentially comprise not just (circuit) resistance, but also inductance and capacitance – an outcome of near universal use of AC supply of varying voltage(s) in various usage. The subject matter of this chapter, therefore, pertains to analysis of electric circuits fed from AC supply.

THE AC EMF

All modern alternating current working – from generation to end-use, that is, utilisation in various forms – ideally presupposes the waveform of all ac quantities, viz. voltage(s), current(s) and flux density to be sinusoidally varying with respect to time. It is helpful to understand how this is achieved from the source, that is, the alternator. As is well known, an alternator is a generator of AC EMF, the induction of the latter being based on the flux-cutting rule.

The Fundamental EMF Equation

A modern, practical alternator comprises 3-phase windings housed in suitably designed slots in the stator, arranged with appropriate spatial distribution, with the 2-pole (or 4-pole) DC field system (providing the necessary excitation) being mounted on the rotor, it being mechanically driven at the appropriate synchronous speed. However, to understand the concepts of the subject matter, consider a simple rectangular coil rotating at uniform speed in a stationary uniform magnetic field, B (tesla), as shown in Fig. 3.1.

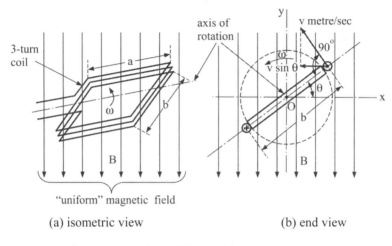

(a) isometric view   (b) end view

Fig.3.1: A rotating coil in a uniform magnetic field

The coil sides b, b, being always parallel to the direction of the field as the coil is rotated, there will be no (motional) EMF induced in these sides and so only the coil sides a, a be considered. Let v m/s be the *peripheral* velocity as shown, then in any angular position θ with respect to Ox, velocity perpendicular to the flux lines will be v sin θ.

Therefore, EMF induced in one side of length a metre = $B a v \sin \theta$ V (applying the "B/v" rule) and EMF in both sides of one turn,

$$e = 2 \, B \, a \, v \sin \theta \text{ V}.$$

If the coil has N turns, then

$$e = 2 \, N \, B \, a \, v \sin \theta \text{ V}$$

Now let the *angular* velocity of the coil be ω rad/s. Then the speed of the coil sides can be expressed

$$n = \frac{\omega}{2\pi} \text{ rps}$$

Since the coil diameter (distance between the coil sides b, b) is b m, the peripheral velocity will be given by

$$v = \pi \, b \, n \text{ m/s}$$

$$= b \, \frac{\omega}{2} \text{ m/s}$$

Hence

$$e = 2 \, N \, B \, a \times \left(b \frac{\omega}{2} \right) \sin \theta \text{ V}$$

$$= N \times (ab) \times B \, \omega \sin \theta \text{ V}$$

or $\qquad\qquad e = \varnothing_m\, N\, \omega \sin\theta\ \ V$

where $\varnothing_m = (ab) \times B$, is the maximum flux (in Wb) linking the coil *when its plane is in the horizontal position.*

If the coil is *driven* at a uniform angular speed, that is, ω is a constant, the induced EMF can be expressed

$$e = E_m \sin\theta\ \ V$$

writing $\qquad\qquad E_m = \varnothing_m\, N\, \omega$

or

$$\boxed{e = E_m \sin\omega t\ \ V}$$

This shows that in the assumed scheme, which is fundamental to a modern alternator with some practical considerations, the induced EMF in the 'machine' is always sinusoidal, that is, following a "sine" variation with time. The graphical variation of the EMF will be as depicted in Fig.3.2.

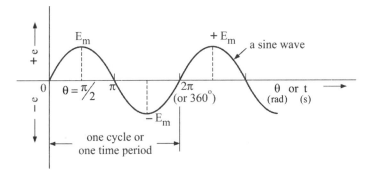

Fig.3.2 : Sinusoidal variation of induced EMF with time in a rotating coil

Root Mean Square (or RMS) and Average Values of the Sinusoidally Alternating EMF

Two quantities of practical importance associated with a *sinusoidally* varying EMF, or any quantity, are:

(a)   The *rms* value, expressed by

E_{rms} (or simply E in MKS system of units)

$$= \sqrt{\frac{1}{2\pi}\int_0^{2\pi} e^2 d\theta}$$

where e is the induced EMF at any instant θ or ωt.

Thus, the subscript "rms" derives from "(square) Root of the Mean of the Square of instantaneous values of the EMF during one cycle."

Substituting $e = E_m \sin \omega t$ in the above expression and simplifying

$$E_{rms} = E = \frac{E_m}{\sqrt{2}} \qquad \text{or} = 0.707 \; E_m$$

Similarly, the RMS value of alternating *current* will be given by[1]

$$I_{rms} \text{ or } I = \sqrt{\frac{1}{2\pi} \int_0^{2\pi} i^2 d\theta} = I_m / \sqrt{2} \; I_m$$

(b) The *average* value, given by

$$E_{av} = \frac{1}{\pi} \int_0^{\pi} e \; d\theta$$

Or, once again, writing $e = E_m \sin \omega t$ in the integral and simplifying

$$E_{av} = \left(\frac{2}{\pi}\right) \times E_m \qquad \text{or} = 0.637 \; E_m$$

The ratio of rms to average value

$$= \frac{0.707}{0.637} \qquad \text{or} \;\; 1.11$$

in the present case is defined as the **form factor,** the value signifying the shape or waveform of a given alternating or time-varying quantity. Thus, for a *rectangular* wave the form factor will be seen to be unity whilst for a *triangular* wave it is 1.15. In general, the more 'pointed' or peaky the wave or wave shape the greater will be the form factor.

Phasor Representation of AC Quantities

An important property of an alternating quantity, as revealed by the mathematical dependence on $\sin \theta$, or $\sin \omega t$, is that its magnitude varies with time or from instant to instant so that the letter the in the expression $e = E_m \sin \omega t$ represents *instantaneous* value of the induced EMF. If θ or ωt is measured with respect to the axis x as shown in Fig.3.1(b), then when

---

[1]In strict sense, the concept of RMS value of current derives from its physical effect, specifically in producing heat in, say, a resistor (where heat is expressed by I^2R t: I the current, R the resistance and t the time). Since the AC is time varying, its heating effect is integrated over t and, in this sense, can be compared to the same heat produced by an 'equivalent' direct current of the same magnitude, that is I. Accordingly, the root-mean-square (or RMS) value of an alternating current is "defined" as that value of steady (or direct) current which would dissipate heat at the same rate in a given resistance. Also, it is the RMS value of an AC quantity – voltage, current or flux – that only can be measured using the conventional instruments.

$\theta = 0$, or the coil plane is in horizontal position[1], the EMF induced in the coil would be zero. In contrast, when $\theta = 90°$, that is when the coil is rotated in the *anticlockwise direction* by one quadrant, the induced EMF will be maximum; decreasing once again as the coil rotates through another quadrant to take the position $\theta = 180°$ when the EMF will again be zero. This is graphically shown in Fig.3.2.

The Instantaneous Value

It is clear that the *magnitude* of the induced EMF is a function of the *angular* position of the coil with respect to the datum or the x-axis. This can be explained by considering the coil side(s) b, positioned with respect to x-axis and its projection on the y-axis, assuming that the maximum induced EMF magnitude can, at the same time, be represented by a 'line' of a given length, such as E_m, as shown in Fig.3.3.

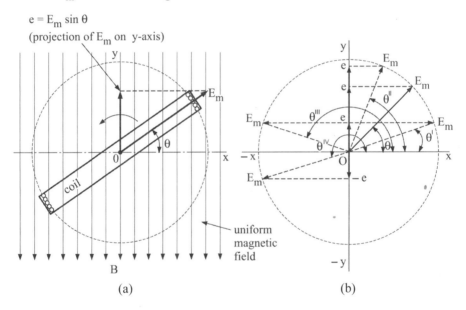

Fig.3.3 : Instantaneous values of induced EMF

Fig.3.3(b) shows that for a constant maximum value of the induced EMF $E_m(= \emptyset_m\, N\, \omega\ V)$, the locus of the latter is a circle (shown dotted). However, its projections on the y-axis for different values of θ, measured

---

[1]In the horizontal position $(\theta = 0)$, the coil is in a position to have 'maximum' *flux* linking with it. However, since it is constant, non-time-varying flux, the induced EMF is zero as these is no 'flux-cutting' action. In the vertical position, there is maximum flux-cutting action and hence the induced EMF is maximum, given by

$$e = E_m\ \sin\frac{\pi}{2}\ \text{or } E_m$$

from the x-axis in anticlockwise direction, represent instantaneous values of E_m, reaching a maximum of E_m for $\theta = \pi/2$ and zero corresponding to $\theta = 0$ and $\theta = \pi$; for $\theta > \pi$ (or 180°) the projections and therefore the induced EMF will be negative as shown.

The Phase and the Phasor

The orientation of the arrow representing the maximum induced value E_m as it 'rotates' anticlockwise[1] (Fig.3.3), and its instantaneous value e corresponding to different values of θ can be identified as its phase[2].

The AC quantity, E_m in this case, will then be identified as a **phasor** similar to a physical quantity like force, called a **vector,** which has a magnitude *and* given direction in *space*.

Thus, all alternating current (AC) quantities – EMFs, PDs, currents – are called phasors having a constant *maximum* value and various instantaneous values; these must not be confused with vectors – a term used to represent only physical quantities such as force, velocity and acceleration.

Phase Difference

Consider two AC quantities such as a voltage and current in a circuit. In practice, it is unlikely that both these will reach their maximum or any given value at the same instant; in general, they may reach the same value at different instants (or angles, θ). Mathematically, these may be expressed

$$e = E_m \sin \theta$$

$$i = I_m \sin (\theta - \phi)$$

indicating that *for the same angular speed*, ω (such that $\theta = \omega t$) or frequency, the current reaches its maximum value *later* than that of the voltage by the angle ϕ, meaning that their instantaneous values maintain a constant angular 'distance' of the angle ϕ. This difference is known as **phase difference** and forms an important aspect of *all* alternating quantities.

Clearly, when $\phi = 0$ so that voltage and current reach their given inst-antaneous values (for example, zero or maximum) *simultaneously*, the phase difference between them is zero; the two quantities are then said to be *in phase*[3]. In general, when ϕ is non-zero, the quantities are said to be *out-of–phase* (by the angle ϕ); when $\phi = \pi/2$, the quantities are said to be in *phase quadrature*.

---

[1]In electrical analyses, the anticlockwise 'rotation' is reckoned 'positive' by convention; clockwise rotation 'negative'.

[2]This concept is somewhat analogous to the movement of celestial bodies in space; for example, the moon which passes through various phases from new moon to full moon.

[3]A special case of AC circuits.

Lagging and leading phase difference

It is logical that of the two quantities being out of phase, one of the two would reach the specific value (e.g. the maximum) *after* the other. The former is then said to be *lagging* the latter (that is, trailing behind it); the latter is said to be *leading* the former. This is depicted in Fig.3.4.

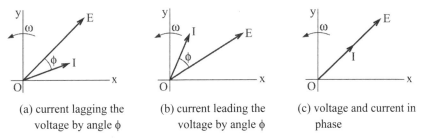

(a) current lagging the voltage by angle ϕ

(b) current leading the voltage by angle ϕ

(c) voltage and current in phase

Fig.3.4 : Phase difference between two AC quantities

These three possible cases are *graphically* illustrated as three sinusoidal waves in Fig.3.5. Note the 'cross-over' points of the two waves on the x-axis in each case; how the *angular* phase difference can be shown along the 'time' axis.

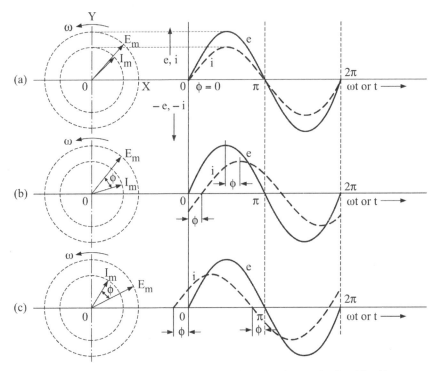

(a) : e and i in phase; (b) : i lagging e by ϕ (or e leading i by ϕ);

(c) : i leading e by ϕ (or e lagging i by ϕ)

Fig.3.5 : Sinusoidal variation of AC quantities with a phase difference

Phasor Addition and Subtraction

If there are two identical AC quantities in a circuit; for example, two potential differences, represented as two phasors[1] $\dot{V}_1$ and $\dot{V}_2$, their resultant by addition or subtraction, that is

$$\dot{V}' = \dot{V}_1 + \dot{V}_2$$
$$\dot{V}'' = \dot{V}_1 - \dot{V}_2 \quad [= \dot{V}_1 + (-\dot{V}_2)],$$

can be obtained by applying the "law of parallelogram" as shown in Fig.3.6.

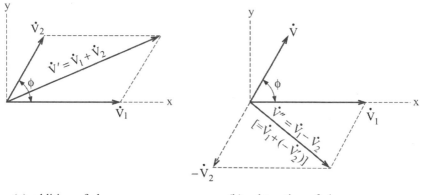

(a) addition of phasors   (b) subtraction of phasors

Fig.3.6 : Phasor addition and subtraction

Then, the *magnitude* of the resultant is given by

$$\left[V_1^2 + V_2^2 \pm 2V_1 V_2 \cos\phi \right], \text{ respectively.}$$

The Reference Phasor

In Fig.3.6, the phasor $\dot{V}_1$ is drawn horizontally (or along x-axis) whilst $\dot{V}_2$ is drawn ahead of $\dot{V}_1$, that is, *leading* $\dot{V}_1$ by an angle ϕ. This is drawn arbitrarily here to demonstrate phasor summation or subtraction. In practice, if the phasors are expressed such that

$$e_1 = \dot{V}_1 \sin \omega t$$

and $\qquad\qquad e_2 = \dot{V}_2 \sin (\omega t + \phi),$

the use of $+\phi$ associated with the second phasor defines its explicit *phasor relationship* with $\dot{V}_1$, showing that it is leading $\dot{V}_1$ by the angle ϕ. Thus, if the two phasors are drawn as in Fig.3.6, $\dot{V}_1$ and $\dot{V}_2$ acquire a special position in the diagram:

---

[1]Note the representation of a phasor quantity, discussed later.

$\dot{V}_1$ is now called the *reference* phasor, since

$\dot{V}_2$ is drawn *with respect to* and ahead of it at an angle ϕ in the *counter-clockwise* direction,

and, of course, $\dot{V}_2$ leading $\dot{V}_1$ by the angle ϕ.

Phasor Diagram

The graphical representation of the two phasors $\dot{V}_1$ and $\dot{V}_2$, showing their relative phasor positions, is a simple case of what is known as a **phasor diagram**. In practice, in an electrical circuit there may be several AC quantities, typically voltages, potential drops and currents in various parts (or branches) of the circuit having different phase differences relative to each other. All these phasors can be graphically drawn, one of them being drawn as the reference phasor whilst others with appropriate phasor relationships with respect to the reference phasor or with respect to each other. This representation is identified as the phasor diagram of the electric circuit in question and forms an important aspect of AC circuit analysis.

The following example will illustrate the concept and method of analysis involving various phasors and the corresponding phasor diagram(s):

Consider the four EMFs as follows:

$$e_1 = 100 \sin \omega t \ \text{V}$$

$[E_{m_1} = 100 \text{ V, and the reference phasor}]$

$$e_2 = 80 \sin \left(\omega t + \frac{\pi}{3} \right)$$

$$\left[E_{m_2} = 80 \text{ V, leading } E_{m_1} \text{ by} \frac{\pi}{3} \right]$$

$$e_3 = 40 \sin \left(\omega t - \frac{\pi}{6} \right)$$

$$\left[E_{m_3} = 40 \text{ V, lagging } E_{m_1} \text{ by} \frac{\pi}{6} \right]$$

$$e_4 = 60 \sin \left(\omega t + \frac{3\pi}{4} \right)$$

$$\left[E_{m_4} = 60 \text{ V, leading } E_{m_1} \text{ by} \frac{3\pi}{4} \right]$$

Calculate the resultant EMF by summation, and its equation.

The EMFs can be expressed as four phasors in the phasor diagram of Fig.3.7(a), *drawn to scale.*

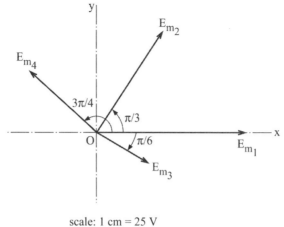

scale: 1 cm = 25 V

Fig.3.7(a) : Phasor diagram of the four EMFs

The resultant EMF can be obtained graphically as shown in Fig.3.7(b), using the various phasors in appropriate relative phasor relationship. The resultant of 'pairs' of EMFs in turn can be worked out by considering (say) first E_{m_1} and E_{m_2}, shown dotted, giving E'_m. The complete polygon can then be obtained by successive summation.

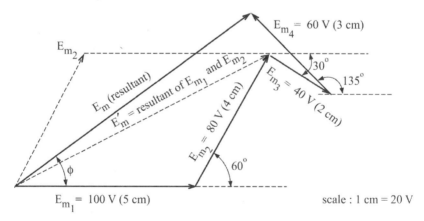

Fig.3.7(b) : To determine resultant of various EMFs

From the phasor polygon and using the scale 1 cm = 20 V, the resultant E_m is found to be 161 V while ϕ is measured to be 34° 44′ or 0.606 rad. Also, it is seen that $\dot{E}_m$ leads the reference phasor $\dot{E}_{m_1}$ so that

$$e = 161 \sin(\omega t + 0.606) \text{ V}$$

Alternative approach

An alternative method to obtain the above result is depicted in Fig.3.7(c) where to-scale *components* of the various phasors along x- and y-axis are drawn, *resolved* from the values of E_{m_1}, E_{m_2}, ... etc.

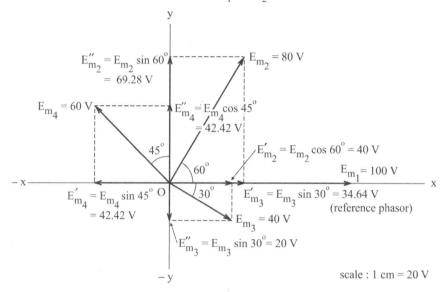

Fig.3.7(c) : Alternative approach to obtain $\dot{E}_m$

Then,

    A. algebraic sum of components along x–axis

$$= E_{m_1} + E'_{m_2} + E'_{m_3} - E'_{m_4}$$

$$= 100 + 40 + 34.64 - 42.42$$

$$= + 132.22 \ \text{V}$$

    B. algebraic sum of components along y–axis

$$= E''_{m_2} + E''_{m_4} - E''_{m_3}$$

$$= 69.28 + 42.42 - 20$$

$$= + 91.7 \ \text{V}$$

Therefore, the resultant $E_m = \sqrt{(132.22)^2 + (91.7)^2}$ $= 160.9 \ \text{V}$

and the 'phase angle' with respect to $\dot{E}_{m_1}$, $\phi = \tan^{-1}\left(\dfrac{91.7}{132.22}\right)$

$$= 34.74° \, (34°44.4')$$

very nearly same as before[1].

---

[1]Clearly, this approach, being completely mathematical, is preferable and more accurate.

Notation of a Phasor

To differentiate from DC or other electrical quantities like power and energy, a phasor (quantity) is denoted by a dot (·) above the letter representing the quantity.

For example: $\dot{E}$, $\dot{I}$, $\dot{B}$. As will be seen, the same also applies to a 'complex' quantity or a *complexor*.

The *magnitude* of the phasor (e.g. an EMF) is then, expressed by $|\dot{E}|$ or simply E^1.

In contrast, a vector, representing a physical quantity such as force is denoted in bold capital or with a bar (-) above the letter.

For example: **F** or $\overline{F}$ for force, or **V** or $\overline{V}$ for velocity.

The Symbolic or "j" Notation

It was stated that when the phase difference between two phasors is 90^{o2}, they are said to be *in quadrature*. This concept is expressed 'mathematically' to also show the *positional* 'location' of a **single phasor** in the x-y plane by using the **"j"** notation.

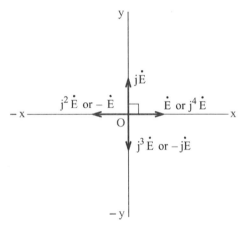

Fig.3.8 : The "j" notation associated with phasors

Imagine a phasor $\dot{E}$ directed along x-axis at a given instant as shown in Fig.3.8. If this is to be 'rotated' in a counter-clockwise direction so as to be directed along the y-axis (that is, a rotation by 90°), a kind of 'operator' may be imagined to achieve this rotation. If the operator be called "j" without going into its nature at the moment, the phasor in its new position along the y-axis can be expressed as $j\dot{E}$, corresponding to, and to denote, the 90° rotation. A further anti-clockwise rotation by 90° will position the phasor again in x-direction, *but in reversed sense*. 'Mathematically', it can be written $j^2\dot{E}$ and equal to $-\dot{E}$ since $\dot{E}$ is now in phase opposition to its first position.

---

[1]This also represents RMS value of the EMF.

[2]Strictly expressed as 90°(electrical) or 90°(E) to differentiate from 'mechanical' degrees, or angular rotation in space.

Therefore, $j^2\dot{E} = -\dot{E}$ or $j^2 = -1$ and $j = \sqrt{-1}$. Thus the operator j is essentially the imaginary quantity, in common use in algebra.

A further rotation of the phasor by "j" will position the phasor along $-y$ and another back to $+x$ as shown.

The Argand Diagram

Refer to the 'alternative approach" used to calculate the value of E_m as depicted in Fig.3.7(c). The four EMF phasors (or rather their maximum values) were resolved along x- and y-axis using $\cos\theta$ and $\sin\theta$ functions.

If the components could be expressed using the "j" notation, then these could be written

$$E''_{m_2} \quad : j\ 69.28\ V$$

$$E''_{m_3} \quad : -j\ 20\ V$$

$$E''_{m_4} \quad : j\ 42.42\ V$$

whilst the phasors along the x–axis will only have algebraic signs.

In terms of the "j" notation, the various EMFs could thus be represented as follows:

$$\dot{E}_{m_1} = 100 + j\ 0\ V \quad \text{[the reference phasor]}$$

$$\dot{E}_{m_2} = 40 + j\ 69.28\ V$$

$$\dot{E}_{m_3} = 34.64 - j\ 20\ V$$

$$\dot{E}_{m_4} = -\ 42.42 + j\ 42.42\ V$$

as shown in the diagram of Fig.3.9.

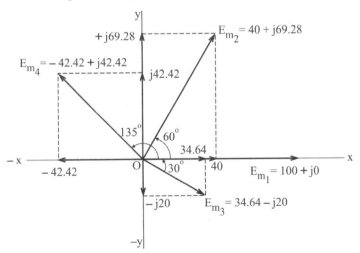

Fig.3.9 : Representation of phasors using "j" notation: the Argand diagram

Such a diagram is referred to as Argand diagram where components of phasors are represented in terms of their real and imaginary parts and their resultants in appropriate quadrants.

Using the "j" notation, it is at times 'easier' to analyse problems involving phasors (and even complexors as shown later). The technique is called the "j method".

The following example illustrates the application of the "j" method to analyse a circuit containing three current phasors:

In a circuit there are three currents which act together. These are individually represented as

$$i_1 = 50 \sin \omega t \ \text{A},$$

$$i_2 = 100 \sin \left(\omega t + \frac{\pi}{4} \right) \ \text{A, and}$$

$$i_3 = 75 \sin \left(\omega t - \frac{\pi}{3} \right) \ \text{A}.$$

Determine the expression for the resultant current.

With i_1 or $\dot{I}_1$ (the maximum value of i_1) chosen as the reference phasor, it can be written

$$\dot{I}_1 = 50 + j0 \ \text{A}$$

Then
$$\dot{I}_2 = 100 \cos \left(\frac{\pi}{4} \right) + j \, 100 \, \sin \left(\frac{\pi}{4} \right) \ \text{A}$$

$$= 70.7 + j \, 70.7 \ \text{A}$$

and
$$\dot{I}_3 = 75 \cos \left(\frac{\pi}{3} \right) - j \, 75 \, \sin \left(\frac{\pi}{3} \right) \ \text{A}$$

$$= 37.5 - j \, 65 \ \text{A}[1]$$

The resultant current can then be worked out as follows:

$$\dot{I} = \dot{I}_1 + \dot{I}_2 + \dot{I}_3$$

$$= (50 + 70.7 + 37.5) + j \, (0 + 70.7 - 65),$$

by (algebraically) adding real and imaginary components

or
$$\dot{I} = 158.2 + j \, 5.7 \ \text{A}$$

and
$$|\dot{I}| = \sqrt{(158.2)^2 + 5.7^2} \ \text{or } 158.3 \ \text{A (approx)}$$

---

[1]Clearly, when so represented, the phasors resemble complex quantities (or complexor) having "real" and "imaginary" parts; that is, of the form $A \pm jB$

The phase angle of $\dot{I}$, $\phi = \tan^{-1}\left(\dfrac{5.7}{158.2}\right)$

$$= 2.06° \text{ or } 0.036 \quad \text{rad (approx)}$$

with respect to the assumed reference.

Therefore, the instantaneous resultant current is expressed by

$$i = 158.3 \sin(\omega t + 0.036) \text{ A}$$

Later, it will be shown that the phasors could be expressed directly in complex form and analysed as above.

Also, the above example and the solution are similar to the "alternative approach" used in the previous example.

Representation of phasors in polar form

As pointed out, the above expressions of phasors using real and imaginary parts or components are similar to complex quantities, typically expressed as

$$A \pm j B$$

Since the letter j associated with phasors is identical to that used in complex algebra, that is, involving the imaginary number $\sqrt{-1}$, it follows that all the rules of complex algebra, viz, addition, subtraction, multiplication and division will apply identically to analyses involving phasors, including rationalisation when dividing one complex number by the other.

Yet another form of representing a phasor, other than a complexor, is the "polar form" where its magnitude and angular relationship with respect to the reference phasor are specified as

$$\dot{V} = |\dot{V}|\,\underline{/\theta}$$

Clearly, this can be resolved into its two components at right angles, thus

$$V_1 = |\dot{V}|\cos\theta, \quad V_2 = |\dot{V}|\sin\theta$$

so that in terms of j notation, it can be written

$$\dot{V} = V_1 + j\,V_2$$

giving $\qquad |\dot{V}| = \sqrt{V_1^2 + V_2^2} \quad$ and $\quad \theta = \tan^{-1}\left(\dfrac{V_2}{V_1}\right)$

The two approaches are therefore interrelated and can be used as desired and helpful.

If there were two phasors, given in polar form such that $\dot{V}_1 = \left|\dot{V}_1\right| \underline{/\theta_1}$ and $\dot{V}_2 = \left|\dot{V}_2\right| \underline{/\theta_2}$, then

$$\dot{V}_1 \times \dot{V}_2 = \left|\dot{V}_1\right| \left|\dot{V}_2\right| \underline{/\theta_1 + \theta_2} \text{ and } \frac{\dot{V}_1}{\dot{V}_2} = \frac{\left|\dot{V}_1\right|}{\left|\dot{V}_2\right|} \underline{/\theta_1 - \theta_2}$$

The exponential form of phasor representation

Consider a phasor

$$\dot{V} = V \cos \alpha + jV \sin \alpha$$

where V is its magnitude and α the angle of lead with respect to an arbitrary reference phasor.

If the angle α were in radians, sin α and cos α could be expressed by the infinite series

$$\sin \alpha = \alpha - \frac{\alpha^3}{\underline{|3}} + \frac{\alpha^5}{\underline{|5}} - \frac{\alpha^7}{\underline{|7}} + \cdots$$

$$\cos \alpha = 1 - \frac{\alpha^2}{\underline{|2}} + \frac{\alpha^4}{\underline{|4}} - \frac{\alpha^6}{\underline{|6}} + \cdots$$

Therefore

$$\dot{V} = V \left[\left(1 - \frac{\alpha^2}{\underline{|2}} + \frac{\alpha^4}{\underline{|4}} - \frac{\alpha^6}{\underline{|6}} + \cdots \right) + j \left(\alpha - \frac{\alpha^3}{\underline{|3}} + \frac{\alpha^5}{\underline{|5}} - \frac{\alpha^7}{\underline{|7}} + \cdots \right) \right]$$

$$= V \left[1 + j\alpha - \frac{\alpha^2}{\underline{|2}} - j\frac{\alpha^3}{\underline{|3}} + \frac{\alpha^4}{\underline{|4}} + j\frac{\alpha^5}{\underline{|5}} - j\frac{\alpha^6}{\underline{|6}} - j\frac{\alpha^7}{\underline{|7}} + \frac{\alpha^8}{\underline{|8}} + \cdots \right]$$

Substituting j^2 for -1 in the above

$$\dot{V} = V \left[1 + j\alpha + \frac{j^2\alpha^2}{\underline{|2}} + \frac{j^3\alpha^3}{\underline{|3}} + \frac{j^4\alpha^4}{\underline{|4}} + \frac{j^5\alpha^5}{\underline{|5}} + \frac{j^6\alpha^6}{\underline{|6}} + \frac{j^7\alpha^7}{\underline{|7}} + \frac{j^8\alpha^8}{\underline{|8}} + \cdots \right]$$

or, simply,

$$\boxed{\dot{V} = V \in^{j\alpha}}$$

since the above series within the square brackets is, in effect, the expansion of $\in^{j\alpha}$, where $\in$ is the base of natural logarithm.

Thus, if there were two phasors such that

$$\dot{A} = A \in^{j\alpha}, \qquad \dot{B} = B \in^{j\beta}$$

then their sum and difference will be expressed by

$$\dot{A} + \dot{B} \quad \text{or} \quad A\in^{j\alpha} + B\in^{j\beta}$$

and $\qquad\qquad \dot{A} - \dot{B} \quad$ or $\quad A \in^{j\alpha} - B \in^{j\beta}$,

respectively,

whilst the product and division will be given by

$$\dot{A} \times \dot{B} = A \in^{j\alpha} \times B \in^{j\beta} \quad \text{or} \quad (A \times B) \in^{j(\alpha + \beta)}$$

and $\qquad\qquad \dfrac{\dot{A}}{\dot{B}} = \dfrac{A \in^{j\alpha}}{B \in^{j\beta}} \quad \text{or} \quad \left(\dfrac{A}{B}\right) \in^{j(\alpha - \beta)}$

Clearly, it would be helpful in multiplication and division of phasors if they were represented in exponential form. This is similar to the treatment when expressed in polar form.[1]

SINGLE-PHASE AC CIRCUITS CONTAINING DIFFERENT ELEMENTS

Potential Drop, Current and Power

1. The 'pure' resistance circuit

This is a circuit which has ohmic resistance only, that is, no magnetic or electrostatic 'flux' is associated with the element.

Consequently, the applied PD is used up to overcome the ohmic drop only. If the instantaneous PD is given by

$$v = V_m \sin \omega t \ \ V$$

and the circuit resistance is R Ω, the current through R will be

$$i = \frac{v}{R} = \left(\frac{V_m}{R}\right) \sin \omega t$$

or, using rms quantities

$$\boxed{I = \frac{V}{R} \ \ A}$$

This shows that the current is *in phase* with the applied voltage or PD at any instant.

The instantaneous power in the circuit is given by

$$p = v \ i \ \ W$$
$$= V_m I_m \sin^2 \omega t$$
$$= \frac{1}{2} V_m I_m - \frac{1}{2} V_m I_m \cos 2\omega t$$

---

[1]Conversely, if a phasor were given as $\dot{P} = P \in^{j\theta}$, it could be expressed as
$$\dot{P} = P \cos \theta + j \, P \sin \theta$$

The expression for power thus consists of two parts:

(a) a constant part $= \dfrac{1}{2} V_m I_m$, and

(b) a periodic part $= -\dfrac{1}{2} V_m I_m \cos 2\omega t$

varying with time at double the supply frequency.

The variations of v, i and p as governed by the above equations are shown in Fig.3.10. Power being a physical entity, and a scalar quantity, its variation with time has no meaningful significance; it is the *average* power consumed in, or supplied from, a circuit over a time period that is to be considered.

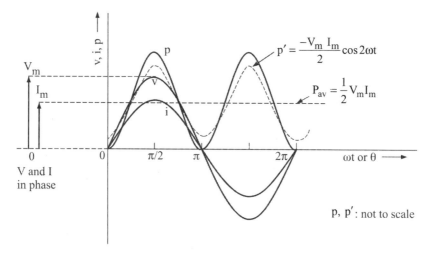

Fig.3.10 : Voltage, current and power in a purely resistive circuit

Now the average value of the double-frequency power term is given by

$$P'_{av} = \frac{1}{2\pi} \int_0^{2\pi} V_m I_m \cos 2\omega t \, d\omega t = \frac{1}{2} \frac{1}{2\pi} V_m I_m \left[\sin 2\omega t \right]_{\omega t=0}^{\omega t=2\pi}$$

$$= 0^1$$

---

[1]This is also clear from the variation of $p' \left(= \dfrac{1}{2} V_m I_m \cos 2\omega t \right)$ shown in the figure. The graph of p' contains two halves of *opposite polarity* in each cycle of p resulting in net power due to p' to be zero.

Hence the net power in the purely ohmic circuit[1] is given by the first term alone, that is

$$P = \frac{1}{2} V_m I_m$$

or

$$= \frac{1}{2} \sqrt{2} V \times \sqrt{2} I \quad \text{or} \quad V \, I \, W$$

that is,

$$\boxed{P = V \times I \quad W}$$

where V and I denote the rms values of the voltage and current in the circuit.

2. The 'purely' inductive circuit

According to the fundamental property of a magnetic field associated with a current, a circuit possesses *self induction* if a magnetic flux is set up when current flows, say in a coil.

Neglecting any saturation effect; for example, considering an air-cored solenoid, the self inductance of the coil is given by

$$L = [\text{flux (Wb) per amp}] \times (\text{no. of turns}) \quad H$$

A purely inductive circuit is one which has zero resistance and, therefore, similar to a pure resistance is unattainable in practice. Thus, if a coil is wound with copper conductor of a given length, *l*, and area of cross section, a, it will always have a resistance given by

$$R = \rho \left(\frac{l}{a} \right) \quad \Omega$$

where ρ = resistivity of copper at the operating temperature.

However, if there is no resistance, there is no ohmic drop and therefore for equilibrium the applied voltage at every instant must be equal and opposite to the induced EMF in the inductive circuit, given by $-L \left(\dfrac{di}{dt} \right)$.

Therefore, when *opposing* the applied potential difference,

$$+L \left(\frac{di}{dt} \right) = v = V_m \sin \omega t$$

---

[1]In practice, it is impossible to obtain a "pure" resistance: a close approximation may be a carbon (or metal oxide) resistance which does not contain a coil wound using a resistance wire. In a resistance consisting of a coil, there will always be a certain amount of magnetic flux and hence an inductive effect. As discussed in Chapters I and II, special winding techniques may have to be used to obviate this phenomenon.

or
$$di = \left(\frac{V_m}{L}\right) \sin \omega t \, dt$$

Integrating both sides,

$$i = -\left(\frac{V_m}{L\omega}\right) \cos \omega t + A, \qquad A = \text{a constant}$$

Under steady-state conditions, A can be assumed to be zero and hence

$$i = -\left(\frac{V_m}{L\omega}\right) \cos \omega t$$

$$= \left(\frac{V_m}{L\omega}\right) \sin\left(\omega t - \frac{\pi}{2}\right)$$

This shows that in terms of phasor relationship, the current at any instant *lags* the applied voltage by $\pi/2$ rad or 90°(E) in a purely inductive circuit. The maximum value of the current is, clearly,

$$I_m = \frac{V_m}{L\omega}$$

and if the applied voltage is in volts and current is in amperes, the quantity $(L\omega)$ must have the dimensions of ohms. Defining this as **inductive reactance**, and denoting it by X_L such that $X_L = L\omega$, the current in rms value is given by

$$\boxed{I = \frac{V}{X_L} \text{ A}}$$

lagging the voltage V by $(\pi/2)$ and is also inversely proportional to frequency, f, since X_L is proportional to f.

The instantaneous power in the circuit is given by

$$p = v \times i = V_m \sin \omega t \times I_m \sin\left(\omega t - \frac{\pi}{2}\right)$$

$$= - V_m I_m \sin \omega t \cos \omega t$$

$$= -\frac{1}{2} V_m I_m \sin 2\omega t$$

Unlike in a purely resistive circuit, there is no constant term in the above expression; only the double-frequency term, the average value of which over a complete cycle is zero. Hence, *net power in a purely inductive circuit is always zero* even though both V and I are finite. This is illustrated graphically in Fig.3.11.

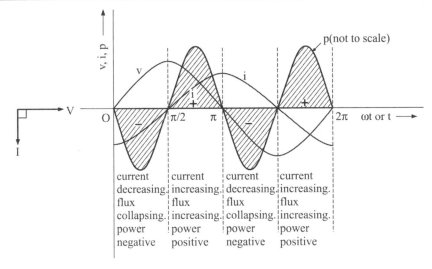

Fig.3.11: Voltage, current and power in a purely inductive circuit

Apart from the mathematical result, depicted in Fig.3.11, with two identical power curves of opposite polarity in each cycle of v or i, the physical meaning of the above phenomenon can be explained in terms of *exchange of power* between the supply and the coil. When the power is 'negative', the current is negative, the flux in the coil is collapsing and the energy stored in the magnetic field of the coil is being returned *to*, instead of being delivered *by*, the supply circuit. The opposite is true when the current is positive (together with the applied voltage): the flux increasing and the energy being now imparted, from the supply circuit to the inductive coil[1]. This is also explained in the figure.

3. The purely capacitive circuit

Consider a pure capacitor having a capacitance of C (farad) connected to an AC supply. Then the charge 'q' stored in the capacitor from instant to instant will be related to the voltage by

$$v = \frac{q}{C} = V_m \sin \omega t, \text{ v being the instantaneous}$$
$$\text{applied voltage}$$

or $\qquad q = C V_m \sin \omega t$

Therefore $\qquad \dfrac{dq}{dt} = \omega C V_m \cos \omega t = i$

where i is the instantaneous current in the circuit given by $\dfrac{dq}{dt}$.

---

[1]As mentioned before, in actual practice there will be a resistance (however small) associated with the coil and hence a net power *drawn* from the supply, equivalent to ohmic power loss in the resistance.

or
$$i = \omega C V_m \sin\left(\omega t + \frac{\pi}{2}\right)$$

This shows that the current in a purely capacitive circuit *leads* the applied voltage by $\pi/2$ rad or $90°$ (E), with the maximum value, I_m, being $\omega V_m C$. In terms of rms values, I_{rms} can be written

$$I = \frac{V}{\left(\dfrac{1}{\omega C}\right)}$$

The quantity $\left(1/\omega C\right)$ must clearly have the dimensions of ohms: it is called the **capacitive reactance**, X_C. (Also, X_C is *inversely* proportional to ω or supply frequency).

Therefore,

$$\boxed{I = \frac{V}{X_C} \text{ A}}$$

For the instantaneous power, as before

$$p = v\, i = V_m \sin \omega t \times I_m \sin\left(\omega t + \frac{\pi}{2}\right)$$

$$= V_m I_m \sin \omega t \cos \omega t$$

$$= \frac{1}{2} V_m I_m \sin 2\omega t$$

Once again, being a double-frequency quantity, the power is pulsating in character, the net power over a cycle being zero[1].

A graphical variation of the applied voltage, current and power from instant to instant in a purely capacitive circuit is shown in Fig.3.12.

The physical meaning of power being alternately positive and negative in the capacitor can be explained similar to the case of a pure inductance. The dielectric flux, being proportional to the applied PD is increasing representing positive power during positive half cycles of v and i, being stored in the capacitor in the form of electrostatic energy [$= \frac{1}{2} Cv^2$].

---

[1]As in the case of a 'pure' resistance and 'pure' inductance, the term 'pure' capacitance (or a capacitor) is a misnomer. A common practice is to use a dielectric (mostly a kind of insulating material) to construct a capacitor which may possess some resistance, and hence ohmic loss, particularly at high frequencies. The nearest to a pure capacitor may be a simple "parallel-plate" capacitor with air as the dielectric where the ohmic loss in the usual sense may be considered to be negligible.

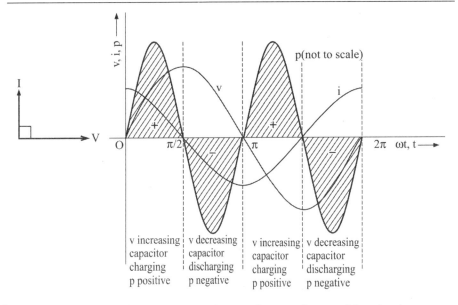

Fig.3.12 : Voltage, current and power in a purely capacitive circuit

When the PD is decreasing, the reverse is the case and the stored energy is 'returned' to the circuit.

4. Circuit containing R and L in series

Consider a simple series circuit comprising a pure resistance in series with a pure inductance across which a sinusoidally varying AC voltage is applied as shown in Fig.3.13[1].

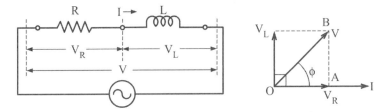

Fig.3.13 : R and L series circuit and phasor diagram

Assuming the common quantity, current, to be the reference phasor, a phasor diagram 'describing' the circuit can be drawn as shown in the figure, in terms of *rms* values of the current and potential drops.

---

[1]The combination can be seen to be a typical, actual case where the inductance part of a practical resistance can be considered as 'clubbed' with pure inductance part of the inductor (say, a coil); whereas the inductance part of the coil can be assumed to be clubbed with the resistance part of a practical resistor.

Clearly, the PD across the resistor will be in phase with the current, I, whereas the one across the inductor will be leading I by 90°(E). The resultant voltage V can then be expressed as

$$V^2 = V_R{}^2 + V_L{}^2 \quad \text{from the triangle OAB}$$

$$= (R\ I)^2 + (L\omega\ I)^2$$

$$= I^2 \times [R^2 + (\omega L)^2]$$

or

$$\boxed{I = \frac{V}{\sqrt{R^2 + X_L^2}}\ A}$$

where $X_L = \omega L$ is the inductive reactance.

The quantity $\sqrt{R^2 + X_L^2}$ must obviously be expressed in ohms: it is called the **impedance**, Z, from the property of impeding the flow of current in the circuit.

It is interesting, and of some importance in circuit analysis, that the 'voltage triangle' OAB drawn in Fig.3.13, formed from the PDs V_R, V_L and V can be similar to an "impedance triangle" in which V_R, V_L and V are replaced by R, X_L and Z, respectively, by dividing the former by the (common) current, I; the phase angle of V with reference to I, marked φ, will also be the 'phase angle' of Z with respect to R, called the "impedance angle".

To calculate the power delivered to the circuit, recall that there is no average power transferred to the inductor and therefore the whole of power intake will be in R, given by $I^2\ R$.

Therefore,

$$P = I^2\ R$$

$$= I \times \left(\frac{V}{Z}\right) \times R$$

$$= V\ I\left(\frac{R}{Z}\right)$$

or

$$\boxed{P = V\ I\cos\phi\ \ W}$$

since

$$\cos\phi = \frac{R}{Z};$$

and

$$\phi = \tan^{-1}\frac{X_L}{R}\left(\text{or }\frac{\omega L}{R}\right)$$

Thus, the net power in such a circuit is no longer given by a simple product V I as would be the case in a simple DC circuit. If the product

V I can be called **volt-amperes (or VA)**, and cos ϕ the **power factor** (p f), the power can be written

$$\text{Power} = (\text{volt-amp}) \times pf^1 \text{ W}$$

or   Power in kW = kVA × p f   where kVA is VA/1000

Variation of Power with Time

With the applied voltage expressed as $v = V_m \sin \omega t$ and current as $i = I_m \sin(\omega t - \phi)$, lagging v by ϕ, the instantaneous power will be given by

$$p = v\,i$$
$$= V_m I_m \sin \omega t \times \sin(\omega t - \phi)$$
$$= \frac{1}{2} V_m I_m [\cos \phi - \cos(2\omega t - \phi)]$$
$$= \frac{1}{2} V_m I_m \cos \phi - \frac{1}{2} V_m I_m \cos(2\omega t - \phi)$$

This expression contains two terms. The first term is non-time-dependent since cos ϕ is a constant for a given load. However, the second term is a periodic quantity of double-the-supply frequency whose average value over a cycle is zero (when integrated from 0 to 2π). Hence, the average power in the circuit is given by

$$P = \frac{1}{2} V_m I_m \cos \phi$$

or

$$P = \frac{V_m}{\sqrt{2}} \times \frac{I_m}{\sqrt{2}} \times \cos \phi$$
$$= V I \cos \phi$$

as before.

The variation of the instantaneous power with time for a given (arbitrary) angle ϕ is plotted in Fig.3.14. It is seen that the power in the circuit alternates between positive and negative quantities, this being due to the presence of the periodic term in the power expression. However, the positive lobes of the variation are much larger (actually, corresponding to a 'small' assumed value of ϕ) than the negative lobes, and the *average*

---

[1]The term power factor plays a prominent role in AC circuits for if the p f of the circuit were poor – ϕ increasing more and more towards 90° – a larger current will be required to deliver the same power at constant applied voltage. This points to great emphasis on the measurement of power and the power factor as shown in later chapters where a more detailed discussion is also given related to other aspects of power factor.

power over the whole period of time is positive and finite as deduced analytically.

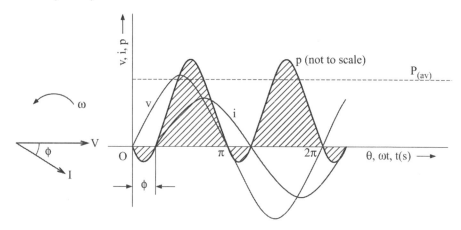

Fig.3.14 : Variation of voltage, current and power in a resistive-inductive circuit

5. Circuit containing R and C in series

This circuit is shown in Fig.3.15. The PD across the capacitor *lags* the current, as well as the voltage drop across the resistor, by 90°. The drop across the resistor itself is in phase with the current, being chosen as the reference phasor. The resultant voltage or the supply voltage is the phasor sum of the two PDs, and *lags* the current by an angle ϕ.

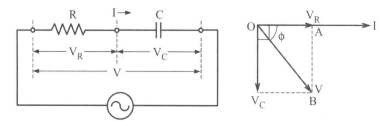

Fig.3.15 : R and C series circuit and phasor diagram

From the voltage triangle OAB

$$V^2 = V_R^2 + V_C^2$$
$$= (R\,I)^2 + (X_C\,I)^2$$

whence

$$I = \frac{V}{\sqrt{R^2 + X_C^2}} \quad A$$

From the phasor diagram ϕ can be deduced as

$$\phi = \tan^{-1}\frac{V_C}{V_R} = \tan^{-1}\frac{X_C}{R}$$

$$= \tan^{-1}\left(\frac{1}{\omega CR}\right) \qquad \text{since } X_C = \frac{1}{\omega C}$$

and the power delivered to the circuit

$$P = V\,I\cos\phi, \text{ as in the case of the R-L series circuit.}$$

The variation of instantaneous power with time can be seen to be similar to that for the R-L circuit with the positive and negative lobes of power, and the net positive power given by the above expression.

6. The general series circuit, containing R, L and C

A circuit containing all the three elements connected in series and the corresponding phasor diagram is shown in Fig.3.16.

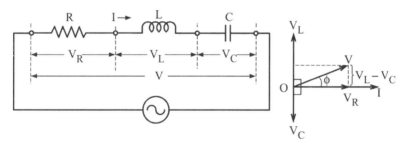

Fig.3.16 : The R-L-C series circuit and phasor diagram

Once again, the current I being 'common' to all the elements can be chosen as the reference phasor. The PD V_R across the resistor will be in phase with I whilst the PDs V_L and V_C across the inductor and capacitor, respectively, will be in phase quadrature with I as shown. Also, both these are in phase opposition with either V_L or V_C being greater than the other[1]. Assuming V_L to be greater than V_C, the phasor diagram will be as shown. The applied voltage is then given by

$$V^2 = V_R^2 + (V_L - V_C)^2$$

or $\qquad (I\,Z)^2 = (I\,R)^2 + (I\,X_L - I\,X_C)^2$

$\qquad\qquad\qquad$ [Z being the *resultant* circuit impedance]

---

[1]The PDs V_L and V_C being equal in magnitude and in phase opposition forms a special case, known as *resonance* condition, discussed later.

from which

$$I = \frac{V}{\sqrt{R^2 + \left(X_L - X_C\right)^2}} \quad A$$

with the resultant angle of phase, ϕ, given by

$$\phi = \tan^{-1} \left[\frac{\left(V_L - V_C\right)}{V_R} \right]$$

or

$$\phi = \cos^{-1} \left[\frac{R}{\sqrt{R^2 + \left(X_L - X_C\right)^2}} \right]$$

or

$$\phi = \cos^{-1} \left(\frac{R}{Z} \right) \quad \text{where} \quad Z = \sqrt{R^2 + \left(X_L - X_C\right)^2}$$

The analysis shows that depending on the relative values of X_L and X_C (or rather ωL and $1/\omega C$, both being frequency dependent), the resultant voltage will either lead or lag the current in the circuit.

In this case, too, the power in the circuit is given by

$$P = V\, I \cos \phi$$

A Parallel, or Branched, Circuit Comprising R, L and C

In such a circuit, the applied *voltage* will be the common quantity whereas each branch current will be different, the total current drawn from the supply then being the *phasor* sum of all these currents.

This condition can be best explained by analysing a numerical problem.

Example.

Consider the parallel circuit consisting of

(a) Z_1: a pure resistance, $R_1 = 5\,\Omega$;

(b) Z_2: a resistance, $R_2 = 3\,\Omega$ in series with an inductive reactance, $X_2 = 4\,\Omega$;

(c) Z_3: a resistance, $R_3 = 2\,\Omega$ in series with a capacitive reactance, $X_3 = 4.5\,\Omega$.

A voltage of 230 V is applied across the three branches. It is required to calculate the individual and resultant current.

The circuit with the various branches and the corresponding phasor diagram are shown in Fig.3.17.

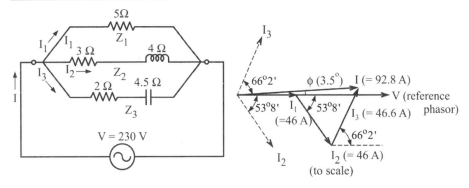

Fig.3.17 : A branched electric circuit and phasor diagram with V as the reference
phasor

From the circuit with applied voltage $\dot{V}$ ($\dot{V} = 230 + j0$) as the reference
phasor and various branch *impedances* as

$$\dot{Z}_1 = 5 + j0$$
$$\dot{Z}_2 = 3 + j4$$
$$\dot{Z}_3 = 2 - j4.5,$$

respectively, the branch currents are given by

$$\dot{I}_1 = \frac{\dot{V}}{\dot{Z}_1} = \frac{230 + j\,0}{5 + j\,0} = \frac{230}{5} = 46 \underline{/0^\circ} \text{ or } 46 + j0 \text{ A}$$

$$\dot{I}_2 = \frac{\dot{V}}{\dot{Z}_2} = \frac{230 + j\,0}{3 + j\,4} = 46\underline{/-53^\circ8'} \text{ or } 27.6 - j36.8 \text{ A}$$

$$\dot{I}_3 = \frac{\dot{V}}{\dot{Z}_3} = \frac{230 + j\,0}{2 - j\,4.5} = 46.6 \underline{/- 66^\circ 2'} \text{ or } 18.9 + j42.6 \text{ A}$$

These currents are shown, to scale, in the phasor diagram of Fig.3.17 with
their relative phasor positions; the polar form of the phasors having been
used. The resultant current is found to be 92.8 A (approx.), leading the
applied voltage by about $3.5^\circ \left(3^\circ 30'\right)$.

Instead of using the graphical solution as above, the problem could be
tackled to obtain the resultant current using the "j" notation as follows:

$$\dot{I}_1 = 46 + j\,0$$

$$\dot{I}_2 = 27.6 - j\,36.8$$

$$\dot{I}_3 = 18.9 + j\,42.6$$

$$\therefore \quad \dot{I} = \dot{I}_1 + \dot{I}_2 + \dot{I}_3 = 46 + j0 + 27.6 - j36.8 + 18.9 + j\,42.6$$
$$= 92.5 + j\,5.8$$

or $I = \sqrt{92.5^2 + 5.8^2} \simeq 92.7 \ A$

and $\phi = \tan^{-1}\left(\dfrac{5.8}{92.5}\right) = 3°35' \ (\simeq 3.58°)$

very nearly same as before.

Comment

A treatment of phasors and electrical circuits in detail is aimed at a clear understanding of analysing circuits containing resistances, inductances and capacitances. For example, in a later chapter, a variety of AC bridge methods are dealt with for the measurement of inductance and capacitance. Unlike the basic Wheatstone bridges extensively used for measuring 'medium' resistances where only the magnitudes are compared, in AC bridges it is essential that not only the magnitudes with correct dimensions, but also the *phasor relationships* of the unknown quantities are balanced at null conditions. For example, considering the simple case of De Sauty bridge for the measurement of capacitance shown below, the balance equations give

$$C_{unknown} = C_{known} \times \left(\dfrac{R_2}{R_1}\right) \ \text{(known ratio)}$$

whilst the phasor diagram shows the balanced *phasor* relationship amongst various PDs and currents.

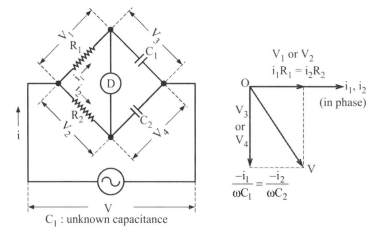

C_1 : unknown capacitance

[The bridge is discussed later in Chapter XI: Alternating-Current Bridge Methods]

"RESONANCE" IN AC CIRCUITS

Oscillatory Systems

Mechanical systems

These systems, in general, can be defined as the ones that have a "to and fro motion" about their mean positions. Typical examples are a pendulum or a turning fork. If left to themselves after initiation of the motion and if there were no damping forces, for example in the form of air friction, the systems will continue to be in motion perpetually, especially in the case of a tuning fork when its frequency of mechanical vibrations coincides with its *natural* frequency of vibration. The fork is then said to be "in resonance". The quantities of importance in such systems are *amplitude* and *frequency* of vibrations. The condition of resonance is generally indicative of "zero energy loss". Similar phenomena, but with a difference, occur in AC electric circuits.

Electrical systems

A. Resonance in the R, L, C *series* circuit

Consider a series circuit comprising a resistance, R, an inductance (say, a purely inductive coil), L, and a capacitance (for example a variable "loss free" air capacitor), C, across which a PD V is applied. [See Fig.13.16]. At a supply frequency, ω (rad/s), the resulting current through the circuit will be

$$I = \frac{V}{Z} = \frac{V}{\sqrt{R^2 + \left(X_L - X_C\right)^2}} \; A \, , \text{ assuming } X_L > X_C.$$

Here, R can normally be assumed to be constant with respect to $f \left(= \frac{\omega}{2\pi} \right)$, the supply frequency in Hz, whilst

(a) $X_L \propto f$, and

(b) $X_C \propto \dfrac{1}{f}$

being given by $2\pi f L$ and $\dfrac{1}{2\pi f C}$, respectively. Clearly, at some frequency f_r, or $\omega_r = 2\pi f_r$,

$$X_L = X_C$$

The current in the circuit will then be limited by R only, given by

$$I = \frac{V}{R} \text{ A}$$

and will be a maximum.

The circuit is then said to be in **resonance.** The various possible variations of the resistance, reactances and resulting current with frequency in this condition are shown in Fig.3.18.

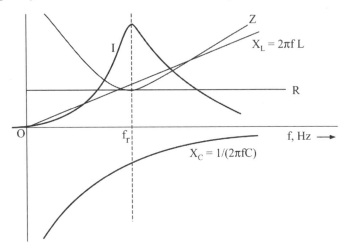

Fig.3.18 : The phenomenon of resonance in R, L, C series circuit

The resonance frequency can be determined from

$$2\pi f_r L = \frac{1}{2\pi f_r C}$$

or

$$f_r^2 = \frac{1}{4\pi^2 LC}$$

and

$$f_r = \frac{1}{2\pi} \times \frac{1}{\sqrt{LC}} \quad \left[\text{or } \omega_r = \frac{1}{\sqrt{LC}} \right]$$

It is seen that whereas the PDs across the inductance and capacitance are equal and opposite and the PD across the resistance is responsible for the very large current, both V_L and V_C are real and non-zero, indicative of energies stored in the inductance and capacitance in electromagnetic and electrostatic form, respectively; for example, the former is given by

$$W_L = \frac{1}{2} L I^2 \text{ J}$$

depending on the magnitudes of L and I.

B. Resonance in a *branched* circuit

This case can be best analysed in terms of the *conductance* and *susceptance* of the branch elements. Thus, conductance, $G = \dfrac{1}{R}$ mho

and total susceptance, $B = \dfrac{1}{2\pi f L} - 2\pi f C$.

B will be zero at a particular frequency, f', defined by

$$\frac{1}{2\pi f' L} = 2\pi f' C$$

or $\qquad\qquad f' = \dfrac{1}{2\pi\sqrt{LC}}$

The circuit admittance will be *minimum*, given by $G = 1/R$ and the *total* current given by $I = G\,V$.

Here, at the 'resonance' frequency, f', the total current is a *minimum*, and for this reason the phenomenon is sometimes called *anti-resonance*. However, the expression for frequency is same as in the case of series resonance.

Comment

The phenomenon of electrical resonance, at times using variations of circuits employing a combination of resistance, inductance and capacitance, forms an important aspect of *tuned* circuits; for example, the ones used extensively as filters in DC converter and inverter equipment to eliminate or suppress certain harmonics. A more important application of such circuits is in High-Voltage DC [HVDC] transmission; a variety of these systems are now in vogue for transmission of huge power over long to very-long distances (1000-2000 km) and at ultra high voltages [± 500 to ± 800 kV].

An interesting, essential application of tuned circuits is in historic radio receivers where circuits in RF stage comprise inductive coils, fitted with adjustable magnetic cores (made of ferrite material), shunted by a ganged (air) capacitor used to tune in the desired radio station.

In all these minute to significant applications, an accurate knowledge of the values of R, L or C is imperative, necessitating the use of good, dependable methods of measurement.

THREE-PHASE ELECTRICITY

Comment

The many-fold advantages of poly-phase as against 1-phase working, esp. in terms of 3-phase generation, transmission, distribution and utilisation are well known and extensively documented in text books. However, in view of the importance of measurement of various 3-phase quantities, and usage of appropriate measuring instruments for the same, it is necessary to review a few aspects and requirements of 3-phase systems as follows.

AC Generators or Alternators

Single-phase alternator

Whilst explaining the principle of induction of EMF in an AC generator by flux-cutting rule, a coil was shown to rotate at a fixed speed in a uniform stationary magnetic field. In practical generators, called alternators, the reverse is true: a typical alternator comprising a (set of) *stationary* coils called the stator and a magnet (or field system) rotating *inside* the stator.

Consider the simplest case as shown in Fig.3.19. Clearly, depending on the relative position of the magnet with respect to the *axial* parts of the coil,

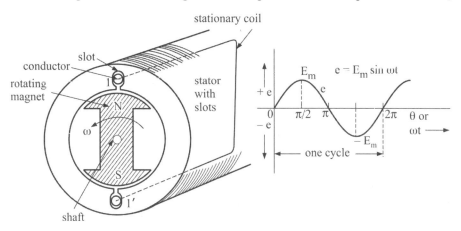

Fig.3.19 : A 'practical' 1-phase alternator and induced EMF

the conductors actually cutting the flux, an EMF will be induced, given by $e = E_m \sin \theta$ (or ωt). Such a machine would be identified as a "single-phase alternator". For simplicity, a 2-pole permanent magnet comprising the rotor is shown in the figure; in practice an electromagnet wound with a suitable number of turns and excited by DC, called the field system, would be used. Also, the rotor must be rotated at 3000 rpm to induce an EMF of 50 Hz according to the basic relation $f = \dfrac{(N \times P)}{120}$ where N is the rotor speed in rpm and P the number of poles in the alternator.

Three-phase alternator

Now consider a similar structure comprising THREE coils, displaced in *space* by $120°$ as shown in Fig.3.20. The magnetic field, to be cut to induce EMFs in the various conductor(s) is still due to a bipolar permanent magnet, capable of being rotated at a constant speed by suitable means. It is easy to visualise that in this case the EMF e_1 in coil $1 \text{-} 1'$ is $120°$ in *advance* or ahead of the EMF e_2 in coil $2 \text{-} 2'$, and this in turn is $120°$ in advance of the EMF e_3 in coil $3 \text{-} 3'$ *at any instant.* If each coil 'supplies' a separate network (or 'load'), this system would be identified as a "3-phase alternator". [Note the direction of rotation of the magnet, being anticlockwise].

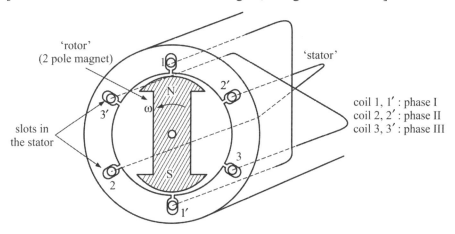

Fig.3.20 : The 'elementary' three-phase alternator

The resulting waveforms of phase-wise induced EMFs and the corresponding phasor diagram is shown in Fig.3.21.

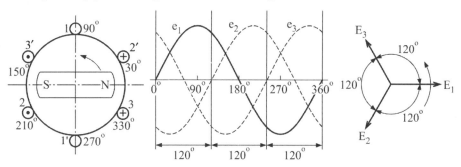

Fig.3.21: EMF waveforms and phasor diagram of a 3-phase alternator

Why three-phase?

Some of the reasons why a three-phase system is preferred to a simple single-phase system in modern-day practice, despite the 'complications' of the former in terms of design and fabrication are enumerated below:

1. As was seen, the power in a single-phase system is *pulsating* since even with unity power factor, there will always be an alternating component of twice the supply frequency. This would be objectionable in many applications, esp. large machines, resulting in non-uniform load control and excessive vibrations.

    With the three-phase working, the power output is very nearly 'constant' as long as the phases are balanced with respect to each other.

2. Single-phase motors, other than the commutator type, possess no starting torque and will require an auxiliary device to enable them to start; in contrast, a commonly employed 3-phase induction motor is self-starting.

3. For a given *frame size*, the power output of a 3-phase machine (an alternator or a motor) is greater than that of a single-phase machine. This means that there is more efficient utilisation of the active material of a 3-phase machine, making them more economical for the same power rating.[1]

4. In general, the 3-phase power transmission may require less 'copper' (in conductors) than single-phase considering the distance, power to be transmitted and given power loss. Thus, with a given voltage across conductors, the 3-phase system may require 3/4th the weight of copper compared to the single-phase system.

5. In single-phase synchronous machines the mmf set up by the armature currents, that is, the armature reaction is pulsating resulting in the induction of eddy currents in the field system with consequent heating and temperature rise.

Voltages, Currents and Power in a Three-phase System

> *Comment*
> Measurement of power in a 3-phase circuit is an important subject, to be dealt with in a separate chapter. It is, therefore, necessary to understand a theoretical basis of how power is related to terminal quantities, viz., voltages and currents in the three phases.

---

[1]Above three phases, the increase in the output is not much and hence the 3-phase system comprising generation (using 3-phase alternators), transmission and distribution lines and utilisation equipment, has been adopted as the 'standard' (polyphase) system throughout the world. In some special cases, it may be necessary to use a 6-phase system; for example, in rectifier units.

The instantaneous voltages in the three phases can be written, assuming phase a as the reference phasor

$$v_a = V_m \sin \omega t$$

$$v_b = V_m \sin\left(\omega t - \frac{2\pi}{3}\right) \qquad \because \ \frac{2\pi}{3} \ \text{rad} \equiv 120°$$

$$v_c = V_m \sin\left(\omega t - \frac{4\pi}{3}\right) \quad \text{or} \quad V_m \sin\left(\omega t + \frac{2\pi}{3}\right)$$

Assuming a lagging power factor, $\cos\phi$, *being same in the three balanced phases*, the phase currents can be expressed as

$$i_a = I_m \sin(\omega t - \phi)$$

$$i_b = I_m \sin(\omega t - 2\pi/3 - \phi)$$

$$i_c = I_m \sin(\omega t - 4\pi/3 - \phi) \text{ or } I_m \sin(\omega t + 2\pi/3 - \phi)$$

Then the total instantaneous three-phase power is

$$p = v_a i_a + v_b i_b + v_c i_c$$

$$= V_m I_m \Big[\sin \omega t \ \sin(\omega t - \phi)$$

$$+ \sin(\omega t - 2\pi/3)\sin(\omega t - 2\pi/3 - \phi)$$

$$+ \sin(\omega t - 4\pi/3)\sin(\omega t - 4\pi/3 - \phi)\Big]$$

Applying for each of the pair of sine terms in the square brackets the trigonometric identity

$$\sin A \sin B = \frac{1}{2}\left[\cos(A - B) - \cos(A + B)\right]$$

and simplifying, the expression for average power would reduce to

$$P = 1.5 \ V_m I_m \cos\phi$$

or

$$P = 3 \times V \times I \times \cos\phi^1$$

$$= 3 \times \text{power per phase, or the total 3-phase power.}$$

Also, since the final form does not contain a term depending on ωt, the *total* power is constant with respect to time.

---

[1] In the expression, V and I are rms values of voltage and current *per phase*. If V were *line* voltage, say across any two phases of a star-connected load, equal to $\sqrt{3} \times V_{phase}$, the total 3-phase power will be given by

$$P = \sqrt{3} \ V_{line} \ I_{line} \cos\phi,$$

I_{line} being line (or phase) current and $\cos\phi$ the power factor of *each* phase.

A graphical representation of power components of a three-phase system with time is shown in Fig.3.22.

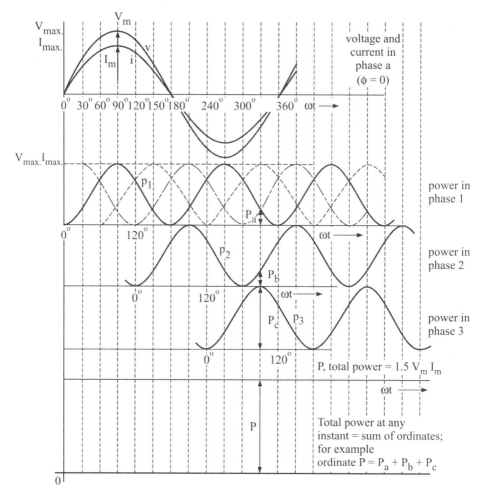

Fig.3.22 : Variation of power in a 3-phase balanced system at unity power factor

Here, the case of unity power factor has been considered for simplicity of the diagram, but the concept would apply for any lagging or leading power factor as brought out in the above analysis. Since a particular point, say a maximum, on each curve is displaced by $120°$ from one phase to the next, the three "power curves" for the individual phases will be as shown in the figure. The maximum instantaneous power in *any one* phase is $V_m I_m$. The graph of total power is obtained by adding ordinates of the three component power curves, showing that the total power is constant and equal to

$$P = 1.5\, V_m\, I_m$$

$$= 3\frac{V_m}{\sqrt{2}}\frac{I_m}{\sqrt{2}}$$

$$= 3 \times V\,I \quad \text{or } 3 \times \text{power per phase}$$

since the power factor is assumed to be unity.

Connections of Three-phase Systems in STAR(Y) or DELTA (Δ)[1]

The three phases, each comprising a phase (or "live") terminal and neutral, of a 3-phase system are seldom operated with individual pairs, but are invariably connected in STAR or DELTA circuits or systems as shown in Fig.3.23.

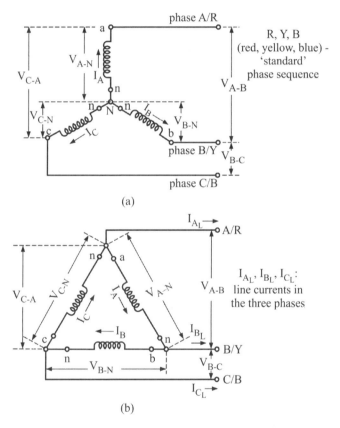

Fig.3.23 : Star and Delta connections in a 3-phase system

It can be seen that the two alternative schemes of connections have some distinct features. These are summarised in Table 3.1.

---

[1]Nomenclature, perhaps derived, respectively, from the shapes which resemble (crudely) a "star" and "delta" (formed, for example, by a river bed near the sea).

Table 3.1: Some distinct features of star and delta connections

Star connection	Delta connection
• The 'start' terminals of all the windings (say, in a 3-phase induction motor) are joined together, to form a "star" or neutral terminal.	• The 'start' terminal of one winding is connected to 'end' terminal of the second winding or phase and so on, in a cyclic manner to form a "closed" circuit, called delta (or mesh).
• The 'line' current in the system is the same as the phase current.	• The 'line' current is the phasor sum of appropriate phase currents, such that $$I_{line} = \sqrt{3} \times I_{ph} \text{ (any)}$$ (in a balanced system)
• The 'line' voltage, e.g., $V_{A\text{-}B}$ is the phasor sum of the appropriate phase to neutral voltages, such that $$V_{line} = V_{A\text{-}B} = \sqrt{3} \times V_{ph} \text{ (any) }^1$$	• The 'line' voltages are the same as the phase-to-neutral voltages, as appropriate.
• The scheme is particularly useful or applicable when choice of two distinct voltage is desired; e.g., in low voltage distribution system(s).	• This scheme is useful when relatively large currents at fixed voltage are to be dealt with; e.g., in transmission systems.
• Under balanced conditions, there is no current through the (common) neutral which can be grounded to meet some system requirements.	• Under some un-balanced conditions encountered in individual phases, some harmonic current components circulate within the closed mesh and do not appear in the line currents.

Transformation of a delta-connected system to a star-connected system and vice-versa

In some electrical measurement procedures, it may be required to transform a delta-connected circuit into an "equivalent" star-connected one or vice-versa.

These transformations can be arrived at as follows:

Delta-to-Star

Consider the two network arrangements shown in Fig.3.24.

---

[1]The relationships $V_{line} = \sqrt{3} \times V_{ph}$ or $I_{line} = \sqrt{3} \times I_{ph}$ are applicable only when the systems are balanced, that is, all the *phase-to-neutral* voltages (or *phase* currents) are equal in magnitude and $120°(E)$ apart with respect to each other.

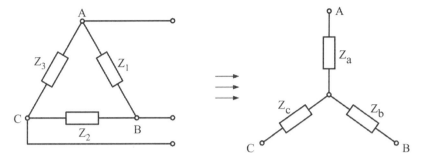

Fig.3.24 : Transformation of a delta network into an 'equivalent' star network

The basis of transformation is that if the two circuits are to be equivalent, then the impedances between the corresponding terminals must be the same in both cases.

Then, considering only the magnitudes,

the impedance between terminals A and B in the delta network is

$$Z_{AB} = \frac{Z_1(Z_2 + Z_3)}{Z_1 + Z_2 + Z_3}$$

and in the star network

$$Z_{AB} = Z_a + Z_b$$

Similar expressions may be derived for the other pairs of terminals, giving

For A-B $\quad Z_a + Z_b = \dfrac{Z_1(Z_2 + Z_3)}{Z_1 + Z_2 + Z_3}$ .....(i)

For B-C $\quad Z_b + Z_c = \dfrac{Z_2(Z_1 + Z_3)}{Z_1 + Z_2 + Z_3}$ .....(ii)

For C-A $\quad Z_c + Z_a = \dfrac{Z_3(Z_1 + Z_2)}{Z_1 + Z_2 + Z_3}$ .....(iii)

To simplify, subtract (ii) from (i) to get

$$Z_a - Z_c = \frac{Z_3(Z_1 - Z_2)}{Z_1 + Z_2 + Z_3}$$ .....(iv)

Add (iii) and (iv), obtaining

$$\boxed{Z_a = \frac{Z_1 Z_3}{Z_1 + Z_2 + Z_3}}$$

and by symmetry

$$Z_b = \frac{Z_1 Z_2}{Z_1 + Z_2 + Z_3} \text{ and } Z_c = \frac{Z_2 Z_3}{Z_1 + Z_2 + Z_3}$$

In practice, the denomenator being common, the expressions can be written by inspection.

Star-to-Delta

The inverse transformation may be deduced using the above expressions (in boxes) as follows:

$$Z_1 + Z_2 + Z_3 = \frac{Z_1 Z_3}{Z_a} = \frac{Z_1 Z_2}{Z_b} = \frac{Z_2 Z_3}{Z_c}$$

from which

$$\frac{Z_1}{Z_3} = \frac{Z_b}{Z_c} \qquad \text{and} \qquad \frac{Z_2}{Z_3} = \frac{Z_b}{Z_a}$$

But

$$Z_a = \frac{Z_1 Z_3}{Z_1 + Z_2 + Z_3} = \frac{Z_1}{\frac{Z_1}{Z_3} + \frac{Z_2}{Z_3} + 1}$$

∴ substituting from above ratios

$$Z_a = \frac{Z_1}{\frac{Z_b}{Z_c} + \frac{Z_b}{Z_a} + 1} = \frac{Z_1 Z_c Z_a}{Z_a Z_b + Z_b Z_c + Z_c Z_a}$$

Hence

$$Z_1 = \frac{Z_a Z_b + Z_b Z_c + Z_c Z_a}{Z_c}$$

Similarly

$$Z_2 = \frac{Z_a Z_b + Z_b Z_c + Z_c Z_a}{Z_a}$$

and

$$Z_3 = \frac{Z_a Z_b + Z_b Z_c + Z_c Z_a}{Z_b}$$

UNBALANCED THREE-PHASE SYSTEMS

Comment

In practice, quite frequently, the 3-phase system may not be balanced, involving one or all of the following conditions:

(a) The supply is balanced, but the (load) impedances are not, giving rise to unbalanced *load* currents, different in each phase, and hence unbalanced potential drops across the loads.

(b) The load is balanced, but the supply is not, thus resulting in unbalanced currents.

(c) Both the supply as well as the load are unbalanced.

In most cases, when the utilisation of electrical energy is from 'stable' source(s), for example the infinite bus, the first situation is usually encountered. In any of the cases of unbalance, measurement of the three important electrical quantities, viz., currents, potential drops and power may have to employ special procedures and these may be rather involved.

Typical Example of Unbalance in Practice

Consider a 3-phase induction motor, fed from a 3-phase supply at appropriate voltage. The stator of the motor may be star or delta connected. When the supply is balanced and the three stator windings of the motor are identical, the machine runs normally.

However, if

(a) a few turns in one of the phase windings are shorted so that the impedances in the three phases are different or/and

(b) one of the line conductors is broken so that the motor is connected to only two phases;

in either case, the *line* currents drawn by the motor will be different, that is, unbalanced.

How is the motor operation going to be affected? This is understood easily and effectively by applying a unique technique of analysis involving what are known as **symmetrical components** of unbalanced electrical parameters, usually voltages and currents.

Symmetrical Components

It was in 1918 that one of the most powerful tools for analysing unbalanced polyphase circuits (specifically the 3-phase circuits) was propounded by

Dr. C.L. Fortescue at a meeting of the American Institute of Electrical Engineers (AIEE) by way of presenting a paper[1].

Fortescue's dictum states that "an unbalanced system of n *related* phasors can be resolved into n systems of *balanced* phasors called the *symmetrical components* of the original phasors." The n phasors of each set of components will be equal in length (magnitude), with equal angles between the adjacent phasors.

Accordingly, three unbalanced phasors of a 3-phase system can be resolved into *three balanced* systems of phasors as follows:

1. a set of *positive-sequence* components consisting of *three* phasors of equal magnitude, displaced from each other by 120°(E) in phase and having the *same phase sequence as the original phasors, R-Y-B*, following anti-clockwise rotation;

2. a set of *negative-sequence* components similar to the above, but having the phase sequence *opposite* to that of the original phasors, that is, R-B-Y, but the rotation being still anti-clockwise;

3. a set of *zero-sequence* components comprising three phasors, equal in magnitude and with *'zero' phase displacement* from each other, that is, in phase with each other.

The three sets representing 3-phase *voltages* of a hypothetical unbalanced system are as shown in Fig.3.25. Component phasors representing currents will be designated in a similar manner with appropriate subscripts.

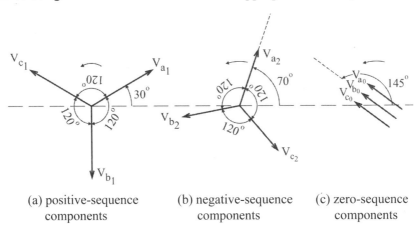

| (a) positive-sequence | (b) negative-sequence | (c) zero-sequence |
| components | components | components |

Fig.3.25 : The three sets of symmetrical components

The original unbalanced phasors will then be expressed in terms of sets of components as follows:

[1]C.L. Frotescue: Method of Symmetrical Coordinates Applied to the Solution of Polyphase Networks, Trans. AIEE, Vol. 37, 1918, pp 1027-1140.

$$\dot{V}_a = \dot{V}_{a_1} + \dot{V}_{a_2} + \dot{V}_{a_0}$$

$$\dot{V}_b = \dot{V}_{b_1} + \dot{V}_{b_2} + \dot{V}_{b_0}$$

$$\dot{V}_c = \dot{V}_{c_1} + \dot{V}_{c_2} + \dot{V}_{c_0}$$

Note that individual set of phasors are themselves symmetrical; hence the name.

To show what the above equations could mean, the three sets of phasors with their *relative* phaser displacements as specified in Fig.3.25 are added together. The result is the three, 'original' unbalanced voltage phasors, $\dot{V}_a$, $\dot{V}_b$, $\dot{V}_c$, of which the three sets of symmetrical components are given by

$$\dot{V}_{a_1}, \dot{V}_{a_2}, \ldots; \dot{V}_{b_1} \ldots, \dot{V}_{b_0}; \ldots, \dot{V}_{c_0}$$

This is illustrated in Fig.3.26 with reference to the three component phasors of Fig.3.25, maintaining *the same phasor relationship*, but slightly different scale of magnitudes. It can be argued that if the original unbalanced voltages were known, it should be possible to derive the three sets of symmetrical components. This is achieved by using a systematic analytical approach as discussed below.

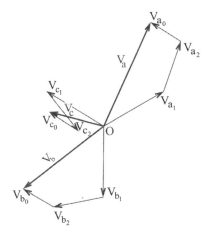

Fig.3.26 : Three hypothetical unbalanced voltages and their symmetrical components

The Operator "a"

An "operator" essential to all analyses involving symmetrical components is identified to be "a", similar to "j" defined previously for dealing with complex quantities whose property was to "rotate" a given phasor by an angle of 90° in anticlockwise direction. To put simply, the operator "a" has the property of "rotating" a phasor in an anticlockwise direction by 120°. Thus, the set of phasors comprising the positive sequence component will be expressed as

$$\dot{V}_{a_1} = \left|\dot{V}_{a_1}\right| \underline{/0°} \ : \text{reference phasor}$$

$$\dot{V}_{c_1} = \left|\dot{V}_{c_1}\right| \underline{/+120°} \ \text{or a} \ \dot{V}_{a_1}, \ \text{leading} \ \dot{V}_{a_1} \ \text{by} \ 120°$$

$$\dot{V}_{b_1} = \left|\dot{V}_{b_1}\right| \underline{/-120°} \ \text{or a}^2 \ \dot{V}_{a_1}, \ \text{lagging} \ \dot{V}_{a_1} \ \text{by} \ 120°$$

These relationships are shown in Fig.3.27(a).

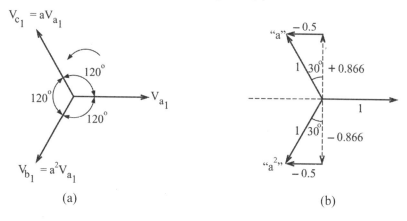

(a)   (b)

Fig.3.27 : Representation of symmetrical components using the operator "a"

Clearly, assuming unit magnitudes of $\dot{V}_{a_1}$, $\dot{V}_{b_1}$, $\dot{V}_{c_1}$, the operator a (and a^2) can be numerically expressed in terms of the operator j as given below, and as shown in Fig.3.27(b).

Thus

$$a = 1 \ \underline{/+120°} = 1 \times (\cos 120° + j \sin 120°)$$

$$= -0.5 + j0.866$$

and

$$a^2 = 1 \ \underline{/+240°} = 1 \times (\cos 240° + j \sin 240°)$$

$$= -0.5 - j0.866$$

As seen, both "a" and "a²" are complex numbers[1].

The other identities using "a" and its 'powers' can be derived as listed below.

$$a^3 = 1 \ \underline{/360°} = 1 + j0 = 1$$

---

[1]A "dot" (·) above a or a^2 is omitted for clarity.

$$a^4 = 1\ \underline{/480^\circ} = 1\ \underline{/120^\circ} = -0.5 + j0.866 = a$$

$$1 + a = 1 + (-0.5 + j\,0.866) = +0.5 + j0.866$$

$$= -a^2 = 1\ \underline{/60^\circ}$$

[it follows that $1 + a + a^2 = 0$]

$$1 - a = 1 - (-0.5 + j0.866)$$

$$= 1.5 - j0.866 = \sqrt{3}\ \underline{/-30^\circ}$$

$$1 + a^2 = 1 + (-0.5 - j0.866)$$

$$= 0.5 - j0.866 = -a = 1\ \underline{/-60^\circ}$$

$$1 - a^2 = 1 - (-0.5 - j0.866)$$

$$= 1.5 + j0.866 = \sqrt{3}\ \underline{/30^\circ}$$

$$a + a^2 = -0.5 + j0.866 + (-0.5 - j0.866)$$

$$= -1 - j0 = 1\ \underline{/180^\circ}$$

$$a - a^2 = -0.5 + j0.866 - (-0.5 - j0.866)$$

$$= 0 + j1.732 = \sqrt{3}\ \underline{/90^\circ}$$

The use of some of these identities could be in effecting rotation of appropriate phasors by angles other than 90° or 120°.

Thus, a set of three unbalanced voltages can be expressed in terms of its symmetrical components as fallows:

$$\dot{V}_a = \dot{V}_{a_1} + \dot{V}_{a_2} + \dot{V}_{a_0} \quad \text{[reference phasor]}$$

$$\dot{V}_b = \dot{V}_{b_1} + \dot{V}_{b_2} + \dot{V}_{b_0}$$

$$\dot{V}_c = \dot{V}_{c_1} + \dot{V}_{c_2} + \dot{V}_{c_0}$$

or, using the relationship between $\dot{V}_{a_1}$, $\dot{V}_{b_1}$, $\dot{V}_{c_1}$ in terms of the operator "a",

$$\dot{V}_a = \dot{V}_{a_1} + \dot{V}_{a_2} + \dot{V}_{a_0} \quad \text{[reference phasor]}$$

$$\dot{V}_b = a^2\,\dot{V}_{a_1} + a\,\dot{V}_{a_2} + \dot{V}_{a_0}$$

$$\dot{V}_c = a\,\dot{V}_{a_1} + a^2\,\dot{V}_{a_2} + \dot{V}_{a_0}$$

And similarly for currents.

It is seen that whilst positive-sequence components are related to each other by the sequence 1, a^2, a, the negative–sequence components follow the sequence 1, a, a^2; the zero–sequence components being in phase with each other, not involving a.

Using matrix form

$$
\begin{bmatrix} \dot{V}_a \\ \dot{V}_b \\ \dot{V}_c \end{bmatrix} = \begin{bmatrix} 1 & 1 & 1 \\ 1 & a^2 & a \\ 1 & a & a^2 \end{bmatrix} \begin{bmatrix} \dot{V}_{a_0} \\ \dot{V}_{a_1} \\ \dot{V}_{a_2} \end{bmatrix}
$$

An important outcome of the analysis using the operator "a", therefore, is the computation of sequence components, given the original unbalanced phasors. Thus, using matrix algebra

$$
\begin{bmatrix} \dot{V}_{a_0} \\ \dot{V}_{a_1} \\ \dot{V}_{a_2} \end{bmatrix} = \frac{1}{3} \begin{bmatrix} 1 & 1 & 1 \\ 1 & a & a^2 \\ 1 & a^2 & a \end{bmatrix} \begin{bmatrix} \dot{V}_a \\ \dot{V}_b \\ \dot{V}_c \end{bmatrix}
$$

from which the other phasors of *same* set follow, giving

$$\dot{V}_{b_1} = a^2 \, \dot{V}_{a_1}, \quad \dot{V}_{c_1} = a \, \dot{V}_{a_1}, \quad \text{and so on.}$$

The following example will demonstrate the procedure employing symmetrical components.

One conductor of a 3-phase supply, feeding a delta-connected load, is "open" as shown in Fig.3.28. The current flowing through the 'healthy' lines is 10 A each. Find the symmetrical components of the line currents.

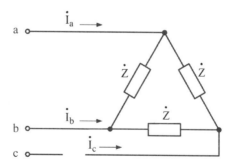

Fig.3.28 : Currents and load for the example

Assume that the current in phase a is the reference phasor.

Then $\qquad \dot{I}_a = |\dot{I}_a| \underline{/0^\circ}$ or $\dot{I}_a = 10 \underline{/0^\circ}$ A

Also, the same current would return via line b.

Therefore, $\qquad \dot{I}_b = -\dot{I}_a$ or $\dot{I}_b = 10 \underline{/180^\circ}$ A

The load is balanced, with each phase impedance being $\dot{Z}$ Ω.

Let line c is open; hence $\dot{I}_c = 0$.

To derive the symmetrical components

$$\dot{I}_{a_0} = \left(\frac{1}{3}\right) \times \left(\dot{I}_a + \dot{I}_b + \dot{I}_c\right)$$

$$= \left(\frac{1}{3}\right) \times (10 \underline{/0^\circ} + 10 \underline{/180^\circ} + 0)$$

$$= \left(\frac{1}{3}\right) \times (10 - 10) = 0$$

$$\dot{I}_{a_1} = \left(\frac{1}{3}\right) \times \left(\dot{I}_a + a\dot{I}_b + a^2 \dot{I}_c\right)$$

$$= \left(\frac{1}{3}\right) \times (10 \underline{/0^\circ} + 10 \underline{/180^\circ + 120^\circ} + 0)$$

(substituting for a in terms of phase displacement of 120°(E) with respect to $\dot{I}_a$)

$$= 5 - j\,2.89 = 5.78 \underline{/-30^\circ}\ \text{A}$$

$$\dot{I}_{a_1} = \left(\frac{1}{3}\right) \times \left(\dot{I}_a + a^2\dot{I}_b + a\,\dot{I}_c\right)$$

$$= \left(\frac{1}{3}\right) \times (10 \underline{/0^\circ} + 10 \underline{/180^\circ + 240^\circ} + 0),$$

(since $a^2 = -120^\circ$ or $+240^\circ$)

$$= 5 + j\,2.89 = 5.78 \underline{/30^\circ}\ \text{A}$$

The same results could be obtained by first representing $\dot{I}_a$ and $\dot{I}_b$ in complex, that is, A $\pm$ j B form, and using the complex forms of a and a^2.

Using appropriate expressions

$$\dot{I}_{b_1} = a^2\, \dot{I}_{a_1} = 5.78\ \underline{/-150^\circ}\ \text{A,}$$

$$\dot{I}_{c_1} = a\, \dot{I}_{a_1} = 5.78\ \underline{/90^\circ}\ \text{A}$$

$$\dot{I}_{b_2} = a\, \dot{I}_{a_2} = 5.78\ \underline{/150^\circ}\ \text{A,}$$

$$\dot{I}_{c_2} = a^2\, \dot{I}_{a_1} = 5.78\ \underline{/-90^\circ}\ \text{A}$$

$$\dot{I}_{b_0} = \dot{I}_{c_0} = 0$$

A graphic representation of various line currents and their components is shown in Fig.3.29.

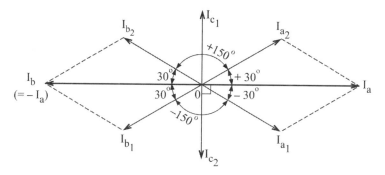

Fig.3.29 : Line currents and symmetrical components for the example

Note that

(a) there is no zero–sequence component of current(s) in the networks;

(b) the components $\dot{I}_{c_1}$ and $\dot{I}_{c_2}$ have finite values, both equal in magnitude (and oppositely directed) although line c is open and can carry no net current;

(c) as would be expected, therefore, the sum of the components in line c is zero, being in phase opposition as shown in Fig.3.29;

(d) the phasor sum of the components in line a is 10 $\underline{/0^\circ}$ A and that in line b as 10 $\underline{/180^\circ}$ A as shown;

(e) both the positive- and negative-sequence components are an appreciable 58% of the line currents in phase a or b.

Comment

If it be assumed that the delta-connected load represents the stator of a 3-phase induction motor (of any rating), there exists a peculiar situation in the machine vis–a–vis the two sequence components of currents and their physical effects in the motor, specifically on the rotor, as follows:

(a) the positive-sequence currents would continue to produce the 'useful' torque in the normal direction, although of reduced magnitude, allowing the rotor to run in the original direction; however, if the developed torque is less than the load, the motor might stall;

(b) the negative–sequence currents would, however, produce an mmf in the airgap rotating at nearly *double the supply frequency* relative to the direction of rotation of the rotor, resulting in induction of eddy currents in all its parts, mainly directed axially on the rotor surface, ultimately leading to excessive heating and temperature rise.

[Even if the motor comes to standstill, the currents will continue to induce eddy currents in the rotor, although at supply frequency, penetrating deeper and still causing heating of the rotor].

In general, any unbalance in a power system, under steadystate or transient operating conditions, would invariably result in some negative-sequence currents which are bound to be variously detrimental, particularly to rotating machines including alternators at power stations.

A special requirement of measurement, or instrumentation, is therefore to provide means to detect and measure negative-sequence currents at appropriate locations in a power system and use this information to be incorporated in providing suitable protection devices; for example a negative-sequence relay.

The significance of the negative–sequence currents, and severity of their effects, would naturally depend on the nature and severity of the unbalance.

WORKED EXAMPLES

1. A circuit consists of a capacitor of 2 μF and a resistor of 1000 Ω connected in series. An alternating EMF of 12 V (rms) at a frequency of 50 Hz is applied across the circuit.

   Calculate (a) the current, (b) the PD across the capacitor, (c) the phase angle between the applied voltage and current, and (d) the power consumed.

   The reactance of the capacitor is, $X_C = \dfrac{1}{2\pi f C}$

   $$= \dfrac{1}{\left(2\pi \times 50 \times 2 \times 10^{-6}\right)}$$

   $$= 1592 \ \Omega$$

   The series resistor being 1000 Ω, the impedance is

   $$Z = \sqrt{(1000)^2 + (1592)^2}$$

   $$= 1880 \ \Omega$$

   Therefore,

   (a) circuit current, $\qquad I = \dfrac{V}{Z} = \dfrac{12}{1880} = 6.4 \times 10^{-3} \ A$

   (b) PD across the capacitor, $\quad V_C = I \times X_C = 10.2 \ V$

   (c) phase angle, $\qquad \phi = \tan^{-1}\left(\dfrac{X_C}{R}\right) = 58^\circ$

   (d) power consumed, $\qquad P = I^2 R = 0.04 \ W$

2. A capacitor of capacitance C, a coil of inductance L and resistance R, and a lamp are connected in series across an alternating voltage, V. Its frequency, f, is varied from a low to a high value while the magnitude of V is kept constant. Describe and explain how the brightness of the lamp would vary.

   When f is varied, the impedance Z of the circuit decreases to a minimum value (the resonance condition) and then increases. Z is a minimum when $X_L = X_C$ and Z is simply equal to R. The current in the circuit will then be maximum and the lamp will be at its maximum brightness.

3. In the parallel branch circuit shown in the figure below, a 1-phase voltage of 230 V (rms) at 50 Hz is applied. Determine the branch currents and show them on a phasor diagram with applied voltage as the reference phasor.

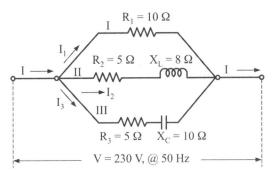

The various branch impedances are

$$\dot{Z}_{I} = R_1 = 10 + j0$$

$$\dot{Z}_{II} = 5 + j8 \ \Omega$$

$$\dot{Z}_{III} = 5 - j10 \ \Omega$$

Then
$$\dot{I}_1 = \frac{\dot{V}}{\dot{Z}_I} = \frac{230}{10} = 23 \pm j0 \text{ or } 23 \ \underline{/0^\circ} \text{ A} \quad [\text{in phase with } \dot{V}]$$

$$\dot{I}_2 = \frac{\dot{V}}{\dot{Z}_{II}} = \frac{230}{(5 + j\,8)} = \frac{230\,(5 - j\,8)}{5^2 + 8^2}$$

$$= 12.9 - j20.7 = 24.4 \ \underline{/-58^\circ} \text{ A}$$

$$\dot{I}_3 = \frac{\dot{V}}{\dot{Z}_{III}} = \frac{230}{(5 - j\,10)} = \frac{230\,(5 + j\,10)}{5^2 + 10^2}$$

$$= 9.2 + j18.4 = 20.6 \ \underline{/63.4^\circ} \text{ A}$$

The phasor diagram is drawn below.

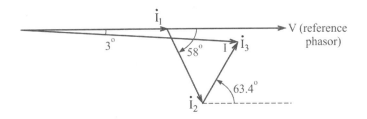

The total current, $\dot{I}$, is given by

$$\dot{I} = \dot{I}_1 + \dot{I}_2 + \dot{I}_3$$

$$= 23 \pm j0 + 12.9 - j20.7 + 9.2 + j18.4$$

$$= 45.1 - j2.3 = 45.2 \underline{/-3^\circ} \text{ A}$$

As indicated in "j" notation or polar form, the current in branch II lags the applied voltage whilst that in branch III leads, being in resistive-inductive and resistive-capacitive circuits, respectively.

4. A 1-phase series-parallel network comprises various resistances, inductances and capacitance as shown in the figure below. Calculate the branch and total current in terms of magnitude and phasor relationship when the network is supplied from a 1-phase, 230 V, 50 Hz supply.

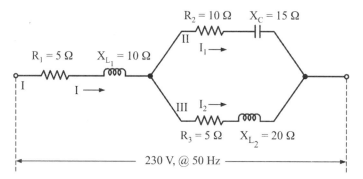

The network impedances are:

branch I $\dot{Z}_1 = R_1 + j\, X_{L_1} = 5 + j\,10$

branch II $\dot{Z}_2 = R_2 + j\, X_C = 10 - j\,15$

branch III $\dot{Z}_3 = R_3 + j\, X_{L_2} = 5 + j\,20$

Resultant of branches II and III in parallel

$$\dot{Z}' = \frac{\dot{Z}_2 \dot{Z}_3}{\dot{Z}_2 + \dot{Z}_3}$$

$$= \frac{(10 - j\,15) \times (5 + j\,20)}{(10 - j\,15) + (5 + j\,20)}$$

$$= 23.5 + j\,0.5 \ \Omega$$

[after rationalisation and simplification]

Observe that the net (physical) effect of the two branches in parallel, one being resistive-capacitive and the other resistive-inductive, is a nearly completely resistive impedance.

The total impedance of the network is

$$\dot{Z} = \dot{Z_1} + \dot{Z'} = 5 + j\,10 + 23.5 + j\,0.5$$

$$= 28.5 + j\,10.5 \ \Omega \text{ and seen to be resistive-inductive.}$$

∴   the total (or input) current in the circuit is

$$\dot{I} = \frac{\dot{V}}{\dot{Z}} = \frac{230 + j\,0}{28.5 + j\,10.5}, \text{ assuming } \dot{V} \text{ to be the reference phasor}$$

$$= 7.1 - j\,2.6 \text{ or } 7.56\underline{/-20^\circ} \text{ A in polar form.}$$

The negative phase angle of the current shows that the net reactive effect of the circuit is inductive, since the parallel branch is nearly resistive and branch I is resistive-inductive.

The PD across the branches in parallel is

$$\dot{V'} = \dot{I} \times \dot{Z'} = (7.1 - j\,2.6)\,(23.5 + j\,0.5)$$

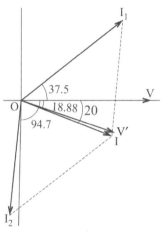

$$= 168.1 - j\,57.5 \text{ or } 177.7\underline{/-19^\circ} \text{ V}$$

∴   the current in branch II is

$$\dot{I_1} = \frac{\dot{V'}}{\dot{Z_2}} = \frac{168.1 - j\,57.5}{10 - j\,15}$$

$$= 7.8 + j\,5.98 \text{ or } 9.8\underline{/+37.15^\circ} \text{ A}$$

and the current in branch III is

$$\dot{I_2} = \frac{\dot{V'}}{\dot{Z_3}} = \frac{168.1 - j\,57.5}{5 + j\,20}$$

$$= -0.7 - j\,8.6^1 \text{ or } 8.6\underline{/-94.7^\circ} \text{ A}$$

[Adding the two branch currents yields $(7.1 - j\,2.62)$, the total or input current as before]

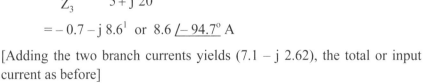

[1]The current $\dot{I_2}$ is seen to be almost in phase quadrature with the supply voltage and lagging. [See the phasor diagram].

5. A 3-phase network containing resistances, inductance and capacitance is connected in star and fed from a 3-phase balanced supply at 400 V line-to-line as shown in the figure below.

Calculate the branch currents $\dot{I}_1, \dot{I}_2$, and $\dot{I}_3$.

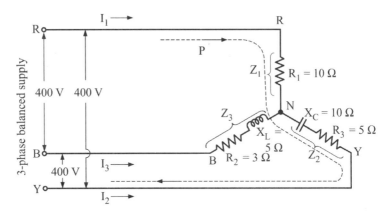

Assume the line voltage across phases R and Y, $\dot{E}_{R-Y}$, to be the reference phasor.

The various applied (line) voltages can then be expressed (using the "j" and "a" operators) as

$$\dot{E}_{R-Y} = 400 + j\,0 = \dot{E}_1 - \dot{E}_2$$

$$\dot{E}_{Y-B} = a^2\,\dot{E}_{R-Y} = 400\,(-0.5 - j\,0.866) = \dot{E}_2 - \dot{E}_3$$

$$\dot{E}_{B-R} = a\,\dot{E}_{R-Y} = 400\,(-0.5 + j\,0.866) = \dot{E}_3 - \dot{E}_1$$

where $\dot{E}_1$, $\dot{E}_2$ and $\dot{E}_3$ represent the supply phase voltages (equal in magnitude).

Writing the various loop equations, for example the one shown by P, in terms of the three line currents and three impedances in phases R, Y and B,

$$\dot{E}_1 - \dot{I}_1\,\dot{Z}_1 + \dot{I}_2\,\dot{Z}_2 - \dot{E}_2 = 0$$

$$\dot{E}_2 - \dot{I}_2\,\dot{Z}_2 + \dot{I}_3\,\dot{Z}_3 - \dot{E}_3 = 0$$

and $$\dot{E}_3 - \dot{I}_3\,\dot{Z}_3 + \dot{I}_1\,\dot{Z}_1 - \dot{E}_1 = 0$$

Simplifying

$$\dot{E}_1 - \dot{E}_2 = \dot{I}_1\,\dot{Z}_1 - \dot{I}_2\,\dot{Z}_2 \qquad\qquad \text{I}$$

$$\dot{E}_2 - \dot{E}_3 = \dot{I}_2\,\dot{Z}_2 - \dot{I}_3\,\dot{Z}_3 \qquad\qquad \text{II}$$

$$\dot{E}_3 - \dot{E}_1 = \dot{I}_3\,\dot{Z}_3 - \dot{I}_1\,\dot{Z}_1 \qquad\qquad \text{III}$$

Also, at the neutral or star point

$$\dot{I}_1 + \dot{I}_2 + \dot{I}_3 = 0 \qquad\qquad \text{IV}$$

To evaluate $\dot{I}_1$, $\dot{I}_2$, $\dot{I}_3$

"$\dot{I}_1$"

From equations I and III

$$\dot{I}_2 = \dot{I}_1\frac{\dot{Z}_1}{\dot{Z}_2} - \frac{\dot{E}_1 - \dot{E}_2}{\dot{Z}_2}$$

$$\dot{I}_3 = \dot{I}_1\frac{\dot{Z}_1}{\dot{Z}_3} + \frac{\dot{E}_3 - \dot{E}_1}{\dot{Z}_3}$$

Substituting in eqn. IV

$$\dot{I}_1 + \dot{I}_1\frac{\dot{Z}_1}{\dot{Z}_2} - \frac{\dot{E}_1 - \dot{E}_2}{\dot{Z}_2} + \dot{I}_1\frac{\dot{Z}_1}{\dot{Z}_3} + \frac{\dot{E}_3 - \dot{E}_1}{\dot{Z}_3}$$

$$= 0$$

or $\quad \dot{I}_1\left(1 + \dfrac{\dot{Z}_1}{\dot{Z}_2} + \dfrac{\dot{Z}_1}{\dot{Z}_3}\right) = \dfrac{\dot{E}_1 - \dot{E}_2}{\dot{Z}_2} - \dfrac{\dot{E}_3 - \dot{E}_1}{\dot{Z}_3}$

or $\qquad\qquad \dot{I}_1 = \dfrac{\dfrac{\dot{E}_1 - \dot{E}_2}{\dot{Z}_2} - \dfrac{\dot{E}_3 - \dot{E}_1}{\dot{Z}_3}}{1 + \dfrac{\dot{Z}_1}{\dot{Z}_2} + \dfrac{\dot{Z}_1}{\dot{Z}_3}}$

$$= \frac{\dot{E}_1 - \dot{E}_2}{\dot{Z}_2\left(1 + \dfrac{\dot{Z}_1}{\dot{Z}_2} + \dfrac{\dot{Z}_1}{\dot{Z}_3}\right)} - \frac{\dot{E}_3 - \dot{E}_1}{\dot{Z}_3\left(1 + \dfrac{\dot{Z}_1}{\dot{Z}_2} + \dfrac{\dot{Z}_1}{\dot{Z}_3}\right)}$$

"$\dot{I}_2$" and "$\dot{I}_3$"

Proceeding similarly

$$\dot{I}_2 = \frac{\dot{E}_2 - \dot{E}_3}{\dot{Z}_3\left(1 + \dfrac{\dot{Z}_2}{\dot{Z}_1} + \dfrac{\dot{Z}_2}{\dot{Z}_3}\right)} - \frac{\dot{E}_1 - \dot{E}_2}{\dot{Z}_1\left(1 + \dfrac{\dot{Z}_2}{\dot{Z}_1} + \dfrac{\dot{Z}_2}{\dot{Z}_3}\right)}$$

and

$$\dot{I}_3 = \frac{\dot{E}_3 - \dot{E}_1}{\dot{Z}_1\left(1 + \dfrac{\dot{Z}_3}{\dot{Z}_1} + \dfrac{\dot{Z}_3}{\dot{Z}_2}\right)} - \frac{\dot{E}_2 - \dot{E}_3}{\dot{Z}_2\left(1 + \dfrac{\dot{Z}_3}{\dot{Z}_1} + \dfrac{\dot{Z}_3}{\dot{Z}_2}\right)}$$

Now the various impedances are:

$$\dot{Z}_1 = 10 + j\,0,\ \dot{Z}_2 = 5 - j\,10,\ \dot{Z}_3 = 3 + j\,5 \quad \text{(in ohms)}$$

To calculate $\dot{I}_1$

$$\dot{Z}_2\left(1 + \frac{\dot{Z}_1}{\dot{Z}_2} + \frac{\dot{Z}_1}{\dot{Z}_3}\right) = \dot{Z}_1 + \dot{Z}_2 + \frac{\dot{Z}_1\,\dot{Z}_2}{\dot{Z}_3}$$

$$= (10 + j\,0) + (5 - j\,10) + \frac{(10 + j\,0)\,(5 - j\,10)}{3 + j\,5}$$

$$= 4.7 - j\,26.2$$

$$\dot{Z}_3\left(1 + \frac{\dot{Z}_1}{\dot{Z}_2} + \frac{\dot{Z}_1}{\dot{Z}_3}\right) = \dot{Z}_1 + \dot{Z}_3 + \frac{\dot{Z}_1\,\dot{Z}_3}{\dot{Z}_2}$$

$$= (10 + j\,0) + (3 + j\,5) + \frac{(10 + j\,0)\,(3 + j\,5)}{5 - j\,10}$$

$$= 10.2 + j\,9.4$$

Therefore,

$$\dot{I}_1 = \frac{\dot{E}_1 - \dot{E}_2}{4.7 - j\,26.2} - \frac{\dot{E}_3 - \dot{E}_3}{10.2 + j\,9.4}$$

$$= \frac{400 + j\,0}{4.7 - j\,26.2} - \frac{400(-0.5\ + j\,0.866)}{10.2 + j\,9.4}$$

$$= -3.65 - j\,13.8 \quad \text{or} \quad 13.88^{1}\ \underline{/-105°}\ \text{A}$$

---

[1]Even while the load is purely resistive, the current I_1 through it is far from being in phase with the voltage.

The current $\dot{I}_2$

$$\dot{Z}_3\left(1+\frac{\dot{Z}_2}{\dot{Z}_1}+\frac{\dot{Z}_2}{\dot{Z}_3}\right)=\dot{Z}_2+\dot{Z}_3+\frac{\dot{Z}_2\,\dot{Z}_3}{\dot{Z}_1}$$

$$=(5-j\,10)+(3+j\,5)+\frac{(5-j\,10)\,(3+j\,5)}{10+j\,0}$$

$$=14.5-j\,5.5$$

$$\dot{Z}_1\left(1+\frac{\dot{Z}_2}{\dot{Z}_1}+\frac{\dot{Z}_2}{\dot{Z}_3}\right)=\dot{Z}_1+\dot{Z}_2+\frac{\dot{Z}_1\,\dot{Z}_2}{\dot{Z}_3}$$

$$=4.7-j\,26.2\quad\text{[deduced earlier]}$$

Therefore,

$$\dot{I}_2=\frac{\dot{E}_2-\dot{E}_3}{14.5-j\,5.5}-\frac{\dot{E}_1-\dot{E}_2}{4.7-j\,26.2}$$

$$=\frac{400\,(-0.5-j\,0.866)}{14.5-j\,5.5}-\frac{400+j\,0}{4.7-j\,26.2}$$

$$=-\,6.6-j\,39.7\quad\text{or}\quad 40.3\,\underline{/-99.4^\circ}\text{ A}$$

The current $\dot{I}_3$

$$\dot{Z}_1\left(1+\frac{\dot{Z}_3}{\dot{Z}_1}+\frac{\dot{Z}_3}{\dot{Z}_2}\right)=\dot{Z}_1+\dot{Z}_3+\frac{\dot{Z}_1\,\dot{Z}_3}{\dot{Z}_1}$$

$$=(10+j\,0)+(3+j\,5)+\frac{(10+j\,0)\,(3+j\,5)}{5-j\,10}$$

$$=10.2+j\,9.4$$

$$\dot{Z}_2\left(1+\frac{\dot{Z}_3}{\dot{Z}_1}+\frac{\dot{Z}_3}{\dot{Z}_2}\right)=\dot{Z}_2+\dot{Z}_3+\frac{\dot{Z}_2\,\dot{Z}_3}{\dot{Z}_1}$$

$$=(5-j\,10)+(3+j\,5)+\frac{(5-j\,10)\,(3+j\,5)}{10+j\,0}$$

$$=14.5-j\,5.5$$

Therefore

$$\dot{I}_3=\frac{\dot{E}_3-\dot{E}_1}{10.2+j\,9.4}-\frac{\dot{E}_2-\dot{E}_3}{14.5-j\,5.5}$$

$$= \frac{400(-0.5 + j\, 0.866)}{10.2 + j\, 9.4} - \frac{400(-0.5 - j\, 0.866)}{14.5 - j\, 5.5}$$

$$= 10.5 + j\, 53.6 \quad \text{or} \quad 54.6^{1}\; \underline{/+\, 79^{\circ}}\; A$$

6. In the load (circuit) of example 5, if $\dot{E}_R$, $\dot{E}_Y$ and $\dot{E}_B$ represent the voltages (or PDs) across the branch impedances $\dot{Z}_1$, $\dot{Z}_2$ and $\dot{Z}_3$, respectively, calculate the branch currents *using symmetrical components*.

The PD equations for the circuit in this case can be written as

$$\dot{E}_R = \dot{I}_1\, \dot{Z}_1 = \dot{I}_1(10 + j\, 0)$$

$$\dot{E}_Y = \dot{I}_2\, \dot{Z}_2 = \dot{I}_2(5 - j\, 10)$$

$$\dot{E}_B = \dot{I}_3\, \dot{Z}_3 = \dot{I}_3(3 + j\, 5)$$

With the currents expressed in terms of their symmetrical components, $I_{a_0}, I_{a_1}, I_{a_2}, \ldots, \ldots \,.\, I_{c_1}, I_{c_2}$, etc., the above equations can be written

$$\dot{E}_R = (\dot{I}_{a_1} + \dot{I}_{a_2})\,(10 + j\, 0)$$

$$\dot{E}_Y = (\dot{I}_{b_1} + \dot{I}_{b_2})\,(5 - j\, 10)$$

$$\dot{E}_B = (\dot{I}_{c_1} + \dot{I}_{c_2})\,(3 + j\, 5)$$

In the above, $\dot{I}_{a_0} = \dot{I}_{b_0} = \dot{I}_{c_0} = 0$ since the "star point" or neutral is 'isolated' and there is no fourth wire.

Using the components of $\dot{I}_a$ alone

$$\dot{E}_R = (\dot{I}_{a_1} + \dot{I}_{a_2})\,(10 + j\, 0)$$

$$\dot{E}_Y = (a^2\, \dot{I}_{a_1} + a\, \dot{I}_{a_2})\,(5 - j\, 10)$$

$$\dot{E}_B = (a\, \dot{I}_{a_1} + a^2\, \dot{I}_{a_2})\,(3 + j\, 5)$$

Also, it is given that

$$\dot{E}_R - \dot{E}_Y = \dot{E}_{R-Y} = 400 + j\, 0$$

$$\dot{E}_Y - \dot{E}_B = \dot{E}_{Y-B} = a^2\, 400$$

$$\dot{E}_B - \dot{E}_R = \dot{E}_{B-R} = a\, 400$$

---

[1]As a check, it is seen that $\dot{I}_1 + \dot{I}_2 + \dot{I}_3 = 0$, very closely, considering the numerous calculations involving complex quantities.

Therefore, substituting $\dot{E}_Y$ from $\dot{E}_R$ from the above expressions and using values of a and a^2,

$$\dot{E}_R - \dot{E}_Y = (\dot{I}_{a_1} + \dot{I}_{a_2})(10 + j\,0) - (a^2\dot{I}_{a_1} + a\dot{I}_{a_2})(5 - j\,10)$$

$$= (21.26 - j\,0.67)\,\dot{I}_{a_1} + (3.84 - j\,9.33)\,\dot{I}_{a_2}$$

after substituting the values of a and a^2 and simplification.

or $(21.26 - j\,0.67)\,\dot{I}_{a_1} + (3.84 - j9.33)\,\dot{I}_{a_2} = 400 + j\,0$

Similarly,

$$\dot{E}_B - \dot{E}_R = (\dot{I}_{c_1} + \dot{I}_{c_2})(3 + j\,5) - (\dot{I}_{a_1} + \dot{I}_{a_2})(10 + j\,0)$$

$$= (a\dot{I}_{a_1} + a^2\dot{I}_{a_2})(3 + j\,5) - (\dot{I}_{a_1} + \dot{I}_{a_2})(10 + j\,0)$$

or $(-15.83 + j\,0.1)\,\dot{I}_{a_1} + (-7.2 - j\,5.1)\,\dot{I}_{a_2} = a\,(400 + j\,0)$

Solving the above two equations for $\dot{I}_{a_1}$ and $\dot{I}_{a_2}$

$$\dot{I}_{a_1} = 25 - j\,11.65$$

and $\qquad \dot{I}_{a_2} = -28.6 - j\,1.7$

From these, the three line/phase currents are obtained as

$$\dot{I}_a = \dot{I}_{a_1} + \dot{I}_{a_2}$$

$$= 25 - j\,11.65 - 28.6 - j\,1.7$$

$$= -3.6 - j\,13.35$$

$$\dot{I}_b = a^2\dot{I}_{a_1} + a\dot{I}_{a_2}$$

$$= -6.83 - j\,39.7$$

and $\qquad \dot{I}_c = a\dot{I}_{a_1} + a^2\dot{I}_{a_2}$

$$= 10.53 + j\,53.1$$

and all three are seen to check closely with the values obtained in Ex. 5.

EXERCISES

1. A coil of inductance L and negligible resistance is connected in series with a resistance R. A supply voltage of 40 V (rms) is applied across the circuit. If the voltage across L is equal to that across R, calculate

    (a) the voltage across each element

    (b) the power absorbed in the circuit if L = 0.1 H and R = 40 Ω

    (c) the frequency of the supply.

    [28.3 V; 20 W; 64 Hz

2. An alternating current of 0.2 A (rms) and of frequency 100/2π Hz flows in a circuit consisting of a resistor R of 20 Ω, an inductor L of 0.15 H and a capacitor C of 500 µF, connected in series. Calculate the AC voltage

    (a) across each component

    (b) across R and L together

    (c) across L and C together

    (d) the total voltage across R, L and C

    [(a) 4 V; 3 V; 4 V (b) 5 V (c) 1 V (d) 4.1 V

3. A lamp which may be regarded as a non-inductive resistor, is rated at 2 A, 220 W. In order to operate the lamp from the 240 V, 50 Hz supply, an inductor is placed in series with it. If the resistance of the inductor is 5.0 Ω what should be the value of its inductance?

    [0.33 H

4. In a series circuit containing a resistance, an inductance and a capacitance, a voltage, V = 0.01 V (rms) is applied. If the values of the components are 10 Ω, 0.4 H and 0.4 µF, respectively, and the circuit is in resonance, calculate

    (a) the resonant frequency

    (b) the maximum current

    (c) the voltage across C at resonance.

    [400 Hz; 0.001 A; 1 V

5. A coil of self inductance 0.2 H and resistance 50 Ω is to be supplied with a current of 1 A from a 240 V, 50 Hz supply. It is desired to make the current in phase with the potential difference of the source. Find the value of the components that must be put in series with the coil.

    [R = 190 Ω; C = 50.6 µF

6. Three non-inductive resistances of 1000 Ω each are star-connected to a 3-phase supply with 200 V across lines. What will be the reading on a voltmeter connected between one of the lines and the star point thus formed if the voltmeter also has a non-inductive resistance of 1000 Ω?

[86.6 V

7. The impedances of the three phases of a star-connected load (no neutral wire) are (5 + j 20) Ω, (12 + j 0) Ω and (1 – j10) Ω, in that order. If the load is connected to a 3-phase supply of 400 V line-to-line voltage, calculate the three line currents.

[0.5 – j 29.65 A; 16.24 – j11.5 A; –16.74 + j 41.15 A

8. Two circuits whose impedances are given by (8 – j 7) Ω and (5 + j 6) Ω are connected in parallel across a 100 V AC supply. Calculate the current passing through each circuit and the total current drawn from the supply.

[9.4 $\underline{/41.2^{\circ}}$ A; 2.8 $\underline{/- 50.2^{\circ}}$ A; 15.7 $\underline{/- 13.4^{\circ}}$ A

IV : Visual Display and Analyses

IV

VISUAL DISPLAY AND ANALYSES

RECALL

In engineering, particularly in electrical engineering and allied fields, invariably various quantities are encountered which are time variant or dependent, expressed as a function of time. Thus

$$X = f(t)$$

where X is the given quantity; for example, a potential difference or a current in a circuit, and f (t) is the expression to provide values of X as a function of time or at various instants. The commonest example in practice is the alternating variation of an EMF, given by

$$e = E_m \sin \omega t$$

which relates to a *sinusoidal* variation of the EMF e with time, with its maximum value, E_m, occurring when $\omega t = \pi/2$. The variation of e is known as a sinusoidal *waveform* as it follows a "sine law" with time. The above is also an example of a *steadystate* variation.

In practice, the variation of electrical quantities in, for example, electric circuits, machines, static equipment and power systems may be much more complex on account of

(a) presence of hormonics of various orders, even during steadystate, to distort the waveform considerably and

(b) transient conditions when the amplitude of the quantity may rise sharply in a few milli- or micro-seconds.

Under these circumstances, it may be essential, and much desirable, to have a *visual* display of the waveforms for observation, detection, measurement and even analyses of the quantities of interest. A variety of instruments have been available to meet the above requirements and, owing to the oscillatory nature of the signals, are called oscillographs in general[1]. A key part of these devices is a suitably designed galvanometer.

---

[1]Although the devices now commercially available can perform nearly all functions, from simple observation of signals to analyses to yield waveform spectrum, it is necessary to choose an instrument with due care with regard to the requirement and affordability; it may not always be necessary to use a sophisticated device. Also, it would be helpful to learn the basic principles of operation of the various available oscillographs and the way these have evolved over decades after considerable research and development.

DUDDELL GALVANOMETER AND ULTRA-VIOLET RECORDER (UVR)

A waveform is essentially a two-dimensional depiction of the electrical quantity, with its amplitude plotted along the y-axis whilst the x–axis usually representing the time variation. Further, the oscillograph must clearly incorporate a moving system that is made to deflect when the voltage, or a PD proportional to the current, under test is applied to the instrument and an appropriate device for the indication or recording of the waveform. It is implied that the moving system must possess negligible inertia in order that it may respond 'instantly' to the changes of voltage or current – a very important requirement when dealing with sharp transients or very-high frequency signals. The earliest form of moving system is the Duddell galvanometer which is based on electromagnetic principle for its operation, and best suited to work at low voltages and comparatively low frequencies, typically up to 300 Hz.

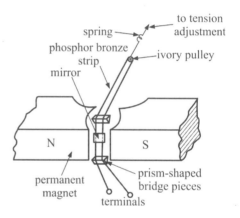

The schematic of such a galvanometer is shown in Fig.4.1. The moving system consists of a single loop of fine phosphor-bronze strip, situated in between the poles of a powerful magnet that could be a permanent or electromagnet type, the latter affording the flexibility of controlling the airgap field strength[1]. The vibrator loop is formed by passing the phosphor-bronze strip round a small ivory pulley which itself is attached to a small spring as shown to adjust the tension of the loop. The loop rests

Fig.4.1 : The schematic of the Duddell galvanometer

on the edges of two prism-shaped bridges located near the top and bottom ends of the pole pieces, and is held sufficiently taut between the bridges. The ends of the strip form terminals for external connections.

The vibrating portion of the loop is thus confined to the section which is situated in the magnetic field. A tiny mirror is attached to the loop at the mid-point. A small piece of soft-iron is usually fitted between the two sides of the loop adjacent to the mirror so that each side has its own airgap. The clearances between the sides of the loop and the pole pieces is of the order of 0.2 mm.

---

[1]In most cases, a 'block' of permanent magnet is used, with circular or semi circular recesses to house 12 to 16 galvanometers and to connect similar number of input signals.

Action

Assuming that a magnetic field B (tesla) exists between the poles and a current I (amp) passes through the loop (being proportional to the PD applied across the terminals), forces B l I (newton) act on each side of the loop, l being the 'active' length of the coil-side in metre. The forces cause one side of the loop to move inward and the other outward making the loop as a whole to deflect in a particular direction depending on the instantaneous direction of the current. Clearly, if the current were alternating, the motion of the loop would be vibrating or oscillatory.

Now if a fine beam of light were cast on the mirror, the beam will be reflected back with oscillations in tune with those of the mirror (or the loop) in a horizontal plane. If the reflected beam is further arranged to fall on a light-sensitive paper, *moving linearly at a given speed* in the horizontal plane, a trace representing the original signal waveform would be produced on the paper as illustrated in Fig.4.2. In practice, the light used is the ultra-violet type, obtained from a special lamp fitted inside the

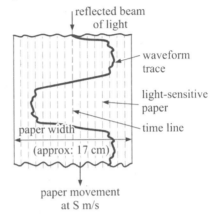

Fig.4.2 : 'Recording' of waveform (trace) on light-sensitive paper

recorder. The paper[1] is specially produced with a microscopic coating which is sensitive to ultra-violet light, such that the trace would appear on the paper after a few seconds and must be "fixed" using a special spray when sufficient/optimum contrast is reached to avoid blackening of the entire paper and the trace. Hence the name: Ultra-Violet Recorder or UVR.

Such UVRs are designed to house several galvanometers or vibrators (as they are called), constructed in long pencil form (about 10 cm long and 5 mm in dia) so as to be easily pushed into position between the holes comprising the magnet poles. Each vibrator differs from the other in terms of its "frequency response" (itself a function of its natural frequency of vibration) and its sensitivity expressed as deflection (in mm/cm) per mV (or mA) input. A fixed mirror, between them or one of the vibrator terminals short circuited, is used to provide a "zero line" (or time line) on the paper about which the signal trace is being recorded.

---

[1]The commonly used UV paper is produced by KODAK Ltd. in the form of rolls of several metres in length and about 17 cm width.

The resistances of the vibrator loops are usually about 4 or 5 ohm, with a sensitivity of about 1 mA being detected, and correspondingly about 4 to 5 mV. The "safe" working current is usually about 100 mA and hence the limiting PD is about 0.4 to 0.5 V. Higher inputs must therefore be attenuated suitably as a safeguard to the vibrators.

The UVR, with each vibrator at a time, can be calibrated by observing the deflection produced by a known direct current of a few mA (measured independently on a DC potentiometer). This would trace a horizontal line along the time axis at the appropriate distance from it.

Advantages and Disadvantages

A UVR, for decades a most versatile means of recording time-dependent electrical signals, has the advantages of

(a) unlimited recording 'time' or length of record on the paper, limited only by the length of paper on the spool and manageability of the recorded paper;

(b) the (time) base of the waveform that can be easily expanded by simply increasing the paper speed;

(c) a number of inputs that can be simultaneously recorded by selecting a suitable galvanometer for each input, thus providing information about relative time variation of the signals, the number being recorded to be restricted only by the clarity of the record;

(d) the recording on the paper that becomes a permanent record for future reference and analysis.

The main disadvantage is the limited frequency response on account of finite inertia of the moving system so that the response to "peaky" waveform or transient inputs suffers from accuracy that can be serious for high- to very-high frequency signals.

Also, unless "fixed" quickly using the right spray, the record may come out to be of poor quality and deteriorate with time.

FIBRE-OPTIC RECORDERS

This is an improved version of the usual ultra-violet recorder in that it eliminates the main disadvantages of a conventional galvanometer-based UVR. In a particular model, a flat fiber-optic strip is used that replaces the pencil-shaped galvanometers of the UVR[1]. The main optical element is still the ultra violet light from a lamp, but it now energises the 'bundle' of thousands of fibre optics the end points of which comprise a strip in the form of a flat surface, in close contact with the light-sensitive paper, arranged to move at a given speed in a horizontal plane similar to that in a UVR.

---

[1]For example, the model 1858, manufactured by Honeywell Inc., USA, called visicorder.

Ordinarily, the strip is kept 'blanked' from the ultra violet light and there is no trace on the paper. In the event of a signal applied to the recorder, a built-in technique allows un-blanking of the fibre optics in a manner that corresponds to the variation of the signal amplitude with time and a trace appears accordingly on the recording paper. The whole process being "static", without any mechanical movement (of optical mirror or lenses etc.), the frequency response of the instrument is very high, limited only by the built-in circuitry. The recorder typically has the following characteristics:

- No. of channels : 18, expandable to 32
- Frequency response : DC – 5 kHz, at up to 18 cm trace amplitude
- Sensitivity : 100 μV per 25 mm
- CMR : 300 V
- numbered trace identification
- 42 push-button selectable paper speeds from 2.5 mm to 300 cm per second
- remote drive and speed control
- automatic record length control
- time-lines recording at five selectable intervals with 0.1% accuracy
- wide choice of plug-ins to suit very low to high level input signals

The recording paper recommended for use with the recorder does not require any fixing and the trace reaches optimum contrast in a few seconds when exposed to mild sun light, and remains unaltered for a long period[1]. Prolonged exposure to sun light must, however, be avoided.

CATHODE RAY OSCILLOSCOPES

A cathode ray oscilloscope, commonly known by the acronym CRO, is the most versatile and effective visual device, many a time unavoidable and forming the "last stage" in various instrumentation for detection, measurement and even analysis of electrical quantities, mostly the potential difference in a circuit as the input. It is called a cathode ray oscilloscope because it traces the required waveform with a beam of electrons, originally called the cathode rays and produced in a "cathode–ray tube" by heating the cathode comprising an indirectly heated tungsten metal component (or filament).

---

[1]Althogh having the drawback of limited frequency response, the UVRs offer an important advantage of providing a long lasting paper record of desired waveforms in *analogue* form for ready reference and analysis.

Clearly, a cathode-ray oscilloscope is essentially an electrostatic instrument the function of which is to trace a v/t graph, that is, a graph of voltage against time, with the voltage along the y-axis and the time along the horizontal or x-axis as in most recorders, the trace being displayed on a specially designed screen of the instrument. When compared to other types of oscillographs, such as the UVR fitted with the vibrators, a CRO has NO mechanical moving parts and hence the electron beam, the main deflecting member, has no inertia. This means that a CRO can theoretically have an infinite frequency response, in practice limited to several mega hertz to serve most practical requirements, and offer extremely high sensitivity as the electron beam can be deflected by a very small voltage.

Whilst the main purpose of a CRO is to be able to observe the waveform of an electrical signal, some modern oscilloscopes can analyse and display the *spectrum* of a repetitive event. In addition, oscilloscopes in modified form are extensively used in a number of applications in science, medicine and engineering; for example, display of the waveform of the heart beat as an electrocardiogram (ECG)[1].

Classification

The modern, "basic" CROs can be broadly classified as:

(a) analogue, and

(b) digital

oscilloscope to serve the various requirements.

The difference between the two types of CRO mainly arises from the manner in which the signal is 'manipulated' between the initial and final stage, apart from the particular application to which the CRO is resorted to. However, there are a number of common features with regard to the construction, applicable to both types.

Constructional Features

The basic entity of a CRO is still the cathode ray tube, even though slightly modified, that was developed in the late 19th century, primarily to demonstrate and explore the physics of electrons, then known as cathode rays. About the time, Karl Ferdinand Braun invented the first CRT (cathode ray tube) oscilloscope, mostly as a physics "curiosity", in 1897. He applied an 'oscillating' signal to electrically charged deflector plates in a phosphor-

---

[1]Countless books and literature abound with detailed description and usage of a variety of CROs. Here, only the essential features are presented, mainly to familiarise the user for proper and effective use of this powerful device.

coated cathode-ray tube. With a reference oscillating signal applied to the *horizontal* deflector plates and a "test" signal to the *vertical* deflector plates, a transient plot of the electrical waveform was produced on the small phosphor-coated screen. There was then a progressive research into the development of modern CROs mostly as a result of its significant uses in countless applications.

The main parts of a typical modern oscilloscope are

- A cathode-ray tube or CRT, comprising
  - electron gun
  - deflection plates
  - fluorescent screen
  - glass envelope
- Amplifiers and controls
  - vertical amplifier(s)
  - horizontal amplifier
  - time base
  - position control
  - focus control
  - intensity control
  - trigger control
  - calibration unit

In addition, a CRO is provided with one or more of the following accessories:

(a) connecting probes and leads for
  - general, and
  - special

  purpose;

(b) plug- ins
  - to provide flexibility of handling input signals or their manipulation (for example, modules for pre-amplification, differentiation or integration of an analogue signal, generally designed for insertion in the slots housing the 'standard' vertical amplifiers);

(c) special-purpose camera(s) for recording a wave form;

(d) leads or chords for remote control.

The front panel of a typical oscilloscope with various controls may look like that shown in Fig.4.3, depicting various key inputs and controls.

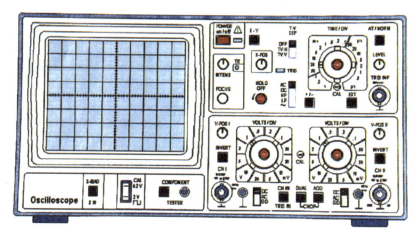

Fig.4.3 : A general-purpose cathode ray oscilloscope

Typically, there are 8 squares or divisions along the vertical axis and 10 along the horizontal (or time) axis on the screen. These squares are usually 1cm apart in each direction.

Cathode-ray tube

The schematic details of a CRT are shown in Fig.4.4.

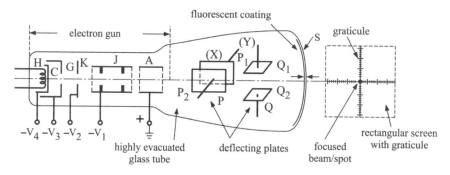

Fig.4.4 : Schematic of a cathode-ray tube

The "heart" of the CRT is the *electron gun* – a device for producing and focusing a concentrated beam of accelerating electrons. The source of the electrons in the Crookes tube, from which the cathode-ray tube was eventually derived, is a piece of metal forming the cathode, C, heated by a heater filament, H, to emit the electrons.

Operation of the CRT

The cathode C is surrounded by a cylindrical *control* electrode, or *grid*, G containing a diaphragm having a negative potential with respect to the cathode. Varying its potential varies the cathode emission current and hence the brilliance or intensity of the electron beam or the spot. The ring-shaped electrode (with a hole), K, is an accelerating electrode to impart a high velocity to the electrons along the tube axis. The electrode J, in a cylindrical form whose potential can be varied, is meant to focus the electron beam to a sharp spot on the screen. This is achieved in association with the high-voltage anode A which also imparts further acceleration to the electrons. All these electrodes are collectively known as the "electron gun".

Electrode potentials

In practice, the screen S, the tube and the anode are "earthed" to avoid danger due to high voltages used in the CRO whilst the other electrodes up to the filament are maintained at increasing negative potentials. Owing to this, touching the outer surface of S with fingers (at ground potential) does not alter the electrostatic field inside the tube and affect its operation.

Focusing

Focusing of the electron beam is important to obtain a sharp trace of the waveform on the screen, particularly when some measurements are also to be made. This is achieved with the help of electrodes J and A using the electrostatic field produced between them as shown in Fig.4.5.

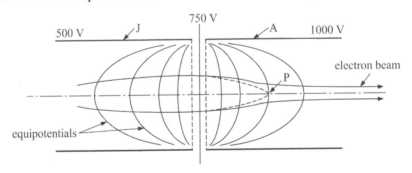

Fig.4.5 : Electrostatic focusing in a CRO

As an example, the schematic of equipotentials in the field for a potential difference of 500 V between the two electrodes is shown in the figure. Electrons entering the field from the filament experience forces from low potential to high at right angles to the equipotential lines. However, they have considerable momentum having been accelerated by a PD of about 500 V, and are travelling fast. Consequently, the field merely deflects them and, because of its cylindrical symmetry, it converges the beam towards the point P. Before they can reach this point, however, they enter the second cylinder (the anode). Here, the potential rises from the axis and the electrons are deflected *outwards*. But, since they are now travelling faster than when they were in the first cylinder, because the potential is *higher* everywhere, their momentum is greater and they are *less* deflected than before. The second cylinder, therefore, diverges the beam less than the first cylinder converged it, and so the beam emerges from the second anode still somewhat convergent. By adjusting the potential of the first anode, the beam can eventually be focused upon the screen to result in a spot, a millimeter or less in diameter (usually about 0.6 mm)[1].

Deflecting plates or electrodes

After leaving anode A, the beam passes in turn between two pairs of deflecting plates P and Q, or (X) and (Y) [see Fig.4.4], the former to cause movement of the beam in the horizontal direction whilst the latter in the vertical direction when voltages are applied across the plates. When the oscilloscope is in use, the varying voltage to be examined is applied between the Q or Y plates. If that were all, then the spot would simply traverse up and down resulting into a vertical line on the screen[2]. Similarly, suppose that a PD is applied to the plates P, with P_1 being more positive than P_2, say, then the spot of light would move towards P_1 tracing a horizontal line from left to right if no other controls were applied.

Time base or sweep

To trace the actual waveform (a variation of voltage from instant-to-instant with time), the X plates must be operated to provide a *"time-axis"*. Frequently, a special circuit is incorporated for this purpose which generates a potential difference across the X plates that rises steadily from zero to a certain value as shown in Fig.4.6(a), and then falls to zero almost instantaneously. It can be made to go through these changes tens, hundreds, thousands, or even millions of times per second (identified as the "sweep frequency"). The rising part of the above trace is called the "sweep" whilst the abruptly falling part the "fly-back", the waveform itself being labeled as "saw-tooth" time-base owing to its shape. Thus, the time-base voltage is applied across the X plates so that the spot, or a given "point" of the waveform under test, is swept steadily to the right, tracing the given

---

[1]As indicated in the figure, electrostatic or electron-focusing devices, are called *electron lenses*. For example, the action of the anodes J and A is roughly analogous to that of a pair of glass lenses on a beam of light, the first lens being converging and the second diverging, but weaker.

[2]This, indeed, is one test to measure the peak-to-peak amplitude of a waveform by reference to the vertical scale in terms of volt/division.

waveform and then flies back swiftly to "zero", and start all over again. The horizontal motion thus provides the needed time-base of the CRO. On it is super-imposed the vertical motion produced by the Y plates, the overall effect being as depicted in Fig.4.6(b). If the sweep frequency, namely the reciprocal of the base of one tooth (the time OT_1 or T_1T_2 in the figure) is an *integral multiple* of the frequency of the voltage whose waveform is being examined, the pattern on the screen will appear stationary[1].

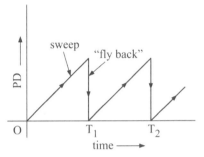 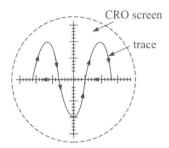

(a) PD across X plates   (b) trace of spot with time on the CRO screen

Fig.4.6 : Time base and trace of spot/waveform on the CRO

In a "double-beam oscilloscope", the two Y plates are joined to terminals labeled Y_1 and Y_2, respectively. An earthed plate between these plates splits the beam into two halves. One half can be deflected by an input voltage connected to Y_1 and the other half being deflected by another input voltage connected to Y_2. With a common time-base applied to the X plates, *two* traces can be displayed simultaneously on the screen and the two different waveforms from Y_1 and Y_2, respectively, can be compared.

Theory of Electrostatic Deflection

The PDs applied across the two pairs of plates set up the electrostatic field between the plates that is responsible for deflection of the beam, with a certain velocity, apart from the manner in which the voltage of the waveform is changing with time.

Consider two parallel plates of an electrostatic deflection system having a PD V volt between them. If the distance between the plates is d metre, the electric intensity will be given by

$$E = \frac{V}{d} \text{ V/m}$$

---

[1]This is one important aspect of the use of a CRO to study unknown waveforms. For the purpose as above, control and adjustments in steps and fine variation of the sweep or time-base form standard features in all CROs.

Let an electron of the beam having a mass m kg and charge e coulomb be directed along the axis at the beginning of the plates (point O) as shown in Fig.4.7.

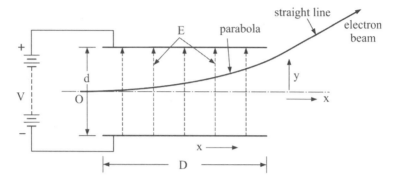

Fig.4.7 : Deflection of an electron (beam) in an electrostatic field

The electron, or the beam, is then deflected upward with a velocity, v, given by

$$e \times V = \frac{1}{2} m v^2$$

where the two sides of the expression give, respectively, the energy acquired by an electron in moving through the PD V, and the kinetic energy of the electron by virtue of its motion. Also, as the electron enters the field between the plates, it is accelerated with an acceleration f given by

$$m f = e \times E$$

The upward velocity of the electron will then be

$$v' = \int_0^t f \ dt = \frac{e \times E}{m} t$$

and its deflection or upward displacement from the axis in time t will be

$$y = \int_0^t v' \ dt = \frac{1}{2} \frac{e \times E}{m} t^2 \qquad .....(a)$$

At the same time, the electron has moved ahead by a distance

$$x = v \times t = \sqrt{\frac{2 e V}{m}} \times t \qquad .....(b)$$

or
$$x^2 = \frac{2 e V}{m} t^2 \qquad .....(c)$$

Substituting for $t^2 \left(= \dfrac{mx^2}{2\,e\,V} \right)$ from eqn. (c) into eqn. (a)

$$y = \frac{1}{2}\left(\frac{E}{V}\right) x^2$$

or
$$y = \frac{1}{(2\,d)}\, x^2, \text{ since } E = \frac{V}{d}.$$

From this, it follows that the path of the electron, or the beam, is a parabola within the distance of the plates along x direction. When the electron just passes the plates, $x = D$, the value of y is then $y = \left(\dfrac{1}{2d}\right) D^2$.

Thus, outside the plates or field, the beam moves in a straight line as shown in the figure.

Example

A beam of electrons moving with a velocity of 1×10^7 m/s enters midway between two horizontal parallel plates P and Q in a direction parallel to the plates as shown in Fig.4.8. P and Q are 5 cm long and 2 cm apart, and have a PD V across them. Calculate V if the beam is deflected so that it just grazes the edge of the lower plate Q as shown.

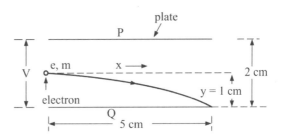

Fig.4.8 : Configuration for the example

Assume e/m = 1.8×10^{11} C/kg.

Electric intensity between plates, $E = \dfrac{V}{D} = \dfrac{V}{2 \times 10^{-2}}$ V/m

Downward acceleration on the electron

$$f = \frac{\text{force}}{\text{mass}} = \frac{E \times e}{m}$$

So, the downward deflection, $y = \dfrac{1}{2}\,f t^2 = \dfrac{1}{2}\,\dfrac{E\,e}{m}\,t^2$

When the beam just grazes the end of lower plate, $y = 1$ cm or 1×10^{-2} m and $x = 5$ cm or 5×10^{-2} m.

Therefore, $t = \dfrac{5 \times 10^{-2}}{1 \times 10^7}$ or 5×10^{-9} s, since $v = 1 \times 10^7$ m/s

(Since the horizontal velocity is not affected by the vertical electric field and is constant).

Therefore,
$$y = 1 \times 10^{-2} = \frac{1}{2} \frac{e}{m} \left(\frac{V}{d} \right) t^2$$

or
$$1 \times 10^{-2} = \frac{1}{2} \times 1.8 \times 10^{11} \times \frac{V}{2 \times 10^{-2}} \times (5 \times 10^{-9})^2$$

whence
$$V = 89 \text{ V (approx)}$$

Essential Features of a Typical CRO

Apart from the electron gun and deflecting plates, a typical CRO will comprise

1. *A time base control* to vary the frequency of the saw-tooth wave or the sweep as mentioned before. This is required to match the frequency of the waveform being examined; for example, from DC to several MHz, so as to make it 'stationary' on the screen. The control operates in steps and also a fine continuously variable one in terms of the time period of the saw-tooth wave. The upper range of the sweep is linked to the bandwidth or highest frequency response of the vertical amplifier, discussed later.

2. *Focus control* to be manually adjusted to vary the CRT grid potential to focus the beam so as to obtain the sharpest possible display of the waveform on the screen. Sometimes, the adjustment may be necessary every time a different waveform is studied.

3. *Intensity or brightness control* being essential to adjust the brightness of the beam (by controlling the cathode potential in the CRT) at level just sufficient to view the trace. Too bright a trace for long might lower the life of, or permanently damage, the fluorescent coating of the screen. Also, a further adjustment of brightness may be necessary when photographing the waveform using a camera.

4. *Vertical and horizontal shift or position control* , usually provided to adjust the position of the beam/trace, and hence the waveform, in a vertical or horizontal direction about the mean or "zero" position marked on the graticule. On the extreme setting, it may be possible to move the trace completely off the screen if so desired.

5. *Vertical amplifier(s)*, very often being essential since a signal to be examined may be so low in magnitude that if applied directly to the deflecting plates, it may not result in a measurable deflection. All CROs are therefore invariably provided with a suitable amplifier to enhance the

signal amplitude/strength before applying to the vertical plates[1]. Conversely, the signal strength may be so large that it may have to be attenuated. The vertical amplifier is thus designed to perform this function, too, and control the level of the applied signal appropriately to provide an acceptable trace of the waveform on the screen.

Amplifier bandwidth: An important requirement of vertical amplifier(s) in practice is its bandwidth or the limiting frequency response. The higher the bandwidth (usually in tens of MHz), the higher the signal frequency that can be faithfully 'processed' in the CRO and displayed on the screen, and relates to the "rise time" of the applied time-varying signal. For example, to view a (perfect) rectangular waveform which has zero rise time theoretically, a 'proper' high bandwidth CRO (say, 25 MHz) will trace it *almost* nearly a rectangular wave; in contrast, a "low" bandwidth CRO will display a "trimmed" waveform, slowly rising and with somewhat rounded corners. This requirement is also related to the design and provision of the time base circuitry of the CRO.

Two important specifications of the vertical amplifier(s), therefore, are

(a) *Sensitivity*: it is expressed as the smallest signal that can be applied in terms of volt or mV/cm. Some CROs may offer a sensitivity as good as 0.1 mV/cm; this, however, may necessitate special measures to handle the signals.

(On the extreme, the highest signal strength that can safely be applied, usually a few hundred volts).

(b) *Bandwidth*: to signify the highest-frequency signals that can be applied, or in a practical sense, the transient signals with extremely small rise time.

6. *Horizontal amplifier*, incorporated in most CROs, to which an external time-varying signal can be connected whilst the time base is de-activated. This is required when some special measurements, for example measuring an unknown frequency as discussed later, are to be made.

7. *"CAL" or calibration output*, provided very often, being a known peak-to-peak alternating output to calibrate the vertical amplifier by adjusting its gain. The signal may be at two levels: one of 0.2 V and the other 2 V and of rectangular waveform. When the amplifier is set at appropriate range (for example, at 0.2 V/cm or 2 V/cm) and the CAL output is connected at the amplifier terminals, the trace on the screen must correspond to the CAL setting in terms of the vertical deflection.

---

[1]The modern CROs are designed to receive at least two signals simultaneously, having dual beam or dual trace function, and are provided with two vertical inputs or amplifiers: VERT 1 and VERT 2.

8. *INVERT control*, often provided on the vertical amplifier to allow swapping of the waveform trace on the screen "up side down", or reverse by 180° in terms of time variation which may be required in some measurements.

Special Features

Dual-trace and dual-beam CROs

In nearly all practical applications, it is usually required to input two signals to the CRO at the same time for simultaneous display or comparison. There are thus two separate vertical amplifiers, each with its own terminal-pair or socket[1]. To meet such a requirement, the single beam from the electron gun is multiplexed electronically to provide two traces to be activated by the two signals, or with one of the traces acting as a "continuous" time base about which is super-imposed the other signal that is being examined. There is only one time base common to both. By a suitable adjustment, the two signals could be viewed in what are known as *chopped* or *alternate* mode, the choice being controlled by a special switch on the front panel. Such dual-trace CROs may, however, have a limitation to high-frequency signals, but are satisfactory at power or relatively low frequencies.

In order to justifiably display high- or very-high-frequency signals *simultaneously*, some elaborate CROs (and, of course, at high cost) are designed to provide two independent *beams*, originating and controlled in basic form, from two separate electron guns having independent focus and intensity controls and pairs of deflection plates as well as their own time bases – somewhat like two single CROs in one CRT.

Clearly, when provided with multiplexing of each beam and splitting in two traces, such CROs are capable of displaying *four* signals simultaneously.

CROs with plug-in modules

A variety of special-purpose "plug-ins" are available with some CROs that can be housed in the slots normally occupied by vertical amplifiers, or even the horizontal amplifier.

These are

1. Differential-input and/or direct-coupled precision amplifier module to respond to extremely low level signals (of the order of a few micro volt) in differential mode that can offer a high CMRR.

2. Operational amplifier module which can be operated as

    (a)   an amplifier with differential input;

---

[1]Most CROs have either "screw-type" terminals or BNC sockets or both, with matching plugs on the leads in the latter case for connecting the signal(s) under test.

(b)   a time integrator with an appropriate gain;

(c)   a (time) differentiator.

depending on the use of appropriate input and output (impedance) components.

3.   A time-base module with "single-stroke" facility to help photographing waveforms directly from the screen without the possibility of overlapping or drifting of waveforms.

4.   "Normal" vertical amplifier module, but with a relatively low band-width (say, about 10 MHz) for power- or low-frequency signals and much high sensitivity, up to 0.1 mV (or 100 μV) per/div (or cm). However, such high sensitivity amplifiers when used at a very low setting may invariably be accompanied by appreciable noise, esp. the "white" noise, which may be undesirable or may have to be eliminated using special filters, or handled by other means.

Special-purpose probes

Open-wire test leads, esp. with very low signals, may invariably pick up electromagnetic interference (EMI) en-route from the surroundings and their capacitance at the probing end is likely to disturb the circuit or device being examined. They are appropriate only for low frequencies and low-impendence devices. For modern high-bandwidth CROs and, in general, when dealing with signals of high frequency, the probe cable is usually a special coaxial cable (with a resistive central conductor to damp out "ringing") with an effective metallic shield.

There are also special "high-voltage" probes, of sufficient length, to deal with high-voltage signals up to a few kilovolts. These probes are designed to attenuate the input signals by ratios such as 10:1 or 100:1. Special earthing arrangement may also be required at the BNC socket terminal in such cases to avoid an accidental shock to the operator.

Storage Oscilloscope

Some analogue CROs are equipped with an extra feature for 'storing' the trace. Storage allows the trace or waveform on the screen to remain "stored" or "frozen" and displayed for several minutes or longer[1]. This is particularly useful when the image is to be photographed using a camera for permanent record. When no longer required, the stored image can be erased from the

---

[1]The storage is accomplished using the principle of secondary emission. When an ordinary electron beam strikes a point on the phosphor-coated surface of the CRO screen, not only does it momentarily cause the phosphor to illuminate, but the kinetic energy of the beam knocks other electrons loose from the surface. This can leave a negative charge that is held on the screen by means of a specially produced positive charge, resulting in the stored image.

screen by pressing a knob on the front panel which activates an electric circuit inside the CRO.

Recording of Waveforms

A permanent record of the waveform(s) displayed on the CRO screen is often required for analysis or study at a future date. On an analogue or the usual CRO, any of the following means may be adopted for the purpose.

(a) Using a "Storage" oscilloscope or a CRO that can operate in storage mode. Once the desired waveform is obtained on the screen by following the normal process, the CRO is switched into storage mode and the waveform is stored using "single-stroke" control provided on all storage oscilloscopes. The image can then be photographed using a suitable camera[1]; for example,

  (i)   an ordinary camera loaded with a high-speed B/W film (say, 400 ASA) and held on a tripod or stand giving up to 35 pictures,

  (ii) a modern digital camera with good resolution, giving a record of hundreds of pictures.

  In both these cases, the pictures can be printed or reproduced later on a photographic or other type of paper(s).

(b) Using a "Polaroid" camera that can be fitted on to a special frame provided on the CRO screen, and loaded with a roll of special film. This can provide an 'instant' print of the waveform. The camera can be operated with open lens and single-stroke control of the CRO. A typical Polaroid camera generally employed in most CROs is shown in Fig.4.9.

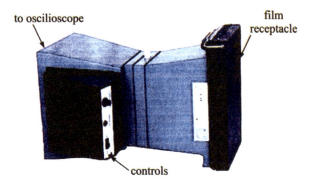

Fig.4.9 : A Polaroid camera

---

[1]The photograph of the waveform is normally obtained as a white trace against a black background. If xeroxed to make a copy (for example, for a report), the time base would be reversed. To avoid it to happen, the waveform can be photographed with the *original time base reversed*, that is, moving from right to left – a feature that may be available in some CROs. The photo-copied picture would then show the waveform correctly.

In general, only about eight pictures can be photographed from a roll using this camera. Also, the developed prints should be "fixed" using a special solution for long-time storage.

A sample Polaroid print of the radial flux density variation in the airgap of a salient-pole hysteresis coupling fitted with a Vicalloy rotor, using a "full-pitch" search coil is shown in Fig.4.10. Note the huge peaks that occur in such a coupling at the trailing pole edge, exhibiting pronounced effect of hysteresis in Vicalloy – a permanent-magnet material (containing vanadium, iron and cobalt).

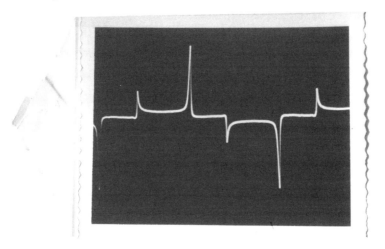

Fig.4.10 : Actual Polaroid print of radial flux-density variation in a salient pole
hysteresis coupling [vertical amplifier sensitivity: 2 mV/cm]

DIGITAL (STORAGE) OSCILLOSCOPE

A later version of the historic *analogue* CRO with some distinct advantages is the *digital* cathode-ray oscilloscope, invariably with storage capability, that was invented by Walter Le Croy after producing the much-needed high-speed digitizers. Starting in the 1980s, digital CROs became prevalent and are now in common use, at times preferred to analogue CROs. Whilst the analogue CRO depends for its operation on continually varying voltages (with time), a conventional digital storage oscilloscope samples the analogue input waveform at a large to very large sample points. These samples are then converted to digital form using appropriate analogue to digital converters (ADCs) and stored in computer memory, being the integral part of the digital CRO. The digital information is repeatedly retrieved and re-converted to analogue form using digital to analogue converters (DACs) for presentation on the screen of a cathode ray tube[1]. The latter element is thus common to both types of oscilloscope.

The digitized samples of the input waveform represent *discrete* points on the waveform, chosen using a given technique. In order to avoid presentation of a discontinuous trace, sometimes not matching the original waveform consisting only of these points, some kind of interpolation mechanism is frequently employed to provide a "continuous" waveform as nearly congruous to the original waveform as possible.

A simplified block diagram of a digital storage oscilloscope is given in Fig.4.11.

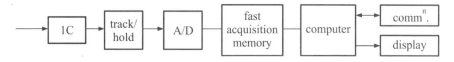

Fig.4.11 : Block diagram of a digital storage oscilloscope

Main Features

- the input circuits (vertical amplifiers etc.) are similar to those for an analogue CRO
- the sampling circuit, track and hold, *tracks* the input signal and switches into the hold state in well-defined instants of time at a given constant value
- A-D converter converts the analogue values into a "number", known as quantisation with typically 8-bits resolution[2]
- the digitised values are stored in the acquisition memory and passed on to the computer for processing
- the "communication" (comm[n].) with the computer, with the signal in digital form, can be maintained through any or all of the modes such as LAN, USB or RS 232.

Some aspects of sampling

When sampling the input signal, samples are picked at fixed intervals, usually chosen in relation to the shape of the input waveform and its time

---

[1]To display the data, CRT with *magnetic field* deflection may be employed instead of the usual electrostatic type. In most modern digital scopes, LCD or LEDs are usually employed for displaying the data.

[2]At the A/D stage, the accuracy of the reconstructed signal at a later stage can essentially be ensured if the Nyquist criterion is fulfilled which states that the sampling frequency must be *at least* twice as high as the highest harmonic frequency of interest in the signal; in practice, it could be much higher than that.

period. At these intervals, the amplitude of the input signal is converted into a number. The accuracy of this number depends on the *resolution* of the device. The higher the resolution, the more accurate the input signal can be reconstructed. Whilst the 'standard' resolution is using 8 bits (the lowest possible), a resolution of 16, 32 or even 64 bits may be employed for more accurate results.

Sample frequency

The rate at which the samples are "taken" is called the sampling frequency, the number of samples per second. Clearly, a higher sampling frequency corresponds to a shorter interval between adjacent samples, leading to a more faithful reconstruction of the waveform later in the analogue form. This is particularly important when dealing with sharply rising transient signals or peaky waveforms (see, for example, Fig.4.10).

Advantages of Digital Oscilloscopes

- With high resolution, it can process extremely low level signals down to 1 µV.

- Using suitable software in the computer, it can faithfully detect peaks of the input signal, esp. in the case of transient signals.

- A much higher bandwidth is possible, as high as 200 MHz.

- Brighter and bigger display with different colours is possible to distinguish multiple traces.

- Specially suited for *quantitative* analyses applications owing to the use of digital technique.

- The memory of the oscilloscope can be arranged not only as a one-dimensional, but even two-dimensional array to simulate a phosphor screen.

APPLICATIONS OF OSCILLOSCOPE(s)

A. Measurement of Amplitude

The amplitude of a given signal, esp. a periodic waveform and containing harmonics, can be easily measured using a well calibrated oscilloscope. For example, when calibrating a search coil wound as secondary of a (long) solenoid, the peak-to-peak value of the induced EMF (essentially sinusoidal if the solenoid is air core) can be measured on the CRO screen in SYN. mode. From this, the constants(s) of the coil can be deduced knowing the number of turns and size etc.

B. Measurement of Time-period

When the signal is synchronised and is stationary on the screen, its time period can be determined from the setting of the time base controls.

Otherwise, time period is simply given by the reciprocal of frequency, or

$$T = \frac{1}{f} \text{ s},$$

for all *periodic* waveforms.

C. Measurement of Frequency

For this measurement, the signal with unknown frequency is applied to the vertical amplifier whilst a signal of *known* frequency is applied to the horizontal plates. The resulting trace on the screen will appear to consist of multiple loops as shown in Fig.4.12. From the count of number of points tangentially to x- and y-axis, the

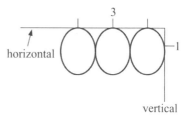

Fig.4.12 : Measurement of frequency

unknown frequency can be determined as follows:

$$\frac{f_H}{f_V} = \frac{\text{number of points of vertical tangency}}{\text{number of points of horizontal tangency}}$$

Referring to the figure

$$\frac{f_H}{f_V} = \frac{1}{3} \qquad \qquad \therefore \quad f_V = 3 \times f_H$$

D. Measurement of Phase Angle

This forms an important application of the CRO when dealing with alternating voltages of sinusoidal waveform. The two signals between which the phase angle is to be measured are connected to vertical and horizontal amplifiers. It is imperative that frequency of both the signals must be *same*. Also, to help making a quick measurement, the two signals should be of same amplitude[1]. The pattern traced on the screen may, in general, be an ellipse as shown in

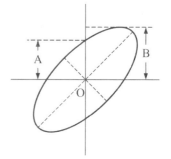

Fig.4.13 :   Measurement of phase angle

Fig.4.13. By knowing the intercept A along the y-axis and the

---

[1]This can be achieved by connecting the two signals independently and adjusting the gains of the amplifiers.

'amplitude' B, the phasor relationship can be given by sin ϕ = A/B. In general, such patterns, called "Lissajous figures", may yield various phase angles in terms of distinct shapes. Some of the Lissajous figures corresponding to 'common' phase angles are illustrated in Fig.4.14.

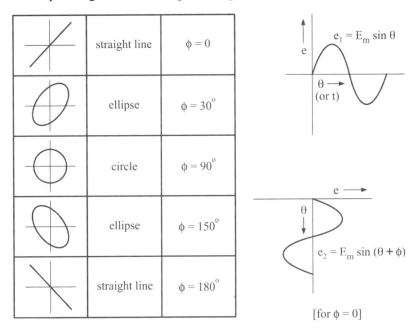

Fig.4.14 : Lissajous figures and phase angles

General Form of Lissajous Figures[1]

Mathematically, the 'generation' of Lissajens figures, in general, is governed by the system of parametric equations

$$x = A \sin(a t + \delta), \quad y = B \sin(b t)$$

where A and B are constants, a and b are inter-related numbers (usually integers) and δ the "phase shift" between x and y. When used in conjunction, the two equations describe complex harmonic motion, such as that produced by an oscillating pendulum as viewed in a plane below it and perpendicular to the vertical 'axis' of oscillations.

The appearance of the generated figure is highly sensitive to the ratio a/b. Thus, for a ratio of 1, the figure would be an ellipse. With other special cases, such as with A = B and $\delta = \pi/2$, the figure would be a circle, and a

---

[1]The family of curves generally known as Lissajous figures (or curves) was first investigated by the scientist Nathaniel Bowditch in 1815, and later in more detail by the French scientist Jules Antoine Lissajous in 1857.

line when $\delta = 0$. [See Fig.4.14]. Another Lissajous figure when $a/b = 2$ and $\delta = \pi/2$ will be a parabola. Other ratios produce more complicated patterns which are 'closed' if a/b is rational. A pattern when $\delta = \pi/2$, $a = 1$ and $b = 3$, recorded on the CRO screen is shown in Fig.4.15. (It would 'rotate' by $90°$ if b were 1 and a equal to 3.

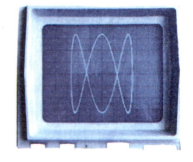

Fig.4.15 : A Lissajous figure on a CRO

OSCILLOSCOPES IN ANALYSES

Apart from their use extensively as end devices in instrumentation and now increasingly as tailor-made instruments in medical applications, for example in recording ECG or observing CT scans, oscilloscopes either in conventional form or somewhat modified find extensive use in various visual analytical instruments. Two important and common applications are

- Spectrum or Spectral Analyser, and
- FFT Analyser

Spectrum Analyser

The most common use of a spectrum analyser is as a test equipment in the design, test and maintenance of radio-frequency circuitry and equipment. A spectrum analyser like an oscilloscope is a basic tool for observing signals, input to the vertical amplifier. However, whilst an oscilloscope displays a signal in the *time domain*, that is signal amplitude v/s time, the spectrum analyser provides the signal display in the *frequency domain*, that is, the amplitude of the signal along the vertical or y-axis, and the frequency content of the signal along the horizontal or x-axis as depicted in Fig.4.16.

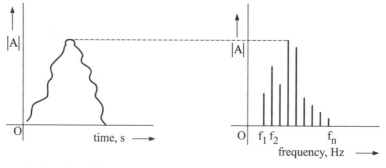

(a) signal oscillogram   (b) frequency spectrum of the signal

Fig.4.16 : A signal in time domain on an oscilloscope and in frequency domain on a spectrum analyser

Usually, the frequencies displayed are chosen as per the requirement.

The concept can be better explained by reference to Fig.4.17 showing two simple examples of waveforms; first in the time domain and then their corresponding spectrum analyser displays. The first figure shows a pure sinusoidal waveform of amplitude A and at frequency of 50 Hz seen on the oscilloscope. On the spectrum analyser screen, only a vertical trace at 50 Hz mark is shown of amplitude A; this may, however, be different if some other scale is chosen.

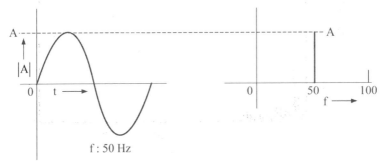

(a) pure sinusoidal signal of 50 Hz

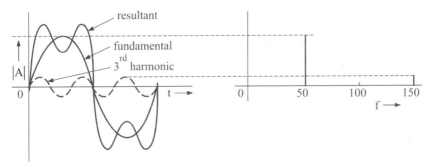

(b) a 'distorted' signal containing fundamental and third harmonic

Fig.4.17 : Displays of waveform on an oscilloscope and on a spectrum analyser

The second figure shows a distorted waveform on the oscilloscope which consists of the fundamental (50 Hz) and the third harmonic (150 Hz). In the frequency domain, on the spectrum analyser, it is traced as two vertical lines at 50 and 150 Hz marks, of corresponding amplitudes. Clearly, if the distorted waveform would include more harmonics of differing magnitudes, all the relevant frequencies would be displayed on the spectrum analyser screen in a similar manner.

The horizontal axis of the analyser is linearly calibrated in frequency with the higher frequency being at the right side, if the frequency spectrum of interest is limited. However, for most applications a logarithmic scale is chosen as it enables signals over a much wider range that can be displayed.

Fig.4.18 shows the display on a spectrum analyser of a voice signal containing some typical audio frequencies and their relative magnitudes. Observe that the figure depicts a continuous spectrum rather than discrete chosen frequencies.

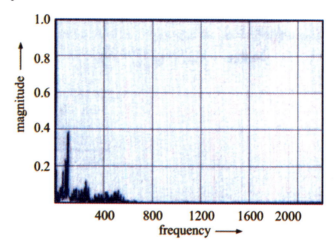

Fig.4.18 : Display of a voice signal on a spectrum analyser

Setting the spectrum analyser frequency

To set the frequency of a spectrum analyser, two selections can be made, being independent of each other. The first method consists of the selection of the "centre" frequency. This sets the frequency of the centre, or 'mid-point' of the scale to the chosen value and is usually where the signal to be monitored would be located. In this way the main signal and the regions on its either side can be monitored. The second selection can be by choosing the span or the extent of the region on either side of the centre frequency that is to be viewed or monitored. The span may be specified as given frequency per division or the total span that is seen on the calibrated part of the screen.

Yet another option that is often available is to set the "start" and "stop" frequencies of the scan. When a spectrum analyser is used in this manner, it is possible to make measurements of the bandwidth of modulated signals that can be checked to determine whether they fall within the desired range. An offshoot of this requirement is in checking and testing the response of filters and networks.

Analogue and digital spectrum analysers

- An analogue spectrum analyser uses either a variable band-pass filter whose mid-frequency is automatically tuned through the range of frequencies of which the spectrum is desired to be measured or use a superhetrodyne technique where a local oscillator is swept through a range of frequencies.

- A digital spectrum analyser computes the discrete Fourier transform, a mathematical process that transforms a waveform into the component of its frequency spectrum.

FFT Spectrum Analyser

This is similar to a spectrum analyser as discussed above, but mainly comprises a *digital* instrument based on digitisation of an input signal which is in time domain.

The words FFT, an acronym for Fast Fourier Transform, essentially relate to Fourier's theorem which states that any waveform in the time domain can be represented by the *weighted* sum of sine and cosine terms with appropriate amplitudes. The FFT (spectrum) analyser samples the input signal, computes the magnitudes of its sine and cosine components and displays the spectrum of these frequency/harmonic components.

An FFT thus presents the complete harmonic analysis of a distorted waveform as viewed on an oscilloscope. Whilst the usual, or analogue, spectrum analyser generally isolates only the frequencies of interest using analogue filters, an FFT analyser resolves the entire frequency content or spectrum of the signal.

The mechanism

In an FFT, the input signal is digitised at a high sampling rate similar to a digital oscilloscope, dependent on the Nyquist criterion. The resulting digital samples record is then mathematically transformed into a frequency spectrum using any of the various available algorithms, implementing the Fourier's theorem.

In terms of obtaining a given frequency spectrum, the FFT analyser has the advantage of great speed as it measures all frequency components at the same time.

V : Measuring Instruments

V

MEASURING INSTRUMENTS

RECALL

Historically, the most commonly used "measuring instrument" in various laboratories to measure or detect the presence of 'electricity' in an electric circuit (mostly DC), irrespective of the cause(s) leading to production of electricity, was the simple galvanometer. A commonest use of the galvanometer was, and still is, to 'detect' an EMF across two terminals in a circuit, or a (small) current flowing through a circuit element. This simple device forms the basis of a variety of measuring instruments even today.

It is, therefore, relevant to study the construction and principle of operation of a "simple" galvanometer[1].

A SIMPLE DC GALVANOMETER

The essential constructional features of a DC galvanometer, also known as D'Arsonval galvanometer[2], are shown in Fig.5.1.

The heart of the moving system in the galvanometer is a rectangular coil of many turns wound using fine insulated copper wire (finer the better, going down to, say, 49 SWG) over an aluminum former, typically about 6 mm in width and about 0.5 mm in thickness. The coil is suitably suspended in a strong magnetic field, provided by a pair of N/S pole pieces of a permanent magnet, and encloses a cylindrical soft-iron core as shown. The core is used to make the magnetic field radial in the airgap through which the coil is free to rotate. The pole pieces are curved to form parts of a cylinder, coaxial with the suspension of the coil. The airgap is typically about 1.5 mm. The coil is provided with tiny spindles at top and bottom ends, supported between two

---

[1]The term "galvanometer", in common use by 1836, referred to the surname of an Italian (electrical) scientist Luigi Galvani.

[2]So called after the inventor Jacques D'Arsonval, a French scientist, circa 1882.

jewel bearings[1], also fitted with a hair spring at both ends as shown. These springs are also used to carry current into and out of the coil. A pointer is attached to the top spindle, capable of moving from left to right over a horizontal, graduated scale to read the 'quantity' being measured.

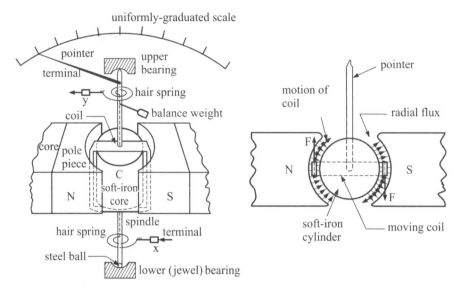

(a) various parts of the galvanometer

(b) plan view of the magnetic field system and moving coil

Fig.5.1 : Constructional features of a DC (or D'Arsonval) galvanometer

As an alternative, in a variation of the galvanometer the pointer is replaced by a small (about 5 mm in diameter), slightly concave mirror attached to a *suspended* moving system carrying the moving coil. The deflection of the coil is then 'magnified' using a suitable lamp–and–scale device to improve accuracy of the reading(s), as discussed later.

Theory of Operation

Let the vertical side of the coil be a and the distance between the sides be b.

Let the number of turns on the coil be N.

When a current of I is flowing along the side, acted upon by a magnetic flux density B at right angles, a force F given by (B a I) will ensue to deflect the coil, the direction being mutually perpendicular to both a and B.

---

[1]These are special 'bearings' shaped and polished to cause negligible friction between the pointed spindle ends (usually fitted with tiny steel balls) and bearing surface(s).

Considering the other side, an equal and opposite force F will result into a couple as shown in Fig.5.2, giving rise to a torque equal to F × b.

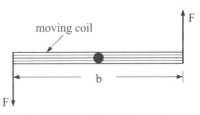

Fig.5.2 : Couple acting on the moving coil

That is, the "deflecting" torque,

$$T_d = F \times b$$
$$= (B\,a\,I) \times b$$
$$= B \times A \times I,$$

where A = 'area' of the coil

Taking into account the number of turns, the total torque will be

$$T_d = N\,B\,A\,I$$

If B is in T, A in m^2 and I in A, the torque will be in Nm.

Control torque

The hair springs are provided, and designed, to produce a control torque, T_c, proportional to the angle of deflection of the pointer such that for final, steady deflection when the pointer comes to rest, the deflecting torque is equal (and opposite) to the control torque.

Hence $T_d = T_c$

or $N\,B\,A\,I = K\,\theta$

where K is the constant of proportionality for the springs.

Therefore, $K\,\theta = (NBA)\,I$

or $\theta = K'\,I$ where $K' = \left(\dfrac{NBA}{K}\right)$

That is, the deflection of the pointer in the galvanometer is proportional to the current being measured, leading to a uniformly graduated scale.

MEASURING INSTRUMENTS

Classification

A common basis of classification of measuring instruments pertains to the way of handling the quantity to be measured, meaning whether the quantity – say, voltage or current – is indicated for an instant or a very brief interval of time, or whether the quantity is connected to the instrument for a very long time-duration, running into hours, days or months. Accordingly, the instruments can be broadly divided into two categories:

    A.   Indicating Instruments

    B.   Recording Instruments

The commonly used instruments such as voltmeters and ammeters belong to the first category, whereas the second category includes instruments such as frequency recorders used to keep a continuous record of system frequency in a utility control room, or an energy meter invariably used to record consumption of electricity in a household or commercial establishment. The latter is based on *time integration* of power over a time duration of interest according to the expression

$$E = \int_0^T P \, dt$$

and forms a very important type of recording or integrating instrument, to be discussed later.

The other broad basis of classification of instruments comprises the categories:

A.   Absolute Instruments

B.   Secondary Instruments

In general terms, *absolute instruments* provide the value of the (electrical) quantity to be measured in terms of the *constants* of the instrument and other basic features, and its 'deflection'. It is not necessary to make comparison of any kind with another instrument or device; in other words these are independent in operation, and a calibration of the instrument may not be required. A classic and important example is the tangent galvanometer that gives the value of the current to be measured in terms of the tangent of the angle of deflection due to the current, the radius and number of turns of the galvanometer coil and the horizontal component of the earth's magnetic field, H.

The tangent galvanometer

The tangent galvanometer, first described by Claude Pouillet in 1837, comprises a circular coil, or an air-core solenoid of short length, wound with N turns and having a mean radius of r (metre) fitted to a horizontal support. A small compass or a short magnetic needle is positioned at the 'centre' of the coil, being capable of rotation in a horizontal plane. The angular position of the needle can be read on a dial around it as indicated in Fig.5.3(a).

The terminals of the coil are brought out on the base for external connections to the circuit through which the current is to be measured, within the prescribed limits.

Initially, the compass needle is set along "zero" marks on the dial, pointing N – S and aligned to the earth's horizontal component of magnetic field. The circular coil, whilst carrying no current, is positioned with its plane along the orientation of the needle.

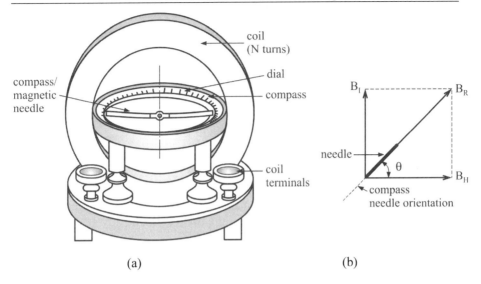

Fig.5.3 : Schematic of a tangent galvamometer

With the settings left undisturbed as above, if now a current is passed through the coil, it sets up its own magnetic field as related to the coil's 'constants' which is at right angles to the earth's field. Under the influence of these two magnetic fields, the needle is deflected, to orient along the resultant magnetic field at some angle of deflection with respect to the original "zero" position. It is important that the leads from the current source to the galvanometer are carefully "twisted" so as to minimise the effect and interaction of field due to lead currents influencing the earth's field [see Appendix VII].

Theory

Let the horizontal component of the earth's magnetic field be B_H T, to be accurately known, and the current through the coil be I A

Then the magnetic field due to the coil

$$B_I = \frac{\mu_0\ I}{2r} \times N \text{ T}$$

The resultant of the two fields is the phasor sum, given by B_R as shown in Fig.5.3(b), being at an angle θ with respect to B_H where θ also represents the angle of deflection of the needle from its original position.

From the phasor diagram

$$\tan\ \theta = \frac{B_I}{B_H}$$

or, substituting for B_I,

$$\frac{\mu_0\,I}{2r} \times N = B_H \tan\theta$$

whence
$$I = \left(\frac{2r \times B_H}{\mu_0\,N}\right)\tan\theta$$

or
$$I = K \tan\theta$$

where K is called the galvanometer "Reduction Factor".

Secondary instruments are so designed and constructed that the quantity to be measured (e.g. current, voltage or power) produces some deflection of the pointer on a dial. However, the latter must be *calibrated* by comparison with either an absolute instrument or one already calibrated and having very good accuracy, before the deflection on the dial can make any sense.

Most indicating instruments in general use belong to this category and themselves may be divided into two classes, viz.,

(a) Sub-standard,

(b) General (or 'common')

depending on their accuracy (or error) and, indirectly, the cost.

The two categories correspond to an important aspect of all measuring instruments that relates to their accuracy, or error(s), associated with the measurements. Clearly, a one hundred percent accurate instrument, that is, the one having zero error is impossible to obtain in practice. Most instruments in common use are designed to achieve an accuracy (or rather error) of about 2 to 5% in general; whilst the one having an accuracy of better than 1% belong to the sub-standard class, a typical figure being 0.5%. Strictly, these figures refer to the level of errors; the general (purpose) instruments being 95 to 98% accurate, used extensively for routine measurements in laboratories, in the form of ammeters, voltmeters and wattmeters etc., whilst the sub-standard instrument being typically 99.5% accurate. Naturally, the latter are designed, manufactured and calibrated with great care and precision and, therefore, cost considerably more. There are, of course, some type of instruments which are inherently quite accurate on account of their principle of operation; for example, an electro-dynamometer instrument, discussed later, and automatically fall in the category of sub-standard instruments[1].

---

[1]The aspect of accuracy, or error, is particularly vital in the case of integrating meters such as conventional energy meters where the error of the basic unit itself gets compounded over the time interval of interest and depending on the error being 'positive' or 'negative', the consumer or the supply company or utility may end up in considerable loss. This phenomenon is discussed in detail later when dealing with energy meters. [See Chapter VIII].

Effects Utilised in Measuring Instruments for their Operation

The electric quantity, be it a current or voltage or any other, when input to a measuring instruments is required to produce a kind of 'effect' that is ultimately related to a deflection torque essential for causing the movement of the pointer over the dial, or lead to recording or integrating with time of the input quantity as appropriate.

The most commonly employed effects are
- A.   the magnetic effect
- B.   the electromagnetic effect
      and, to some extent,
- C.   the electrostatic effect.

The other effects of historic importance and usage are
- D.   the heating effect, used in the form of elongation of a wire due to heat produced when carrying a current, employed in the "hot-wire" ammeter, and
- E.   the chemical effect, based on electrolysis process, employed in earlier energy meters.

Each of the first three effects are discussed in detail in the following sections, in relation to various instruments using any one of the effects for their operation.

INDICATING INSTRUMENTS

The commonest instruments belonging to this category are the ammeters, voltmeters and wattmeters for AC or DC application(s). There are, of course, milli-ammeter/micro-ammeter, and milli-voltmeters/micro-voltmeters which are also basically indicating instruments, but in lower range. The other (special) indicating instruments are the power-factor meters and frequency meters which are dealt with separately; although these, too, may have quite a few constructional and operational features similar to common voltmeter or ammeter.

Essential Forces or Torques

In *most* indicating instruments it is essential to have THREE distinct forces, or torques, acting *simultaneously* on the moving system that also carries the pointer, for the instrument to operate satisfactorily and indicate correctly the quantity being measured. In some instruments, however, one or even two of the forces/torques may not be present.

The three essential torques are the
1.   deflecting torque
2.   controlling torque
3.   damping torque

The deflecting torque

The deflecting, or operating, torque is the 'main' torque that is responsible for causing the moving system of the instrument to move or deflect from its "zero" position (usually the left most, effected by special means), corresponding to zero input condition. Clearly, this torque comes into play by virtue of the input electric quantity, a current or voltage (or a current proportional to the voltage) and its magnetic or other effect that forms the basis of working of the instrument. Also, the magnitude of this torque will correspond to the magnitude or value of the input quantity.

The controlling torque

It is easy to visualise that under the influence of the deflection torque alone, the movement or deflection of the moving system will, in theory, be continual; that is, it would continue to move unidirectionally unless restrained by suitable means. This restrain is provided by a *control torque* which comes into play as soon as the moving system is deflected, will rise as the magnitude of deflecting torque increases and will reach a final value such that it would be *equal and opposite* to the deflecting torque under final, steady condition so as to bring the pointer to rest. Accordingly, in general, if T_d is the deflecting torque corresponding to a given electric input, current or voltage, and T_c the control torque, then for steady position of the pointer

$$T_c = T_d,$$

always.

Means of providing control torque

Spring Control

The controlling torque in all modern indicating instruments is *almost invariably* provided by a spiral spring, or a pair of them, commonly called a "hair" spring (similar to those used in old-type watches or clocks). In spring control, one end of the spiral or hair spring made of phosphor bronze, is attached to the spindle (or shaft) of the moving system in the instrument as shown in Fig.5.4 whilst the other is fixed, fitted suitably to the support frame. The control torque is produced when the spring is tightened as the moving system rotates (almost always clockwise when viewed from top) and the pointer moves forward from left to right, such that the more the spring is tightened, or compressed, more is the (control) torque. It can be shown that, for a given spring, and neglecting the effect of temperature variation, the controlling torque is proportional to the angle of deflection of the moving system, that is,

$$T_c \propto \theta$$

or
$$T_c = K\,\theta$$

where K is a constant depending on the design, construction and material used for the spring as well as on the operating temperature.

It is important that, to give a control torque actually proportional to the angle of deflection, the number of turns on the spring should be fairly large so that the 'deformation' per unit length is minimal. Also, the spring should be stressed within limits so as to avoid a permanent set.

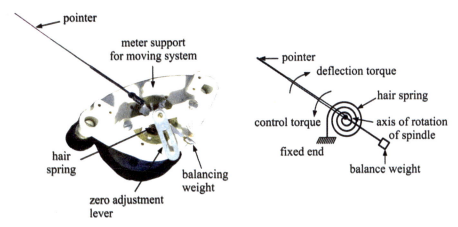

Fig.5.4 : Hair spring in an indicating instrument and control torque

The other requirements of the control spring are that

(i)   its material should be non-magnetic;

(ii)   the spring should not be subject to excessive fatigue;

(iii) its resistance should be low, for the spring may also form a part of the current path; and

(iv) the temperature coefficient of the spring material should be extremely small to eliminate changes in its properties with temperature.

Two springs are sometimes fitted to the spindle, one at the top and the other at the bottom, arranged to be coiled in opposite directions so as to eliminate the temperature effect on the length of the spring when the moving system in the instrument is deflected; one spring is extended while the other is compressed. This also affords the springs to carry the current to be measured to the coil in a moving-coil instrument.

Gravity Control

In this method of providing controlling torque, used in old days with some advantage, a small weight is attached to the moving system in such a way that it produces a restoring torque when the system is deflected as illustrated in Fig.5.5. By virtue of the process of restoration, the controlling torque, for a deflection of θ, is $w\,l\sin\theta$, where w is the control weight and l its distance from the axis of rotation of the moving system as shown. Although sturdy compared to a delicate hair spring, the controlling torque obtained from gravity suffers from two disadvantages:

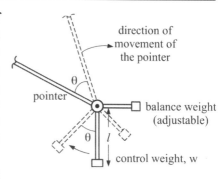

Fig.5.5 : Principle of gravity control

(a) a gravity-controlled instrument must obviously be used in a vertical position in order that the control may operate;

(b) the controlling torque is proportional to sine of the deflection, θ, of the moving system and not the deflection itself as in the case of spring control, thereby resulting in a non-uniform scale, even when the deflecting torque is proportional to the input quantity.

Damping torque

A damping force, and torque, is essential in an indicating instrument in order to bring the moving system to rest quickly in its final deflected position. Without such damping, the pointer of the instrument would continue to oscillate about its final position before coming to rest. This is owing to the inertia of the moving parts. The time of oscillations may be considerable for large inputs and a waste in taking reading on the scale. The oscillatory motion is governed by a second order differential equation and with the effect of damping taken into account, the movement of the pointer can be graphically depicted as in Fig.5.6, depending on the effectiveness of damping.

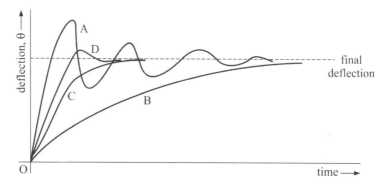

Fig.5.6 : Types of damping in indicating instruments

Referring to the figure, curve A shows the condition of under damping when the pointer would oscillate for quite some time before reaching the final deflection. Curve B shows the opposite trend; the pointer rises slowly from zero to its final deflection: there are no oscillations about the final position – a case of overdamping. Both these examples represent the cases of time wasted. Curve C corresponds to the condition when the pointer rises 'quickly' to its final deflected position without oscillations and the damping is said to be "critical", or perfect, and the instrument is called "dead beat". In practice, it may be desirable that the damping is slightly less than critical such that the pointer overshoots the final position by a very small amount and then 'quickly' settles to the final deflected value as shown by curve D[1].

It is important that the damping force, or torque, must come into play only while the moving system is actually in motion. Thus, the final deflection of the instrument must in no way be affected by damping. This feature has an important bearing on the method of damping utilised in a given type of instrument.

Methods of damping

The three methods that have been commonly employed for decades to provide damping are

   (i)   air friction damping

   (ii) fluid friction damping

   (iii) eddy current, or electromagnetic, damping.

Of the above, the fluid friction damping though robust and efficient is more of an academic interest and no longer used in instruments these days. This is being briefly discussed first.

Fluid Friction Damping

In this method, a light vane, for example the one made of aluminium is attached to the spindle of the moving system and is dipped into a trough or container filled with the damping liquid, which may be a particular oil of appropriate viscosity, and should be completely submerged as depicted in Fig.5.7. The damping force is produced by virtue of frictional drag on the disc due to the liquid and would always be in the direction opposite to the motion due to deflecting torque, the friction force being zero when there is no motion.

---

[1]A simple and interesting example of under-damped oscillations, or rather mechanical vibrations, is the tinkling bicycle bell that is manually activated; the sound stopping if the bell cover is pressed with hand – a case of over-damping.

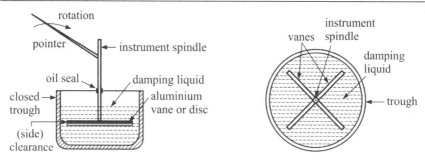

(a) cross section of damping trough
    using single, horizontal vane

(b) an alternative arrangement using
    four vanes at right angles

Fig.5.7 : Schematic of fluid friction damping

An alternative arrangement of design is shown in Fig.5.7(b) to provide increased damping by using four vanes joined at right angles, and immersed in a vertical plane (the plan view is shown in the figure, the spindle being shown in cross section). The damping fluid plays an important role in the quality of damping and must

(a) not evaporate quickly (that is, the volume of the liquid should not change with time);

(b) be free from temperature effect on viscosity;

(c) not have corrosive action on the vanes;

(d) be a good insulator.

The entire assembly must be leak proof. A serious drawback of the method is that the instrument must be used in a vertical position so that the trough containing the liquid is held vertically, and must be handled carefully during transportation to avoid creeping of fluid through the oil seal, if any, and the resulting mess.

Air Friction or Pneumatic Damping

A common method of providing effective damping torque and which is widely in use at present in a variety of instruments is by utilising friction due to air, or rather making use of air *pressure differential*. A scheme usually employed comprises a rectangular, thin aluminium vane attached to the moving-system spindle and just free to move through a range of about 90° inside a closed, sector shaped chamber as illustrated in Fig.5.8.

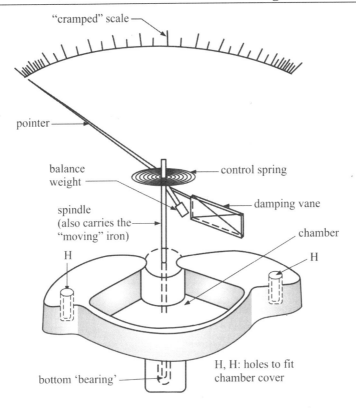

Fig.5.8 : A common arrangement of air friction damping

The production of damping torque can be explained as follows:

Assume a condition when the pointer is at rest and the vane occupies some stationary position inside the chamber. The air pressure on the two side of the vane is same, causing no force to move the vane.

As the pointer is deflected on either side of this position, and hence the vane inside the chamber, the air pressure on one side of the vane is increased whilst on the other side it is reduced resulting in a slight 'vacuum' in this side, causing a pressure differential and thereby producing a force and torque on the vane so as to restore the original condition. The time involved is only a few milli-seconds and thus the clearance (only a mm or so) between the edges of the vane and chamber sides does not have much effect on restoring the equilibrium. The system is designed such that the damping force and torque always acts in the direction opposite to the deflection (torque) and only during the vane movement.

Eddy Current, or Electromagnetic, Damping

In many instruments, esp. when related to their design and operational features, eddy currents can be made use of to produce the required damping torque; this being a very effective and elegant means to achieve the desired damping.

A classic example of an instrument designed to produce eddy-current damping by virtue of its design is the PMMC or a permanent magnet, moving coil instrument, itself a modified version of the D'Arsonval galvanometer. In such instruments, the moving coil, free to rotate inside the gap between the magnet poles and a soft iron core, is wound on a thin, aluminium former of given dimensions. The basis of production of damping torque is the induction of eddy EMF (and hence eddy currents in the 'closed circuit') in the former as it moves due to deflection torque, cutting the airgap flux. The process can be explained as follows:

Refer to the diagrams shown in Fig.5.9.

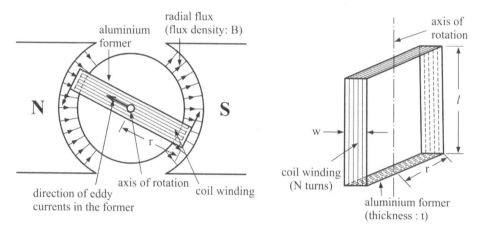

Fig.5.9 : Eddy-current damping in a moving coil instrument

(a) Eddy-current damping torque in a COIL

Consider a coil of N turns, closed on itself, wound on a *non-metallic* former or frame and free to rotate in the airgap having a flux density of B T. Then any rotation of the coil would induce an EMF and current in the coil which, interacting with the magnetic field of the magnet, would give rise to forces acting on each coil side. These must, by Lenz's law, oppose the rotation. The torque so produced thus automatically fulfils the requirements of a 'perfect' damping torque, viz.,

-   the induction effect, and hence the torque, is zero when there is no rotation, and

-   the torque would vary directly as the speed of rotation.

Torque expression

Let the angular velocity of the coil be $d\theta/dt$ rad/s at a deflection of θ so that at a radius r (or half the coil side), the velocity of the coil side will be

$$v = r \left(\frac{d\theta}{dt} \right) \text{ m/s, } r \text{ being in m.}$$

By "$B\,l\,v$" rule, the induced EMF in the coil, considering both the coil sides, will be

$$e = 2 \times B\,l\,r \left(\frac{d\theta}{dt} \right) \times N \ \ V$$

and as the coil is short-circuited, the induced current will be given by

$$i = \frac{e}{R} = \frac{2\,Bl\,N\,r}{R} \left(\frac{d\theta}{dt} \right) \ A$$

where R is the coil resistance in ohm.

The force on each coil side will be given by

$$F = B\,l\,i\,N \ \ N$$

or

$$F = B\,l\,N \times \frac{2\,Bl\,N\,r}{R} \left(\frac{d\theta}{dt} \right) \ N \text{, substituting for } i$$

and the torque

$$T = F \times 2\,r$$

$$= \frac{4\,B^2 l^2 r^2 N^2}{R} \left(\frac{d\theta}{dt} \right) \ Nm$$

Also, in the full equation of motion of the instrument, D expressed as

$$D = \frac{4\,B^2 l^2 r^2 N^2}{R}$$

would constitute the *damping constant* due to eddy current damping alone.

(b) Eddy-current damping torque due to (aluminium) FORMER

The aluminium former may be considered as a single-turn coil and if its dimensions are l_1, and r_1, (in m) and its resistance R_1, (in Ω), the damping torque will be given by

$$T_D = \frac{4\,B^2 l_1^2 r_1^2}{R_1} \left(\frac{d\theta}{dt} \right) \ Nm, \text{ since } N = 1$$

$$= D_1 \left(\frac{d\theta}{dt} \right) \ Nm$$

where D_1 now represents the damping constant in the movement equation due to induced currents in the aluminium (or any metallic) *former.*

If the former has a width of w (m) and thickness t (m), its resistance can be worked out as follows:

total length $= (2 \times r_1 + l_1) \times 2$ m

area of cross section $= w \times t$ m^2

$$\therefore \qquad R_1 = \frac{2 \times (2 \times r_1 + l_1)}{w \times t} \times \rho \ \Omega$$

where ρ is the resistivity of aluminium.

Important: Other than the advantage of being the most effective means of producing the required damping torque, note that the instrument does not require any additional part or a fixture, thus improving torque/weight ratio and resulting in improved performance.

The other category of instruments incorporating eddy-current damping are the induction-type to be discussed later.

Categorisation of Indicating Instruments Based on Design, Construction and Principle of Operation

These can be commonly classified as

(i)   Moving coil type

(ii) Moving iron type

(iii) Induction type

The moving-coil type can be further sub-divided into

(a) permanent magnet type

(b) dynamometer type

whilst the moving-iron type instrument can be of

(a) attraction, or

(b) repulsion

type.

In terms of *usage*, a further consideration may be whether the instrument is to be used in

(a) horizontal position; for example, in a laboratory, or

(b) panel mounted; for example, on the control panel when some design features such as the bearings and support for the moving system acquire a special significance.

General considerations in the selection of instruments

These include

- measurement requirement, that is, to measure current, voltage, power etc., and the range of the instrument;

- accuracy, desired or acceptable for the purpose;
- errors of measurement : inherent to the instrument or the possibility of being caused by human factors;
- cost, involved or affordable, esp. when a number of instruments are in use.

Ammeters and Voltmeters

Ideal, theoretical requirements

Ammeters

These are always connected in *series* with the load in an electric circuit and carry the actual load current, or a part of it, through their internal circuitry. In theory, no current should flow *through* the ammeter so as to avoid any potential drop across its terminals or any power loss and temperature rise within. This is, however, impossible to achieve in practice; nevertheless, the current through the ammeter to produce required deflecting torque should be as small as possible, as also the internal resistance of the instrument which should ideally be zero.

Voltmeters

These are always connected *across* the load or the supply and, depending on the internal resistance, would draw a current from the circuit or the supply for their operation. Ideally, the internal resistance should be infinite so that the current drawn is zero and the voltmeter measures true potential drop across the load without altering the circuit configuration, or the supply.

Once again, in practice it is not possible to avoid a current being drawn by the voltmeter owing to its finite, although 'very large', internal resistance – a phenomenon known as "loading" – but can be kept to a minimum by using a suitable design. These requirements directly or indirectly play an important role when making measurement of *power* in a circuit as dealt with later.

MOVING COIL INSTRUMENTS

Permanent Magnet Moving Coil, or PMMC, Instruments
Ammeters and voltmeters

A permanent magnet moving coil type ammeter (or even a voltmeter) is very similar to a D'Arsonval galvanometer in terms of design, construction and principle of operation, described earlier. [See Fig.5.1].

A PMMC ammeter (or in general, the instrument) will comprise

- a permanent magnet, with the provision of a soft-iron, cylindrical core, fitted concentrically between the poles;[1]
- a rectangular coil of N turns, wound on an aluminium former, using fine insulated (enamelled or cotton-covered) wire;
- a moving system consisting of a spindle, carrying a pointer and balancing weight, if any;
- control (hair) springs usually made of phosphor-bronze and in a pair;
- two bearings, one each at top and bottom to support the spindle, the bottom one being a well-designed jewel bearing to offer minimum friction;
- a graduated scale, usually fitted with a flat mirror beneath it to avoid parallax;
- a pair of terminals, marked with polarity;
- casing[2].

Design Criteria

In a typical moving coil instrument the 'operating' current to give full-scale deflection may vary from about 0.1 to 50 mA, flowing in a coil of about 20 to 100 turns. The gap flux density may be about 0.1 to 0.5 T, with the radial gap length of about 1 to 1.5 mm. The full-scale deflection (angle) is generally about $120°$ and the torque to produce the same may be of the order of 0.5 g-cm with the above coil data.

Working

A nearly uniform and radial flux, achieved by the provision of the cylindrical soft-iron core, exists in the airgap of the magnet system. The vertical coil sides, free to rotate within the gap, experience a force (oppositely directed on the two coil sides) when a current passes through the coil, thus producing the deflecting torque and causing the pointer to deflect in a clock-wise direction (when viewed from the top). A steady deflection is obtained when the deflecting torque is equal to the controlling torque provided by the spiral springs which also act to carry the current to the coil. The damping torque is invariably provided by the eddy currents induced in the aluminium former as discussed earlier.

A schematic of the working of the instrument is shown in Fig.5.10.

---

[1]This arrangement may be 'reversed' in many designs.

[2]In some cases, a suitable shunt may also be provided inside the casing to increase the range of the ammeter as discussed in a later chapter.

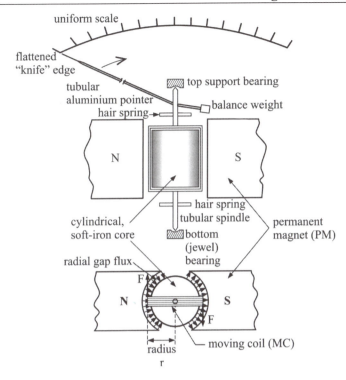

Fig.5.10 : Schematic of the constructional details of a typical PMMC instrument

Note that, as a feasible alternative to the 'basic' design used in some modern instruments, it is possible to "reverse" the construction of a PMMC instrument comprising a magnet *inside* the coil which is surrounded by an annular soft-iron core on the outside. Such an arrangement is illustrated in Fig.5.11.

Fig.5.11 : A PMMC instrument of reverse construction

Deflecting and controlling torques

Assuming that there are N turns on the coil, the airgap flux density is B (T) and each turn of the coil is carrying a current I (A), the deflecting torque developed in the coil will be

$$T_d = B\,N\,A\,I\ Nm$$

where A represents the area of the coil in m^2, as in the case of a D'Arsonval galvanometer.

Or $T_d = K\,I\ Nm,$

where K = BNA is a constant for a given instrument.

This shows that the deflecting torque in a PMMC instrument is *proportional* to the current through the coil at any instant[1].

The controlling torque provided by the spring(s) is proportional to deflection θ, or

$$T_c = K'\,\theta, \qquad K'\ \text{being the "spring constant"}$$

For steady deflection,

$$T_c = T_d$$

or $K'\,\theta = K\,I$

or $\theta = \left(\dfrac{K}{K'}\right) I = K_1\,I,\ \text{say}$

That is, $\theta \propto I$

showing that the scale is *uniformly* divided in this type of instrument – an important property.

Moving coil voltmeter

A moving coil voltmeter is essentially a PMMC milli-ammeter with a high resistance connected in series with the moving coil to limit the current to its operating value for the full-scale deflection when connected to the voltage of the system as shown in Fig.5.12. Thus, if the operating current is 15 mA to produce full-scale deflection as a mille-ammeter and the internal resistance

---

[1]Alternatively, the torque can be written as $T_d = \emptyset\,NI\ Nm$ where Ø denotes the flux linking the coil, assumed constant in the above expression. If, however, the flux may be changing with deflection in a particular position of the coil, for example in the extreme positions, the torque will be given by

$$T_d = N\,I\frac{d\emptyset}{d\theta}\ Nm$$

and this may have a bearing on the calibration of the instrument and graduations on the scale.

of the coil and springs (through which the current passes) is 1 Ω, the internal potential drop will be 15 mV. If the instrument is to read 100 V on full-scale as a voltmeter, the additional resistance to be connected in series will be

$$R = \frac{100 - 15 \times 10^{-3}}{15 \times 10^{-3}} \quad \text{or} \quad 6665.667 \ \Omega$$

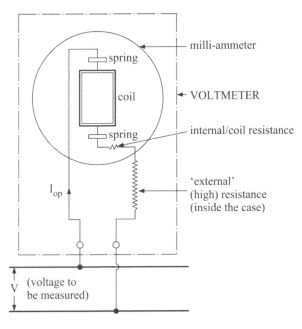

Fig.5.12 : A PMMC milli-ammeter converted to a voltmeter

General Equation of Motion of a Moving Coil Instrument

Assuming the deflecting torque in a moving-coil instrument being given by G i where i is the coil current and G a constant related to the size of the coil, number of turns and gap flux density, all in appropriate units, the motion of the moving system of the instrument will be governed by

$$Gi - C\,\theta - D\frac{d\theta}{dt} \ = \ J\frac{d^2\theta}{d\,t^2}$$

corresponding to a deflection θ,

where

J is the moment of inertia of the moving system,

C θ representing the controlling torque, and

$D\left(\dfrac{d\theta}{dt}\right)$, the damping torque

both acting in opposition to the deflecting torque. $J\left(\dfrac{d^2\theta}{dt^2}\right)$ denotes the 'accelerating' torque, actually responsible for the motion in a particular manner.

The above equation can be written

$$J\frac{d^2\theta}{dt^2} + D\frac{d\theta}{dt} + C\theta = Gi$$

This equation being of great importance, defines the relationship between deflection and time for any type of moving coil instrument. If the right-hand side of the equation is equated to zero, that is, G i = 0,

or
$$J\frac{d^2\theta}{dt^2} + D\frac{d\theta}{dt} + C\theta = 0,$$

the solution of this equation (dealt with in Chapter XII) would lead to the motion of the coil (the effect) when carrying *no current*; or the current having already produced the deflection (the cause). Depending on the relative values of the various constants, the resulting motion can be under-damped, over-damped or critically-damped, related to a variety of instruments.

Merits and Limitations of PMMC Instruments

Merits

(a) The deflecting torque being proportional to coil current, and deflection (θ) being simply proportional to the current, the scale of the instrument is uniform throughout as shown in Fig.5.13

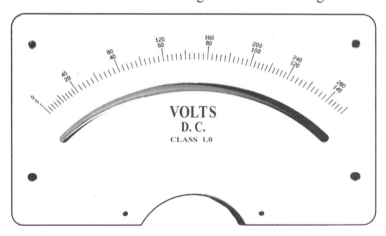

Fig.5.13 : Scale of a typical PMMC instrument

(b) The instrument possesses a relatively high torque/weight ratio owing to simple, light-weight construction of the moving system and use of polished jewel bearings for negligible friction

(c) The angle or span of deflection is as much as 120°, and can even be extended to 300° in instruments using special design of magnet system

(d) Low power consumption since a very small current is capable of producing ample deflecting torque and the coil itself having a very low resistance

(e) Perfect damping, simply afforded by eddy currents induced in the aluminium (or sometimes copper) former, inherently used for winding the coil

(f) High accuracy of measurement which can be 1% or better

(g) Freedom from errors due to hysteresis and stray magnetic field, esp. if used with a magnetic screen

(h) By the use of interchangeable shunts and voltage multipliers, a single instruments may be used as an ammeter or voltmeter, even of high ranges

Limitations

(a) The principal limitation of a PMMC instrument is that it can be used for measurement of DC quantities alone[1]

(b) The error caused by heating and temperature rise, esp. in a voltmeter where some ventilation may have to be provided

(c) The error due to ageing of the permanent magnet resulting in change (usually reduction) of airgap flux

(d) Some error may also occur due to thermo-electric EMF(s), esp. when (internal) shunts are provided to increase the range of an ammeter

(e) The instruments are rather delicate in construction, necessiating careful handling

(f) Relatively high cost

Dynamometer-type Moving Coil Instruments

Compared to a permanent magnet moving coil instrument, a dynamometer moving coil instrument – also known as electro-dynamometer type – differs in that the permanent magnet system is 'replaced' by a pair of fixed coils carrying a current, to provide the required magnetic field. This field when interacted with by the current in the moving coil produces the desired deflecting torque; the actual process may be somewhat complicated. Thus, the instrument comprises a set of fixed and moving coils as shown schematically in Fig.5.14(a). A cut-away view of the instrument is shown in Fig.5.14(b) whilst a symbolic representation of fixed and moving coils is

[1]Nowadays, many firms manufacture a range of moving-coil ammeters and voltmeters for use in AC systems by incorporating a suitable (bridge) rectifier between the supply and the instrument, and by appropriate calibration of the scale.

depicted in Fig.5.14(c). A variation of the fixed and moving coils, nearly rectangular in shape, in an actual instrument is illustrated in Fig.5.15.

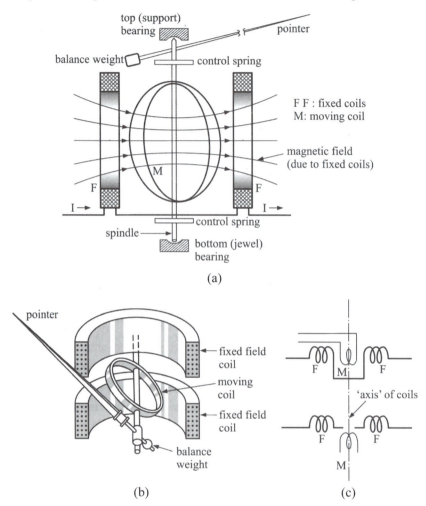

(a)

(b)   (c)

Fig.5.14 : Schematic and cut-away view of a dynamometer instrument

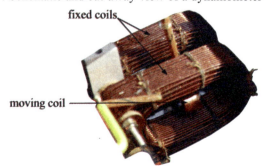

Fig.5.15 : Fixed and moving coils in an actual instrument

Referring to Fig.5.14, the two fixed coils, arranged co-planer, are in the form of Helmholtz coils[1]. This ensures the magnetic field due to the current in the fixed coil being very nearly uniform between the coils as indicated[2]. The moving coil is positioned to move in between the fixed coils and, carrying a current by itself, produces a magnetic field that interacts with the magnetic field due to fixed coils. The torque of the instrument is thus dependent on the strengths of the magnetic fields of both fixed and moving coils at the given instant. The spindle, fitted with the moving coil carries the pointer, balancing weight and (two) control springs as shown, similar to a PMMC instrument. Two polished jewel bearings support the moving system to allow its free movement. The instrument is provided with air damping in the form of a vane moving in a closed chamber. A wooden casing with appropriate screening inside, to minimise the effect of stray magnetic field is used to house the entire assembly.

The coils are almost invariably air-cored; the use of iron, although desirable to enhance the magnetic field, is avoided in a dynamometer instrument to eliminate errors due to non-linearity of magnetic field(s) owing to saturation as also due to eddy currents and hysteresis, esp. when the instrument is used for AC measurements.

Theory of Torque Production

A. A single coil carrying a current I

Consider first a single coil of N turns as shown in Fig.5.16, carrying a current I (A) and having a self inductance L (H). The coil would produce a magnetic field which if 'undisturbed' by, say, the presence of a magnetic material in the vicinity of the coil, would affect its inductance.

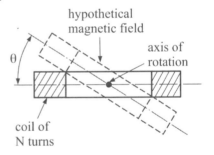

Fig.5.16 : A rotating coil and production of torque

Further, if the coil were free to rotate about an axis as shown, the inductance will also be a function of the angle of deflection, θ. Also, for a given inductance and current in the coil, there will be an energy stored in the system given by

$$E = \frac{1}{2} L I^2 \; J$$

---

[1]See Appendix II for details.

[2]The important requirement of a uniform magnetic field would call for a careful design of fixed coils in terms of their dimensions and lateral spacing.

Now let I change to (I+δI), δI being a small change of current, and, as a result, let θ change to (θ+δθ) and L to (L+δL).

Then the increase in energy stored in the "magnetic" field will be

$$\Delta E = \delta\left(\frac{1}{2}L\,I^2\right)$$

$$= \frac{1}{2}L\,2\,I\,\delta I + \frac{1}{2}I^2\delta L$$

$$= L\,I\,\delta I + \frac{1}{2}I^2\delta L \quad J$$

If T Nm is the value of the control torque exerted by the springs corresponding to the deflection θ, the extra energy stored in the control springs due to the change δθ is T δθ J.

Then, total increase in the stored energy

$$= L\,I\,\delta I + \frac{1}{2}I^2\,\delta L + T\,\delta\theta \quad J \quad \text{in time } \delta t \text{ s.}$$

The flux linked with the coil when carrying current I, having an inductance L and number of turns N is

$$\varnothing = \frac{L\,I}{N}$$

and the induced EMF in the coil will be

$$e = \frac{N\,\delta\varnothing}{\delta t} \quad V$$

or
$$e = \frac{\delta\,L\,I}{\delta t} \quad V$$

This must be balanced by an equal and opposite *supply* EMF and energy drawn from the supply, given by

$$e\,I\,\delta t = \frac{\delta(LI)}{\delta t} \times I \times \delta t \quad J$$

$$= I\,\delta(L\,I)$$

$$= I\,(L\,\delta I + I\,\delta L)$$

$$= L\,I\,\delta I + I^2\,\delta L$$

and this must equal the extra energy stored as obtained above.

Hence

$$L\,I\,\delta I + \frac{1}{2}I^2\,\delta L + T\,\delta\theta = L\,I\,\delta I + I^2\,\delta L$$

from which

$$T \, \delta\theta = \frac{1}{2} I^2 \delta L$$

or

$$T = \frac{1}{2} I^2 \frac{\delta L}{\delta\theta} \; Nm$$

when I is in ampere and the inductance is in henry.

Even though T was assumed to be control torque due to springs, under steady condition this must equal the deflecting torque produced by the coil.

B.   The dynamometer instrument

As mentioned, such an instrument typically comprises two sets of coils. A simplified schematic arrangement of the fixed and moving coils of a dynamometer instrument is shown in Fig.5.17.

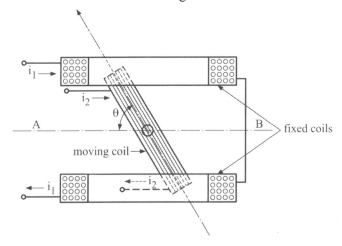

Fig.5.17 : Relative position of fixed and moving coils in a dynamometer instrument

Let i_1 and i_2 be the currents in the fixed and moving coils, respectively, and let the plane of the moving coil in its resulting deflected position make an angle θ with the planes of the fixed coils as shown. Let M be the mutual inductance between the coils in this position[1].

Then the flux density of the magnetic field produced by the fixed coils will be proportional to i_1 whilst, by interaction, the force on the moving coil will be related to i_2. Hence, in simple terms, the torque on the (moving) coil will be proportional to $i_1 \times i_2$. If the two sets of coils are connected in series, i_1 will be equal to i_2, say i, and the torque will be proportional to i^2.

---

[1]Clearly, the value of M will depend on the relative positions of the coils; in other words, on the value of angle θ, and will be maximum when $\theta = 0$.

The whole set-up can then be regarded as made of "one" coil of inductance L where

$$L = L_1 + L_2 + 2M,$$

L_1 and L_2 being the self inductances of the two coils. Mathematically, the torque in the system will be given by

$$T = \frac{1}{2} i^2 \frac{\delta L}{\delta \theta} \ \text{Nm}$$

as derived earlier for the single coil.

Considering independently, L_1 and L_2 will be constant and, therefore,

$$\frac{\delta L_1}{\delta \theta} = \frac{\delta L_2}{\delta \theta} = 0$$

Hence, the torque for the electrodynamic instrument will be given by

$$T = \frac{1}{2} i^2 \frac{\delta}{\delta \theta}(L_1 + L_2 + 2M), \qquad \text{substituting for L.}$$

or

$$T = i^2 \left(\frac{\delta M}{\delta \theta} \right)$$

And, if the two currents are unequal,

$$\boxed{T = i_1 \ i_2 \ \frac{\delta M}{\delta \theta} \ \text{Nm}}$$

The maximum value of M, M_{max}, would occur when θ is 180°, corresponding to full deflection of the moving system representing maximum flux linkage. When $\theta = 0$, $M = -M_{max}$. Between these two limits, M will be proportional to the 'projection' of $-M_{max}$ on the axis AB^1, or

$$M = -M_{max} \cos \theta$$

and hence,

$$\boxed{T = i_1 \ i_2 \ M_{max} \ \sin \theta \ \text{Nm}}$$

The above expression shows that with M_{max} a constant, the developed torque is proportional to the product of currents in the two coils and $\sin \theta$. The usual range of θ in the instrument may be about 45° to 135° and, considering the values of sin 45° to sin 135°, the actual instrument may be appropriately calibrated for a particular function.

---

[1]Provided, of course, the magnetic field produced by the fixed coils is uniform between them: a Helmoltz's coil effect.

Electro-dynamometer Instrument as Ammeter, Voltmeter or Wattmeter

With the instrument being able to achieve sub-standard accuracy (0.5% or better), a dynamometer instrument can be designed and used as *precision* ammeters and voltmeters, and invariably as all-purpose wattmeter.

A. Use as an ammeter

For use as an ammeter, the fixed and the moving coils are connected in series so that the *same* current flows through them or

$$i_1 = i_2 = i \ A$$

and the torque is given by

$$T = K \, i^2, \quad K \text{ being a constant,}$$

and is unidirectional *irrespective of the direction (or polarity) of current at any instant*. This shows that a dynamometer ammeter can be used with the same accuracy in DC as well as AC measurements[1]. Since the moving coil, wound using fine wire, cannot carry a large current, a shunt is essentially used in actual practice and the (internal) connections as an ammeter will be as shown in Fig.5.18.

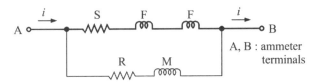

Fig.5.18 : Connection of a dynamometer instrument as ammeter

Here, FF are the fixed coils, M the moving coil, S a small *non-inductive* resistor in series with the fixed coils and R the "swamping" coil in series with M, the purpose of such a coil/resistance being discussed in a later chapter. S and the two fixed coils act as a shunt for the moving coil[2].

B. Use as a voltmeter

The connections for use of a dynamometer instrument as a voltmeter are shown in Fig.5.19 where, once again, all the three coils are connected in series together with a high to very high resistor, depending on the voltage to be measured.

---

[1]Together with high accuracy, this forms an important feature of a dynamometer ammeter, used with advantage in various precision measurement of low EMFs involving potentiometers as described in Chapter IX.

[2]With the type of connections shown, the currents through the fixed and moving coils, that is, i_1 and i_2 may be somewhat different.

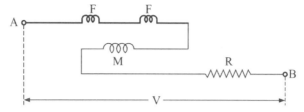

Fig.5.19 : Connections as a voltmeter

With the small current limited by R, the torque may be rather small as in the case of the ammeter and the instrument has to be carefully calibrated. Also, the torque being unidirectional, it is independent of the direction or polarity of voltage being measured.

C. Use as a wattmeter

By far the most important and common use of a dynamometer instrument is as a wattmeter for measurement of power, essentially in single-phase circuit, but with suitable connections in multi-phase circuits, too. [See Chapter VII: "Measurement of Power"].

For use as a wattmeter, the fixed coils are connected in series and designed to carry even a large load current directly, whilst the moving coil in series with a high, non-inductive resistance, is connected across the supply or load as shown in Fig.5.20.

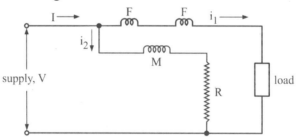

Fig.5.20 : Connections of a dynamometer instrument as a wattmeter

As seen, the current i_1 through the fixed coils is the load current (or proportional to it in some cases) whilst the current i_2 through the moving coil is proportional to the supply voltage, very nearly equal to the voltage across the load and limited by the external (high) resistance R which may include the resistance of the moving coil itself, given by

$$i_2 = \frac{V}{R} \text{ A}$$

The developed torque will then be

$$T = K\, i_1\, i_2$$
$$= K\, i_1 \times \frac{V}{R} \quad \text{or}\quad K'\, V\, i_1 \quad (\text{where } K' = K/R),$$

that is, proportional to power in the circuit *at any instant*.

For a DC circuit or load, the developed torque and deflection is truly proportional to power in the circuit. However, in an AC circuit the torque is proportional to *instantaneous* power and the deflection is proportional to the mean torque owing to inertia of the moving system, thus proportional to *mean* power. For sinusoidally varying voltage and (load) current in an AC circuit, the mean power is given by VI cos ϕ, where V and I are rms values of the supply voltage and load current and cos ϕ the load power factor. Thus, the instrument connected as above works as a wattmeter and can be calibrated to read the power directly in a DC or AC circuit.

The moving coil carrying a current proportional to supply voltage is called *voltage* (or *pressure*) coil whilst the set of fixed coils as a whole carrying load current (or part of it) are together termed *current* coil. Without external accessories, a typical wattmeter can be designed for voltage coil to be connected across a supply of up to 750 V whereas the current coil to carry up to 50 A[1].

Advantages and Limitations of Dynamometer Instruments

Advantages

(a) The main advantage of a dynamometer or electrodynamic instrument is that it can be used with equal accuracy with DC or AC networks since the torque is proportional to *square* of the input quantity

(b) It can be used as an ammeter, voltmeter or a wattmeter, the latter being almost universally in use

(c) In the absence of any magnetic parts or a magnet, the instrument is free from the effects of hysteresis and eddy currents when used with AC

(d) For the same reason, the instruments can be designed with sub-standard accuracy and therefore extensively used in precision measurements

(e) The developed torque, when used in AC network, is free from waveform errors

Disadvantages

(i) Owing to the use of air-cored coils, fixed as well as moving, the magnetic field responsible for production of torque is very small, typically about 20 mT. This results in a small deflecting torque and low torque/weight ratio

---

[1]For voltages in excess of 750 V and load current more than 50 A, special means/accessories may have to be employed, still using the same wattmeter. See Chapter VI.

(ii)   The torque being proportional to square of current in an ammeter, or current proportional to voltage in the case of a voltmeter, the scale is relatively cramped in the initial part[1]

(iii)   The low magnetic fields set up by the coils are susceptible to stray or external magnetic field; to minimise these effects, the working system is usually shielded using appropriate means

(iv)   The moving coil, wound using thin wires to accommodate large number of turns to produce appreciable magnetic field even at low currents, has high resistance which can be affected by ambient temperature

(v)   The instruments are as a rule very expensive compared to other types of instruments of the same range; these are therefore not used for routine measurements

(vi)   In AC measurements, the instruments can be employed in the power-frequency range only on account of the factors related to self- and mutual-inductance

MOVING-IRON (MI) INSTRUMENTS

These constitute the most widely used indicating instruments mainly as ammeters and voltmeters, and to a lesser extent as some special purpose instruments, for their distinct advantages. The deflecting torque in a moving-iron (abbreviated MI) instrument is due to magnetic force(s) on a small piece of iron that is free to move in a field produced by a coil. The magnetic force can be in the form of *attraction* or *repulsion*. Accordingly, the instruments can be classified as

(a) attraction type, and

(b) repulsion type.

In both types, the current, or a (small) current proportional to the voltage, to be measured is passed through a coil, usually cylindrical in form, wound on a suitably designed bobbin or former with a given number of turns to produce optimum mmf corresponding to the current passing through the coil.

Attraction-type MI Instruments

The attraction type moving-iron instrument utilises the force of attraction which a solenoid (or a coil of that configuration) would exert on an iron core/part. When the current is passed through the coil the magnetic field thus produced would attract a piece of iron, attached to the spindle, towards the 'hollow' of the coil as shown in Fig.5.21, thereby causing the pointer to move clockwise.

---

[1]However, the scale may be fairly uniform in the case of a wattmeter where the deflecting torque is proportional to mean power in the circuit.

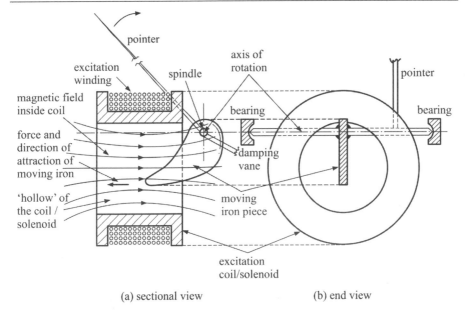

Fig.5.21 : Schematic of an attraction type moving iron instrument

The spindle also carries

(i)   the pointer

(ii) (air) damping vane

(iii) control spring(s)

(iv) balancing weight

and is supported between two jewel bearings. As shown, the iron piece is *not* of a 'regular' shape; rather the shape is especially designed to achieve as nearly uniform scale as possible[1].

Repulsion-type Instruments

These instruments also employ an "excitation" coil similar to the attraction type instruments and other accessories such as the pointer, control springs, (air) damping vane and balancing weight. In addition, the special constructional features of a repulsion-type moving-iron instrument is the presence of *two* iron pieces (or 'rods') inside the coil; one fixed to the inside of the coil, the other movable and attached to the spindle. When the coil is excited by the flow of current in it, the two iron pieces are *similarly* magnetised at any instant (that is, with the same polarity) causing repulsion of one piece with respect to the other, resulting in the rotation of the spindle

---

[1]As shown by the expression of torque later, the scale would otherwise follow the "square" law; the torque being proportional to square of current through the coil.

and pointer attached to it. Two typical constructions of such an instrument are illustrated in Fig.5.22(a) and (b). An isometric, "cut-away" of an instrument is shown in Fig.5.22(c).

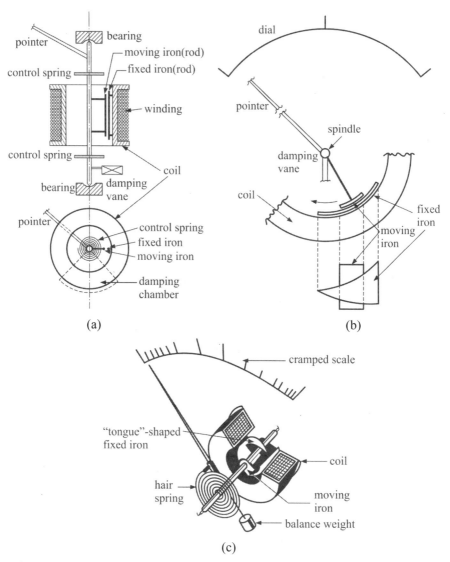

Fig.5.22 : Constructional details of typical repulsion-type moving-iron instruments

In the second variety, the iron pieces are flat, one of these being "tongue"-shaped and fixed to the inside of the coil, bent in the form of an arc. The other is rectangular in shape, attached to the spindle as shown. Both pieces are bent so as to form a concentric assembly with a small airgap between the two. Once again, as the coil is excited, the two pieces are identically magnetised to repel each other, causing rotation of the moving system. The movement of the iron piece attached to the spindle is towards the narrower end of the fixed iron, resulting in clockwise deflection of the pointer over the dial. The tongue-shaped fixed iron is designed to yield a nearly uniform scale.

A variation of construction of an MI instrument of repulsion-type, depicting the moving iron in two halves, is shown in Fig.5.23.

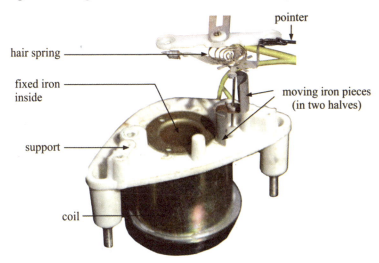

Fig.5.23 : Dismembered view of a moving-iron instrument of repulsion type

Design criteria

A typical, basic moving-iron instrument, whether an ammeter or a voltmeter, is designed to have a 'working' mmf of about 300 A: say a current of 1 A in a coil having 300 turns or a similar combination, care being taken not to exceed the number of turns beyond limit; nor use of very thin wire to keep the self inductance and internal coil resistance to a minimum. In contrast, for a given (maximum) mmf, only a few turns may suffice when the instrument is designed as an ammeter and a large current, say 50 A, is to be measured. With the mmf of the above order and suitably designed "irons", the full-scale deflection torque may be about 0.2 g-cm and a torque/weight ratio of up to 0.1 may be achieved[1]. Depending on the use as an ammeter or voltmeter, the power loss in the instrument may typically be about 4 to 8 W, the latter

---

[1]To improve the effectiveness of the force of attraction or repulsion and to maximise the deflection torque for a given mmf, the iron pieces are made of mu-metal which has very high relative permeability: of the order of 10,000. This also results in higher torque/weight ratio as only a small moving (iron) piece may be necessary.

accounting for loss in the internal resistance of the coil plus the external, series resistance of the voltmeter. As in a typical indicating instrument, the full-scale deflection may extend to nearly 120°.

Torque expression

An exact expression for the production of deflection torque in terms of the input, basically the current in the coil, is nearly impossible on account of the complexity of the instrument: the design using iron piece(s) and the B/H characteristic(s) of the iron material. However, a 'working' expression may be derived by consideration of the *energy* relations when there is a small *increment* in the current supplied to the instrument, a process similar to that used for deriving torque expression for the dynamometer type instrument. The derived expression is strictly applicable to a repulsion-type instrument, but also applies in practice to the attraction type[1].

Thus,

let the 'initial' current through the coil be I A,

the instrument inductance be L H, and

the corresponding deflection due to the deflection torque be θ *rad*.

If the current increases by a small amount dI, then the deflection would change by dθ and the inductance by dL.

In order to effect the increase of current, there must be an increase in the applied voltage given by

$$e = \frac{d}{dt}(L\,I) = I\frac{dL}{dt} + L\frac{dI}{dt} \quad V$$

and the electrical *energy* supplied will be

$$e\,I\,dt = I^2\,dL + I\,L\,dI \quad J$$

In the same interval, the stored energy would have changed from

$$\frac{1}{2}L\,I^2 \text{ to } \frac{1}{2}\left[(L+dL)(I+dI)^2\right] \quad J$$

Hence the *change* in stored energy is given by

$$\frac{1}{2}\left(I^2 + 2\,I\,dI + dI^2\right)(L+dL) - \frac{1}{2}L\,I^2 \quad J$$

Neglecting second- and higher-order terms, the increase in stored energy becomes

$$dE = I\,L\,dI + \frac{1}{2}I^2\,dL \quad J$$

---

[1]The key to the satisfactory operation of either type of instrument is the "hand" calibration of the dial or scale, in terms of proportionality of the torque to the square of input/coil current.

If the control torque due to spiral spring(s) proportional to the deflection θ is T Nm, the extra energy stored in the control springs when θ changes to $(\theta + d\theta)$ will be T $d\theta$ J.

From the principle of the conservation of energy

Electrical energy supplied = increase in stored energy + mechanical work done (that is, the energy stored in control springs)

or
$$I^2 dL + I\,L\,dI = I\,L\,dI + \frac{1}{2}I^2\,dL + T\,d\theta$$

giving
$$T\,d\theta = \frac{1}{2}I^2 dL$$

or
$$\boxed{T = \frac{1}{2}I^2\,\frac{dL}{d\theta}}$$

at the deflection θ.

T will be in Nm if I were in ampere, L in henry and θ in radian. The above expression shows that

- the developed torque is proportional to *square* of the current input to the coil, and also,
- the torque depends on the change of the instrument inductance with deflection[1].

If $dL/d\theta$ were 'constant' within the limits of θ_{min} and θ_{max}, the developed torque will simply be proportional to square of the current, that is,

$$\boxed{T = K\,I^2}, \qquad \text{K being a constant.}$$

It is clear that the above expression for developed torque will apply to *any* type of moving iron instrument.

Variation of Inductance

1. In the attraction type, the increase of inductance with increase in deflection is due to the change in the extent to which the iron piece is pulled within the coil.

2. In the repulsion type, it is due to the increase of flux produced by the decrease in demagnetising effect of one iron piece on the other as the two pieces move apart, or away from each other.

---

[1]There is considerable difficulty in applying this result to the design of instruments as the estimation of L for different values of deflection is a complex matter; usually this is achieved by experience and trial.

Example

A moving iron ammeter has the following relationship between its inductance with respect to angular deflection:

Deflection, θ (degrees) :   15   30   45   60   75

Inductance, L (μH)   : 410   430   448   463   474

If a current of 2 A produces a deflection of $50°$, determine the constant of control springs in the instrument.

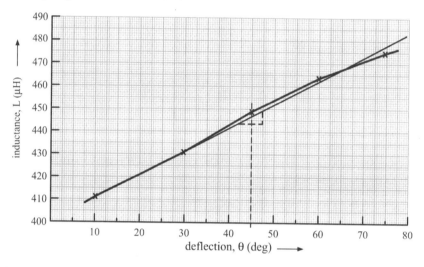

Fig.5.24 : Inductance v/s deflection for the moving-iron instrument of the example

A graph of inductance and deflection is plotted as in Fig.5.24, from which the slope of the curve, or $\dfrac{dL}{d\theta}$, at $\theta = 50°$ is

$$\frac{dL}{d\theta} = \frac{7.5}{10} \text{ or } 1.05 \ \mu\text{H/deg}$$

$$= 60.2 \ \mu\text{H/rad}$$

$\therefore$
$$T = \frac{1}{2}I^2\left(\frac{dL}{d\theta}\right)$$

$$= \frac{1}{2}\times(2)^2 \times 60.2\times10^{-6}$$

$$= 120.4\times10^{-6} \ \text{Nm}$$

This must equal the control torque at that deflection, that is,

$$K\,\theta = 120.4\times10^{-6}, \quad K = \text{control spring constant}$$

$$\therefore \quad K = \frac{120.4 \times 10^{-6}}{\left(\dfrac{50}{180}\right) \times \pi} \quad \text{(by converting angle of deflection into radians)}$$

$$= 138 \ \text{Nm/rad}$$

When the instrument is well-designed, the inductance is very nearly proportional to deflection within the working range. If a straight line approximation is used, $dL/d\theta$ is found to be $1.02 \ \mu H/deg$ at $\theta = 50°$ and the developed torque given by

$$T = \frac{1}{2} \times 4 \times 58.5 \times 10^{-6} \ \text{or} \ 116.96 \times 10^{-6} \ \text{Nm}$$

that is, an error of about 2.8%.

With a straight-line approximation for L v/s θ, $dL/d\theta$ will be a constant for all values of θ and developed torque will be proportional only to I^2.

Calibration and scale

Moving iron instruments are invariably hand calibrated, that is, graduations are marked with hand on a blank dial while the instrument is being calibrated with the help of a sub-standard instrument, either as an ammeter or voltmeter. The developed torque being proportional to *square of the current* through the coil, it is independent of the polarity or direction of current in the circuit. An important implication of this is that the instruments are calibrated and used to measure *rms* current or voltage. This being numerically equal to direct current, the instrument(s) will read with equal accuracy for DC measurement, too, without any change or modification of the calibration.

Owing to the "square law", the scale is always cramped at the beginning, close to the zero mark, and somewhat at the end as indicated (qualitatively) in Fig.5.25[1].

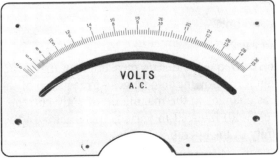

Fig.5.25 : Scale of a typical MI instrument

---

[1]If an elaborate design is used, the L v/s θ variation can be so achieved as to somewhat 'improve' the scale for low values of current/voltage.

Sources of error in MI instruments

A. Errors with both DC and AC

(i) *Hysteresis error*

This is a serious error in these instruments, esp. when used for DC measurements. Owing to hysteresis in the iron pieces of the operating system, the readings are likely to be higher when descending values of current or voltage are measured than when ascending values are observed. This is due to a higher value of B for a given H as a result of hysteresis effect when traversing back the loop. The error can be reduced by

(a) making the iron parts small so that they demagnetise quickly;

(b) operating the instrument at a low value of flux density, along the linear part of the B-H curve (this may, however, result in a reduction of deflection torque);

(c) using a very high permeability and small hysteresis-loop material, such as mu-metal for designing the iron pieces – the usual practice in modern instruments.

(ii) *Stray magnetic field*

As the operating coil in the instruments is essentially "air-core" type and the magnetic field due to the coil correspondingly rather weak, stray magnetic fields can seriously affect the operation/reading of the instrument. The usual means of minimising these effects is by suitable screening of the instrument (inside the casing); alternatively the screen may be positioned immediately surrounding only the coil, being more effective and small in size.

(iii) *Miscellaneous*

These comprise the effect of temperature, friction and ageing of control springs as in the case of moving coil instruments.

B. Errors with AC only

(i) *Frequency errors*

Variation of frequency during measurement may produce error in the deflection torque owing to changes of reactance of the operating coil as well as changes in the magnitude of eddy currents (and hence the 'reactive' field) induced in various metal parts of the instruments in the vicinity of the operating coil.

(ii) *Waveform errors*

A distorted waveform of the current (or the current proportional to voltage) will have a direct bearing on the reading of the quantity on the scale which is (hand) calibrated in terms of rms value(s), assuming a sinusoidal variation of the quantity. Hence, when used for AC measurements, the limit of frequency is up to 100 Hz and the waveform should generally be sinusoidal for reliable readings on the scale.

Advantages and limitations

Advantages

Moving iron instruments possess several advantages over other types so as to qualify for general use in laboratory and elsewhere.

(i)   The instruments can be used with equal accuracy in AC or DC circuits. This follows from the fact that the deflecting torque is proportional to square of the current through the operating coil, and by virtue of the process of calibration

(ii) The instruments are robust owing to simple construction of the moving parts – there being no moving coil and the fact that there are no current leads to the coil

(iii) The stationary part being simply a coil carrying a given number of turns, it can be designed for any suitable current/mmf. For working as an ammeter, the coil can directly carry the current to be measured, up to 50 A with only 3 or 4 turns[1]. Alternatively, the designed value of ATs can be achieved by using large number of turns carrying a small current, say about 20 mA. The latter construction is suitable for the instrument working as a voltmeter when the internal resistance of the coil, being a few hundred ohms or more, would form part of the voltmeter resistance for a given range

(iv) The simplicity of construction invariably results in low cost and affordability of the instrument for general use

(v)   The basic construction with suitable modification is useful for some special-purpose instruments such as frequency or power factor meters, discussed later

Limitations

(i)   As the developed torque is proportional to square of the coil current, the scale is invariably cramped at the lower range and the best results are obtained for the readings being taken in the middle part of the scale. [See Fig.5.25]

(ii) The dependability of the instrument on iron piece(s) for its working makes them susceptible to a variety of magnetic and frequency errors as discussed

---

[1]For measuring circuit current higher than 50 A, a suitable internal or external shunt, or in the case of AC measurement a CT may be employed.

(iii)   The instruments are inherently not as accurate as moving coil or dynamometer instruments owing to the presence of iron; the accuracy may usually be about 2% or so

(iv)   Since the magnetic field is rather weak, the developed torque, and hence the torque to weight ratio, is low as compared to moving coil instrument of the same range

(v)   The damping being pneumatic type, it is not as effective as eddy-current damping in moving coil instruments; the damping chamber itself may develop minor defects if the instrument is not carefully handled

ELECTROSTATIC INSTRUMENTS

By virtue of their principle of operation, such instruments are essentially used as voltmeters[1]. Their main advantages are:

(a)   They give equally accurate readings when used in AC or DC systems/circuits

(b)   Since no current is drawn by the instrument, they offer highest accuracy compared to any other indicating instrument

(c)   There is no iron present in their working system, hence the instruments are free from all errors related to magnetic fields and stray magnetic effects

(d)   There are no errors due to variation of waveform and frequency of input voltage

(e)   The power loss in these instruments is negligible

Their chief disadvantage is that the operating force/deflecting torque is very small in general, esp. for low voltages; their most useful range being from about 500 V to several hundred, or thousand, volts.

The deflecting torque of an electrostatic instrument is derived from the force of attraction or repulsion between charged conductors or elements used in the instrument; the action of the latter thus depends directly on the potential difference in the system. Accordingly, higher the potential difference, higher the deflecting torque.

Types of Electrostatic Voltmeters

Two types of electrostatic voltmeter are in general use:

1. The quadrant type
2. The attracted disc type

---

[1]However, in rare cases, esp. in very high voltage systems, these instruments may also find their application in the measurement of current and power, incorporating precision, non-inductive standard resistors to provide a potential drop proportional to the current in the system.

In general, the former types are meant for voltages up to about 20 kV whilst the latter for voltages above that.

Quadrant electrostatic voltmeter

Also known as an *electrometer*, the principle of operation of this type is illustrated schematically in Fig.5.26. A light, rigid metal vane, V, usually made of aluminium is attached to the instrument spindle, S, also carrying the pointer, P, and is situated within the hollow quadrant Q, both being sector shaped as shown. The spindle is held between jewel bearings, B.

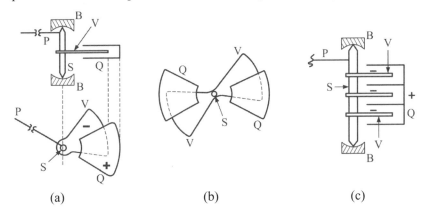

(a)   (b)   (c)

Fig.5.26 : Schematic of a quadrant-type electrostatic instrument and its variation

When the vane and the quadrant are oppositely charged to the potential difference that is to be measured, the vane is attracted or pulled into the quadrant, resulting in the rotation of the spindle S and deflection of the pointer P. The deflecting torque may be doubled by using a 'double' vane and *two* quadrants as shown in Fig.5.26(b). Alternatively, the torque can be increased many-fold by using a 'multi-cellular' construction as in Fig.5.26(c). The control torque in the voltmeter is provided by hair springs and damping by using an air chamber.

Theory of operation

As in the case of a moving-iron instrument, the expression of deflecting torque in an electrostatic instrument can best be derived by consideration of stored energy. Assuming that the quadrant and the vane are connected across a source of potential difference V, let the 'effective' *capacitance* between the vane and the quadrant be C in the given relative position of the two as shown in Fig.5.26 and the corresponding deflection of the vane/pointer be θ.

Then the charge Q on the instrument will be CV. Now let V increase to (V+δV) and correspondingly θ to (θ+$\delta\theta$), C to (C+δC) and Q to (Q+δQ), respectively.

Under these conditions, the increase in energy stored in the electrostatic field will be

$$\delta\left(\frac{1}{2}CV^2\right) = \frac{1}{2}V^2\,\delta C + CV\,\delta V \quad J$$

If T is the control torque (in Nm) due to hair springs corresponding to the angle θ (in *rad*), the extra energy stored in the control system will be $T\,\delta\theta$ J, and the total extra energy stored in the instrument will be

$$= T\,\delta\theta + \frac{1}{2}V^2\,\delta C + CV\,\delta V \quad J$$

However, during this change the source supplies a charge δQ at a potential V; that is, it supplies the energy of value

$$V\,\delta Q = V\,\delta(CV) = V^2\,\delta C + CV\,\delta V$$

Clearly, the energy supplied by the source should equal the (total) energy stored in the instrument.

$$\therefore T\,\delta\theta + \frac{1}{2}V^2\,\delta C + CV\,\delta V = V^2\,\delta C + CV\,\delta V$$

whence

$$T\,\delta\theta = \frac{1}{2}V^2\,\delta C$$

or

$$\boxed{T = \frac{1}{2}V^2\frac{\delta C}{\delta\theta} \quad Nm}[1]$$

where V is in volt and C in farad.

This expression shows that the deflecting torque in an electrostatic instrument is proportional to the square of the applied voltage, independent of its polarity. The voltmeter is thus equally suited for DC or AC voltage measurement. For the latter, the voltmeter will read the rms value.

Heterostatic and ideostatic connections

A practical form of electrometer, particularly suited to DC voltage measurement, is shown in Fig.5.27. There are four, fixed metal (double) quadrants, arranged to form (hollow) circular 'boxes' with short airgaps between the quadrants as shown. Inside the quadrants, an aluminium, sector-shaped (double) vane is free to rotate, suspended by means of a phosphor-bronze string.

---

[1]As a matter of interest, compare this expression with that derived for a MI instrument where the deflection torque was obtained to be $T_d = \frac{1}{2}I^2\left(\frac{dL}{d\theta}\right)$.

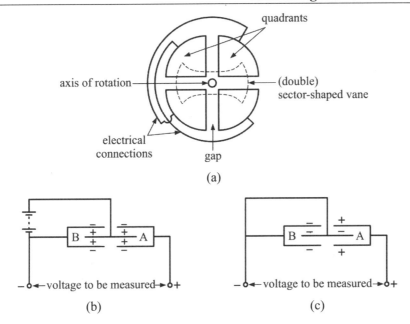

Fig.5.27 : A quadrant electrometer showing two types of connection

With reference to the polarity of the DC voltage to be measured, it can be input to the voltmeter in two different ways as shown. In Fig.5.27(b) is shown what is known as the "heterostatic" connection. In this, a battery of high voltage is used to charge the vane to a potential considerably above that of the quadrant to which the negative of the voltage to be measured is connected. The deflection/rotation of the vane then depends on the difference of the (constant) potential of the vane and the variable (according to the applied voltage) potential of the quadrant.

The other type of connection, called "ideostatic" and generally used in commercial instruments is shown in Fig.5.27(c). In this scheme, the vane is connected directly to one pair of the quadrants as shown, the other pair being connected to positive polarity of the voltage to be measured.

Action

Referring to Fig.5.27(b), with the polarities as marked, the end A of the vane is *repelled* by the fixed quadrant adjacent to it, whilst the end B is *attracted* by its adjacent quadrant resulting in the deflection of the vane. For the ideostatic connection scheme, end B of the vane is repelled by the adjacent quadrant whilst end A is attracted by the fixed quadrant adjacent to it.

Attracted-disc type voltmeter

The schematic of an "attracted-disc" type electrostatic voltmeter is shown in Fig.5.28. F and H are two parallel (circular) plates; F being fixed whilst H, suspended by a (double) leaf spring S is free to rotate. The spring itself is supported at the top by a micrometer device M used for initial adjustment of the "zero" position of the instrument. The movable disc H is surrounded by a "guard ring" G, electrically connected to the disc H, the purpose of which is to render the field between the moving and fixed discs uniform. When in use, the voltage to be measured is applied across the two discs.

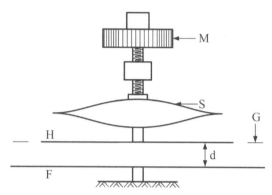

Fig.5.28 : Attraction-disc type electrometer

The moving disc is attracted downward and is brought back to its zero position by turning the micrometer head M, the movement required to return the disc to zero being observed. The spring and the micrometer head are "calibrated" by first shorting the two discs of the instrument, setting the moving disc to its "zero" position and adding known weights to the disc. The movements of the micrometer required to bring the disc back to zero are observed for different known weights, the calibration thus being achieved. Hence, in reality, the force of attraction produced by a certain PD across the discs is measured in terms of the movement of the micrometer and the PD itself is determined from the dimensions of the instrument and measured force.

Theory of attracted-disc type voltmeter

The two discs acting together form a parallel-plate capacitor with the area of plate/disc H being common with the fixed disc F.

Let the area of the plate H be A^1 m^2, the vertical distance between the plates be d m, and the PD across the plates be V volts.

---

[1]With the provision of the guard ring separated from H by an airgap, the effective area of the disc/plate H for computation of capacitance may be taken as its actual area plus half the area of the gap between H and G with minimal error.

Let the 'effective' capacitance between the plates be C F.

Then for the air capacitor,

$$C = \frac{\varepsilon_0 A}{d} \text{ F}$$

ε_0 being the absolute permittivity.

Following the derivation of deflecting torque for the quadrant electrometer, the force F between the plates will be given by

$$F = \frac{1}{2} V^2 \frac{\delta C}{\delta d} = \frac{1}{2} V^2 \left(-\frac{\varepsilon_0 A}{d^2} \right)$$

$$= -\frac{\varepsilon_0 A \, V^2}{2 \, d^2} \text{ N}$$

the negative sign indicating that the force on disc H acts downward.

A commercial electrometer

A commercial form of an electrometer due to Lord Kelvin and capable of measuring even low voltages (100 – 1000 V) is illustrated in Fig.5.29.

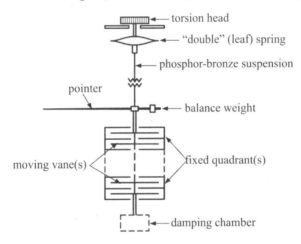

Fig.5.29 : Schematic of the Kelvin electrometer

This instrument is essentially a quadrant-type electrometer with a number of moving vanes and quadrants cascaded vertically as shown. This results in producing an appreciable deflecting torque even at low voltages as the torque is multiplied by virtue of the presence of various combinations of vanes and quadrants.

By using a phosphor-bronze suspension instead of (jewel) bearings, the friction torque is considerably reduced. The torsion head capable of being moved very slowly, is for zero adjustment. The damping may be by a vane dipping into an (enclosed) air-dashpot.

Vacuum-enclosed electrostatic voltmeters

A modern development is to enclose the working parts of the (usual) electrostatic voltmeter within a highly evacuated chamber to bring down considerably the frictional error produced by the surrounding air on the moving system. The superior dielectric strength of the high vacuum in comparison with that of air at normal pressure also enables the clearance between the plates to be reduced thereby resulting in increased force for a given applied voltage.

SPECIAL-PURPOSE INDICATING INSTRUMENTS

Power Factor Meters

Power factor, or cosine of the angle by which current lags or leads the potential drop across the load or an element, is an important quantity in all AC systems or circuits.

Indirectly, the power factor can be deduced from the knowledge of the current, PD and active power consumed by the load, obtained by way of independent measurements. If these quantities are I, V and W, respectively, in a given system of units, the power factor, abbreviated **pf**, is given by

$$\mathrm{pf} \text{ or } \cos \phi = \frac{W}{V \times I},$$

its numerical value being between 0 and 1.

However, if the measured quantities are rms values and the phasor relationship of V and I is not known, the pf deduced as above cannot reveal whether the current is lagging behind or leading the applied voltage or PD across the load.

Power-factor meters are designed to indicate the power factor of a circuit directly without the need of measuring circuit current, PD or the power AND, what is more important, point out whether the power factor is lagging or leading.

Dynamometer-type single-phase power factor meter

This type of power factor meter is similar to a dynamometer wattmeter, comprising a current and a voltage circuit. The current circuit consists of two fixed coils and carries the current in the circuit whose power factor is be measured, or a definite fraction of this current (by incorporating a shunt, for example).

The voltage circuit is in the form of two *identical* coils fitted together with their planes at right angles to one another as shown in Fig.5.30. One of the coils is connected to the supply (also across the load) through a non-inductive resistance whilst the other has a 'pure' inductance in series with this coil and connected across the same supply. The two coils thus carry the *same* current in magnitude and produce magnetic fields of equal strength, *but in phase quadrature.*

The deflection of the instrument depends on the phase difference between the main, or load, current and the currents in the two branches of the voltage circuit, that is, upon the power factor of the load.

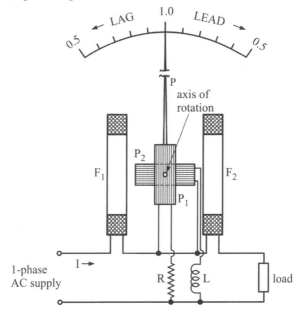

Fig.5.30 : Schematic of a single–phase dynamometer-type power factor meter

In the diagram of Fig.5.30, F_1, F_2 are two fixed coils carrying the circuit/load current, the magnetic field of these coils being proportional to the load current. The coils are so positioned, similar to Helmholtz coils, as to produce a near uniform field in the space between them as depicted in Fig.5.31. Pivoted between them at the central point are the set of two moving coils P_1 and P_2, rigidly fixed at an angle of 90° with respect to each other.

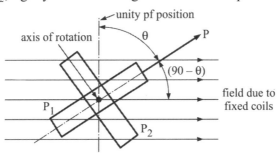

Fig.5.31 : Principle of operation of the power factor meter

These coils rotate together and carry a pointer which moves over a graduated scale to indicate the power factor of the circuit directly. The dial of the meter is calibrated to indicate whether the p f is "lag" or "lead", the 'limits' in general being 0.5 lag to 0.5 lead[1].

The physical dimensions and shape, as well as the number of turns on each of the two coils, are the same so that they produce equally strong magnetic field individually, but displaced by 90° in space and in time phase when equal currents having a time phase difference of 90°(E) are passed through them. This is achieved by connecting a resistance and a pure inductance in the circuits of the two coils as shown. From the point of producing a phase difference of 90°, the inductance may be replaced by a capacitance to allow the same current. Note that the magnitude of current in the circuit would also depend on the supply frequency. No mechanism is required to produce any control torque in these instruments; slight air damping may, however, be provided.

Principle of Operation and Production of Deflection Torque

The action of the meter can be explained by reference to Fig.5.31, and the basic principle of production of torque by interaction of two magnetic fields, or one magnetic field interacting with currents(s) in movable coils. Thus, torques are produced due to interaction of main field due to F_1, F_2 and that due to 'pressure' coils P_1 and P_2. The torque developed due to either P_1 or P_2 is maximum if the axis of the coil is parallel to the main field and zero if the two are at right angles. For any 'intermediate' position of the set, torques will be produced due to both P_1 and P_2 and under the impact, the set of coils would begin to rotate and come to rest when the two torques are equal (and opposite) and the net torque is zero.

Theory

Refer to Fig.5.31. Consider coil P_1 first. It is like pressure coil of a dynamometer wattmeter, the current through which being in phase with the supply, controlled by resistance R. It therefore produces a torque corresponding to "active" power in the circuit, that is, $V\,I\cos\phi\left(\dfrac{dM}{d\theta}\right)$ where $\cos\phi$ is the load power factor and M is the mutual inductance between P_1 and F_1, F_2 corresponding to any deflection θ, as derived earlier for a dynamometer instrument.

---

[1]This is because in practice a power factor less than 0.5 lag (or lead) is seldom encountered in most circuits.

Now the coupling, and hence the mutual inductance M, will be maximum when $\theta = 0$ with the plane of P_1 aligned with the planes of F_1 and F_2.

Therefore,

$$M = M_{max} \cos \theta \quad \text{or} \quad \frac{dM}{d\theta} = - M_{max} \sin \theta$$

That is,

$$\frac{dM}{d\theta} \propto \sin \theta, \quad \text{disregarding } -\text{ve sign,}$$

θ being measured with reference to the vertical axis, corresponding to unity p f position. [See Fig.5.31].

Then the torque acting on P_1 can be expressed as

$$T_{P_1} = K \, V \, I \cos \phi \sin \theta$$

Considering the coil P_2 with the current in it lagging the supply voltage by $90°(E)$, the torque produced due to P_2 can be written

$$T_{P_2} = K \, V \, I \cos (90 - \phi) \cos \theta^1$$

At equilibrium

$$T_{P_1} = T_{P_2}$$

or $K \, V \, I \cos \phi \sin \theta = K \, V \, I \sin \phi \cos \theta$

or $\tan \theta = \tan \phi$

That is, $\boxed{\theta = \phi}$

Hence, the angular position taken by the coils or the deflection of the moving system is a measure of power factor angle of the load, or the power factor itself if the instrument is so calibrated. It is easy to visualise that the movement of the pointer will be to one or the opposite side of the vertical depending on whether the power factor is lagging or leading.

The meter has the advantage of being unaffected by the *magnitudes* of current and voltage unless these are very small. However, owing to the use of an inductor or capacitor in series with one of the coils to produce the required $90°(E)$ phase difference, the reading may be affected by the frequency and waveform of the supply.

---

[1]$dM/d\theta$ for coil P_2 will be proportional to $\cos \theta$ since the two coils are fixed at right angles in space.

Dynamometer-type power factor meter for balanced 3-phase load

To measure power factor in a *balanced* 3-phase load in which the power factor is the same in each phase or having the same angle of lag or lead, a 3-phase power factor meter, essentially a dynamometer-type (that is, incorporating a set of fixed and moving coils) is best suited and has the following distinct advantages:

(a)   there is no necessity for phase splitting by artificial means using an inductor or capacitor, since the required phase displacement between the currents in the two moving coils results from the 3-phase, balanced supply itself;

(b)   the instrument indicates the load power factor, *per phase*, independent of frequency and waveform of the supply.

The construction of the meter is similar to that of a 1-phase instrument, having two fixed coils and a pair of identical coils which form part of the moving system. However, the two moving coils are fixed with their planes 120° apart in space as shown in Fig.5.32, which also shows the connections of the various coils. The two fixed coils, F F, are connected in series in phase R whilst the moving coils are connected across phases R and B and R and Y, respectively, each in series with a non-inductive resistance, R as shown.

There is no control-torque mechanism provided in the meter.

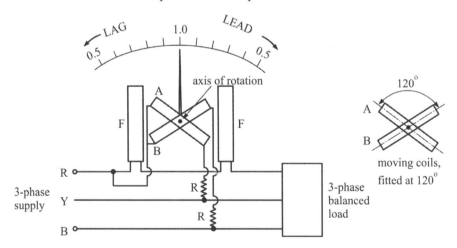

Fig.5.32 : Schematic of the 3-phase dynamometer-type power factor meter

Theory and Operation

Refer to the schematic of coils and phasor diagram in Fig.5.33.

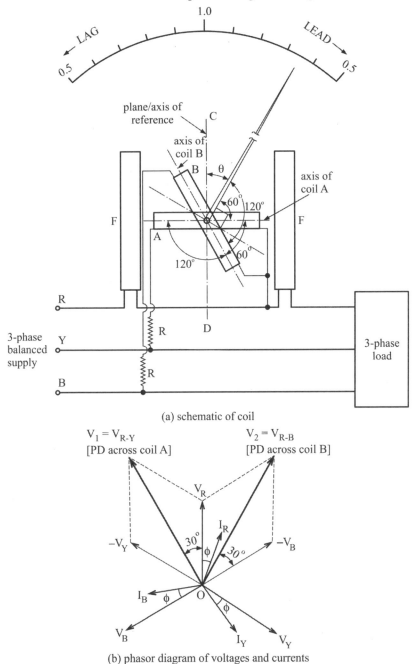

(a) schematic of coil

(b) phasor diagram of voltages and currents

Fig.5.33 : Operation of the meter for a power factor angle ϕ and corresponding phasor diagram

Following the analysis for the 1-phase power factor meter, if M_A is the mutual inductance between the fixed coils and moving coil A, then its variation corresponding to a deflection of θ will be

$\dfrac{dM_A}{d\theta} \propto$ sine of the angle between 'reference' axis CD, the position corresponding to unity power factor, and axis of coil A at equilibrium [see the phasor diagram in Fig.5.33]

that is,

$$\dfrac{dM_A}{d\theta} \propto \sin\left(\theta + 60^0\right)$$

Similarly,

$\dfrac{dM_B}{d\theta} \propto$ sine of the angle between the reference position and axis of coil B

that is,

$$\dfrac{dM_B}{d\theta} \propto \sin\left(\theta + 120^0\right)$$

Also, the phase angle between the current through fixed coils, I_R, and the PD across coil A, V_{R-Y} or V_1, is $(30 + \phi)$ as shown in the phasor diagram of Fig.5.33(b).

Therefore, torque acting on coil A

$$T_A = K_A V_1 I_R M_{max_A} \cos(30^0 + \phi) \sin(60^0 + \theta)$$

where K_A is a constant and M_{max_A} denote the maximum value of the mutual inductance between coil A and fixed coils, and $(30^0 + \phi)$ the phase angle between the current through the fixed coils and the PD across coil A, V_{R-Y} or V_1.

Similarly, the torque acting on coil B

$$T_B = K_B V_2 I_R M_{max_B} \cos(30^0 - \phi) \sin(120^0 + \theta)$$

where K_B is another constant, M_{max_B} the maximum value of mutual inductance between coil B and fixed coils, and $(30^0 - \phi)$ the phase angle between the current through the fixed coils and the PD across coil B, V_{R-B} or V_2

But for identical coils A and B

$$K_B = K_A = K \quad \text{and} \quad M_{max_B} = M_{max_A} = M_{max}$$

Also, for a balanced supply $\left| V_2 \right| = \left| V_1 \right| = \left| V_L \right|$

Hence

$$T_A = K\, V_L\, I_R\, M_{max}\, \cos(30° + \phi)\, \sin(60° + \theta)$$

and

$$T_B = K\, V_L\, I_R\, M_{max}\, \cos(30° - \phi)\, \sin(120° + \theta)$$

For steady deflection,

$$T_A = T_B$$

or

$$\cos(30° + \phi)\, \sin(60° + \theta) = \cos(30° - \phi)\, \sin(120° + \theta)$$

On expanding and simplifying the trigonometric expressions and substituting for $\sin 30°$, $\cos 30°$, etc.

$$\boxed{\theta = \phi}$$

This shows that the deflection of the moving system in the instrument is a measure of load power factor angle[1].

Moving-iron type power factor meter for three-phase load

This type of meter differs from the moving-coil type by the use of a moving-iron system in place of the set of coils to which the pointer is attached. The instruments are mostly used for measurement in 3-phase balanced circuits and have two main advantages:

(a)   larger working force or deflection torque owing to presence of "iron" which enhances the magnetic field;

(b)   a scale that can theoretically extend to $360°$, thus improving the accuracy of reading.

Rotating-field Type

A power factor meter of this type was developed by the Westinghouse Company and is shown schematically in Fig.5.34. It comprises a set of three *identical* coils, marked A, positioned $120°$ apart in space. A rotating magnetic field is thus produced when the coils are excited from a 3-phase *balanced* supply similar to that by the 3-phase (distributed) winding of an induction motor. The excitation currents to the coils are drawn from secondary windings of three current transformers, one each in the balanced 3-phase supply. Another coil, B, with hollow circular interior is placed at the 'centre' of the system of coils, A, and is connected in series with a resistance, R, across any two phases of the supply. At the centre of hollow of coil B is pivoted a short iron rod or spindle, S, carrying a pointer and (air) damping vanes towards the top. Also attached to the spindle are two sector-shaped iron pieces I_1 and I_2, one above and one below the coil B. No controlling mechanism is provided.

---

[1]The theory, and hence the deflection expression, will not apply if the load were unbalanced.

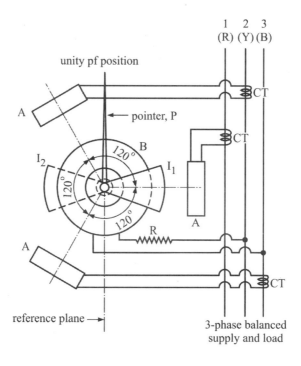

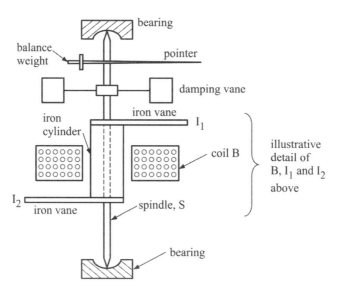

Fig.5.34 : Moving-iron type 3-phase power factor meter

Action

The currents through coils A, or the rotating magnetic field produced by them, interact with the alternating or pulsating flux induced in the iron vanes by the current, and magnetic field thereof, in coil B to produce a torque and hence rotation of the moving system in a particular direction. The pointer comes to a steady deflection when the resultant torque on the moving system is zero.

The angular deflection, θ, of the pointer is then a measure of the power factor angle of the load, the scale being calibrated to read the power factor directly as in the case of the dynamometer type instrument.

Theory of Operation

Referring to Fig.5.35, let the current through the coils be I_1, I_2, I_3, lagging the respective applied voltages (V_1, V_2, V_3) by power factor angles ϕ_1, ϕ_2, ϕ_3. However, for balanced supply and identical coils, A,

$$V_1 = V_2 = V_3 = V$$

$$I_1 = I_2 = I_3 = I$$

and

$$\phi_1 = \phi_2 = \phi_3 = \phi$$

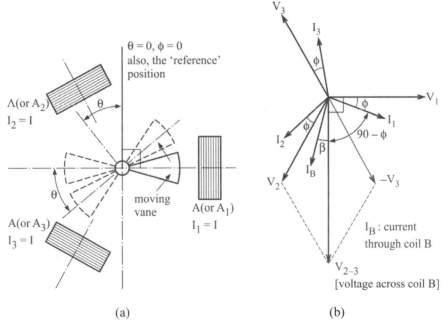

(a)

(b)

Fig.5.35 : Theory of operation of the moving-iron power factor meter

Let the moving system acquire a steady deflection of θ under the above excitation conditions. It is assumed that a small current I_B flows through coil B, lagging the applied voltage (the line voltage V_{2-3} in this case) by a small angle, β, owing to inductance of the coil. However, for the purpose of analysis β can be assumed to be zero.

Then, referring to the phasor diagram in the figure and proceeding as in the case of the dynamometer-type instrument, in terms of expression for power in each phase and that for $(dM/d\theta)$ for each of the coils, A, the deflection torque can be written

$$T_D = K \, [V_{2-3} \, I_1 \, \cos (90° - \phi) \sin (90° + \theta) +$$
$$V_{2-3} \, I_2 \, \cos (330° - \phi) \sin (120° + 90° + \theta) +$$
$$V_{2-3} \, I_3 \, \cos (210° - \phi) \sin (120° + 120° + \theta)]$$

where K is a constant.

For the "power terms" -

the phase angle between V_{2-3} and $I_1 = 90° - \phi$

the phase angle between V_{2-3} and $I_2 = 330° - \phi$

the phase angle between V_{2-3} and $I_3 = 210° - \phi$

and for the $(dM/d\theta)$ terms -

the spatial phase displacement between axis of coil A_1 and deflection $= 90° + \theta$

the spatial phase displacement between axis of coil A_2 and deflection $= 120° + 90° + \theta$

the spatial phase displacement between axis of coil A_3 and deflection $= 120° + 120° + 90° + \theta$

For a steady deflection, the net torque T_D must be zero. Then, substituting

$$I_1 = I_2 = I_3 = I$$

and $$V_{2-3} = V_L$$

and simplifying the above expression by trigonometric expansions and substituting for $\cos 90°$, $\sin 90°$ etc.

$$\boxed{\theta = \phi}$$

That is, the deflection of the pointer is a direct measure of ϕ, the load power factor angle[1].

---

[1]Clearly, whether the power factor is lagging or leading will still be indicated by the deflection, being on the opposite sides of the vertical position.

Errors

Although these meters have the advantages cited above and further listed later, they also suffer from various basic errors on account of the use of iron in the moving system, that is, those caused by hysteresis, eddy currents, frequency and waveform. Stray magnetic fields such as the ones caused by presence of cables in the vicinity carrying very large currents, typically in a testing bay can also affect the working. Hence, for better accuracy, the dynamometer-type power factor meters are always preferable.

Single-phase moving-iron type power factor meter

A single-phase, moving-iron power factor meter is based on the same principal as the 3-phase meter, with similar construction, but some difference. The working of the meter is explained with reference to the schematic shown in Fig.5.36.

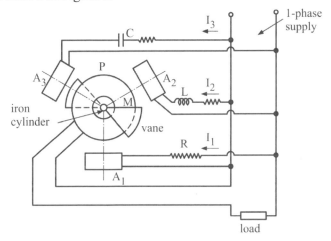

Fig.5.36 : Single-phase MI power factor meter

Thus, referring to the figure

A_1, A_2, A_3: three identical, fixed coils, displaced $120°$ in space

P   : a circular coil, connected in series with the load

M   : moving system comprising soft-iron cylinder, sector-shaped vanes (attached to the cylinder, one above and one below the coil P as in the three-phase meter); and, the spindle, pointer and damping vanes (not shown) to provide air damping.

The three fixed coils are connected to the supply through a resistor, an inductor and capacitor, respectively, as shown, such that the

- current in coil A_1 is *in phase* with the supply voltage, V;

- current in coil A_2 *lags* the supply voltage by $60°$; and

- current in coil A_3 *leads* the voltage by $60°$.

The connections of coil A_1 are (then) reversed with respect to connections of the other two coils so that currents in the three coils are finally displaced by $120°$ with respect to each other as shown by the phasor diagrams of Fig.5.37.

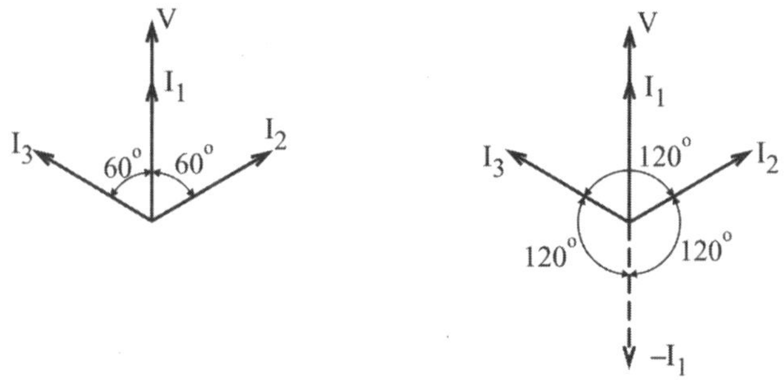

"normal" currents in coils A_1, A_2, A_3 current in coil A_1 reversed

Fig.5.37 : Currents in coils A_1, A_2 and A_3 of the 1-phase power factor meter

The mutually displaced currents in the three coils, being of *equal* magnitude, produce a rotating magnetic field. Once that is achieved, the rest of the theory and production of deflection torque is identical to that of the 3-phase MI power factor meter.

Advantages of MI Power Factor Meters

- The working forces are appreciably large
- All the coils are fixed; therefore robust construction
- The scale may extend to nearly $300°$ (theoretically $360°$); hence better reading accuracy
- Relatively in-expensive compared to the dynamometer type

Disadvantages

- Errors may be introduced in the meters mainly owing to iron losses in cylinder and vanes; also, due to stray magnetic fields
- The calibration of the meter may be affected by frequency and waveform errors
- In general, moving-iron power factor meters are bulky and less accurate than those of the dynamometer type

FREQUENCY METERS

Frequencies in AC systems generally relate to power frequencies, viz., 50 or 60 Hz although, in electromagnetic phenomenon, there is the vast range from audio frequencies (limited to about 15 kHz) to GHz. The meters dealt with here are essentially meant to measure power frequencies.

In the AC power net-works or systems, the tolerance on the variation of frequency is limited to ± 3% as per the national (or international) standards;

that is, for a 'base' frequency of 50 Hz, the allowed deviation is 48.5 to 51.5 Hz. An accurate measurement of system frequency therefore acquires special significance[1].

For the purpose, a variety of frequency meters are available. Their working depends on one of the following principles:

(a) mechanical resonance

(b) electrical resonance

(c) movement of 'iron' in an electromagnetic field

Mechanical Resonance Type

[or vibrating–reed frequency meter]

This type of frequency meter is commonly used as "panel-mounted" meter for visual *indication* of supply frequency at a glance.

The main details of the meter are shown in Fig.5.38.

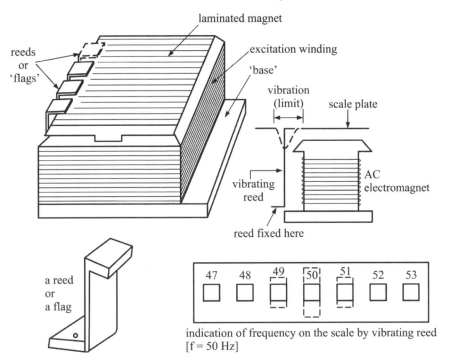

Fig.5.38 : Constructional features of a vibrating-reed frequency meter

---

[1]An accurate knowledge of supply frequency is also essential in a number of measurements; for example, AC bridge methods discussed later.

Also, proper measurement and/or monitoring of power system frequency is important since the load flow, and even the system stability, is crucially dependent on frequency.

The meter comprises a laminated electromagnet, carrying a winding excited by the supply of which the frequency is to be measured. On one side of the magnet are arranged a row of thin, steel strips, called reeds or flags, as shown. Each of the reeds is 4 to 5 mm in width and about 0.3 to 0.5 mm thick, *slightly differing in cross-sectional dimensions from each other.* The reeds are bent at the two ends as shown with one end rigidly fixed and the other free to vibrate laterally, through suitable rectangular 'cuts' in the face plate of the meter as indicated. Owing to difference in individual size (and hence weight), each reed has a unique natural frequency of vibration[1].

On excitation, the electromagnet produces a force of attraction on the reeds during each half cycle and hence the frequency of force having *twice* the frequency of the supply. Under the action of the force, all the reeds start vibrating. However, the reed of which the natural frequency matches the supply frequency (and the force owing to it) – a condition of resonance –, vibrates more intensely, causing a 'band' of vibrations, as seen on the scale, pointing to *the* frequency of the supply.

The scale is marked in the range of, say, 47 Hz to 53 Hz as the supply frequency will seldom cross these limits. The indication, even though not very accurate, serves the purpose to an observer to note the supply frequency and the trend of its variation. At times two of the reeds might vibrate with equal amplitude, thus indicating a frequency approximately equal to the mean value of the adjacent frequencies. Since the force due to electro-magnet current largely depends on the supply *frequency*, the operation of the meter is independent of the supply waveform.

Frequency Meter Based on Electrical Resonance

The schematic of this type of meter is shown in Fig.5.39. The main part is the laminated iron core of *varying* area of cross section, carrying at one end a magnetising coil of appropriate number of turns. The latter is connected across the (single-phase) AC supply and thus draws a current of the frequency that is being measured.

The current I in the magnetising coil lags the supply voltage V by an angle depending on the coil impedance although this is immaterial. There is also a moving coil M threading the protruding core as shown, horizontally suspended at the pivot and thus capable of lateral movement inside and out of the core. A capacitor, C, is connected across the coil as shown. The coil M also carries a pointer, arranged to move over a graduated scale which is calibrated to read the supply frequency, the central, vertical position corresponding to 'normal' frequency as shown.

---

[1]This property is somewhat akin to a tuning fork. Also, the reeds are arranged along the magnet surface in ascending order of their natural frequencies.

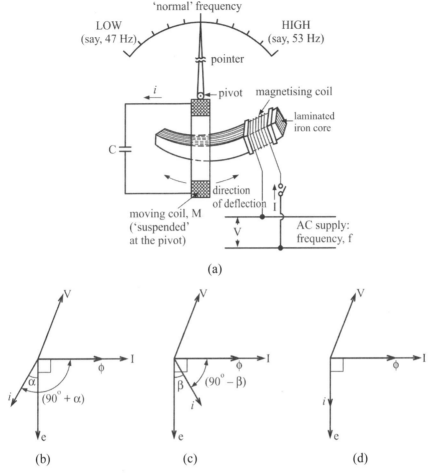

Fig.5.39 : Electrical resonance type frequency meter
(a) constructional details; (b), (c), (d) phasor diagrams

Principal of operation

The principle of operation of the meter depends on the relative magnitudes and phasor relation of currents in the two coils: the magnetising coil and the moving coil. The current in the former is constant for a given supply voltage and given frequency, producing a flux, Ø, being *in phase with the current*. Assuming that this flux is linked fully with the moving or suspended coil, an EMF e is induced in the coil, lagging the flux by 90°. The current in the moving coil, however, is the result of net reactance in its circuit which consists of

(i) a capacitive reactance, $1/\omega C$, for a given (supply) frequency; and

(ii) an inductive reactance, ωL, which for a given frequency also depends on the inductance L, itself a function of the lateral position of the moving coil with respect to the core[1].

For a better understanding of the operation of the meter, consider the phasor diagram of Fig.5.39(b) when the inductance and hence the inductive reactance, ωL, is high owing to high supply frequency compared to the capacitive reactance, that is $X_L > X_C$. The current, i, in the moving coil is thus largely inductive and lags the induced EMF, e, by an angle α as shown. The 'torque' developed[2] is then given by $T_1 = K \, I \, i \, \cos(90°+\alpha)$, that is, $- K \, I \, i \, \sin \alpha$. In contrast, when the supply frequency is low, the capacitive reactance would be greater than the inductive reactance, the current in the moving coil would be largely capacitive and hence leading the induced EMF, e, by an angle β as shown in Fig.5.39(c). The 'torque' in the instrument will now be given by $T_2 = K \, I \, i \, \cos(90°-\beta)$, i.e., $K \, I \, i \, \sin \beta$ and will be in the 'opposite' sense. In figure (d), the inductive reactance is equal to the capacitive reactance so that i is in phase with e and the torque being given by $K \, I \, i \, \cos 90°$, i.e., zero.

Now the two types of reactance in the moving coil being equal shows the condition of (series) resonance. Hence, for any supply frequency the moving coil, and the pointer, comes to rest corresponding to zero torque condition when the two reactances are equal. For this, the moving coil occupies a position about the core to change the value of inductance to result in X_C being equal to X_L, that is, every time leading to an electrical resonance.

The value of the capacitor C is so chosen that the pivoted coil takes up a convenient central (or vertical) position when the (supply) frequency is at its 'normal' value (for example, 50 or 60 Hz).

An advantage of this type of frequency meter is that if the inductance of the moving coil changes gradually as it moves about the core, great sensitivity can be achieved, with the meter capable of measuring even a slight change of the frequency as shown on the scale.

Moving-iron Type Frequency Meter

The action of this type of frequency meter depends on the relative variation of currents with frequency between two parallel circuits, one inductive and

---

[1]The inductance being higher when the moving coil is closer to the magnetising coil (that is, to the right in the figure) and small as it moves away to the left.

[2]Following the usual expression based on the 'standard' principle of torque production in terms of an interaction of the flux and resulting current in a coil.

the other resistive, connected across the supply of which frequency is to be measured.

The instrument, developed by the Weston company, comprises two sets of *identical* coils A and B made into two halves and mounted with their planes perpendicular to each other as shown in Fig.5.40. At the 'centre' is pivoted a long and thin soft-iron needle free to rotate within the (hollow) space of the coils and attached to a spindle which carries a pointer and (air) damping vane (not shown). There is no controlling-torque mechanism.

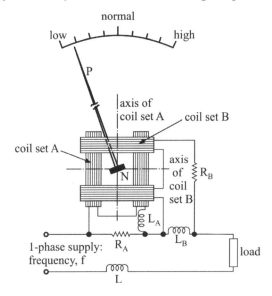

Fig.5.40 : Moving-iron type or Weston frequency meter

The connection scheme of the two coil sets is as shown in the figure. Two halves of each coil set are connected in series. Coil A, in series with a 'pure' inductance L_A, is connected across the resistor R_A in series with the load. Coil B, in series with a non-inductive resistor R_B, is connected across a 'pure' inductance L_B, again in the series circuit of the load. The series inductor L is included to damp out harmonics, if any, in the waveform of the current through the circuit.

Action or principle of operation

The soft-iron needle, and hence the pointer, takes up a position, dependent on the currents through the coils A and B, and is best explained by reference to 'phasor' diagrams of Fig.5.41.

(a) Frequency lower than the normal

The potential drop across L_B decreases due to lower reactance. Current in coil set A, and therefore the force due to it, decreases while that in coil set B increases due to lower reactance of L_A.

Or $I_A < I_B$

and $F_A < F_B$

The resultant force on the needle is such as to rotate it in counter-clockwise direction.

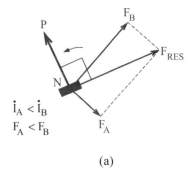

$i_A < i_B$

$F_A < F_B$

(a)

(b) Frequency higher than the normal

The current through coil A is increased whilst that through B decreases.

Or $I_A > I_B$

and $F_A > F_B$

The resultant force on the needle makes it, and the pointer, to rotate in clock-wise direction.

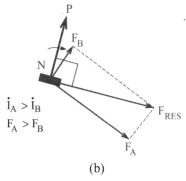

$i_A > i_B$

$F_A > F_B$

(b)

(c) Normal frequency

The circuit elements are so designed that both the currents are equal and the resultant force on the needle makes it take the vertical position.

Or $I_A = I_B$

and $F_A = F_B$

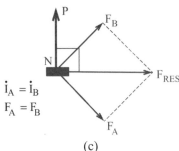

$i_A = i_B$

$F_A = F_B$

(c)

Fig.5.41: Phasor diagrams of the Weston frequency meter

SYNCHROSCOPES

These are special-purpose instruments meant/used to synchronise or parallel a 3-phase alternator to a 3-phase bus or another 3-phase alternator. Synchroscopes indicate the right instant when the synchronising (or paralleling) switch can be closed safely as depicted in Fig.5.42.

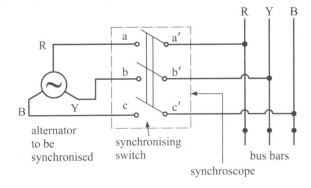

Fig.5.42 : The principle of synchronisation of an alternator

For satisfactory synchronisation, it is imperative that
- voltages (generally L-L) of the two 'supplies' be exactly equal in *magnitude*
- the voltages must be exactly *in phase* on the respective sides

The second requirement implies that
- the frequency of the two supplies, must be identical, and
- the phase sequence of the voltages on the two sides must be same.

A synchroscope ensures all these.

Simple Laboratory Methods

These are commonly known as the

(a) dark-lamp method,   (b) bright lamp method

or a combination of the two.

In the first method, two lamps (rated at 230 V each, for the 3-phase lab supply of 400 V between lines, and 60 W), joined in series, are connected across switch terminals in each phase; for example, across a, a′ ; b, b′ ; c, c′. [See Fig.5.42]. The correct instant of synchronisation is when all the six lamps are 'completely' dark. At this instant the switch can safely be closed.

In the second method, the lamps are connected criss-cross; for example, a, b′ ; b, c′ ; c, a′ (with reference to the same figure). The correct instant of synchronisation in this case is when all the lamps are at their brightest when the synchronising switch can be closed.

In both these cases, the requirements for correct synchronisation can be explained by reference to simple phasor diagrams showing various instances of mis-match of voltages on the two sides. Both the methods, however, may be prone to subjective errors.

A Dynamometer-type Synchroscope

(Also known as Weston synchroscope)

This instrument is similar to a single-phase dynamometer wattmeter, having a pair of fixed coils and a moving coil, both designed to carry only small currents as against a wattmeter in which the fixed coils might carry the full 'load' current.

The instrument also incorporate the essence of the "bright lamp method" by the use of a small (meaning low-voltage) lamp which is at full brightness at the instant of synchronism. The schematic of the instrument is shown in Fig.5.43.

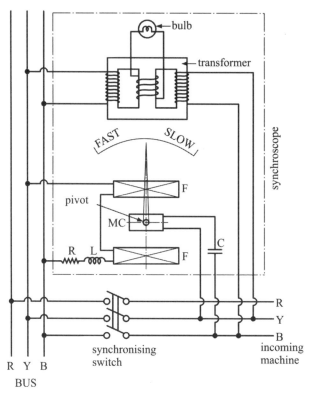

Fig.5.43 : Weston (or dynamometer-type) synchroscope

Operation

The fixed coils F, F in series with a resistor and inductor are connected across two of the phases of the *existing* supply, whilst the moving coil in series with a capacitor is connected across corresponding phases of the incoming machine as shown. This means that the currents in the two sets of coils will be in phase quadrature and equal, indicating that the two supplies are *balanced* with respect to each other. At this instant, the net torque on the moving coil will be zero, the pointer being vertically up and indicating the correct moment for closing the synchronising switch.

Since the same situation may occur when the two voltage sets may be $180°$ out of phase rather than in phase, a special transformer is included in the instrument, connected across the two supplies as shown. The central limb of the transformer carries a winding, the EMF induced in which feeds a lamp. Now the correct instant of synchronisation is when torque on the moving coil is zero (that is, the pointer is in vertical position) AND the lamp is at its brightest.

In the practical form of the instrument, the lamp is provided directly behind the opal glass, marked "SLOW" and "FAST" as shown. This is to avoid any confusion about observing both the pointer *and* brightness of the lamp at the same time, and to judge the instant for synchronisation correctly. Clearly, if the frequencies of the two supplies are different, the pointer will keep oscillating about the 'mean' position and the lamp would keep flickering.

PHASE SEQUENCE INDICATOR

This device is useful to ascertain current phase sequence of a 3-phase supply when phase-sequence sensitive equipment is to be connected to the supply; for example, a 3-phase induction motor where the correct direction of rotation is important.

Induction-type Phase Sequence Indicator

This is based on the induction-motor action vis-à-vis a three phase supply. In its simplest form, the 'meter' comprises a set of three Y-connected, identical coils, spaced (or space-distributed) $120°$ apart. When excited by a balanced (or even 'slightly' unbalanced) supply, a rotating magnetic field is produced. A circular aluminium disc is mounted above the coils, with its axis of rotation coinciding with the 'centre' of the coils. Under the action of the rotating magnetic field, eddy currents are induced in the disc which interact with the field to produce a rotational torque.

Depending on the supply connections, the disc rotates in clockwise or counter-lockwise direction – an indication of the phase sequence of the supply. The outward appearance of the device is shown in Fig.5.44. The whole assembly is enclosed in a case, with the three supply terminals appropriately marked and the direction of rotation of the disc seen through a glass cover. The disc is marked with an arrow to indicate the direction of rotation.

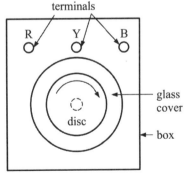

Fig.5.44 : Phase sequence indicator

WORKED EXAMPLES

1. A moving-coil instrument has a coil of width 25 mm, depth 30 mm and number of turns 150. The airgap flux density is 0.15 T. Calculate the deflection torque when carrying a current of 8 mA. Also, calculate the deflection if the control spring constant is 3 micro Nm/deg.

At 8 mA or 0.008 A,

the deflecting torque is, $T_d = N \times B \times l \times w \times I$

Substituting,

$$T_d = 150 \times 0.15 \times 0.03 \times 0.025 \times 0.008$$

$$= 135 \times 10^{-6} \ Nm$$

Let this produces a deflection of θ (deg).

The control torque is

$$T_c = 3 \times 10^{-6} \times \theta$$

At steady deflection

$$3 \times 10^{-6} \times \theta = 135 \times 10^{-6}$$

whence

$$\theta = 45°$$

2. The inductance of a given moving-iron ammeter is approximated to be $\left(8 + 4\theta - \dfrac{1}{2}\theta^2\right)$ $\mu H/rad$ of deflection from the zero position. The control spring constant of the ammeter is $12 \times 10^{-6} \ Nm/rad$. Calculate the deflection in the ammeter for a current of 5 A.

Given $\qquad L = 8 + 4\theta - \dfrac{1}{2}\theta^2$

$\therefore \qquad \dfrac{dL}{d\theta} = (4 - \theta) \times 10^{-6} \ H/rad$

For a moving-iron ammeter, the deflecting torque is given by

$$T_d = \frac{1}{2} I^2 \frac{dL}{d\theta}$$

Therefore, at 5 A,

$$T_d = \frac{1}{2} \times 5^2 \times (4 - \theta) \times 10^{-6} \text{ Nm}$$

and the control torque for a deflection θ (rad) corresponding to T_d,

$$T_c = 12 \times 10^{-6} \times \theta \text{ Nm}$$

For a steady deflection

$$\frac{1}{2} \times 5^2 \times (4 - \theta) \times 10^{-6} = 12 \times 10^{-6} \times \theta$$

or $\qquad\qquad 24.5\ \theta = 50$

from which

$$\theta = 2.04 \text{ rad} \quad \text{or } 117° \text{ [since } \pi \text{ rad} = 180°]$$

3. The table below describes the relationship between deflection and inductance of a MI instrument:

deflection, θ (deg)   : 20 30   40   50   60   70 80   90
inductance, L, micro H : 335 345 355.5 366.5 376.5 386 391.3 396

Find the current and torque required to give a deflection of 45 deg, given the control spring constant as 0.4×10^{-6} Nm/deg .

The deflecting torque is given by

$$T_d = \frac{1}{2} I^2 \left(\frac{dL}{d\theta} \right) \text{ Nm,} \quad \theta \text{ in radians}$$

Also, the control torque is

$$T_c = 0.4 \times 10^{-6} \text{ Nm/deg}$$

At 45°,

$$T_c = 0.4 \times 45 \times 10^{-6}$$
$$= 18 \times 10^{-6} \text{ Nm}$$

At steady deflection, $T_d = T_c$

$$\therefore \qquad \frac{1}{2} I^2 \left(\frac{dL}{d\theta} \right) = 18 \times 10^{-6}$$

Using the data in the above table, a graph is plotted of inductance v/s deflection as shown below.

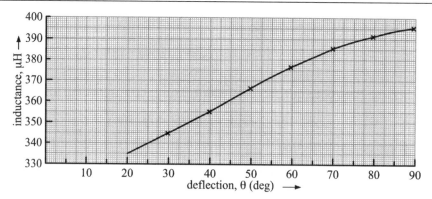

From the graph at $\theta = 45°$, the slope of the curve is obtained as

$$\frac{dL}{d\theta} = \frac{366.5 - 355.5}{50 - 40}$$

$$= 1.1 \times 10^{-6} \quad \text{H/deg}$$

$$= 1.1 \times 10^{-6} \times 57.3 \quad \text{H/rad} \quad [\text{since } 180° = 3.14 \text{ rad}]$$

$$= 63.06 \times 10^{-6} \quad \text{H/rad}$$

$$\therefore \frac{1}{2} I^2 \times 63.06 \times 10^{-6} = 18 \times 10^{-6}$$

giving

$$I = 0.755 \text{ A}$$

Then

$$T_d = \frac{1}{2} \times (0.755)^2 \times 63.06 \times 10^{-6}$$

$$= 18 \times 10^{-6} \text{ Nm}$$

$$= T_c \text{ (as expected)}$$

4. The relationship between the inductance of a moving-iron ammeter, the current and the deflection of the pointer is as follows:

current, A	1.2	1.4	1.6	1.8
deflection (deg)	36.5	49.5	61.5	74.5
inductance (µH)	575.2	576.6	577.8	578.8

Calculate the deflecting torque when the current is 1.5 A.

The deflecting torque, $T_d = \dfrac{1}{2} I^2 \dfrac{dL}{d\theta}$

Referring to the given table, two graphs are plotted: I v/s deflection (deg) and I v/s inductance as shown below.

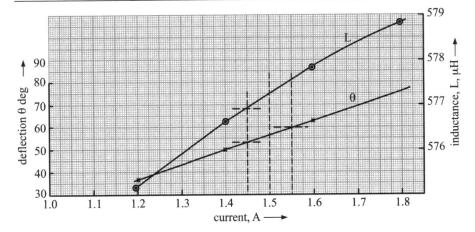

From these, the value of $\dfrac{dL}{d\theta}$ at the current of 1.5 A can be obtained as follows

$$L_{1.45} = 576.9 \ \mu H; \qquad L_{1.55} = 577.55 \ \mu H$$

$$\theta_{1.45} = 53°; \qquad \theta_{1.55} = 59°$$

$$\therefore \qquad \Delta L = 0.65 \ \mu H; \qquad \Delta\theta = 6°$$

and

$$\frac{dL}{d\theta} = \frac{\Delta L}{\Delta\theta}$$

$$= \frac{0.65}{6}$$

$$= 0.108 \ \mu H/rad$$

$$= 0.108 \times 57.3 \ \text{or} \ 6.188 \ \mu H/rad$$

Hence, at $\qquad I = 1.5 \ A,$

$$T_d = \frac{1}{2} \times (1.5)^2 \times 6.188 \times 10^{-6} \ \text{or} \ 6.96 \times 10^{-6} \ Nm$$

Without taking recourse to plotted graphs, if the given values of L and θ are used for

$I = 1.4$ and 1.6, about 1.5 on either side, then

$$\frac{dL}{d\theta} = \frac{577.8 - 576.6}{61.5 - 49.5} = \frac{1.2}{12} = 0.1 \times 10^{-6} \ H/deg$$

or $\qquad 0.1 \times 57.3 = 5.73 \times 10^{-6} \ H/rad$

and $\qquad T_d = \dfrac{1}{2} \times 1.5^2 \times 5.73 \times 10^{-6} = 6.44 \times 10^{-6} \ Nm$

5. A dynamometer ammeter is arranged so that only one hundredth of the total (input) current passes through the moving coil and the remainder through the fixed coils. The mutual inductance between the two coils varies with the angle of displacement of the moving coil from its zero position as follows:

angle, θ :	0	15	30	60	90	105	120
M (μH) :	-336	-275	-192	0	$+192$	$+275$	$+336$

If a torque of 1.05×10^{-5} Nm is required to give a full-scale deflection of $120°$, calculate the current at the full-scale deflection.

What will be the current for half the full-scale deflection?

The connections of the dynamometer instrument as an ammeter and current distribution is as shown.

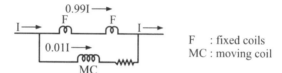

From the given data, the plot of M v/s θ is shown below

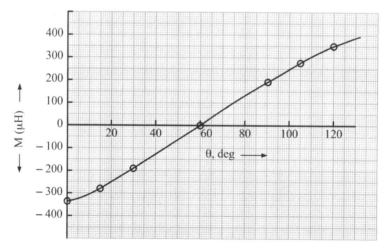

The expression for the deflecting torque as an ammeter is

$$T_d = i_1 i_2 \frac{dM}{d\theta} \text{ Nm}$$

where i_1 and i_2 are the currents through the fixed and moving coils, respectively.

(i)   Full-scale deflection : $120°$

At $120°$ deflection, the value of $\dfrac{dM}{d\theta}$ is estimated from the graph in terms of slope of the curve at $\theta = 120°$.

Thus

$$\Delta M = 370 - 310 = 60 \ \mu H$$

$$\Delta\theta = 130 - 110 = 20 \ \text{deg}$$

$$\therefore \qquad \frac{dM}{d\theta} = \frac{\Delta M}{\Delta\theta} = \frac{60}{20} = 3 \ \mu H/deg$$

$$= 3 \times 57.3 \text{ or } 172 \ \mu H/rad$$

The full-scale deflection is given as 1.05×10^{-5} Nm.

$$\therefore \ i_1 \ i_2 \times 172 \times 10^{-6} = 1.05 \times 10^{-5}$$

Substituting for i_1 and i_2 from the distribution

$$0.01I \times 0.99I \times 172 \times 10^{-6} = 1.05 \times 10^{-5}$$

or $\qquad\qquad I^2 = \dfrac{1.05 \times 10^{-5}}{0.01 \times 0.99 \times 172 \times 10^{-6}} = 6.166$

whence

$$I = 2.48 \ A$$

(ii) At half full-scale deflection, $\theta = 60°$

Here, from the graph

$$\Delta M = 70 + 60 = 130 \ \mu H$$

$$\Delta\theta = 70 - 50 = 20 \ \text{deg}$$

$$\therefore \qquad \frac{dM}{d\theta} = \frac{\Delta M}{\Delta\theta} = \frac{130}{20} = 6.5 \ \mu H/deg$$

$$= 372.45 \ \mu H/rad$$

The deflecting torque will be 5.25×10^{-6} Nm.

$$\therefore \ 0.01I \times 0.99I \times 372.45 \times 10^{-6} = 5.25 \times 10^{-6}$$

or $\qquad\qquad I^2 = \dfrac{5.25 \times 10^{-6}}{0.01 \times 0.99 \times 372.45 \times 10^{-6}}$

whence $\qquad\qquad I = 1.193 \ A$

6. The following measurements were made on a voltmeter of the dynamometer type having a range of 60 V and a (series) resistance of 780 Ω.

applied voltage, V	:	30	40	50	60
deflection (deg)	:	14	24	37	54
inductance (mH)	:	74.8	78.5	82.8	88.6

Calculate the deflecting torque when the applied voltage is 45 V.

The connections for a dynamometer voltmeter are as shown.

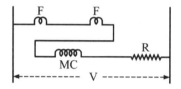

FF : fixed coils
MC : moving coil
R   : series resistance

The expression for the deflecting torque is

$$T_d = V^2 \frac{dM}{d\theta}$$

or, rather,

$$T_d = \left(\frac{V}{R}\right)^2 \frac{dM}{d\theta}$$

Refer to the graph of V v/s M and θ shown below.

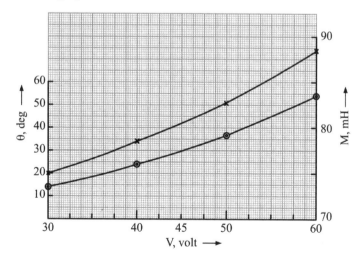

V, volt ⟶

At θ = 45°, the slope of V v/s M curve is

$$\Delta M = 81 - 80; \quad \Delta\theta = 2.5°$$

$$\frac{dM}{d\theta} = \frac{\Delta M}{\Delta \theta} = \frac{1}{2.5} = 0.4 \text{ mH/deg}$$

$$= 22.92 \text{ mH/rad}$$

Hence
$$T_d = \left(\frac{45}{780}\right)^2 \times 22.92 \times 10^{-3} \text{ Nm}$$

$$= 76.3 \times 10^{-6} \text{ Nm}$$

7. The torsional constant of the control springs of the movement of a 10 A dynamometer wattmeter is 10.5 μNm/rad. The variation of mutual inductance with angular position of the moving system is practically linear over the operating range, the rate of change being 0.078 mH/rad. If the full-scale deflection of the instrument is 83°, calculate the current in the voltage coil at full-scale deflection.

The deflection torque of a dynamometer wattmeter is given by

$$T_d = I_1 I_2 \frac{\delta M}{\delta \theta} \text{ Nm}$$

where I_1 and I_2 are the current in fixed (current) coil(s) and in the moving (voltage) coil, respectively, and $\delta M/\delta \theta$ is the slope of the M v/s θ curve at the given deflection. Also, at steady deflection,

$$T_d = T_c, \text{ the control springs torque}$$

Here,
$$I_1 = 10 \text{ A}, T_c = 10.5 \times 10^{-6} \text{ Nm/rad}$$

$$T_d = 10 \times I_2 \times 0.078 \times 10^{-3} \text{ Nm}$$

At 83° or $\left(\frac{83 \times \pi}{180}\right)$ rad $= 1.4479$ rad,

$$T_c = 10.5 \times 1.4479 \times 10^{-6} \text{ Nm}$$

$$\therefore \quad 10 \times I_2 \times 0.078 \times 10^{-3} = 10.5 \times 1.4479 \times 10^{-6}$$

whence
$$I_2 = 19.5 \text{ mA}$$

8. An electrostatic voltmeter of the double-vanes and quadrant-type construction has six identical 'units', cascaded vertically. The radius of the outer edges of the vanes is 3 cm and the distance or separation between each vane surface and the corresponding, adjacent quadrant surface is 2 mm. A PD of 600 V produces a deflection of 30°. Determine the deflecting torque of the instrument and control-torque constant. Assume that capacitance of each of the capacitors formed out of the vane surface and *any* of the quadrant surface near it is given by

$$C = \frac{\varepsilon_0 \frac{1}{2} r^2 \theta}{d} \quad \text{farad}^1$$

with the usual notation.

Capacitance of *one* vane surface and corresponding quadrant surface, for an overlap angle, θ

$$C = \frac{\varepsilon_0 \frac{1}{2} r^2 \theta}{d}$$

$$= \frac{8.854 \times 10^{-12} \times 3^2 \times 10^{-4} \ \theta}{2 \times 2 \times 10^{-3}}$$

$$= 1.99 \times 10^{-12} \times \theta \ \text{F}$$

Altogether there are 24 active vane surfaces

$\therefore$ total capacitance of the system

$$C = 24 \times 1.99 \times 10^{-12} \times \theta \ \text{F}$$

$$= 47.76 \times 10^{-12} \times \theta \ \text{F}$$

$$\left[\text{and} \ \frac{dC}{d\theta} \left(\text{or} \ \frac{\delta c}{\delta \theta} \right) = 47.76 \times 10^{-12} \ \text{F/rad} \right]$$

and deflecting torque for 600 V PD

$$T_d = \frac{1}{2} V^2 \times \frac{\delta C}{\delta \theta} = \frac{1}{2} (600)^2 \times 47.76 \times 10^{-12}$$

$$= 8.59 \times 10^{-6} \ \text{Nm}$$

This produces a deflection of 30° or $\dfrac{\pi}{6}$ rad.

$\therefore$ control torque constant

$$= \frac{T_d \left(= T_c \right)}{\theta}$$

---

[1]In general, for a single 'unit' quadrant-type instrument with a double-ended vane and corresponding two quadrants, the capacitance will be

$$C = \frac{2 \times 2 \times \varepsilon_0 \times \frac{1}{2} \times r^2 \times \theta}{d} \quad \text{for an overlap angle } \theta \text{ (rad)}.$$

$$= \frac{8.59 \times 10^{-6}}{\left(\dfrac{\pi}{6}\right)}$$

$$= 16.4 \times 10^{-6} \quad \text{Nm/rad}$$

9. The capacitance of a $0-2000$ V electrostatic voltmeter increases uniformly from 42 to 54 μF, from zero to full-scale deflection. It is required to increase the range of the instrument to 20 kV by means of an external capacitor. Calculate the capacitance required for the purpose.

For the increased range, the range multiplying factor

$$m = \frac{20,000}{2,000} = 10$$

Capacitance of the voltmeter at full scale,

$$C_V = 54 \ \mu F$$

The multiplying factor with an external capacitance, C_s, at the full-scale value would be

$$m = \frac{V}{N} = \frac{Z_1}{Z} = \frac{C_S + C_V}{C_S} = 1 + \frac{C_V}{C_S}$$

$\therefore$ $\qquad 1 + \dfrac{C_V}{C_S} = 10$

from which $\qquad C_S = 6 \ \mu F$

EXERCISES

1. The coil of a moving-coil voltmeter has 100 turns and has an effective height of 3 cm and width of 2.5 cm. The voltmeter gives full-scale deflection with a current of 5 mA. The controlling torque of the control spring is 0.5 gcm for full-scale deflection. Estimate the airgap flux density.

    [0.133 T

2. The dimensions of the coil of a PMMC voltmeter are 4 cm × 2.6 cm. It has 80 turns and the airgap flux density is 0.15 T. The resistance of the voltmeter is 15,000 Ω, including the external series resistance. Calculate the deflecting torque in the voltmeter when a voltage of 300 V is applied across its terminals.

    [2.5×10^{-6} Nm

3. The coil of a moving-iron voltmeter has a resistance of 300 Ω and an inductance of 1 H. It has an external series resistance of 2,200 Ω. The meter reads 250 V when measuring a DC voltage of 250 V. What will be the reading of the voltmeter when 250 V AC at 50 Hz is connected across it?

    [246 V

4. The inductance of a moving-iron ammeter is given by $L = (20 + 10\,\theta - 3\,\theta^2)$ mH, where θ is the angle of deflection in radians. Determine the deflection in the meter for a current of 8 A if the control spring constant is 10×10^{-6} Nm/rad.

    [1.584 rad or 90.776°

5. The relationship between the inductance of a 2 A moving-iron ammeter, the current and deflection of the moving system is as follows:

current, A	:	0.8	1.0	1.2	1.4	1.6	1.8	2.0
deflection, deg	:	16	26	36.5	49.5	61.5	74.5	86.5
inductance, mH	:	573.2	574.2	575.2	576.6	577.8	578.8	579.5

    Calculate the deflecting torque of the ammeter at input currents of 1 A and 2 A.

    [2854.7×10^{-6} Nm; 6621.75×10^{-6} Nm

6. The inductance of a moving-iron instrument is given by $L = (0.01 + x \times \theta)^2$ mH where θ is the deflection from zero position in degrees. The angular deflections of the instrument corresponding to the currents of 1.5 A and 2 A are 90° and 120°, respectively. Determine the value of x.

    [0.0475×10^{-3}

7. The mutual inductance of an electro-dynamic/dynamometer ammeter varies uniformly at the rate of 0.0035 µH/rad. The full-scale current of the ammeter is 25 A. The control spring constant of the instrument is 1×10^{-6} Nm/deg. Determine the angular deflection of the ammeter for full scale.

    [125.4°

8. The current coil of a dynamometer wattmeter is rated at 20 A whilst its pressure coil has a resistance of 20 kΩ. The variation of mutual inductance with deflection of the moving system is nearly linear, being 0.4584 mH/deg. The torsional constant of the instrument is 0.1 gcm/deg. If the pressure coil is fed from a supply at 250 V, calculate the deflection of the instrument when working at full load.

    [91.7°

9. An attracted disc type electrostatic voltmeter has two flat parallel plates, each having an effective area of 15 cm^2. The separation between the plates is 2 mm. If a voltage of 1000 V is applied across the plates, what is the force of attraction. Permittivity of free space, $\varepsilon_o = 8.854 \times 10^{-12}$.

    [16.6×10^{-4} N

10. In a quadrant–type electrostatic voltmeter, the variation of capacitance (between fixed and moving vanes/parts) with the angle of deflection from zero is given as follows:

deflection, θ (deg) :	0	10	20	30	40	50	60	70	80	90	100
Capacitance, µF :	37	55	71	86	100	112	124	134	144	152	160

    The torsional constant of the control springs in the voltmeter is 5.5×10^{-6} Nm/rad.

    Calculate the values of PD across the voltmeter to give a deflection of (a) 50°, (b) 100°.

    [164.35 V; 280.7 V

11. The torsional constant of the control spring of a 3000 V electrostatic voltmeter is 0.072 gcm/rad. The full-scale deflection of the instrument is 80°. Assuming the rate of change of capacitance with angular deflection to be constant over the entire operating range, calculate the total change of capacitance from zero to full scale.

    [3.054 pF

QUIZ QUESTIONS

1. Instruments can be broadly classified as

   (a) _____ (b) _____

2. Apart from accuracy, the other important feature of an instrument is_____

3. A "secondary" instrument is the one that depends on

   □ absolute quantities   □ calibration

   □ recording the quantity to be measured   □ none of these

4. A typical indicating instrument incorporates

   □ a deflecting torque   □ a control torque

   □ a damping torque   □ all of these

5. A "common" ammeter/voltmeter basically measures

   □ current passing through it   □ power supplied to it

   □ energy consumed in it   □ none of these

6. A PMMC instrument can measure

   □ AC or DC quantity   □ only DC quantity

   □ only AC quantity   □ none of these

7. The deflecting torque of a PMMC instrument depends on

   □ only the current   □ only airgap flux

   □ only moving coil size   □ all of these

8. The power consumption of a typical PMMC instruments is in

   □ microwatt   □ milliwatt   □ watt   □ kilowatt

9. A typical PMMC voltmeter is nothing but a galvanometer with a high resistance in series.

   □ true   □ false

10. The damping in a PMMC instrument is provided by

    □ eddy currents induced in the coil former   □ fluid friction

    □ gravity   □ none of these

11. The scale of a PMMC instrument is

    □ cramped in the beginning   □ uniform

    □ cramped in the middle   □ cramped at the end

12. The control torque in a typical spring-controlled instrument is proportional to

    □ θ □ θ^2 □ $\sqrt{\theta}$ □ $1/\theta$

    [θ = angle of deflection]

13. Main source(s) of error in typical PMMC instruments are

    □ effect of temperature   □ weakening of magnet

    □ weakening of control spring   □ all of these

14. The PMMC instrument can have

    □ poor accuracy   □ sub-standard accuracy

    □ good accuracy   □ none of these

15. The pointer of an indicating instrument comes to rest when the deflecting torque is equal to control torque.

    □ true   □ false

16. The control spring in the indicating instruments consists of

    □ iron   □ phosphor-bronze   □ aluminium   □ brass

17. The moving-iron instrument in common use are

    (a) _____   (b) _____

    type

18. A MI instrument is suitable for measuring

    □ AC quantity only   □ DC quantity only

    □ AC and DC quantity   □ none of these

19. The deflecting torque in a MI ammeter is proportional to

    □ current   □ (current)^2 □ 1/current   □ $\sqrt{\text{current}}$

20. A typical MI instrument is best used for the supply frequency in

    □ audio-frequency range   □ radio-frequency range

    □ power-frequency range   □ none of these

21. If an ammeter is used as a voltmeter, it will

    □ indicate a very high value   □ give a very low value

    □ indicate no reading   □ burn out in all probability

22. A dynamometer instrument has

     ☐ one fixed coil and two moving coils

     ☐ two fixed coils and one moving coil

     ☐ one fixed and one moving coil

     ☐ three fixed coils and two moving coils

23. A dynamometer instrument can be designed to work as

     ☐ an ammeter   ☐ a voltmeter

     ☐ a wattmeter   ☐ any of these

24. The damping in a dynamometer instrument is generally

     ☐ air-friction damping   ☐ eddy-current damping

     ☐ fluid-friction damping   ☐ damping by gravity

25. A low power-factor wattmeter of dynamometer type has _____ pressure coil inductance.

26. The common types of indicating instrument in the lab should be used in

     ☐ vertical position   ☐ horizontal position

     ☐ tilted position   ☐ inverted position

27. The two types of electrostatic voltmeters are

     (a) _____   (b) _____

28. The two "standard" connections of a quadrant-type electrostatic voltmeter are called

     (a) _____   (b) _____

29. An electrostatic voltmeter is best suited to measure very high voltages.

     ☐ true   ☐ false

30. An electrostatic voltmeter can be used to measure

     ☐ AC PD alone   ☐ DC PD alone

     ☐ AC or DC PD   ☐ none of these

31. With special measures, an electrostatic instrument can be used to measure PDs, current and power.

     ☐ true   ☐ false

32. The typical expression for deflecting torque in an electrostatic voltmeter is given by

$$T_d = \text{_____}$$

33. The range of an electrostatic voltmeter can be best extended by

    □ a high resistance in series   □ an inductance in series

    □ a capacitance in series   □ a capacitor in series whose value
                                         is smaller than the voltmeter
                                         capacitance

34. An electrostatic voltmeter has highest accuracy because it draws little
    current.

    □ true   □ false

35. The control torque in a 1-phase power factor meter is provided by

    □ spiral spring   □ gravity

    □ eddy currents   □ none of these

36. The 'normal' range of measurement of a typical power factor meter is

    □ 0.5 lag to unity   □ 0.5 lead to unity

    □ 0.5 lag to 0.5 lead   □ 0 to unity

37. A typical dynamometer power factor meter has

    □ one pair of coils   □ two pairs of coils

    □ three pairs of coils   □ any number of coils

38. A reed-type frequency meter works on the principle of

    □ resonance of natural frequency

    □ simple oscillations

    □ linear movement of reeds

    □ none of these

39. A synchroscope is useful for

    □ measuring current   □ measuring voltage

    □ measuring frequency   □ synchronising AC machines

40. An alternator is being synchronised to the bus bars of 50 Hz frequency.
    The bulbs of the synchroscope flicker at a frequency of 10 Hz. The
    frequency of the alternator is

    □ 60 Hz   □ 40 Hz

    □ 60 Hz or 40 Hz   □ none of these

VI : Instrument Transformers

VI

INSTRUMENT TRANSFORMERS

RECALL

The earliest form of an instrument commonly in use was the DC, permanent-magnet, moving-coil galvanometer, based on the principle of mechanical force produced in a moving coil carrying a direct current and placed in a magnetic field of a permanent magnet. The current, usually a few mA, produced just enough torque so that the instrument could measure the current directly, working as a milli-ammeter. Connected across a small PD, it could measure small voltages as milli-voltmeter, or act simply as a (centre-zero) galvanometer to 'detect' current in a circuit. The instrument could be appropriately modified to read large currents as ammeters, or even measure high voltages. It still forms the basic part of a modern permanent magnet moving coil (or PMMC) instrument, used as an ammeter or a voltmeter essentially in DC circuits or applications. By incorporating a bridge rectifier between the input and deflection system, these instruments could be used in AC measurements, too.

EXTENSION OF INSTRUMENTS RANGE

Very often, it is desirable to 'convert' a simple milli-ammeter, itself being a modification of the basic (D'Arsonval) galvanometer designed to read a few mA, into an ammeter having a range of, say, 5 or 10 A and further to extend the range of such an ammeter to 50 or 100 A using suitable devices. Similarly, the range of a voltmeter designed to read up to 300 V may be required to read up to 1000 V, without having to make drastic changes in its design. The argument may also apply to a wattmeter meant to measure power in a circuit where an extension of range of the wattmeter from 10 A/300 V to 50 A/600 V may be called for. In all these requirements, special means or devices may have to be employed to serve the purpose. The type of a device and principle of operation will depend on the given instrument, that is, an ammeter or voltmeter etc. as well as whether the instrument is being used for DC or AC measurements.

Extension of Range of Ammeters

Use of a shunt

The range of an ammeter, from measuring a few amperes to measuring tens of amperes, can be extended by simply connecting a suitable small resistance, called a **"shunt"**, in *parallel* with the instrument, esp. the PMMC type ammeters used for direct-current measurements. The schematic of connection of a shunt in parallel to the ammeter of which the range is to be extended is shown in Fig.6.1.

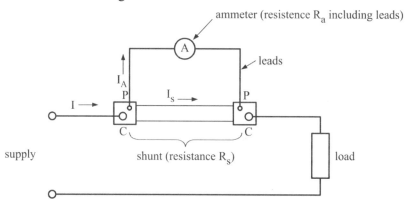

Fig.6.1 : Schematic of an ammeter shunt and division of current

In the figure

PP :   "potential" terminals

CC :   "current" terminals

I_A :   ammeter current for full deflection
(in the existing range)

I_S :   current through the shunt during measurement
(for extended range)

I :   *total* current to be measured ($= I_A + I_S$)

The Shunt

A shunt connected across the ammeter either internally or externally, is almost always known as a "4-terminal resistor". In its 'basic' form, it comprises two solid copper (or brass) blocks, joined by a number of manganin strips as shown in Fig.6.2. The strips constitute the desired resistance; the resistance

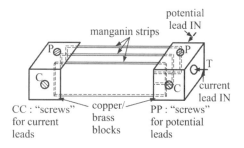

Fig.6.2 : A 4-terminal resistance as a shunt

of blocks being negligible. The blocks have a pair of terminals each. The large ones, known as "current terminals", are used for (series) connection to the load whilst the other two, called "potential terminals" are used to connect suitable leads (at times supplied along with the shunt) to connect to the ammeter. The photograph of an actual 4-terminal resistance for laboratory use is shown in Fig.6.3.

Referring to Fig.6.1

voltage drop across shunt = voltage drop across the ammeter

or $\qquad I_S R_S = I_A R_A$

or $\qquad R_S = \dfrac{I_A}{I_S} \times R_A = \dfrac{I_A}{I - I_A} \times R_A$

$$= \frac{1}{\left(\dfrac{I}{I_A} - 1\right)} \times R_A$$

Fig.6.3 : A 4-terminal resistance for laboratory use

R_S is thus the value of the shunt resistance.

The ratio I/I_A, or the ratio of total current to the ammeter current, is usually defined as "multiplying factor" of the instrument or "multiplier power" of the shunt. Denoting this by K, the shunt resistance can be expressed as

$$R_S = \left(\frac{1}{K - 1}\right) \times R_A$$

or $\qquad K = 1 + \dfrac{R_A}{R_S}$

Use of shunt in AC circuits

When used in alternating-current circuits, it may be necessary to consider inductance (and therefore inductive reactance at supply frequency) of the instrument as well as that of the shunt in addition to the resistances. The division of currents in the ammeter and the shunt will then be according to their *impedances*.

Thus, let R_A and L_A be the resistance and inductance of the ammeter and R_S and L_S be that of the shunt. Then the two currents will divide in the ratio of

$$\frac{\dot{I}_A}{\dot{I}_S} = \frac{\dot{Z}_S}{\dot{Z}_A} = \frac{\sqrt{R_S^2 + \omega^2 L_S^2}}{R_A^2 + \omega^2 L_A^2}$$

and in order that the ratio of impedances be independent of ω, or rather the supply frequency, $\dfrac{L_S}{R_S}$ must equal $\dfrac{L_A}{R_A}$; that is, the time constants of the shunt and ammeter 'circuits' be the same. If that condition is fulfilled, let

$$\frac{L_S}{R_S} = \frac{L_A}{R_A} = k\;;$$

then

$$\frac{i_A}{i_S} = \frac{\sqrt{R_S^2 + \omega^2 L_S^2}}{\sqrt{R_A^2 + \omega^2 L_A^2}} = \frac{\sqrt{R_S^2 + \omega^2 k^2 R_S^2}}{\sqrt{R_A^2 + \omega^2 k^2 R_A^2}} = \frac{R_S}{R_A}$$

or *independent of even inductances*[1].

Also, the multiplying power of the shunt will be given by

$$K = \frac{R_A + R_S}{R_S} = 1 + \frac{R_A}{R_S}$$

same as before.

Main requirements of shunts

For satisfactory, error-free operation, it is essential that

(a) the temperature coefficient of the shunt and instrument shall be very low, and nearly the same, so that considering the ratio R_A/R_S, the multiplying power shall be independent of temperature, either ambient or that due to internal heating;

(b) the resistance of the shunt should not vary with time, this being ensured by proper, careful annealing during manufacturing[2];

(c) the thermoelectric EMF effects must be as small as possible, esp. at large (load) currents that might result in appreciable heating;

(d) the leads used for connecting the shunt to the ammeter should be as small as possible, made of thick, stranded (and flexible) copper wires, so that their resistance is negligible compared to R_A or R_S; at times special leads are provided with the shunts to make the external connections when the leads resistance is not quite negligible and is included in R_A.

---

[1] In practice, this is achieved by designing the two 'circuits' such that the two ratios are at least approximately equal. Indeed, it may well be that the inductances may be negligibly small compared to resistances.

[2] Both these requirements are nearly fully met by the use of manganin for making shunts, in the form of thin strips soldered or brazed 'rigidly' into the copper blocks.

Extension of Range of Voltmeters

Voltmeter multipliers

Resistance Multiplier

In general, the range of a voltmeter can be extended by simply connecting a non-inductive resistance of appropriate value in series with the voltmeter, so that the full-scale current, responsible for producing full-scale deflection corresponding to non-extended range is kept the same.

Thus, if i is the current to produce full-scale deflection originally, r the 'internal' resistance of the voltmeter and V is the enhanced value of the voltage to be measured, then

$$i = \frac{V}{R + r}$$

where R now represents the resistance to be connected in series[1]. See Fig.6.4.

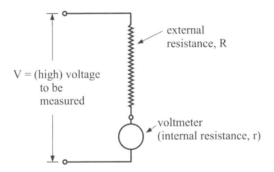

Fig.6.4 : A simple resistance multiplier

Or

$$R = \frac{V - ir}{i} = \frac{V}{i} - r$$

When used in the above simple form with DC voltmeters, it is essential that their resistance must be constant; this means that the temperature coefficient of the resistance material should be very small. Also, if the resistance is very large, appreciable heating may result even if the operating current is small. Therefore, ample provision must be made for cooling or ventilation.

Usually, multipliers may be supplied for external connection to the given voltmeter.

---

[1]In this 'capacity', R acts as a current limiter.

Operation with AC Circuits

When used in AC circuits, it is necessary that the total *impedance* of the voltmeter circuit including the external resistance must remain nearly the same for any variation of frequency. The added resistance should, therefore, have a very small inductance. For this reason, the resistance coil is wound upon flat mica strips to keep the area enclosed by the turns of wires to a minimum, reducing the enclosed flux for a given current.

A form of series resistance used for a voltmeter and housed within its case is shown in Fig.6.5.

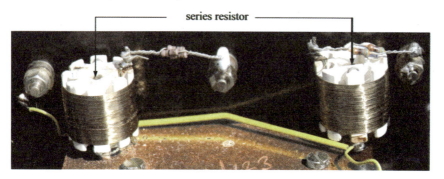

Fig.6.5 : Series resistance for a voltmeter

Multipliers using capacitors

This class of multipliers is normally suited for measurement of extremely high voltages, typically several hundred kV, and is generally described in relation to extension of range of electrostatic voltmeters which themselves can be represented by an equivalent capacitance and not just an internal (high) resistance. Even though 'simple' resistance potential dividers as shown in Fig.6.4 could still be used in (extremely) high voltage measurement, their use would suffer from the following serious drawbacks:

(a)   Whilst the ratio of the resistances in inverse proportion to the applied voltage v/s the voltmeter range will hold good for DC voltages, the same may be erroneous in the case of high AC voltage due to the effect of stray capacitance(s), usually dependent on the voltage being measured, in parallel with the resistance divider. Although it may be possible to counter the effect by using suitable shielding, the task may be quite elaborate and not fully effective.

(b)   The continuous leakage current, even if only a few micro-ampere, through the resistance may result in appreciable heating and temperature rise that may affect the reading of the voltmeter[1].

---

[1]For example, a 100 MΩ resistance across a 400 kV voltage to be measured will end up with a continuous power loss of 1.6 kW.

Capacitance Divider

The schematic of a simple capacitance potential divider to increase the range of an electrostatic voltmeter is depicted in Fig.6.6. Here, too, there may be some shunting effect due to stray capacitances, as also due to leakage resistances of the various capacitors, once again resulting in i^2R loss and heating, but the effects may usually be negligible[1].

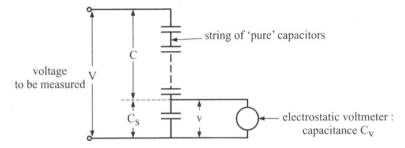

voltage to be measured V

C — string of 'pure' capacitors

C_S v   electrostatic voltmeter : capacitance C_V

Fig.6.6 : Connection of an electrostatic voltmeter in a capacitor multiplier

Theory

Let the capacitance of the electrostatic voltmeter be C_V, the voltmeter being connected across a capacitance of C_S, forming a small part of the string of capacitors such that C represents the capacitance of the remaining string.

Then, with C_S and C_V in parallel, the equivalent capacitance of the string across V is

$$C_{eq} = \cfrac{1}{\cfrac{1}{C} + \cfrac{1}{C_S + C_V}} \quad \text{or} \quad \frac{C(C_S + C_V)}{C_S + C_V + C}$$

and the reactance of the string at angular frequency of ω will be given by angular

$$X = \frac{1}{\omega C_{eq}} = \frac{C_S + C_V + C}{\omega C(C_S + C_V)} \ \Omega$$

The equivalent reactance of C_V and C_S in parallel is given by

$$X' = \frac{1}{\omega(C_S + C_V)} \ \Omega$$

---

[1]If the effect of leakage resistance(s) is not negligible and the capacitances being "lossy", the theory may be developed considering the resistance appropriately in parallel with capacitances, assuming both to be independent of the range of voltage to be measured.

The ratio of the applied to the voltage across the voltmeter is then given by

$$\frac{V}{v} = \frac{X}{X'} = \frac{C_S + C_V + C}{C} \quad \text{or} \quad 1 + \frac{C_S + C_V}{C}$$

Since C_V is variable, the voltmeter may have to be calibrated along with the multiplier. However, if the voltmeter capacitance is negligible, the ratio of the voltage will simply be

$$\frac{V}{v} = \frac{C_S + C}{C} \quad \text{or} \quad 1 + \frac{C_S}{C}$$

It is thus desirable to use a value of C_S much greater than C_V to avoid the necessity of having to calibrate the voltmeter along with the multiplier.

Comment

Extension of instrument range using shunts and multipliers has limited applications, mainly for low voltages and currents. Shunts for ammeters (DC and AC) can be useful for measuring current up to, say, 100 to 150 A. Similarly, external resistances as a form of voltage multiplier, find extensive use in pre-designed and in-built devices for DC and AC (MI type) voltmeters. For measurement of large quantities – current, voltage and power – esp. in AC systems these devices are of only limited practical use.

INSTRUMENT TRANSFORMERS

According to Indian Standards an instrument transformer is one that is intended to supply measuring instruments, meters, relays and other similar apparatus. These constitute the most common and important electromagnetic devices that have been in use in all AC power systems for measurement of (load) currents up to hundreds or thousands of ampere and (terminal) voltages of several hundred kilovolts, and even for a variety of protection requirements. Broadly divided in two categories, the *current* transformers are essentially for current measurement and *voltage* or *potential* transformers for measuring voltages, abbreviated as CTs and PTs, respectively. In both cases, the primary of the transformer carries the load or primary current, or connected across the high, terminal voltage, whilst the secondary is connected to an appropriate ammeter or voltmeter[1]. The actual current or voltage to be measured is then obtained in terms of the readings of the instruments connected across the secondary and the turns ratio of the windings, in the simplest manner.

---

[1]It may not always be that the measuring instruments are connected across the secondary; rather the output(s) of the instrument transformer secondary may be appropriately wired for a variety of purposes.

Primarily designed and developed for measurement of main current and voltage, the instrument transformers find their use in measurement of a number of other electrical quantities where 'direct' measurement of the quantity is not possible.

Chief Advantages of the Use of Instrument Transformers

1. Single-range instruments (ammeters, voltmeters, wattmeters, energy meters etc.) can be used to measure a large range of inputs.

2. The 'basic' instruments can be designed for, and be of, low, convenient full-scale range; for example, 5A for ammeters and 110 V or 230 V for voltmeters.

3. The indicating instruments can be located at a distance, remote from the sources of high voltage and current (and hence the associated EM fields), convenient for general instrumentation and measurement. This affords greater safety to operators/personnel[1].

4. Instrument transformers are now available in various range and accuracy to suit a given requirement and usually do not need much maintenance.

5. The combination of an instrument transformer and a suitable-range instrument may yield a better economy compared to the use of a single instrument of high range even if it were available.

THE CURRENT TRANSFORMER

The design, specification and other important aspects of current transformers are governed by IS 2705 (various parts), itself having its genesis in IEC Pub 185 (1987) and BS 3938:1973. The CT essentially comprises a magnetic core of laminated sheet construction and two windings wound over it. The primary winding designed to carry the main or load current is connected in series with the load circuit whilst the secondary is designed for connection to the ammeter of appropriate range as shown in Fig.6.7(a).

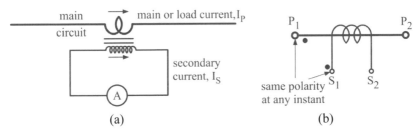

Fig.6.7 : Basic connections of a CT

---

[1]This distance, however, should not be too great to avoid common problems associated with analogue signals, viz., attenuation and pick-up en-route affecting or distorting the basic signals.

The markings of the terminals, as depicted in Fig.6.7(b), are in accordance with the IS.

When designed, constructed and used as in Fig.6.7(a), the CT is known as a "wound type" current transformer. This is commonly employed for "low-level" current measurement for use in labs etc. The CT has two primary and two secondary terminals, with proper marking with respect to polarity of the supply at a given instant.

"Bar type" CTs

This type, commonly designed to measure large to very large currents in a power system and usually permanently fitted, surrounding the bus bars which form the single-turn (bar) primary (winding), as shown schematically in Fig.6.8. The magnetic coupling of the secondary in this case is thus due to the magnetic field inside the core which itself is the result of the circular field produced by the main current around the bus bar. The "single" turn can be considered to be closing at 'infinity', electromagnetically linking with the secondary winding which is in the form of a toroid as shown. The

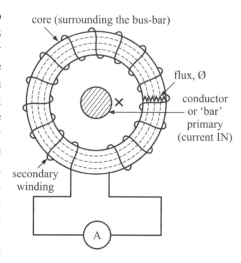

Fig.6.8 : Schematic of a bar primary CT

relatively large spacing between the bar and the secondary winding in the figure is for clarity; in practice the two may fit snugly.

Constructional Features

Core

This constitutes the most important part of a current transformer and needs much consideration. As will be seen later, the two requirements of a 'good' core are

(a) to have as low magnetising mmf as possible;

(b) the core material should be of lowest iron loss (comprising both eddy-current and hysteresis loss).

Clearly, the above two requirements point to the necessity of a very small 'no-load' current for the CT core. For this reason

   (i) the magnetic path (and thus the size) of the assembled core should be small;

(ii) the core material should be of lowest eddy-current loss coefficient;

(iii) the core must be operated at a reasonably low flux density to ensure a high relative permeability, no saturation and negligible hysteresis loss.

The flux density commonly used, corresponding to rated primary current, is about 0.1 T, a very low value that may necessiate an appreciable cross-section, increased size and cost of the CT, but necessary to minimise excitation mmf. The laminations comprising the core may be sheared from Stalloy or, now more commonly, mu-metal – a material with the relative permeability being as high as 10,000 or more and extremely low losses, but very expensive. However, when choosing CT core material, its cost or even the overall cost is seldom the criterion. The initial high cost, for a high-accuracy CT, is more than justified when considering its performance over the many years of service for which the CT is continuously connected in the system. After shearing, the laminations are usually annealed to relieve any residual mechanical stress and further bring down the iron losses.

Forms and Shapes of Core

The assembled core-form or shape may depend on whether the CT is "wound" type or bar (primary) type. For the former, two shapes are commonly employed:

(a) rectangular, using "L"-shaped laminations, with both primary and secondary windings arranged on the side limbs;

(b) shell-type, using "E" and "I" laminations, with the windings being arranged on the middle limb in this case.

For bar-primary CTs, the core is essentially circular, assembled using round, ring-shaped laminations. These variations are illustrated in Fig.6.9.

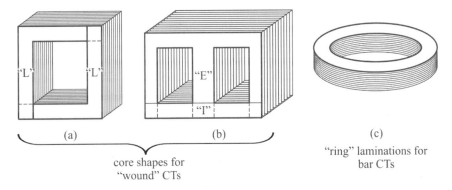

core shapes for
"wound" CTs

"ring" laminations for
bar CTs

Fig.6.9 : Types of core for wound- and bar-type current transformers

The ring-type has the important advantage of there being no (over lapping) airgap joints, a truly continuous magnetic circuit/path and hence requiring minimum possible magnetising mmf. However, the process of winding the secondary, consisting of several hundred turns and requiring looping the conductor all around the core cross-section may be tedius and time consuming[1].

In some manual measurements, by using a split core, or what is known as a "clip-on" ammeter, working on the same principle, current in the main circuit can be measured anywhere without the need of a permanent current transformer and having to disconnect the circuit for the purpose. A typical clip-on ammeter is shown in Fig.6.10.

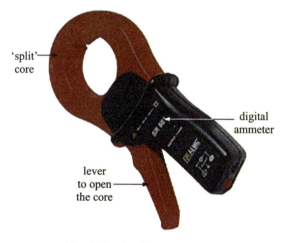

Fig.6.10 : A clip-on ammeter

Winding(s) and insulation

It is important that windings, esp. the secondary winding consisting of several hundred turns, are tightly wound to minimise leakage inductance (and hence reactance). Also, the conductor size should be adequate to keep the winding resistance and hence the copper loss as low as possible. Accordingly, No. 14 SWG copper wire with double enamel insulation or silk covering is frequently used for the secondary winding and copper strips for primary in the wound-type CTs. As revealed later in the theory of a current transformer, it is essential to achieve low impedance (and impedance angle) of the winding. Thus, a ring-type core would appear ideally suited for a bar-primary CT and is generally preferred for the purpose. During short circuits in power systems, the currents may rise by several times the value corresponding to normal operation. This may result in very high electro-mechanical forces, stressing the CT windings enormously under severe fault conditions. The CT windings must, therefore, be designed to

---

[1]However, special winding machines are now available for making toroidal coils or windings, but manual winding may still be preferred.

withstand such forces and stresses without causing any physical damage and maintaining the original electromagnetic performance. The bar-primary, ring-core construction is considered superior/satisfactory from this point of view, too, as compared to other types.

Insulation

For CTs designed for use in low-voltage systems (say, 6.6 kV or less), the windings are insulated with cotton or special insulation tape and varnish, and are baked in special ovens. For much higher voltages, the CTs may be compound-filled and housed in special tanks filled with oil for the purpose of insulation as well as cooling/ventilation. CTs designed for operation at EHV may have an appropriately designed porcelain bushing to provide insulation against the system voltage and mechanical support for conductors.

Typical Terms Associated with Current Transformers

I_p : Rated primary current of the CT, also the upper limit of the main or load current of the power system circuit in which the current is intended to be measured

I_s : Corresponding secondary current which also implies the full-scale range of the ammeter, or the device; for example, the current coil of a wattmeter or an energy meter, to be connected across the secondary winding

n_p : Number of turns of the primary winding; equal to just one for a bar-primary

n_s : Number of turns of the secondary winding

n : *Turns* ratio of the CT

$$= \frac{\text{no. of secondary turns}}{\text{no. of primary turns}} \text{ or } \frac{n_s}{n_p}$$

[Note the definition]

N : *Nominal* ratio of the CT

$$= \frac{\text{rated primary current}}{\text{rated secondary current}}$$

[both primary and secondary currents are specified on the CT 'name' plate: for example, 1000/5]

R : *Actual* current transformation ratio

$$= \frac{I_p}{I_s}$$

[In a 'perfect' CT, that is, one with no errors this would equal the nominal ratio]

RE : The ratio error of the CT

$$= \frac{\text{Nominal ratio} - \text{Actual ratio}}{\text{Actual ratio}}$$

that is,

$$= \frac{N - R}{R} \times 100\%$$

θ   : Phase angle (or *phase angle error*) of the CT, usually in degrees

[This is the angle by which the primary current (phasor) deviates from the *reversed* secondary current (phasor): is reckoned positive if the former *lags* the latter; else negative[1]].

B   : Burden of the CT

This is expressed as the volt-ampere output, VA in short, on the secondary side at the rated secondary current and specified power factor. It would thus relate to the *rated* impedance, at rated current, of the devices connected in *series* across the secondary winding[2]. This can be explained with reference to various possible connections across the secondary winding as shown in Fig.6.11

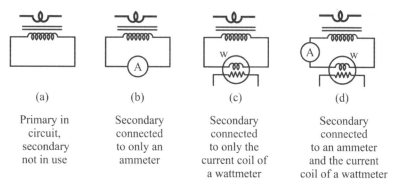

(a)   (b)   (c)   (d)

Primary in   Secondary   Secondary   Secondary
circuit,   connected   connected   connected
secondary   to only an   to only the   to an ammeter
not in use   ammeter   current coil of   and the current
                                     a wattmeter   coil of a wattmeter

Fig.6.11 : Various possible connections of secondary winding of a CT

Clearly, at a given secondary current, the burden in case (d) is much more.

Note also the case (a) where, as shown, the secondary winding *must* be shorted when not in use and primary winding is carrying current.

The ISS has defined standard burdens for which CTs are invariably designed as discussed later.

---

[1]Once again, for a 'perfect' CT, the two phasors drawn in a phasor diagram of the CT will be in phase, the phase angle error thus being zero.

[2]As will be seen, the actual burden of a CT, along with the no-load current will influence the ratio and phase-angle errors of the CT.

Class and Accuracy of CTs

Current transformers are classified on the basis of their accuracy which primarily refers to the error arising from the ratio of *actual* secondary current for a given primary or load current, known as the "ratio error" as discussed later, and the other due to the phase difference between the primary and secondary which must ideally be 180°, but is not so in practice, leading to what is called the "phase-angle error". Both BSS and ISS have specified these errors that depend on the "class" of CTs, in general pertaining to their application(s). This is further discussed later.

Based on the above, the specification of a typical CT will mainly contain the following information (not necessarily in that order) :

Rated primary current

Rated secondary current

Frequency

Rated burden

Class

and, of course, the serial no etc.

[The "construction" would suggest the type of primary]

Theory of Operation

In dealing with the theory of the current transformer, it is important to understand that the primary current of the CT *is determined by the circuit to which it is connected in series*, and not as a reflection of the secondary current which itself is determined by the load as in the case of an ordinary 1-phase transformer. A very small part of the current in the primary is used up to magnetise the core, set up a mutual flux and hence electro-magnetically induce a 'small' EMF and hence a *current* in the secondary winding which forms a 'closed' circuit, the induced EMF depending on the turns ratio[1].

---

[1]This is a kind of "reverse" action compared to an usual 1-phase transformer where a *voltage* is induced in the secondary, resulting in a current flow depending on the load.

Actual current transformation ratio

Let

I_0 = no-load current of the CT, *referred to primary*, having components

I_m = magnetising component, responsible for producing the mutual flux, Ø

I_e = component corresponding to *total* iron loss

r_s = resistance of the secondary winding

x_s = reactance of the secondary winding

E_s = induced secondary EMF

E_t = terminal voltage at secondary winding, corresponding to total impedance in series of the connected instruments[1], at a given current

δ = phase angle of the secondary current to secondary induced EMF

Δ = angle between secondary current and the terminal voltage; also the impedance angle of the secondary burden

α = angle by which I_0 leads the mutual flux

θ = phase angle (error) of the transformer

I_p = primary or load current at any time

[being the phasor sum of I_0 and reversed secondary current, taking into account the turns ratio]

I_s = secondary current (say, as read on the ammeter)

The various quantities are shown in the phasor diagram of Fig.6.12. All angles as also the current I_0 in comparison with I_p are shown much larger for clarity. All currents are rms, complex quantities and angles

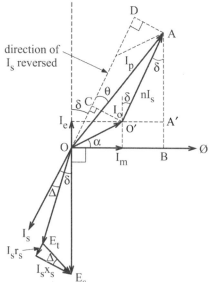

Fig.6.12 : Phasor diagram of a current transformer [not to scale]

---

[1]For example, deflection coil(s) of ammeter(s), current coil(s) of wattmeter(s) and series coil of an energy meter etc, all connected in series.

are in radians. To obtain the *actual* current transformation ratio, it is required to derive the primary current in terms of the secondary current and other quantities.

Refer to the phasor diagram

From the Δ O′A′A

$$O'A' = n\,I_s \sin\delta, \quad AA' = n\,I_s \cos\delta$$

Then, from the Δ OBA

$$I_p^2 = (I_e + n\,I_s \cos\delta)^2 + (I_m + n\,I_s \sin\delta)^2$$

$$= (I_0 \sin\alpha + n\,I_s \cos\delta)^2 + (I_0 \cos\alpha + n\,I_s \sin\delta)^2$$

Expanding and neglecting terms containing I_0^2 since $I_0 \ll I_s$ or I_p

$$I_p^2 = n^2 I_s^2 (\cos^2\delta + \sin^2\delta) + 2\,n\,I_s\,I_0 \times$$

$$(\cos\delta \sin\alpha + \sin\delta \cos\alpha)$$

or
$$I_p = \sqrt{n^2 I_s^2 + 2\,n\,I_s\,I_0 \sin(\alpha+\delta)}$$

which, to a close approximation, is

$$= n\,I_s + I_0 \sin(\alpha + \delta)$$

[since $\{n I_s + I_0 \sin(\alpha+\delta)\}^2 = n^2 I_s^2 + 2\,n\,I_s\,I_0 \sin(\alpha+\delta) + I_0^2 \sin^2(\alpha+\delta)$, the last term being too small as $I_0 \ll I_s$]

Thus, the current transformation ratio of the CT is

$$R = \frac{I_P}{I_s} = \frac{nI_s + I_0 \sin(\alpha+\delta)}{I_s}$$

or
$$\boxed{R = n + \frac{I_0}{I_s} \sin(\alpha+\delta)}$$

and differs from the (CT) turns ratio, n, by the term

$$\frac{I_0}{I_s} \sin(\alpha+\delta)$$

In terms of the magnetising and iron loss components of the no-load current, the ratio R may also be expressed as

$$R = n + \frac{I_0 \sin\alpha \cos\delta + I_0 \cos\alpha \sin\delta}{I_s}$$

$$= n + \frac{I_e \cos \delta + I_m \sin \delta^1}{I_s}, \quad \begin{array}{l} \text{since } I_0 \sin \alpha = I_e \\ \text{and } I_0 \cos \alpha = I_m \end{array}$$

Phase angle (error)

In the phasor diagram of Fig.6.12, note the position of I_s reversed with respect to I_p. The angle θ between the two, caused by

    (a)   finite magnetising or no load current, and

    (b)   the secondary burden of given power factor

is a measure of *phase* or **"phase angle"** error of the CT; for in the absence of these, both I_p and I_s (reversed) will be in phase.

From the $\triangle OCO'$, $\underline{/COO'} = 90 - (\alpha + \delta)$

and $\qquad\qquad O'C = I_0 \sin\{90 - (\alpha + \delta)\}$

$$OC = I_0 \cos\{90 - (\alpha + \delta)\}$$

Also, from the $\triangle OAD$,

$$\tan \theta = \frac{AD}{OD} = \frac{O'C}{OD},$$

since in the rectangle $CO'AD$, $O'C = AD$.

Substituting,

$$\tan \theta = \frac{I_0 \sin \{90 - (\alpha + \delta)\}}{n I_s + I_0 \cos \{90 - (\alpha + \delta)\}}$$

since in the rectangle $CO'AD$, $CD = n I_s$

or $\qquad\qquad \tan \theta = \dfrac{I_0 \cos (\alpha + \delta)}{n I_s + I_0 \sin (\alpha + \delta)}$

or, writing $\qquad\qquad \theta = \tan \theta \qquad$ since θ is very small

$$\boxed{\theta = \frac{I_0 \cos (\alpha + \delta)}{n I_s + I_0 \sin (\alpha + \delta)}} \qquad \text{in radians}$$

Expanding the numerator,

$$\theta = \frac{I_0 \left[\cos \alpha \cos \delta - \sin \alpha \sin \delta\right]}{n I_s + I_0 \sin(\alpha + \delta)}$$

---

[1]This expression immediately brings out the importance of the two components of the no load current of a CT in determining the extent of error introduced in the actual current ratio.

Since $I_0 \ll nI_s$ and $(\alpha + \delta)$ may be very small, neglecting the term $I_0 \sin (\alpha + \delta)$,

$$\theta = \frac{I_m \cos \delta - I_e \sin \delta}{n\ I_s}$$ rad, very approx[1]

and, in degrees,

$$\theta = \frac{180}{\pi} \times \frac{I_m \ \cos \delta - I_e \sin \delta}{n\ I_s} \ \text{deg}$$

The phase angle may also be expressed in terms of the primary current, I_p, as follows:

From the Δ OAD,

$$\sin \theta = \frac{AD}{OA} = \frac{I_0 \ \sin \{90 - (\alpha + \delta)\}}{I_p}$$

$$= \frac{I_0 \ \cos (\alpha + \delta)}{I_p}$$

$$= \frac{I_0 \ [\cos \alpha \cos \delta - \sin \alpha \sin \delta]}{I_p}$$

or $$\theta = \sin \theta = \frac{I_m \cos \delta - I_e \sin \delta}{I_p} \ \text{in radians}$$

Again, if the secondary (burden) power factor is near unity, δ may be very small.

Then, $I_m \cos \delta \simeq I_m$ and $I_e \sin \delta \simeq 0$

and $$\theta = \frac{I_m^2}{I_p} \ , \ \text{very approximately}$$

or, similarly, $$\theta = \frac{I_m}{nI_s}$$

---

[1]Again, for a given burden condition, the phase angle is dependent on the components of the no-load current. Note the two relative terms in the numerator and negative sign compared to the expression for current transformation ratio.

[2]This expression shows an obvious effect of I_m on θ. For I_e to be very nearly zero, an increasing value of I_m will "shift" the I_p phasor [see Fig.6.12] to the right of I_s (reversed) and hence an increase in the phase angle, θ.

Characteristics of Current Transformers

General

If a current transformer is meant to measure (the load/primary) current only, it is essential that the secondary current will be a definite and known fraction of the former, and this would simply depend on the turns ratio, n. However, in practice an error is invariably introduced on account of the magnitude of the exciting (or no load) current, however small, for a given secondary current and its power factor in the secondary circuit as given by the expression

$$R = n + \frac{I_0}{I_s} \sin(\alpha + \delta).$$

The actual current ratio is, therefore, *not constant* under all conditions of load (the secondary/primary current and burden) and frequency.

In power measurements, the important requirement is that the phasor of the secondary current shall be exactly $180°$ out of phase with respect to the primary. Again, this condition is not fulfilled since a CT, in practice, does have a phase angle (error).

In general, it can be seen that the error in the actual ratio, and hence a factor affecting the ratio error, is largely dependent upon the value of the *iron-loss* component, I_e, of the exciting/no load current, I_0, whilst the phase angle (error) depends on the *magnetising* component, I_m.

This is also clear from the simplified expressions

$$R = n + \frac{I_e}{I_s}$$

and

$$\theta = \frac{I_m}{n\, I_s}$$

if the angle δ which is usually very small, is assumed to be zero[1].

Turns compensation

In most current transformers having fairly large number of secondary turns (for example a 1000/5 A bar-primary CT with 200 secondary turns), a scheme called "turns compensation" is an easy method to make the *actual* current transformation ratio more nearly equal to the *nominal* ratio than would obtain if the *turns ratio* were equal to the nominal ratio. Thus, in the above example, the secondary turns would be equal to, say, 198 instead of

---

[1]As discussed for θ, this aspect is graphically clear from the position of phasor I_p in the phasor diagram relative to that of I_e and I_m [see Fig.6.12].

200 to achieve the desired result. Whilst there is no rigorous, theoretical basis to optimise the reduction in number of secondary turns, an indication can be obtained from the ratio. I_e/I_s for standard/rated secondary current and an 'estimate' of I_e for the same. It is observed that the phase angle error is only a little affected by a change of one or two turns in the number of secondary winding turns[1].

Effects of variation of CT parameters

A. Effect of Variation of Secondary Load Current, I_s

The expressions for both RE and PA have I_s in the denominator in the "error" terms. Hence, as I_s increases, for a given burden, 'load' power factor and constant frequency, and assuming no significant change in the no-load current, I_0, the error terms *decrease* in magnitude. This shows that, in general, both RE and PA (error) *reduce* with increasing value of CT secondary (or rather the primary) current with the former becoming less negative and the latter less positive. This trend is qualitatively depicted in Fig.6.13. The phase angle error may typically vary from $-2°$ to $+2°$ whilst the ratio error from about -1% to 1% for the secondary current variation of zero to 100%.

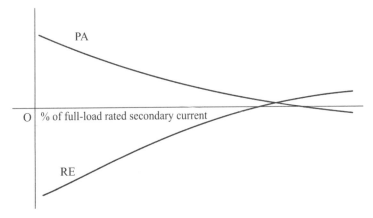

Fig.6.13 : Effect of secondary current variation on RE and PA

B. Effect of Variation of Secondary Burden (in VA)

For a given secondary current, *at a given power factor angle*, δ (or Δ), the effect of increasing burden means the 'necessity' to increase the secondary terminal voltage and, in turn, a corresponding increase of the induced EMF and consequently the core flux and flux density. The

---

[1]Clearly, it may be trickier to achieve optimum compensation if the nominal, or turns ratio, is relatively small, e.g. just 20.

exciting current, I_0, is accordingly increased, with relatively more increase of I_m as well as I_e. Also, an increase of I_0 would lead to

(i) an increase in the magnitude of I_p, thereby resulting in an increase of RE;

(ii) a 'clockwise' shift of the I_p phasor [see Fig.6.12], and thus an increase of PA.

Hence, in general, an increase of the secondary burden will result in an increase of *both* RE and PA (error), the former becoming more negative and the latter more positive. The *trend* is graphically shown in Fig.6.14 for two arbitrarily chosen values of secondary burden for a typical current transformer with respect to secondary current.

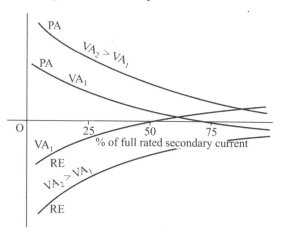

Fig. 6.14 : Effect of variation of secondary burden on RE and PA

C. Effect of Variation of Power Factor of the Secondary Burden

Assuming a given primary (or load), and corresponding secondary current, unchanged exciting current, I_0, and given burden, the effect of increasing power factor angle δ (or Δ), that is, reducing lagging power factor, is to 'move' the I_s phasor more to the left (refer to phasor diagram of Fig.6.12). This amounts to "$-nI_s$" phasor moving more and more in phase with the I_0 phasor, resulting in an increase of the I_p magnitude. The actual current ratio, I_p/I_s, is thus increased and the RE made less positive (or more negative).

The phase angle error is clearly *reduced* with reduction of power factor (that is, increase of angle δ or Δ) since $(-nI_s)$ moves more into phase with I_0. The order may be reversed in the case of leading power factor; however, leading power factor burdens are seldom encountered in practice, if ever.

D. Effect of Variation of Supply Frequency

This is more the case of academic interest since a current transformer is very rarely called upon to operate at a frequency very different from the one for which it is designed[1]. Thus, errors introduced in a CT on account of slight variation of frequency can be considered negligible.

However, consider a reduction in the supply frequency to study the effect on a CT's RE and PA. The reduction, for a given secondary current and power factor, would necessitate an *increase* in the core flux to 'maintain' a constant induced EMF, being proportional to the product of B_{max} and f, and hence the terminal voltage for a given burden. The magnetising component of I_0 would therefore increase to account for the increase in induced EMF. This would correspond to

(i) an increase in the magnitude of I_p and the actual transformation ratio, I_p/I_s, and the RE becoming less positive or more negative;

(ii) a "clockwise' shift of I_p phasor with reference to $(-nI_s)$, resulting in an increase of PA (error).

Thus a reduction of the supply frequency has nearly the same effect on the RE and PA of the CT as in the case of increased secondary burden.

Example

A current transformer of nominal ratio 1000/5 A has a total secondary impendence of $0.4 + j\,0.3\,\Omega$. At rated current, the (equivalent) primary magnetising and core loss components are 6 A and 1.5 A, respectively. The CT primary has 4 turns. Calculate the ratio error and phase angle at rated primary current if the secondary has (a) 800 turns, (b) 795 turns.

The nominal ratio of the CT $= \dfrac{1000}{5}$ or $\,200$

The (primary) magnetising current, $I_m = 6$ A

and iron-loss component, $I_e = 1.5$ A

The burden impedance is $0.4 + j\,0.3\,\Omega$

$\therefore$ the angle, $\delta = \tan^{-1}\left(\dfrac{0.3}{0.4}\right)$ or $36.87°$

If n is the turns ratio, the actual current ratio of the CT is given by

$$R = n + \frac{I_e \cos\delta + I_m \sin\delta}{I_s}$$

---

[1]Invariably, the system (or supply) frequency is never allowed to vary more than $\pm 3\%$ or less for stringent reasons such as system stability and load flow control.

and the phase angle $\quad \theta = \dfrac{I_m \cos\delta - I_e \sin\delta}{n\, I_s}$

Here

$$I_e \cos\delta + I_m \sin\delta = 1.5 \cos 36.87° + 6 \sin 36.87°$$

$$= 1.2 + 3.6 \text{ or } 4.8$$

$$I_m \cos\delta - I_e \sin\delta = 6 \cos 36.87° - 1.5 \sin 36.87°$$

$$= 4.8 - 0.9 \text{ or } 3.9$$

(a) when $\qquad n = \dfrac{800}{4}$ or 200

$$R = 200 + \left(\dfrac{4.8}{5}\right) \quad \text{or} \quad 200.96$$

$\therefore \qquad RE = \dfrac{200 - 200.96}{200.96} \times 100 \quad \text{or} \quad -0.48\%$

and $\qquad PA = \dfrac{3.9}{200 \times 5} = 0.0039 \text{ rad or } 0°13.4'$

(b) when $\qquad n = \dfrac{795}{4}$ or 198.75

$$R = 198.75 + 0.96 \text{ or } 199.71$$

$\therefore \qquad RE = \dfrac{200 - 199.71}{199.71} \times 100 \quad \text{or} \quad +0.145\%$

and $\qquad PA = \dfrac{3.9}{198.75 \times 5} = 0.00392 \text{ rad or } 0°13.5'$

Thus, the RE is considerably 'improved' using turns compensation. However, there is little effect on the PA.

Open-circuiting of the Current Transformer Secondary

It has been stressed that a current transformer is an example of a very special electromagnetic device and differs from an ordinary (1-phase) transformer in that the magnetisation of its core is as a result of *current* (or the mmf) due to the primary current which also constitutes the load current to be measured. When the secondary also carries a current, as a result of EMF induced by electormagnetic induction, the mmf of the secondary current balances the primary (winding) mmf, and this is essential. If, therefore, the secondary circuit of a current transformer is opened[1], when 'normal' current may be flowing in the primary, a very high flux density will suddenly ensue in the

---

[1]This may happen accidentally or inadvertently, esp. in a lab by students.

CT core owing to the absence of the balancing secondary mmf as the primary mmf, being a fixed quantity, is not reduced when the secondary current is zero. The result would be

(a) the high flux density, possibly driving the CT core into saturation, will induce a very high ("open circuit") voltage across the secondary winding, greatly straining the CT insulation and even posing a danger to the operator if he accidently comes in contact with the winding terminals;

(b) the high magnetising forces acting upon the core may, if suddenly removed, leave behind considerable residual magnetism in the core[1], so that the ratio and phase angle errors, obtained after such an open circuit may be far from those actually associated with the CT during normal operation.

It is thus essential that the secondary winding of a CT is neither left open nor accidentally opened *when the primary is carrying current.*

As illustrated in Fig.6.11(a), the secondary winding may simply be short circuited even if not in use for making any measurements, as it is practically a closed circuit during normal use.

Demagnetisation after an open circuit

It is important that the CT core should be demagnetised completely before putting to use in case of an open circuiting and resulting in (appreciable) residual magnetism.

There are two methods that are normally used for the purpose:

1. The first method consists of passing through the primary winding a current at least equal to (but ideally slightly greater than) that which was passing through it at the time of open circuit, the secondary winding being left open[2], and then gradually reducing this current to zero. This can ideally be arranged using a motor driven, 1-phase alternator the output of which is connected to the primary winding as shown in Fig.6.15

   After the initial setting, the motor driving the supply alternator is shut down with the alternator field still excited. As the set slows down, the alternator voltage falls gradually to zero[3], and the CT core

---

[1]This is true even in the case of high-permeability material such as Stalloy or mu-metal which have a characteristic "rectangular" B-H loop.

[2]But making sure to avoid any personnel coming in contact with the winding terminals to eliminate the risk of electric shock from the high induced EMF.

[3]Of decreasing frequency corresponding to speed of the motor at any instant.

material is made to pass through a large number of cycles, or loops, of magnetisation of gradually decreasing amplitude, finishing at zero magnetisation as shown in Fig.6.15(b).

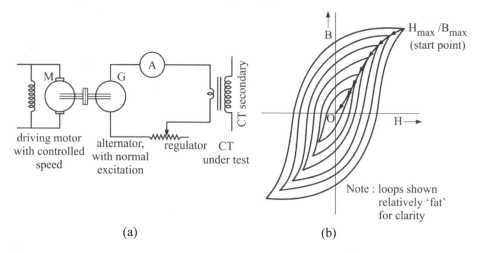

(a)   (b)

Fig.6.15 : Demagnetisation of CT core using an alternator

2. The second method consists of connecting across the CT secondary winding a resistance which is sufficiently high (several hundred ohms) so as to very nearly amount to an open circuit as at the time of its open circuiting. The "full" current, or rather the current at open circuiting, is passed through the primary winding, obtained from any suitable regulated source of AC supply, whilst the secondary resistance is gradually reduced to zero as uniformly as possible. The process means that the magnetisation of the CT core is reduced from a very high value to its near normal value in a continuous, gradual manner, resulting in complete demagnetisation of the core. The schematic of the method is shown in Fig.6.16.

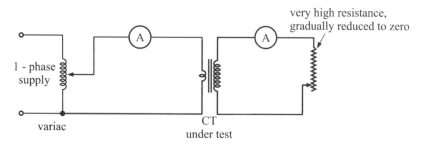

Fig.6.16 : Demagnetisation of CT core using high resistance across the secondary

'Standard' Ratios, Ratings and Permissible Errors

Ratios and ratings

Instrument transformers are designed and manufactured to conform to standard ratios, that is, the primary and secondary current values, as specified by Indian Standards Specification, the ISS, or BSS otherwise. This also applies to current transformers which are specified to be "wound" type for 'low' primary currents, say up to 100 A, but invariably "bar" (primary) type for high to very high currents.

Primary Current

CTs may be designed for primary current ratings varying from 5 A to 6000 A. The IS 2705 specifies rated primary currents for single-ratio CTs to be 10, 12.5, 15, 20, 25, 30, 40, 50, 60, 75 A, with the underlined values being preferred, *and their decimal multipliers or fractions.*

Secondary Current

The rated secondary current is specified as 1 A or 5 A.

Marking

The terminals of the primary and secondary windings are marked as depicted in Fig.6.7(b). It is specified that the terminals P_1 and S_1 will be of same polarity at any instant when the CT is connected in the circuit.

Burden

The burden of a current transformer is expressed as its output in volt-ampere (denoted by VA) *at the rated secondary current*[1], Thus, for a (specified) burden of 15 VA, the rated secondary current being 5 A, the secondary terminal voltage is to be 3 V and the (total) impedance of the connected load must not exceed 0.6 Ω. Likewise, for the rated current of 1 A, the terminal voltage can be 15 V for the same burden and the secondary impedance can be 15 Ω.

---

[1]As illustrated in Fig.6.11, the terminal voltage at the secondary will depend on the number of instruments (or rather the impedances of their 'operating' coils) connected in series; this, or their total impedance in series, then constitutes the CT burden.

Rated burdens for various classes of current transformers as per the IS 2705 are: 2.5, 5.0, 7.5, 10, 15, and 30 VA.

Errors

The CTs are, in general, rated with respect to their two main errors, viz. RE and PA, with no reference being made to *thermal rating* or permissible heating and temperature rise. The latter may become important and to be considered for very high-ratio CTs, used in 'large' power systems and constricted CT designs or installations. Limits of Error for Special Application Accuracy Classes 0.2S and 0.5S [as per IS 2705] Errors and Accuracy Permissible errors (and hence the levels of accuracy) of CTs are expressed as the "class" of the CT: the 'higher' the class, smaller must be the ratio and PA errors (and hence better the accuracy).

In the BSS, this is denoted by the letters AL, BL, A, B, C, D, implying decreasing accuracy or increasing (ratio and phase angle) errors; class AL being of highest accuracy (error not exceeding 0.1%), being suitable for testing work of the highest precision. The IS 2705 defines the standard accuracy classes as 0.1, 0.2, 0.5, 1, 3, and 5 percent., the accuracy class 0.2S and 0.5S being specifically recommended for use with energy meters designed for load current range of 50 mA to 6 A, that is, from 1% to 120% of the rated current of 5 A.

The two errors for the "S" class CTs as per standard are specified in the following tables:

Ratio error

Accuracy class	Percentage RE (in %) at percentage of rated current				
	1	5	20	100	120
0.2S	0.75	0.35	0.20	0.20	0.20
0.5S	1.50	0.75	0.50	0.50	0.50

Phase angle (error)

Accuracy class	Phase displacement in minutes at percentage of rated current				
	1	5	20	100	120
0.2S	30	15	10	10	10
0.5S	90	45	30	30	30

> **Comment**
>
> The above discussion/description essentially relates to current transformers as used in *AC* power systems for measurement, monitoring, metering and protection. This is because their basic design, working and operation is due to the laws of electromagnetic induction and EMFs induced by virtue of 'transformer action', that is the EMF given by $e = dB/dt$. There is, however, a different class of *transductors*, increasingly being required and used in DC applications, for example now prominently in HVDC transmission systems, mainly to sense the *change* of direct current in the event of faults and also for control of power flow.
>
> These devices, called transductors as against transformers are also based on the principle of electromagnetism (if not on electromagnetic induction). Since these form an important requirement of measurements, a typical such transductor for DC is described in Appendix IV, bringing out the basic principle of operation.

VOLTAGE OR POTENTIAL TRANSFORMERS

A potential transformer is characterised by two features:

(a) the design, operating principle and performance is much like a typical 1-phase (or 3-phase) 'ordinary' transformer;

(b) the extent of errors is relatively less when compared to that of a typical current transformer.

However, a potential transformer, too, has both the errors, viz. the ratio and phase angle errors, owing to the presence of a finite excitation current accounting for magnetisation of core and iron losses therein, and the secondary burden depending on its level and power factor etc.

Theory of a PT

The phasor diagram of a potential transformer under normal working condition is shown in Fig.6.17. The main quantities of interest are the secondary terminal voltage, corresponding to a given

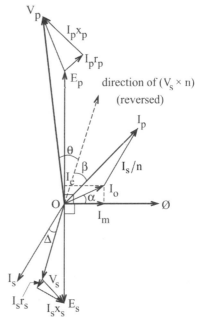

Fig.6.17 : Phasor diagram of a potential transformer

burden/secondary current and the resultant primary terminal voltage, V_p. As in the case of a current transformer, the phasors in the diagram in terms of their relative magnitude and phasor relationship are considerably exaggerated for clarity.

Let

$\emptyset$ =   mutual or working flux in the PT core

I_o =   no-load or exciting current, with

         I_m as its magnetising component, and

         I_e as the iron-loss component

n   =   turns ratio of the PT

         (= no. of primary turns/no. of secondary turns)[1]

E_s =   induced voltage in the secondary winding due to $\emptyset$

E_p =   corresponding induced voltage in the primary winding

I_s =   secondary winding current for a given burden

$I_s r_s$ =   resistive potential drop in the secondary

$I_s x_s$ =   reactive potential drop in the secondary

         (assuming a lagging power factor)

V_s =   secondary terminal voltage $\left[= E_s - \left(I_s \times \sqrt{r_s^2 + x_s^2} \right) \right]$

I_p =   primary current

         [phasor sum of I_0 and I_s reversed, being I_s/n in magnitude]

r_p, x_p =   resistance and reactance of the primary winding, with corresponding potential drops $I_p r_p$ and $I_p x_p$ in phase and phase quadrature of I_p, respectively.

V_p =   terminal voltage of the primary[2] $\left[= E_p + \left(I_p \times \sqrt{r_p^2 + x_p^2} \right) \right]$

Δ =   phase angle of the secondary load circuit, that is, the angle by which secondary current lags the secondary terminal voltage

α =   angle by which the no-load current leads the mutual flux

β =   angle between the phasors I_p and V_s (reversed)

θ =   angle between the phasors V_p and V_s (reversed)

         ["phase angle error" of the PT]

---

[1]The turns ratio of a PT is more conventionally defined as compared to a CT.

[2]Clearly, the construction and position of terminal voltages follows the pattern of the usual transformer.

Ratio expression

The expression for the actual voltage ratio V_P/V_S of a PT can be best worked out by considering the phasor diagram on a much enlarged scale, referring all the phasors on the *same* side, that is, by 'turning back' the secondary phasors on the primary side. This provides the necessary ease and clarity to obtain the various quantities. To derive the primary voltage, refer to the (enlarged) phasor diagram shown in Fig.6.18[1].

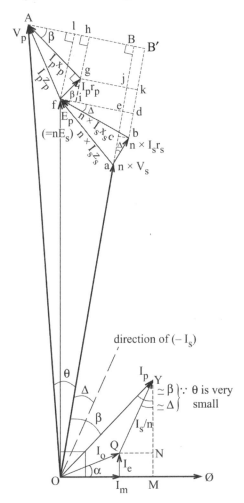

Fig.6.18 : Much enlarged phasor diagram of the PT with all phasors drawn/referred to primary side

---

[1]The secondary voltages when referred to primary side become n times; currents are *divided* by n as in an ordinary transformer. Similar reasoning holds for the winding impedance drops.

Constructional details of the phasor diagram

(a) The phasor $n \times V_s$ is extended to point B

(b) The line bB' is drawn parallel to a B

(c) The lines f d and g k are perpendiculars to bB'

(d) The lines g h and f l are perpendiculars to AB'

The rest of the construction is self-explanatory.

Further, in the triangle

1. a b c, a b is parallel to QY.

    $\therefore$ the angle c a b $= \Delta$ and a c $=$ a b cos $\Delta = n\, I_s\, r_s$ cos Δ

2. b d f, b d is parallel to a e and f d is perpendicular to c e or b d

    $\therefore$ the angle between a c and a b $=$ the angle between f b and f d, that is, Δ

    $\therefore$ $b\, d = b\, f \sin \Delta = n\, I_s\, x_s \sin \Delta = c\, e$

3. f g i, f g is parallel to OY or current I_p and i g is parallel to OB

    $\therefore$ the angle between OB and OY (or I_p), that is,

    $\beta =$ the angle between f g and i g or the angle f g i

    $\therefore$ $i\, g = f\, g \cos \beta = I_p\, r_p \cos \beta = e\, j$

4. A g h, g h $=$ j B, A g is perpendicular to f g and A h is perpendicular to i g h

    $\therefore$ the angle between f g and i g, that is, β is also the angle between A g and A h, or the angle g A h $= \beta$

    $\therefore$ $g\, h = A\, g \sin \beta = I_p\, x_p \sin \beta = j\, B$

Finally, in the triangle OAB,

$$OB = O\, a + a\, c + c\, e + e\, j + j\, B$$

and substituting from above expressions

$$= nV_s + n\, I_s\, r_s \cos \Delta + n\, I_s\, x_s \sin \Delta + I_p\, r_p \cos \beta + I_p\, x_p \sin \beta$$

$$= V_p \cos \theta$$

Now in the triangles OYM and QYN, since θ is very small, of the order of a few *minutes* in a typical potential transformer, the angle OYM is very nearly equal to β and the angle QYN almost same as Δ, with both V_p and nV_s being almost at right angles with the flux, Ø.

Therefore,

$$I_p \cos \beta = I_e + \left[\frac{I_s}{n}\right] \cos \Delta$$

and $\quad I_p \sin \beta = I_m + \left[\frac{I_s}{n}\right] \sin \Delta$ $\quad\Bigg\}$ from Δ OYM

Also $\cos \theta = 1$ (very nearly)

Hence,

$$V_p \cos \theta = V_p = nV_s + nI_s (r_s \cos \Delta + x_s \sin \Delta) + I_p(r_p \cos \beta + x_p \sin \beta)$$

and, substituting for $I_p \cos \beta$ and $I_p \sin \beta$ from above

$$V_p = nV_s + nI_s (r_s \cos \Delta + x_s \sin \Delta)$$

$$+ r_p \left[I_e + \left(\frac{I_s}{n}\right) \cos \Delta \right] + x_p \left[I_m + \left(\frac{I_s}{n}\right) \sin \Delta \right]$$

$$= nV_s + I_s \cos \Delta \left[n\, r_s + \left(\frac{r_p}{n}\right) \right]$$

$$+ I_s \sin \Delta \left[n\, x_s + \left(\frac{x_p}{n}\right) \right] + I_e r_p + I_m x_p$$

$$= nV_s + \left(\frac{I_s}{n}\right) \cos \Delta \left(n^2 r_s + r_p \right)$$

$$+ \left(\frac{I_s}{n}\right) \sin \Delta \left(n^2 x_s + x_p \right) + I_e r_p + I_m x_p$$

Dividing by V_S, the actual voltage transformation ratio of the PT is thus given by

$$\frac{V_p}{V_s} = n + \frac{\left(\dfrac{I_s}{n}\right)\left[R_p \cos \Delta + X_p \sin \Delta \right] + I_e r_p + I_m x_p}{V_s}$$

where

R_p = equivalent resistance of the PT referred to the primary = $n^2 r_s + r_p$

and X_p = equivalent reactance of the PT referred to the primary = $n^2 x_s + x_p$

Alternatively, V_p may be written

$$V_p = nV_s + n\, I_s \cos \Delta \left[r_s + \left(\frac{r_p}{n^2}\right) \right] + n\, I_s \sin \Delta \left[x_s + \left(\frac{x_p}{n^2}\right) \right] + I_e r_p + I_m x_p$$

and

$$\frac{V_p}{V_s} = n + \frac{nI_s[R_s \cos\Delta + X_s \sin\Delta] + I_e r_p + I_m x_p}{V_s}$$

where R_s and X_s now represent the equivalent resistance and reactance, respectively, of the PT referred to its secondary. Depending on the manner in which the resistances and reactances are expressed, the error in the voltage transformation of the PT is given by

$$\frac{\left(\dfrac{I_s}{n}\right)[R_p \cos\Delta + X_p \sin\Delta] + I_e r_p + I_m x_p}{V_s}$$

or

$$\frac{nI_s[R_s \cos\Delta + X_s \sin\Delta] + I_e r_p + I_m x_p}{V_s}$$

and depends, in addition to the primary and secondary impedances, on the secondary current (or the burden at given terminal voltage, V_s), and the two components of the exciting current of the PT[1].

Phase angle (error)

Referring to the (enlarged) triangle OAB of the same phasor diagram in Fig.6.18,

$$\sin\theta = \frac{AB}{OA}, \quad \text{where} \quad OA = V_p$$

Now, the side AB comprises $A\,l + l\,h + h\,B$

or $AB = (Ah - lh) + (lB' - BB')$

Further

$$Ah = Ag \cos\beta = I_p x_p \cos\beta, \text{ from the } \Delta\,A\,g\,h$$

$$lh = fi = fg \sin\beta = I_p r_p \sin\beta, \text{ from the } \Delta\,f\,g\,i$$

$$lB' = fd = fb \cos\Delta = n\,I_s\,x_s \cos\Delta, \text{ from the } \Delta\,f\,b\,d$$

$$BB' = bc = ab \sin\Delta = n\,I_s\,r_s \sin\Delta, \text{ from the } \Delta\,a\,b\,c$$

---

[1]It is seen that the expressions for the (ratio) error in a PT are somewhat 'complex' compared to the corresponding forms for a CT. This is on account of finite impedance of the primary winding that may, in fact, be considerable in proportion to that of the secondary.

$\therefore \qquad AB = I_p x_p \cos \beta - I_p r_p \sin \beta + n I_s x_s \cos \Delta - n I_s r_s \sin \Delta$

and $\quad \sin \theta = \dfrac{I_p x_p \cos \beta - I_p r_p \sin \beta + n I_s x_s \cos \Delta - n I_s r_s \sin \Delta}{V_p}$

$\qquad \qquad = \theta$, in rad since θ is very small

Alternatively, the phase angle can be expressed as

$\tan \theta = \dfrac{AB}{OB}$

$\qquad = \dfrac{I_p x_p \cos \beta - I_p r_p \sin \beta + n I_s x_s \cos \Delta - n I_s r_s \sin \Delta}{n V_s + n I_s (r_s \cos \Delta + x_s \sin \Delta) + I_p (r_p \cos \beta + x_p \sin \beta)}$

using the expression for OB derived earlier.

Neglecting all terms of OB other than $n V_s$, being too small in comparison, and since θ is very small

$\tan \theta = \theta = \dfrac{I_p x_p \cos \beta - I_p r_p \sin \beta + n I_s x_s \cos \Delta - n I_s r_s \sin \Delta}{n V_s}$, in rad

Or, substituting for $I_p \cos \beta$ and $I_p \sin \beta$ terms from the triangles OYM and QYN of Fig.6.18, as before,

$$\theta = \dfrac{x_p \left[I_e + \left(\dfrac{I_s}{n} \right) \cos \Delta \right] - r_p \left[I_m + \left(\dfrac{I_s}{n} \right) \sin \Delta \right] + n I_s x_s \cos \Delta - n I_s r_s \sin \Delta}{n V_s}$$

introducing the magnetising and iron-loss components of the no-load current.

Or, following some simplification,

$$\theta = \dfrac{\dfrac{I_s \cos \Delta}{n} X_p - \dfrac{I_s \sin \Delta}{n} R_p + I_e x_p - I_m r_p}{n V_s} \text{ in rad}$$

where R_P and X_P are the equivalent resistance and reactance of the PT referred to primary as before.

Similarly, using R_s and X_s, the equivalent resistance and reactance referred to secondary,

$$\theta = \dfrac{n I_s \cos \Delta \times X_s - n I_s \sin \Delta \times R_s + I_e x_p - I_m r_p}{n V_s}$$

$$= \dfrac{I_s}{V_s} (X_s \cos \Delta - R_s \sin \Delta) + \dfrac{I_e x_p - I_m r_p}{n V_s} \text{ rad}$$

similar to the expression(s) of actual voltage transformation ratio derived earlier[1].

Turns Compensation

It is clear from the actual voltage ratio expression that, if the PT is primarily being used for voltage measurement, the secondary voltage (or the actually measured quantity) cannot be related to the primary voltage by the turns (or nominal) ratio, n, on account of the error term

$$\frac{nI_s\left[R_s\cos\Delta + X_s\sin\Delta\right] + I_e r_p + I_m x_p}{V_s}$$

that can be compensated by adjusting the turns ratio as in the case of a CT. This can be applied with relative ease and more successfully for a PT owing to the large to very large number of primary turns (running into several thousand for PTs used in EHV systems). By adjusting *primary* turns, the actual ratio can be made almost equal to the nominal ratio for all practical values of the primary voltage.

Design Considerations for PTs

Windings

Design aspects of potential transformers are less critical and straightforward as compared to CTs owing to their conventional principle of operation, connection and usage[2]. However, since the error term depends predominantly on the exciting current, the core has to be of low reluctance, the working flux density being necessarily less than 1 tesla[3]. Since both the windings comprise large number of turns, it is vital to wind the turns as close as possible and the two windings to be placed very close to each other to minimise leakage reactance. It is customary that PTs used for voltages below 220 kV are conventional or "wound" type, that is, having both primary and secondary windings of n_1 and n_2 turns to suit the respective voltages. However, for voltages of 220 kV and 400 kV, the PTs are usually

---

[1]Once again, it is seen that the phase angle (error) of a PT typically depends on secondary burden and components of the primary no load/exciting current for a given turns ratio. With this, it is possible to make the actual ratio more nearly equal to the nominal ratio for all practical values of the primary voltage.

[2]For example, the transformer core would be a simple rectangular or shell type and not a ring type, and the windings thereby easy to wind.

[3]The core material is usually Stalloy; mu-metal being not warranted. Also, the flux density is relatively higher than used in a CT on account of higher induced voltages involved.

"capacitive" type, that is, using string of capacitors connected in series; thus, more like a capacitor-type potential divider. This affords better insulation level, low loss (and temperature rise), lesser possibility of faults and better performance.

Insulation

Owing to the high voltages encountered, esp. on the primary side, insulation, both inter-turn and overall, plays an important role in a PT. The windings of voltage transformers designed for very high voltages on the primary side (say, 6.6 kV and above) are almost invariably oil-immersed. The terminals of the primary windings are brought out through specially designed porcelain bushings which may be enormous in size protruding for metres for, say, a PT used on a 400 kV system since the entire primary voltage is borne by the bushing. [See the sketch in Fig.6.19].

PTs designed for lower voltages may simply be air insulated; that is, the windings being simply housed in a tank, or just a box.

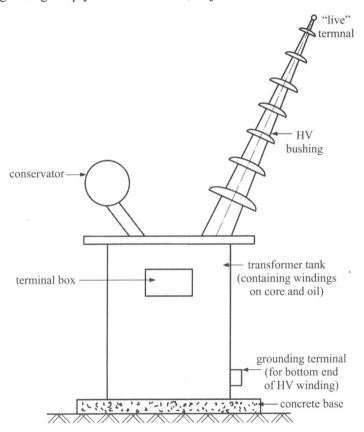

Fig.6.19 : A 1-phase PT for very high system voltage

1-phase or 3-phase

Potential transformers may be designed to operate as 1-phase units, usually for HV and EHV requirements, or three identical primary and secondary windings using a "3-phase" core as an ordinary 3-phase transformer, but with rather 'un-definable' errors on account of magnetic complexity of a 3-phase core. These may be better suited for 'low'-voltage applications.

PTs usually have small to very-small currents flowing through their windings. Hence, even with the burden comprising a high resistance coil of a voltmeter or pressure coil of a wattmeter, the heat produced due to "I^2R" loss and the consequent temperature rise may not be as critical as in a CT. The role of oil in the tank (esp. for HV PTs) is therefore more to provide insulation than cooling and ventilation.

Connections

In contrast to CTs, potential transformers must never 'experience' a short circuit and the secondary winding can be left open-circuited when not in use[1]. The total, equivalent impedance of all the instruments connected *in parallel* constitutes the PT burden, as shown in Fig.6.20, and should be within the rated or specified limits. The marking of terminals of the primary and secondary windings as recommended by the ISS is depicted in Fig.6.20(d). The terminals A and a on the primary and secondary, respectively, are of the same polarity at any instant.

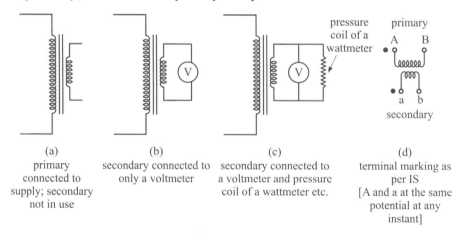

(a)	(b)	(c)	(d)
primary connected to supply; secondary not in use	secondary connected to only a voltmeter	secondary connected to a voltmeter and pressure coil of a wattmeter etc.	terminal marking as per IS [A and a at the same potential at any instant]

Fig.6.20 : Connections of a typical potential transformer and terminal marking

---

[1]Although it is not advisable for the PT secondary to remain open for long owing to the requirement of mmf 'balance'!

Standard Ratios and Burdens

PTs are commercially available in a variety of standard ratios as per IS 3156:1992, in accordance with IEC Pub 186(1987) and BS 3941:1975, and IS 12360:1988. The primary voltage for a 1-phase PT may be just 230 V, but may vary as 400 V, 3.3 kV, 6.6 kV, 11 kV, 33 kV, 66 kV, 110 kV, 132 kV, 220 kV, 400 kV. The standard also recognises 100 kV and 110 kV as primary voltages where these are already in existence, but these are not the preferred values[1]. The secondary voltage is standardised at 110 V. The IS defines rated burdens or outputs of PTs as 10, 15, 25, 75, 100, 150, 200, 300, 400, and 500 VA, with the underlined values as the preferred ones, at 0.8 p f (lag) and rated system frequency of 50 Hz.

Accuracy

As per the BS, PTs are divided into six classes – A, B, C, D, AL, and BL – with respect to their accuracy and performance. The limits of accuracy range from 0.5% ratio error and 20 minutes phase angle (error) to 2% and 30 minutes for class A, B, and C; for class D, the errors are defined as 5% ratio error and 30 minutes phase angle. For classes AL and BL, ratio error is limited to 0.25% and 0.5% whilst the phase angle is allowed to be not more that 10 minutes and 20 minutes, respectively[2].

According to the IS, for measuring voltage transformers, the accuracy class is designated by the highest permissible percentage voltage error at *rated voltage and with rated burden* prescribed for the accuracy class concerned. Thus, the standard accuracy classes for PTs are specified to be 0.1, 0.2, 0.5, 1.0, and 3. The limits of the ratio and phase-angle errors for these classes are given in Table 6.1.

Table 6.1: Limits of voltage errors and phase displacement for various PTs

Class	Percentage voltage (ratio) error	Phase displacement (minutes)
0.1	± 0.1	± 5
0.2	± 0.2	± 10
0.5	± 0.5	± 20
1.0	± 1.0	± 40
3	± 3.0	–

[1]With the proposed up-gradation of the AC transmission voltage to 765 kV in the country, the PTs will have to be designed for a (3-phase) primary voltage of 765 kV, the construction being capacitive type.

[2]Even with an error of just 0.25%, the absolute voltage error in a 400 kV/110 V PT may be as high as 1000 V on the primary side.

PT Characteristics

The limits (of variation) of ratio error and phase-angle errors in PTs are less critical and 'overall' performance more satisfactory as compared to typical CTs. Fig.6.21 illustrates this feature, showing qualitative plots of typical ratio and phase angle for a voltage transformer, with respect to various values of secondary voltage. Clearly, there is very little change in the errors with change of (secondary) voltage[1].

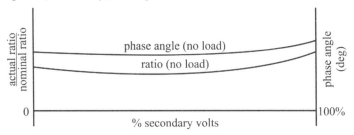

Fig.6.21 : Typical, general variation of ratio and phase angle errors of a potential transformer at 'no load', ($I_s = 0$)

[The actual to nominal ratio may vary from about 1.0 to 1.006 whilst the PA limits may be 0 to 0.15^0]

Effect of Variation of Secondary Voltage: no load and on load [$I_s = 0$; $I_s \neq 0$]

At no load, the secondary terminal voltage will be the same as the induced EMF. Now, for example, if the applied primary voltage is reduced, the flux density in the core will correspondingly be reduced, with some reduction in the exciting current, I_0. This means a reduction in the primary current, the primary impedance drop and ultimately primary terminal voltage and hence the ratio and phase angle errors. The effect is depicted in the phasor diagram of Fig.6.22.

With the $I_p x_p$ drop being at right angles with I_0, the primary terminal voltage, V_p, is very nearly in phase with $-nV_s$, or $-nE_s$, as shown and therefore resulting in a very small phase angle (error). Similar reasoning can be

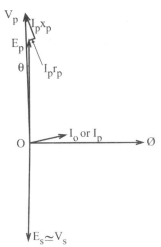

Fig.6.22 : Phasor diagram of a PT when $I_s = 0$

---

[1]Cf. the corresponding curves of errors for a CT, plotted against secondary current [see Fig. 6.13].

considered when the secondary is carrying a current, I_s (usually very small on account of high impedance of the burden) and when the secondary terminal voltage varies as indicated by Fig.6.21.

Effect of Variation of Secondary Burden

With the full rated voltage applied to the primary, the secondary voltage may also be proportionally same for a given burden in VA. Therefore, any *increase* in the PT burden would mean an increase of the secondary current[1]. Thus, if the initial burden were 15 VA and the rated secondary voltage 110 V, then the secondary current will be (15/110) or 0.136 A. If the burden is increased to, say, 50 VA, the secondary current will increase to (50/110) or 0.45 A.

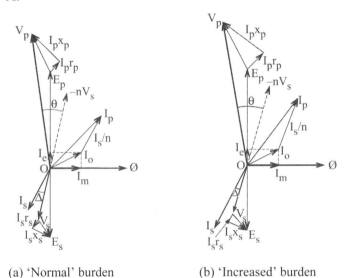

(a) 'Normal' burden   (b) 'Increased' burden

Fig.6.23 : Phasor diagrams corresponding to increase in a PT burden

[Not to scale]

V_s, Δ, I_e	: assumed unchanged
I_s, E_s, $\emptyset$, I_m, $I_s Z_s$, I_p, $I_p Z_s$	: increased with increase of burden
Net result	: both actual voltage ratio and phase angle increase

Hence, with increased burden and a corresponding increase of the secondary current and thereby the I_p current, assuming exciting current to

---

[1]Since the burden in a PT comprises voltmeter and wattmeter pressure coil etc. *connected in parallel*, the more such parallel branches, the more the *total* secondary current.

remain unaltered, the primary impendence drop and therefore V_p will increase in magnitude. The effect will therefore be to increase the actual voltage ratio[1], V_P/V_S.

The increased primary impedance drop will result in the 'shift' of the V_p phasor to the left with respect to the nV_s (reversed) and hence an increase of the phase angle error. This is illustrated by the phasor diagrams of Fig.6.23 (a) and (b).

Effect of Change of Power Factor of Secondary Burden

If the power factor of the secondary burden were reduced, that is, the power factor angle Δ increased (assuming a lagging power factor), I_p would become more nearly in phase with I_0, assuming I_s to remain unchanged, thereby resulting in an increase of I_p and the primary impedance drop. This would increase the magnitude of V_p for a given secondary voltage and hence an increased voltage ratio, V_P/V_S. However, the I_pZ_p voltage drop as well as V_p are slightly shifted to right owing to a downward shift of I_p. This results in a reduction of the phase angle. Similarly, for an increase of the power factor, the phase angle may show an increase[2].

Effect of Variation of Frequency

For a given applied or primary voltage, a *reduction* of supply frequency, for example, may necessitate an increase of core flux with a corresponding increase of I_m and therefore I_0. However, since a slight increase of exciting current (owing to only a slight permissible variation of frequency) does not influence a PT errors appreciably, the effects of variation of supply frequency are not so significant as in the case of a current transformer. In general, a slight increase of I_0, and hence I_p, may be compensated by a reduction in the value of primary winding reactance and therefore the impedance drop and the magnitude of V_p, thus maintaining a nearly constant voltage ratio.

---

[1]If it is assumed that secondary terminal voltage, V_s, were the same, an increase of secondary current will increase the induced EMF, E_s, and also the primary induced EMF, E_p, leading to enhanced V_p. Also, the exciting current may increase slightly to account for increased induced EMFs and hence I_p.

[2]Even though not encountered in practice, the effect of change of a *leading* power factor of the burden may naturally be 'reverse' of that for the lagging power factor.

Comment

'DC voltage transformer'

Measurement of high to very high DC voltages is commonly encountered in HVDC transmission systems, similar to high direct current measurement. A device usually employed for the purpose is DCVT, or Direct Current Voltage Transductor. A typical such device and its principle of operation is described in Appendix IV.

TESTING OF INSTRUMENT TRANSFORMERS

The accuracy, or rather the errors and inaccuracy, of current and potential transformers can seriously influence the measurement of primary current (which may be in kilo-ampere) and primary voltage (of the order of several thousand or hundred thousand volts in large power systems) in terms of secondary quantities which are *actually* measured using ammeters and voltmeters of given accuracies[1]. Similar arguments hold when instrument transformers are utilised for control and protection of huge inter-connected power systems.

It is, therefore, essential that CTs and PTs are tested rigorously following the manufacture and later at regular intervals and their performance ascertained for 'accurate' measurement of primary quantities.

Testing of Current Transformers

As seen, CTs have a wide variation of errors, both ratio and phase angle, over the range of secondary load and burden and hence require frequent testing and calibration. The actual transformation ratio is required to be determined within one-tenth of 1 per cent. and phase angle to within one twentieth of a degree (or a few minutes). This would call for elaborate means of testing.

There are two types of method of testing CTs:

(a) absolute methods

(b) comparison methods

In the absolute methods, the transformer errors are determined in terms of physical constants of the testing circuit, viz., resistance, inductance or capacitance. In contrast, in the comparison methods the errors of the CT under test are compared with those of a 'standard' current transformer of precision type, having a high degree of accuracy.

---

[1]When CTs and PTs are used to measure *integrated* electrical quantities; for example, energy using energy meters over a long period, or for relaying in power systems for protection, the errors of instrument transformers used in conjunction with the meters and relays become all the more significant.

An absolute method[1]

The test set-up for this method is shown in Fig.6.24.

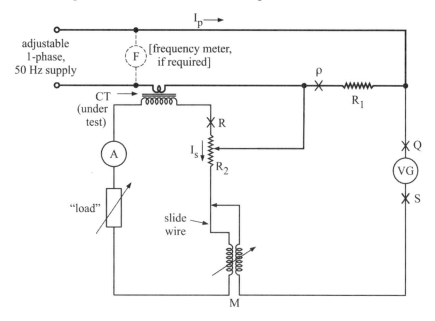

Fig.6.24 : Circuit for testing of current transformer using mutual inductance

Referring to the figure

CT   : current transformer under test

R_1, R_2: non-inductive, known, precision resistors

M   : adjustable, known, air-core precision-type mutual inductance

VG   : vibration galvanometer, tuned to supply frequency

The resistor R_1 is chosen so as to 'safely' carry the primary current. For satisfactory results, it is desirable that the resistor R_2 is such as to make R_2/R_1 approximately equal to nominal ratio of the CT (under test).

Test Procedure

With the resistor R_1 carrying a chosen primary current (rated or a fraction of it), adjust (in steps, if necessary) the resistor R_2, both the fixed and slide-wire part, and the mutual inductance M such that null is obtained in the vibration galvanometer. Note the values of R_1, R_2 and M, as also the supply frequency.

---

[1]Also known as "Mutual Inductance Method" owing to the use of a standard mutual inductance and its inclusion in the results.

Theory

At balance, or null in the VG, the PD on one side of the galvanometer, $PD_{(P-Q)}$, must equal the PD across the variable resistor R_2 and M in series, that is, $PD_{(R-S)}$ *both in magnitude and phasor relationship.*

or $$\dot{PD}_{(P-Q)} = \dot{PD}_{(R-S)}$$

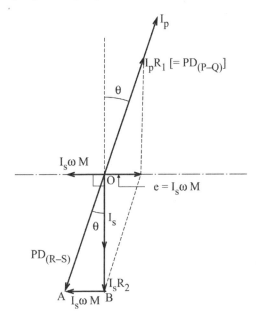

Fig.6.25 : Phasor diagram of voltages for CT testing

Referring to phasor diagram of Fig.6.25,

$$\dot{PD}_{(P-Q)} = \dot{I}_p R_1$$

and $$\dot{PD}_{(R-S)} = I_s R_2 + j\, I_s\, \omega\, M$$

where $I_s\, \omega\, M$ is the EMF induced in the secondary of the mutual inductance due to current I_s in its primary, and in phase quadrature with the current.

From the triangle OAB

$$\tan\theta = \frac{AB}{OB} = \frac{I_s\omega M}{I_s R_2} = \frac{\omega M}{R_2}$$

or θ, the phase angle of the CT $= \tan^{-1}\left(\dfrac{\omega M}{R_2}\right)$.

Note that if there were no phase angle error the potential drop across M will be zero and both phasors I_s and I_p will be in phase opposition.

Also, from the same triangle,

$$\cos\theta = \frac{OB}{OA} = \frac{I_s R_2}{PD_{(R-S)}} = \frac{I_s R_2}{PD_{(P-Q)}} = \frac{I_s R_2}{I_p R_1}$$

Since θ is very small, $\cos\theta = 1$ (approximately) and

$$\frac{I_s R_2}{I_p R_1} = 1$$

and the (actual) current ratio

$$\frac{I_p}{I_s} = R = \frac{R_2^1}{R_1}$$

Thus, the CT errors are obtained in terms of the test circuit constants R_1, R_2 and M and hence an accurate knowledge of these is essential. Also, the supply frequency figures in the calculation of phase angle and must be measured/known accurately. The impedance of the secondary of the CT, shown as a variable load, is actually the burden of the CT (and variable "load" current) for which the test is performed. This impedance, however, comprises the total resistance R_2 and impedance of the primary of M, in addition to the set value of the impedance in "load". These values must, therefore, be known to "define" the actual burden of the CT under test.

Method by comparison

Silsbee's Deflection Method

This method, due to Silsbee is based on comparison of the performance of the CT under test with that of a standard current transformer *having the same nominal ratio* such that the two errors of the CT being tested are obtained in terms of those of the standard CT and must be accurately known. The schematic of the connection for the testing is shown in Fig.6.26.

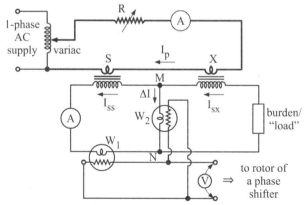

Fig.6.26 : Circuit diagram for testing of a CT by comparison method

---

[1]This emphasises the choice of R_2 v/s R_1 or the ratio R_2/R_1.

In the circuit, S and X represent the standard and CT under test, respectively. The primary windings of the two CTs are connected in series, together with an adjustable resistor R and an ammeter to measure the primary or "load" current, to an adjustable single-phase AC supply. Both the CTs thus carry a common primary current. The two secondaries are also connected in series as shown along with a precision ammeter and current coil of a precision-type wattmeter, W_1. This series circuit also carries a 'suitable' burden. The current coil of the other wattmeter, W_2, is connected across nodes M and N in a differential mode so that the current through this coil is, ΔI, the *difference* of the two secondary currents, I_{sx} and I_{ss} as shown. The pressure coils of the two wattmeters are connected to the rotor of a phase shifter[1] so that the phase of the voltage with respect to the current I_{ss} can be adjusted as required.

Test Procedure

After adjusting the primary current the two steps of testing are as follows:

A. Adjust the phase shifter supply (by rotational movement of the rotor clockwise or anticlockwise) such that wattmeter W_1 reads zero. Note the reading of wattmeter W_2. Let this reading be W_{2_q}.

B. Again adjust the phase shifter supply such that now the wattmeter W_1 reads a maximum, given by W_{1_p}. Note the reading of the wattmeter W_2.

Let this be W_{2_p}.

Theory

In the first case when wattmeter W_1 reads zero, the current I_{ss} through its current coil must be in phase quadrature with V, the voltage across its pressure coil. Then the reading W_{2_q} of wattmeter W_2 is given by

$$W_{2_q} = V \times \Delta I_q$$

---

[1]A phase shifter, a special electromagnetic device discussed later in detail (see Chapter VII: Measurement of Power), essentially comprises a stator having a 3-phase distributed winding (similar to an induction motor) excited from a 3-phase balanced supply. The rotor, wound with a single-phase winding, is capable of rotation through 360°, its position being able to be read (with reference to a datum) on a graduated scale, fitted on the stator. The EMF induced in the rotor is *constant for a given excitation of the stator*, but its *phase angle* with reference to the applied voltage (the 1-phase supply above) would depend on the angular position with respect to the datum marked on the scale.

where ΔI_q is the in-phase component of ΔI, the 'difference' current through current coil of wattmeter W_2 at any time. Therefore, from the phasor diagram of Fig.6.27,

$$W_{2_q} = V \times \Delta I_q = V \times CA \text{ (from the triangle OCA)}$$

or
$$W_{2_q} = V \times I_{sx} \times \sin(\theta_x - \theta_s)$$

[from the triangle OAC in which $OA = I_{sx}$ and angle $AOC = (\theta_x - \theta_s)$]

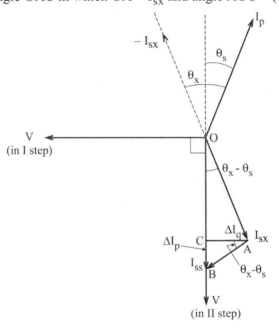

Fig.6.27 : Phasor diagram pertaining to CT testing by comparison method

Referring to the test results of the second step, the voltage V being in phase with the current I_{ss}, the two wattmeters readings are given by[1]

$$W_1 = W_{1_p} = V \times I_{ss}$$

and
$$W_2 = W_{2_p} = V \times \Delta I_p$$

where ΔI_p is now the in-phase component of the difference current ΔI.

Substituting for ΔI_p with reference to the phasor diagram in which
$$\Delta I_p = OB - OC$$
$$= I_{ss} - I_{sx} \cos(\theta_x - \theta_s) \text{ from the } \Delta \text{ OAC,}.$$

---

[1]In both the cases the voltages across the pressure coils of wattmeters W_1 and W_2 are same in *magnitude*, but in phase quadrature themselves as shown in Fig.6.27.

$$W_{2_p} = V \times [I_{ss} - I_{sx} \cos (\theta_x - \theta_s)]$$

$$= V \times (I_{ss} - I_{sx}),$$

writing $\cos (\theta_x - \theta_s) = 1$

since $\theta_x - \theta_s$ is very small

$$= V \times I_{ss} - V \times I_{sx}$$

$$= W_{1_p} - V \times I_{sx}$$

or $\quad V \times I_{sx} = W_{1_p} - W_{2_p}$

If R_x and R_s denote the actual current ratios of the CT under test and the standard CT, respectively, then

$$R_x = \frac{I_p}{I_{sx}} \quad \text{and} \quad R_s = \frac{I_p}{I_{ss}}$$

and

$$\frac{R_x}{R_s} = \frac{I_{ss}}{I_{sx}} = \frac{V \times I_{ss}}{V \times I_{sx}} = \frac{W_{1_p}}{W_{1_p} - W_{2_p}} = \frac{1}{1 - \dfrac{W_{2_p}}{W_{1_p}}}$$

or

$$\frac{R_x}{R_s} = 1 + \frac{W_{2_p}}{W_{1_p}} \quad \text{(approximately)}$$

$\therefore$

$$R_x = R_s \times \left[1 + \frac{W_{2_p}}{W_{1_p}}\right]$$

So that, knowing R_s, R_x can be deduced.

Phase angle

From the results of the first step

$$\sin (\theta_x - \theta_s) = \frac{W_{2_q}}{V \times I_{sx}}$$

and from those of the second step

$$\cos(\theta_x - \theta_s) = \frac{V \times I_{ss} - W_{2_p}}{V \times I_{sx}}$$

$$\therefore \qquad \tan(\theta_x - \theta_s) = \frac{W_{2_q}}{W_{1_p} - W_{2_p}}$$

or $\qquad (\theta_x - \theta_s) = \dfrac{W_{2_q}}{W_{1_p} - W_{2_p}}$ since $(\theta_x - \theta_s)$ is very small

or, neglecting W_{2_p} being much smaller[1],

$$(\theta_x - \theta_s) = \frac{W_{2_q}}{W_{1_p}}$$

Hence $\qquad \theta_x = \theta_s + \dfrac{W_{2_q}}{W_{1_p}}$ rad

$$= \frac{180}{\pi}\left(\theta_s + \frac{W_{2_q}}{W_{1_p}}\right) \text{ deg}$$

In this method, too, the burden in the secondary of the current transformer under test includes, in a rather complex manner, the impedances comprising the current coils of both the wattmeters and the ammeter. However, the burden is common to both the current transformers.

Testing of Potential Transformers

Tests for the determination of the two errors of a potential transformer may also be classified into absolute or comparison methods. However, potential transformers, by and large, are associated with relatively less degrees of both the ratio and phase angle errors as compared to current transformers. Their testing and calibration is thus not as critical as for the latter. Further, owing to extremely high voltages encountered on the primary side, conduction of the tests is not easy and great skill is necessary. The apparatus connected on the high voltage side is also prone to introducing errors resulting from distributed capacitive effects.

An absolute method

As an illustration, a method in which the voltage of the transformer secondary is compared with a suitable fraction of the applied primary voltage by the use of a special resistance potential divider is described here.

---

[1]It is assumed that the standard CT specification would be similar; for example, its nominal ratio to be the same as that of the CT under test. Thus, the secondary currents I_{ss} and I_{sx} would nearly be equal and their difference, ΔI, or the in-phase component of it, would be very small.

The magnitudes and phase of the difference between these two voltages are measured and the PT errors are derived therefrom. The circuit arrangement for the method is shown in Fig. 6.28.

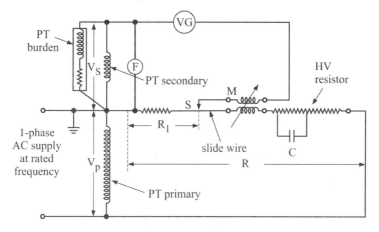

Fig.6.28 : Circuit for PT testing

The burden for which the PT is to be tested is connected across the secondary winding, comprising a resistance and inductive reactance and thus a certain power factor. The primary is connected to the applied voltage at rated frequency. One of the secondary winding terminals is joined to one of the primary terminals which is grounded. A non-inductive, very high resistance is connected across the primary winding as shown to constitute a potential divider comprising a fixed resistor in series with a slide wire (resistance), S. The upper end of the potential divider consists of a *shielded* high voltage resistor in series with the primary of a variable mutual inductor, M. The inductance of the primary of the mutual inductor may introduce a phase error; to compensate this, a small part of the HV resistor is shunted by a capacitor. VG is a vibration galvanometer, tuned to the supply frequency. The resistances are so chosen such that the nominal ratio of the PT is equal to R/R_1, R_1 being its value at balance.

At balance, as indicated by null in the vibration galvanometer, the values of R_1 and M are recorded. It is also necessary to accurately note the supply frequency.

A phasor diagram corresponding to the balanced condition is shown in Fig.6.29, in which $I = V_p/R$ is the current in the potential divider.

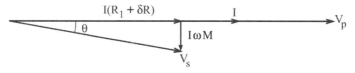

Fig.6.29 : Phasor diagram for PT testing

The transformer secondary voltage V_s is very approximately given by

$$V_s = I (R_1 + \delta R)$$

neglecting the mutual inductor drop, $I \omega M$, since θ is very small and where δR represents the small part of the slide wire resistor in series with R_1 at the time of balance. Then, substituting for I,

$$V_s = \frac{V_p}{R}(R_1 + \delta R)$$

and the actual voltage ratio of the PT

$$\frac{V_p}{V_s} = \frac{R}{(R_1 + \delta R)}$$

$$\text{The ratio error} \quad = \frac{\text{nominal ratio} - \text{actual ratio}}{\text{actual ratio}}$$

$$= \frac{\dfrac{R}{R_1} - \dfrac{R}{(R_1 + \delta R)}}{\dfrac{R}{(R_1 + \delta R)}}$$

$$= \frac{(R_1 + \delta R)}{R_1} - 1 = \frac{\delta R}{R_1}$$

Thus, the slide wire can be calibrated directly in terms of the ratio error.

Phase angle

From the phasor diagram

$$\tan \theta \simeq \theta = \frac{I \omega M}{I \times (R_1 + \delta R)} \simeq \frac{\omega M}{R_1}$$

and the mutual inductor can be calibrated directly in terms of phase angle of the PT.

WORKED EXAMPLES

1. Calculate the resistance of the shunt which when connected across a milliammeter designed for a current of 5 mA for full-scale deflection to be converted into an ammeter of 0-10 A. The (moving) coil has an internal resistance of 20 Ω.

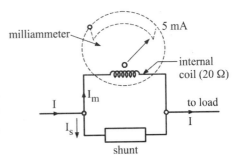

The configuration of the circuit is as shown in the figure.

For meter to show full-scale deflection with shunt when carrying 10 A load current, the current through the meter I_m will still be only 5 mA or 0.005 A. The current diverted to the shunt, I_s, will be

$$I_s = 10 - 0.005 \text{ or } 9.995 \text{ A.}$$

However, the PD across the shunt will be the same as that across the meter coil.

PD across the meter $= 20 \times 0.005$ or 0.1 V

This must equal $R_s \times 9.995$ where R_s is the shunt resistance.

$$\therefore \qquad R_s = \frac{0.1}{9.995} = 0.010005 \ \Omega$$

2. The circuit of a milliammeter consists of a swamp resistance of 3.8 Ω in series with a former-less coil. The coil has sides 1.5 cm long and is 1 cm wide, consisting of 60 turns of a thin wire having a resistance of 0.4 Ω/m. If a current of 5 mA produces full-scale deflection, calculate the value of a shunt required to convert the instrument into an ammeter of full-scale range of 0.1 A

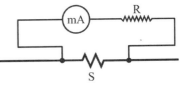

The circuit for the instrument is as shown where R denotes the swamp resistance and S the required shunt.

Coil resistance, $\qquad r = \dfrac{2 \times (1.5+1) \times 60}{100} \times 0.40 = 1.2 \ \Omega$

PD across the milliammeter in series with the swamp, R, with 5 mA through it

$$= 5 \times 10^{-3} \times (1.2 + 3.8) = 25 \text{ mV}$$

When the shunt is connected as shown, this is also the PD across the shunt when it is carrying a current of $(100 - 5)$ or 95 mA.

$$\therefore \text{ shunt resistance, } S = \frac{25 \times 10^{-3}}{95 \times 10^{-3}} = 0.263 \ \Omega$$

3. A moving-coil instrument has a resistance of $5 \ \Omega$ between terminals, and its full-scale deflection is obtained with a current of 0.015 A. The instrument is to be used with a manganin shunt to measure 100 A at full scale. Calculate the error caused by a 10°C rise in temperature when

(a)   the internal resistance of $5 \ \Omega$ is due to copper only;

(b)   a $4 \ \Omega$ manganin swamping resistor is used in series with a copper resistor of $1 \ \Omega$.

Assume the temperature coefficient of copper as 0.004 ohm/ohm/$^\circ$C and that of manganin 0.00015 ohm/ohm/$^\circ$C.

(a)   When the whole of the $5 \ \Omega$ constitutes the resistance in the copper of the instrument coil and leads as shown in the figure,

instrument current   = 0.015 A

$\therefore$ shunt current $= 100 - 0.015$

= 99.985 A

PD across the shunt = 0.015×5

= 0.075 V

$\therefore$ shunt resistance $= \dfrac{0.075}{99.985}$

= 0. 00075 Ω (manganin wire)

Then the shunt resistance at 10°C temperature rise

$$= 0.00075 \times (1 + 10 \times 0.00015)$$

$$= 0.000751 \ \Omega$$

The instrument resistance after 10°C rise in temperature

$$= 5 \times (1 + 10 \times 0.004)$$

$$= 5.2 \ \Omega$$

The instrument current corresponding to 100 A input, in inverse proportion to the resistances forming the parallel circuit, A and B will be,

$$= \frac{0.000751}{5.200751} \times 100 = 0.01444 \text{ A}$$

and the instrument reading $= \dfrac{100}{0.015} \times 0.01444 = 96.27 \text{A}$

∴ percentage error due to temperature rise

$$= 100 - 96.27 \text{ or } 3.73\% \text{ (low)}$$

(b) The connections in this case are as shown

Resistance of the instrument circuit after 10°C rise in temperature

$=1 \times (1 + 10 \times 0.004) +$

$\quad 4 \times (1 + 10 \times 0.00015)$

$= 5.046 \ \Omega$

The shunt resistance after 10°C rise in temperature is still the same, that is, 0.000751 Ω

Instrument current corresponding to 100 A input

$$= \frac{0.000751}{5.046751} \times 100 = 0.01488 \text{ A}$$

and the instrument reading $= \dfrac{100}{0.015} \times 0.01488 = 99.2 \text{ A}$

∴ percentage error due to the temperature rise
$$= 100 - 99.2 \text{ or } 0.8\% \text{ (low)}$$

[This shows the great advantage of using a swamping resistance in parallel with the shunt].

4. A bar-primary CT with 300 turns on the secondary has a secondary burden comprising a non-inductive resistance of 1.5 Ω and a reactance of 1Ω, including the transformer winding. With 5 A secondary current, the magnetising mmf is 100 A and the iron loss is 1.2 W, referred to primary.
   Determine the ratio error and phase angle (error) of the CT.

Secondary burden impedance is

$$Z_s = \sqrt{1.5^2 + 1^2}$$

$$= 1.8 \ \Omega$$

∴ for the secondary,

$$\cos \delta = \frac{1.5}{1.8} = 0.833$$

and

$$\sin \delta = \frac{1}{1.8} = 0.555$$

Induced EMF in the secondary for the current of 5 A,

$$E_s = 5 \times 1.8 = 9 \text{ V}$$

and the corresponding EMF in the primary

$$E_p = \frac{E_s}{n} = \frac{9}{300} = 0.03 \text{ V}$$

The loss component of the no-load current

$$I_e = \frac{\text{iron loss}}{E_p} = \frac{1.2}{0.03} = 40 \text{ A} \quad \text{(referred to primary)}$$

and the magnetising component

$$I_m = \frac{\text{primary mmf}}{\text{primary no. of turns}}$$

$$= \frac{100}{1} \quad \text{or } 100 \text{ A (referred to primary)}$$

Now the actual current transformation ratio is given by

$$R = n + \frac{I_e \cos \delta + I_m \sin \delta}{I_s}$$

$$= 300 + \frac{40 \times 0.833 + 100 \times 0.555}{5}$$

$$= 317.6$$

∴ % ratio error, RE $= \dfrac{\text{nominal ratio} - \text{actual ratio}}{\text{actual ratio}} \times 100$

$$= \frac{300 - 317.6}{317.6} \times 100 = -5.54\%$$

The phase angle (error) is given by

$$\theta = \frac{180}{\pi} \left(\frac{I_m \cos \delta - I_e \sin \delta}{n I_s} \right) \text{ deg}$$

$$= \frac{180}{\pi} \left(\frac{100 \times 0.833 - 40 \times 0.555}{300 \times 5} \right) \text{ deg}$$

$$= 2.34°$$

5. An 8/1 current transformer has an accurate ratio when the secondary is short circuited. The inductance of the secondary is 60 mH and its resistance is 0.5 Ω. When the secondary load has a resistance of 0.4 Ω and inductance of 0.7 mH, estimate the current ratio and phase-angle error of the CT. Assume the supply frequency as 50 Hz, no iron loss and magnetising current to be 1% of the rated primary current.

The net burden on the CT is

$$R = 0.5 + 0.4 = 0.9 \ \Omega$$

$$L = 0.06 + 0.0007 \ \text{or} \ 0.0607 \ \text{H}$$

and

$$X = 0.0607 \times 2\pi \times 50 = 19 \ \Omega$$

∴

$$\delta = \tan^{-1}\left(\frac{19}{0.9}\right) = 87.3°$$

$$\cos \delta = 0.047 \ \text{and} \ \sin \delta = 0.9988$$

Using the expression for actual current ratio

$$= n + \frac{I_e \cos \delta + I_m \sin \delta}{I_s}$$

$$= 8 + \frac{0 \times 0.047 + 0.08 \times 0.9988}{1}$$

[since $I_e = 0$ and $I_m = 1\%$ of 8 A or 0.08 A]

$$= 8.08$$

Phase angle

$$\theta = \frac{180}{\pi} \times \frac{I_m \cos \delta - I_e \sin \delta}{n \ I_s} \ \text{deg}$$

$$= \frac{180}{\pi} \times \frac{0.08 \times 0.047}{8}$$

$$= 0.027° \quad \text{or} \quad 0° \ 1.62'$$

6. A current transformer of nominal ratio 1000/5 A is operating with a total secondary impedance (or burden) of $(0.4 + j \ 0.3) \ \Omega$. At rated current, the components of the primary current associated with the core magnetisation and core-loss effects are 6 A and 1.5 A, respectively. The primary has 4 turns. Calculate the ratio error and phase angle at rated primary current if the secondary has (i) 800 turns, (ii) 795 turns.

This problem is best solved by referring to the phasor diagram of the CT as shown.

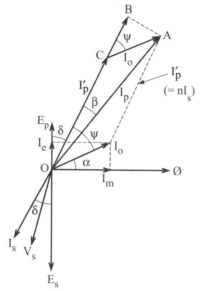

The burden of the CT is $0.4 + j\,0.3\ \Omega$

$\therefore \qquad\qquad \delta = \tan^{-1}\left(\dfrac{0.3}{0.4}\right) = 36.87°$

$\therefore \qquad\qquad \cos\delta = 0.8 \ \text{ and } \ \sin\delta = 0.6$

Also $\qquad\qquad \alpha = \tan^{-1}\left(\dfrac{1.5}{6}\right) = 14°$

and the angle $\qquad \psi = 90 - (\alpha + \delta)$
$$= 90 - 50.87 \text{ or } 39.13°$$

The no-load current
$$I_0 = \sqrt{1.5^2 + 6^2}$$

$$= 6.186\ \text{A} = \text{CA (in } \triangle \text{ ABC)}$$

and $\qquad\quad \text{BC} = 6.186 \times \cos 39.13 = 4.8\ \text{A};$

$$\text{AB} = 6.186 \times \sin 39.13 = 3.903\ \text{A}$$

Assuming $\text{OB} \simeq \text{OA} = I_p = 1000\ \text{A}$, since β is very small

$I_p' = \text{OC} = \text{OB} - \text{BC} = I_p - \text{BC} = 1000 - 4.8 = 995.2\ \text{A}$

(i)   Hence the ratio error at $n_s = 800$ or $n = 200$

$$\text{RE} = \frac{nI_s - I_p}{I_p} = \frac{I_p' - I_p}{I_p} = \frac{995.2 - 1000}{1000} = -\,0.0048 \text{ or } -\,0.48\%$$

and the phase angle (error)

$$\beta = \tan^{-1}\left(\frac{AB}{OB}\right)$$

$$= \tan^{-1}\left(\frac{3.903}{1000}\right) = 0.224°$$

(ii) When the secondary turns are 795 and the primary current is to be 1000 A, I_p' is still 995.2 A.

$$\therefore \qquad I_s = 995.2 \times \frac{4}{795}(1/n)$$

With the nominal ratio still being $\dfrac{1000}{5}$, the 'apparent' primary current will be

$$= \frac{1000}{5} \times 995 \times \frac{4}{795} = 1001.4 \text{ A}$$

and the ratio error

$$RE = \frac{1001.4 - 1000}{1000} = 0.0014 \text{ or } 0.14\%$$

The value of β, being very small, may be the same, that is $0.224°$

7. A current transformer with 5 primary turns has a secondary burden given by $(0.16 + j\,0.12)\,\Omega$. When the primary current is 200 A, its magnetising current component is 1.5 A and iron-loss component 0.4 A. Determine the number of secondary turns needed to make the CT ratio 100/1, and also calculate the phase-angle error of the CT.

Let the desired *ratio* under the given conditions be n.

With the burden given as $(0.16 + j\,0.12)\,\Omega$, the angle

$$\delta = \tan^{-1}\left(\frac{0.12}{0.16}\right) \quad \text{or} \quad 36.87°$$

$$\therefore \qquad \cos\delta = \cos 36.87 = 0.8 \quad \text{and} \quad \sin\delta = \sin 36.87 = 0.6$$

The actual current transformation ratio

$$R = n + \frac{I_e \cos\delta + I_m \sin\delta}{I_s}$$

$$= n + \frac{0.4 \times 0.8 + 1.5 \times 0.6}{2}$$

$$[\text{for the desired ratio of } 100/1]$$

$$= n + 0.61 = 100 \text{ (as desired)}$$

$\therefore$ $\qquad\qquad\qquad$ n $= 100 - 0.61$ or 99.39

Hence no. of secondary turns

$$n_s = 5 \times 99.39 \quad \text{or} \quad 496.95 \text{ or } 497$$

The phase angle error

$$= \frac{180}{\pi} \times \frac{I_m \cos \delta - I_e \sin \delta}{n I_s}$$

$$= \frac{180}{\pi} \times \frac{1.5 \times 0.8 - 0.4 \times 0.6}{99.39 \times 2}$$

$$= \frac{180}{\pi} \times \frac{0.96}{99.39 \times 2}$$

$$= 0.276°$$

8. A ring-core current transformer with a nominal ratio of 500/5 and a bar primary has a secondary resistance of $0.5\ \Omega$ and negligible reactance. The resultant of the magnetising and the iron-loss components of the primary current associated with a full-load secondary current of 5 A in a burden of non-inductive type of $1.0\ \Omega$ is 3 A at a power factor of 0.4.

Determine the correct ratio and phase angle error of the CT. Calculate also the total flux in the core, the supply frequency being 50 Hz.

Total secondary impedance, *including* burden

$$Z_S = (0.5+1.0) + j\,0 \quad \text{or} \quad 1.5 + j\,0\ \Omega$$

$\therefore$ $\qquad\qquad\qquad$ $\delta = 0; \quad \cos \delta = 1, \quad \sin \delta = 0$

The no-load current,

$$I_0 = 3\,\text{A} \ @ \ \text{pf of } 0.4$$

or $\qquad$ $\cos \alpha = 0.4$, giving $\alpha = 66.4°$

and $\qquad$ $\sin \alpha = \sin 66.4 = 0.916$

$\therefore$ $\qquad\qquad$ $I_e = I_0 \cos \alpha = 1.2\ \text{A}$ and $I_m = I_0 \sin \alpha = 2.75\ \text{A}$

The actual current ratio is given by

$$R = n + \frac{I_e \cos \delta + I_m \sin \delta}{I_s}$$

$$= 100 + \frac{1.2 \times 1.0 + 2.75 \times 0}{5},$$

$\qquad\qquad$ assuming turns ratio, n, equal to

$$\text{nominal ratio} = \frac{500}{5}$$

$$= \frac{500+1.2}{5} \text{ or } \frac{501.2}{5}$$

Phase angle error $\theta = \dfrac{180}{\pi} \times \dfrac{I_m \cos\delta - I_e \sin\delta}{n\, I_s}$

$$= \frac{180}{\pi} \times \frac{2.75 \times 1.0 - 1.2 \times 0}{100 \times 5}$$

$$= 0.315°$$

With secondary impedance of $1.5\ \Omega$ at full secondary current of 5 A, the secondary induced EMF is 5×1.5 or 7.5 V.

$\therefore$ $4.44\, f\, \varnothing_m\, n = 7.5$

giving

$$\varnothing_m = \frac{7.5}{4.44 \times 50 \times 100}$$

$$= 337.8 \times 10^{-6}\ \text{Wb}$$

9. A current transformer has a bar primary and a 200-turns secondary. The secondary supplies a current of 5 A to a non-inductive burden of $1.0\ \Omega$ resistance. The requisite flux is set up in the core by 80 AT in the primary winding. The frequency is 50 Hz and the cross-sectional area of the core is $10\ \text{cm}^2$. Neglecting the effects of magnetic leakage and the iron and copper losses, calculate

(a) the ratio and phase-angle of the transformer;

(b) the flux density in the core.

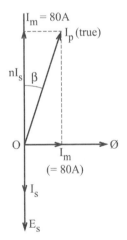

Since the burden is non-inductive, $\delta = 0$

$\therefore$ $\cos\delta = 1$ and $\sin\delta = 0$

Also, since the core loss is neglected,

$$I_e = 0 \text{ and } I_m = \frac{80}{1} = 80\ \text{A}$$

(referred to primary)

To determine the actual current ratio, refer to the phasor diagram shown on the right.

For $I_s = 5$ A and $n = 200,$

$$n\, I_s = 1000\ \text{A}$$

$$\therefore \qquad I_{p(true)} = \sqrt{(n\,I_s)^2 + I_m^2}$$

$$= \sqrt{(1000)^2 + (80)^2}$$

$$= 1003.2 \text{ A}$$

$\therefore$ correct transformation ratio of the CT $= 1003.2/5$

Also, phase angle, $\beta = \tan^{-1}\left(\dfrac{80}{1000}\right) = 4.574°$

Now secondary induced EMF, $E_s = 5 \times 1.0$ or 5 V

$$= 4.44 \times f \times \emptyset_m \times n$$

$$\therefore \qquad \emptyset_m = \frac{5}{4.44 \times 50 \times 200}$$

$$= 0.0001126 \text{ Wb} \quad \text{or} \quad 112.6 \times 10^{-6} \text{ Wb}$$

$$\therefore \quad \text{core flux density} = \frac{112.6 \times 10^{-6}}{10 \times 10^{-4}} = 0.113 \text{ T}$$

10. A 1000/100 V potential transformer has the following constants:

primary resistance and reactance : 94.5 Ω; 66.2 Ω

secondary resistance : 0.86 Ω

total equivalent reactance referred to primary : 110 Ω

no-load current (primary side) : 0.02 A @ 0.4 pf

Calculate

(a) PA of the PT at no-load

(b) burden in VA at unity pf at which the phase angle will be zero.

Given

$$r_p = 94.5 \ \Omega \ ; \quad x_p = 66.2 \ \Omega$$

$$r_s = 0.86 \ \Omega \ ; \quad X_p = 110 \ \Omega$$

$$I_0 = 0.02 \text{ A @ pf of } 0.4; \quad \text{or } \phi = \cos^{-1} 0.4 = 66.42°$$

$$\therefore \qquad I_e = 0.02 \times 0.4 = 0.008 \text{ A}$$

and $\qquad I_m = 0.02 \times \sin 66.42 = 0.0183 \text{ A}$

Nominal ratio $= \dfrac{1000}{100} = 10$; assume also to be the turns ratio, n

(i) At no load, $I_s = 0$

and the phase angle (PA) is simply given by

$$PA = \frac{180}{\pi} \times \frac{I_e x_p - I_m r_p}{n \, V_s} \text{deg}$$

$$= \frac{180}{\pi} \times \frac{0.008 \times 66.2 - 0.0183 \times 94.5}{10 \times 100}$$

$$= \frac{180}{\pi} \times \frac{-1.19975}{1000}$$

$$= -0.06877^\circ \quad \text{or} \quad -4.126'$$

(ii) At unity power factor, $\Delta = 0$

$\therefore \quad \cos \Delta = 1.0;\ \sin \Delta = 0$

Hence

$$PA = \frac{180}{\pi}\left[\frac{I_s}{V_s}\left(X_s \cos \Delta - R_s \sin \Delta\right) + \frac{I_e x_p - I_m r_p}{n V_s}\right]$$

$$= 0 \text{ [as specified]}$$

Using the alternative expression

$$PA = \frac{I_s}{n} \times X_p + I_e x_p - I_m r_p = 0$$

or $\quad \dfrac{I_s}{n} \times X_p = I_m r_p - I_e x_p$

or $\quad I_s = \dfrac{n}{X_p}\left(I_m r_p - I_e x_p\right)$

Substituting

$$I_s = \frac{10}{100}\left(0.0183 \times 94.5 - 0.008 \times 66.2\right)$$

$$= 0.119975 \text{ A}$$

$\therefore$ secondary burden under specified condition

$$= 100 \times 0.119975 \quad \text{or} \quad 11.9975 \quad \text{or} \quad 12 \text{ VA (approx)}$$

11. A single-phase voltage transformer has a ratio of 3810/63 V. The nominal secondary voltage is 63 V and the total equivalent resistance and leakage reactance of the PT referred to the secondary are 2 Ω and 1 Ω, respectively. Calculate the ratio and phase-angle errors of the transformer when supplying a burden of $(100 + j\, 200)$ Ω. State any assumptions made.

Since no mention is made, assume that the no-load current of the transformer is negligible.

Hence,

$$I_e = I_m = 0$$

The secondary burden is $100 + j\, 200$ Ω

∴ $$Z_s = \sqrt{100^2 + 200^2} = 223.6 \ \Omega$$

and

$$I_s = \frac{63}{223.6} = 0.281 \ \text{A}$$

Also, assuming that V_s and E_s are in phase,

$$\Delta = \tan^{-1}\left(\frac{200}{100}\right) = 63.43°$$

∴ $\cos \Delta = 0.447$ and $\sin \Delta = 0.894$

The PT turns ratio, $n = \dfrac{3810}{63} = 60.47$ $\begin{bmatrix}\text{assumed to be equal to the}\\ \text{nominal ratio}\end{bmatrix}$

Neglecting the no-load current components, the actual primary voltage is given by

$$V_p = n\, V_s + n\, I_s\, [R_S \cos \Delta + X_S \sin \Delta]$$
$$= n\, \{V_s + I_s\, (R_S \cos \Delta + X_S \sin \Delta)\}$$

Substituting,

$$V_p = 60.47\, [63 + 0.281\, (2 \times 0.447 + 1 \times 0.894)]$$
$$= 60.47 \times 63.502$$
$$= 3840 \ \text{V}$$

The ratio error,

$$RE = \frac{nV_s - V_p}{V_p} \times 100\%$$

$$= \frac{3810 - 3840}{3840} \times 100$$

$$= -0.78\ \%$$

The phase-angle error is given by

$$PA = \frac{180}{\pi} \times \frac{I_s}{V_s}\left(X_s \cos \Delta - R_s \sin \Delta\right)$$

$$= \frac{180}{\pi} \times \frac{0.281}{63}\left(1 \times 0.447 - 2 \times 0.894\right)$$

$$= \frac{180 \times 0.281 \times \left(-1.341\right)}{\pi \times 63}$$

$$= -0.343^{\circ}$$

12. Two CTs of the same nominal ratio 500/5 A are tested by Silsbee's method. With the current in the secondary of the standard CT at its rated value, the "difference" current, $\Delta\dot{I}$, is $0.05 \underline{/-125^{\circ}}$ A with respect to the CT secondary current. The standard CT has a "Ratio Correction Factor" (RCF) of 1.002 and phase angle error of $10'$. Determine the RCF and phase angle error of the CT under test.

$$\left[RCF\ of\ a\ CT = \frac{actual\ ratio}{nominal\ ratio} \right]$$

Nominal ratio of each CT

$$K_n = \frac{500}{5} = 100$$

Secondary current of the standard CT

$$I_{ss} = 5\ A$$

The difference current with respect to the above is

$$\Delta\dot{I} = 0.05\ \underline{/-125^{\circ}}$$

$$= -0.02868 - j\,0.041\ A$$

∴

$$\dot{I}_{sx} = \dot{I}_{ss} - \Delta\dot{I}$$

$$= (5 + j\,0) - (-0.02868 - j\,0.041)$$

$$[with\ \dot{I}_{ss}\ as\ reference]$$

$$= 5.0286 + j\,0.041$$

$$= 5.0289\ \underline{/+0.467^{\circ}}\ A$$

Actual ratio of the standard CT

$$= RCF \times nominal\ ratio$$

$$= 1.002 \times 100 = 100.2$$

∴ actual primary current of the standard CT

$$I_p = 100.2 \times 5 = 501\ A$$

For the CT under test

$$actual\ ratio = \frac{I_p}{I_{sx}} = \frac{501}{5.0289}$$

$$= 99.6242$$

∴ the Ratio Correction Factor or RCF of the test CT

$$= \frac{actual\ ratio}{nominal\ ratio}$$

$$= \frac{99.6242}{100}$$

$$= 0.9962$$

Phase angle between I_{sx} and I_{ss} is $0.467°$ or $28.02'$

Phase angle of the standard CT $= 10'$

∴ phase angle (error) of the CT under test

$$= 28.02 + 10$$

$$= 38.02'$$

EXERCISES

1. A 100/5 A CT, at its rated burden of 25 VA, has an iron loss of 0.2 W and magnetising current of 1.5 A. Determine its ratio error and phase angle when supplying rated burden having a ratio of resistance to reactance of 5.

    [− 1.075 %; 0.75°

2. A 1000/5 A, 50 Hz current transformer having a bar primary has a rated secondary burden of 12.5 VA. The secondary has 196 turns and a leakage inductance of 0.96 mH. With a purely resistive burden at rated full load, the magnetisation mmf is 16 A whilst the loss component of no-load current is 12 A, both referred to the primary.

    Determine the ratio and phase-angle errors of the CT.

    [+ 0.08 %; 0.327°

3. A 1000/5 A, 50 Hz current transformer has a secondary burden comprising a non-inductive impedance of 1.6 Ω. The primary is bar type (1 turn). Neglecting leakage reactance, assuming magnetising mmf of 100 A and iron loss in the core to be 1.5 W at full load, calculate the ratio error of the CT. Determine also the core flux density if the core has a cross-sectional area of 10 cm^2.

    [− 3.61 %; 0.18 T

4. A 6900/115 V nominal-ratio potential transformer has 22,500 turns in the primary and 375 turns in the secondary. The open-circuit primary current is 0.005 A at a power factor angle of 73.7°. With a given burden connected across the secondary, the primary current is 0.0125 A, lagging the voltage by 53.1°. The PT has the following impedance data.

    Primary winding resistance and reactance   : 1200 Ω and 2000 Ω
    Secondary winding resistance and reactance : 0.4 Ω and 0.7 Ω

    Determine

    (i)   the secondary current, terminal voltage and secondary burden;
    (ii) the actual transformation ratio and phase angle of the PT.

    [(i) 0.48 A; 114.5 V; 55 VA; (ii) 60.3; 0.90°

5. Two CTs of the same nominal ratio 500/5 A are tested by Silsbee's method. With the current in the secondary of the standard CT adjusted at its rated value, the "difference" current is $\Delta \dot{I} = 0.05 \,\underline{/-127°}$A, expressed with respect to the secondary current of the standard CT as the reference. If the Ratio Correction Factor and phase angle error of the standard CT are 1.0015 and +8′, respectively, calculate the RCF and phase-angle error of the CT under test.

    [0.9955; 35.3′

QUIZ QUESTIONS

1. A CT is a
   □ Step-up transformer   □ Step-down transformer   □ neither

2. A CT can be used to measure current in AC or DC circuits.
   □ true   □ false

3. A CT with a bar primary has
   □ one turn   □ no turn
   □ 10 turns   □ any number of turns
   in the primary.

4. A CT can also be used in the REVERSE manner.
   □ true   □ false

5. A clip-on ammeter is a form of CT.
   □ true   □ false

6. The THREE important ratios of a CT are
   a. _____   b. _____   c._____

7. The burden of a CT is expressed as
   □ rated primary current   □ rated secondary current
   □ VA rating of the CT   □ none of these

8. Name the most common two errors of a CT
   (i) _____   (ii) _____

9. Name the most commonly used core materials for CTs
   (i) _____   (ii) _____

10. A CT is rated at 1000/5 A. Its nominal ratio is _____.

11. The secondary of a bar-primary CT has 200 turns. Its turn ratio is _____.

12. The ratio error in a typical CT can be controlled by
    □ adjusting turns ratio   □ reducing excitation current
    □ controlling the burden   □ all of these

13. A CT rated at 100/5A has loss component of the exciting current as 0.6 A, the secondary winding burden being purely resistive.
    Its transformation ratio is
    □ 20   □ 20.12
    □ 20.2   □ 20.6

14. When in use, the secondary of a CT is always kept shorted even without burden.

    □ true   □ false

15. Name the typical circuit elements that may constitute burden of a CT

    a. _____ b. _____ c. _____ .

16. A PT is a form of very accurate

    □ step-up transformer   □ step-down transformer

    □ power transformer   □ distribution transformer

17. Name the typical circuit elements that may constitute burden of a PT

    a._____ b. _____ c. _____ .

18. The phase-angle error of a typical PT is less serious than that of a CT.

    □ true   □ false

19. Ratio-error correction in a PT is easier to apply because

    _____ .

20. Capacitor-type PTs are generally preferred in power systems with voltages exceeding 200 kV.

    □ true   □ false

21. Testing of instrument transformers is important to check

    □ ratio error   □ phase angle

    □ both   □ neither

22. Name some important advantages of using instrument transformers.

    _____

    _____

    _____

VII : Measurement of Power

VII

MEASUREMENT OF POWER

RECALL

What is power?

From the earliest days, power is associated with "work done", or rather the rate of work done (work done itself being equal to the energy), so that if energy, E, is spent in, say, moving an object to a distance D in time t by the use of a force F, then E can be given by

$$E = F \times D \times t$$

and the power being used given by

$$P = F \times D$$

One form of energy would obviously be heat, exercising the law of conservation of energy. In electrical systems, the basic quantity, viz., electrical current I, is associated with heat produced when the current flows in a resistance R so that by a given combination of I and R, the heat produced can be expressed as

$$H \propto I^2 R$$

according to the Joule's law[1].

Naturally, the other factor to be added to the above law is the *time* to account for all the heat produced in the given resistance. Accordingly, the expression for heat can be written

$$H = I^2 R t$$

In MKS system of units, if I is in ampere, R in ohm and t in second,

$$H = I^2 R t \text{ joule}[2]$$

that is, $\qquad H = I \times (IR) \times t$

$$= V \times I \times t,$$

where V is the PD across the resistance.

---

[1] Postulated by James Prescott Joule as early as circa 1841.

[2] The unit of *energy* named after Joule.

If the quantity $(V \times I)$ is expressed as power then, indirectly, power is rate of work done or energy in time t, or

$$P = V \times I$$

the unit of which in MKS system is watt, denoted by W.

Then, (heat) energy,

$$H = W \times t \quad \text{joule or watt-sec.}$$

POWER IN DC CIRCUITS

In the simplest case of a direct current, I, being delivered by a source of EMF E volt, to an external resistance R, the source itself having an *internal* resistance r ohm[1], the power appearing as heat will be

$$P = I^2 R \ W$$

However, I itself will be given by

$$I = \frac{E}{R + r} \ A$$

and hence

$$P = \frac{E^2 R}{(R + r)^2} \ W$$

Here, neither the current through the resistance nor the power consumed in it is time dependent; that is, it is constant at all times[2].

---

[1]Ideally, this should be zero; however, in practice all sources of EMF have an actual or equivalent internal resistance.

[2]However, an interesting aspect is to realise when the power in the circuit would be a maximum. For this,

let the power,

$$P = \frac{E^2 R}{(R + r)^2} = E^2 R (R + r)^{-2} \text{, as above.}$$

Differentiating with respect to R,

$$\frac{dP}{dR} = E^2 \left[(R + r)^{-2} + R(-2)(R + r)^{-3} \right] = E^2 \left[\frac{1}{(R + r)^2} - \frac{2R}{(R + r)^3} \right]$$

$$= E^2 \left[\frac{R + r - 2R}{(R + r)^3} \right]$$

Equating the numerator to zero for P to be a maximum gives R = r; that is, the 'load' resistance to be equal to the "internal" resistance of the 'source' – a concept of great importance in practice, better known as "impedance matching".

POWER IN AC SYSTEMS OR CIRCUITS

Power in Single Phase Circuits

In AC systems, now universally employed at all levels from generation to utilisation, the power at a given *instant* is still expressed as the product of voltage (or potential drop across a resister), and current in the circuit, but *varies with time since both the voltage and the current themselves are time dependent*, and is not a constant quantity in time as in the case of DC. Thus, in a circuit, assumed resistive-inductive, let v is the applied voltage and i the resulting current at any instant.

Then, let

$$v = V_m \sin \omega t$$

and

$$i = I_m \sin(\omega t - \phi)$$

where

V_m = maximum value of the applied EMF/PD

I_m = maximum value of current in the circuit

ω = angular frequency of the supply

ϕ = the angle of the impedance comprising the resistance R in series with an inductance L, such that $\phi = \tan^{-1}(\omega L/R)$

Then, the *instantaneous* power is given by

$$p = v\,i$$

$$= V_m \sin \omega t \times I_m \sin(\omega t - \phi)$$

$$= V_m I_m \left[\sin \omega t \, \sin(\omega t - \phi) \right]$$

$$= \frac{1}{2} V_m I_m \left[\cos \phi - \cos(2\omega t - \phi) \right]$$

or, writing θ for ωt,

$$p = \frac{1}{2} V_m I_m \left[\cos \phi - \cos(2\theta - \phi) \right]$$

and the mean or *average* power over one cycle of variation of v, i and p with time (or θ)

$$P = \frac{1}{2} V_m I_m \frac{1}{2\pi} \int_0^{2\pi} \left\{ \cos \phi - \cos(20 - \phi) \right\} d\theta$$

$$= \frac{V_m I_m}{4\pi} \left[\theta \cos\phi - \frac{\sin(2\theta - \phi)}{2} \right]_0^{2\pi}$$

$$= \frac{V_m I_m}{2\pi} \cos\phi$$

Since the average value of the second term in the square brackets, containing a double-frequency component, is zero over the cycle.

A graphical variation of v, i and p is shown in Fig.7.1, with p as instant-to-instant product of v and i. The magnitude of p is negative during $\theta = \phi$ periods and is subtracted from the positive variation of p, such that average value of the power will be a maximum for $\phi = 0$ (no negative component) and gradually reducing for $\phi > 0$.

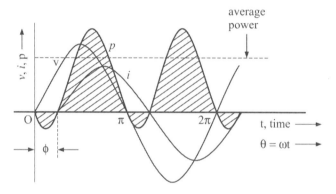

Fig.7.1 : Variation with time of v, i and p in an AC circuit

The expression for average power can be written

$$P = \frac{V_m}{\sqrt{2}} \times \frac{I_m}{\sqrt{2}} \times \cos\phi$$

or

$$\boxed{P = V\,I\,\cos\phi}$$

where V and I now represent the *rms* values of the potential difference and current, respectively. Hence, in AC circuits the average power depends not only on the product of PD and current, but also the factor cos ϕ, which itself is related to the resistance and reactance combination[1], and is identified as **power factor** (abbreviated p f) in all AC systems.

---

[1]In the above discussion, a resistive-inductive 'load' has arbitraily been chosen so that current *lags* the PD by an angle ϕ. In a resistive-capacitive load, the current will *lead* the PD, but the power will still be given by the same expression and the numerical value of power factor will still be cos ϕ.

[See Chapter III for more cases related to power in AC systems].

Comment

Implications and practical importance of power factor

In practice, the *total* power in an AC circuit consists of *two* components:

  (a) "active power", given by $V I \cos\phi$, and

  (b) "reactive power", expressed as $V I \sin\phi$

V and I being the supply voltage and *total* current drawn by the load.

Whilst it is the active power that is truly utilised, the reactive power only adds to the total *current* requirement of the circuit, such that

$$I_{active} = I \cos\phi,$$

$$I_{reactive} = I \sin\phi$$

and

$$I = \sqrt{\left(I_{active}\right)^2 + \left(I_{reactive}\right)^2}$$

Also, from

$$P = V I \cos\phi,$$

$$I \propto \frac{1}{\cos\phi},$$

for given 'active' power and supply voltage, so that poorer the power factor, higher the total current (demand).

  Accordingly, power factors of circuits at various stage of an AC system have an important bearing on

(a) rating, design and performance of a variety of power system equipment like

  - alternators

  - transformers

  - switchgear

  - transmission lines and distribution networks (involving supports and conductors)

(b) electricity tariffs and overall economy of the utility

since all the above depend on the magnitudes of *total* currents in the circuits and NOT merely the active components. The poorer the power factor, worse will be the situation with regard to the above parameters; it is, therefore, important to "maintain" a 'healthy' power factor throughout the system at all times!

Power in Polyphase Systems/Circuits

The 3-phase (balanced) system[1]

In the expression for power in a 1-phase circuit considered above, the double-frequency term is indicative of fluctuations in the magnitude of power from instant to instant and, clearly, may not be acceptable in a number of applications. This drawback is eliminated considerably in a circuit or system fed from a poly-phase supply. As an example, a 3-phase balanced supply is considered here, feeding into a *balanced* star-connected, 3-phase, resistive-inductive load, having a power factor $\cos\phi$ in *each phase*.

Let the three applied voltages in the circuit be

$$v_a = V_m \sin\omega t \text{ [assumed to be the 'reference' phasor]}$$

$$v_b = V_m \sin\left(\omega t - \frac{2\pi}{3}\right), \text{ displaced from } V_a \text{ by } 120^\circ(E)$$

$$v_c = V_m \sin\left(\omega t - \frac{4\pi}{3}\right), \text{ displaced from } V_a \text{ by } 240^\circ(E)[2]$$

where the maximum value of PD in each phase is V_m volt.

Assuming the three (phase or load) currents to be lagging the respective voltages by ϕ (the power factor angle), the instantaneous currents are

$$i_a = I_m \sin\left(\omega t - \phi\right)$$

$$i_b = I_m \sin\left(\omega t - \frac{2\pi}{3} - \phi\right)$$

$$i_c = I_m \sin\left(\omega t - \frac{4\pi}{3} - \phi\right)$$

I_m being the maximum current per phase.

The 'total', instantaneous power in the load is, therefore,

$$p = v_a i_a + v_b i_b + v_c i_c$$

$$= V_m I_m \times \left[\sin\omega t \times \sin\left(\omega t - \phi\right) + \sin\left(\omega t - \frac{2\pi}{3}\right) \times \sin\left(\omega t - \frac{2\pi}{3} - \phi\right) \right.$$

$$\left. + \sin\left(\omega t - \frac{4\pi}{3}\right) \times \sin\left(\omega t - \frac{4\pi}{3} - \phi\right) \right]$$

---

[1]A poly-phase (3-phase in the present case) *balanced* system is characterised by positive-sequence voltages across the phases, equal in magnitude and identical phase difference (120° E here), on the supply side and *identical* impedances in each phase on the load side.

[2]Following the standard 'rotation' of the positive-sequence voltages to be anti-clockwise.

Each term in the (square) bracket consists of a product of two sine terms which can be expanded using the identity

$$\sin A \, \sin B = \frac{1}{2}\left[\cos(A-B)-\cos(A+B)\right]$$

Then, proceeding similar to the case of a "single-phase" circuit, obtaining the average value of p over the cycle spanning $\theta = \omega t$ from 0 to 2π, and disregarding the three double frequency terms whose average value when integrated over a time period will be zero, the expression for power reduces to

$$P = \frac{3}{2}V_m I_m \cos\phi$$

or

$$P = 3 \times \frac{V_m}{\sqrt{2}} \times \frac{I_m}{\sqrt{2}} \, \cos\phi$$

or

$$P = 3 \, V \, I \cos\phi$$

that is,

$$P = 3 \times \text{power/phase.}$$

To illustrate how the average power contains much less fluctuations in the 3-phase system, the variations of load voltage and current in phase a and power in each phase is shown in Fig.7.2 where for simplicity the power factor is assumed to be unity; this being not a limitation.

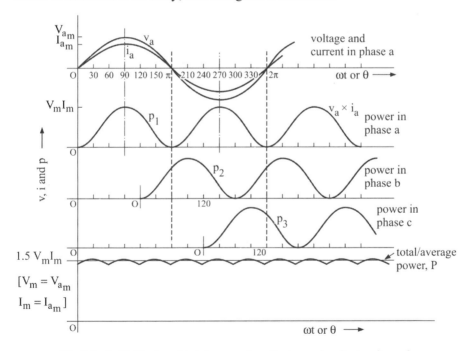

Fig.7.2 : Individual and total power in a 3-phase circuit at unity p f

(not to scale)

As shown by the envelope of superposed power curves, there are little 'ripples' and the average power is very nearly constant with time[1]. Note that the above analysis is strictly applicable to a balanced supply feeding into a balanced load, having the *same* power factor in each of the three phases. The expressions of power derived above establish that the (load) power factor, cos ϕ, constitutes the third important quantity, in addition to the applied voltage and circuit current. Thus, the power in AC circuits cannot be obtained by simply measuring voltage and current and multiplying the two, but an instrument is essential that can account for the power factor. A wattmeter is such an instrument[2].

Power Measurement Using Wattmeters

Use of single-phase electrodynamic/dynamometer wattmeter[3]

Wattmeter Connections in Single-phase AC Circuits

Fig.7.3 shows a wattmeter connected in a 1-phase circuit for measurement of power. The "current" coil is connected in 'series' and carries the load current drawn from the supply whilst the "voltage" or "pressure" coil is connected across the supply/load[4]. A suitable high resistance is invariably connected internally in series with the pressure coil to limit the current through the moving coil and to make the pressure-coil circuit of the instrument largely resistive.

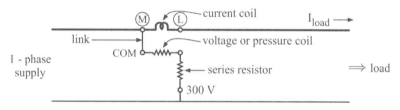

Fig.7.3 : A wattmeter connected in a 1-phase circuit

[Terminals Ⓜ, Ⓛ and COM, 300 V, as shown are marked on the wattmeter]

---

[1]The power output in a 1-phase v/s 3-phase circuit would thus appear to be somewhat analogous to a single-cylinder v/s multi-cylinder engine of a car and the torque appearing at the shaft in the two situations.

[2]However, it is implied that if there were an instrument to measure the power factor, used in association with a voltmeter and ammeter to measure the PD and current, respectively, the 'true' power could be obtained by the product of all the three quantities.

[3]See chapter "Measuring Instruments" for a detailed description and principle of operation of a dynamometer wattmeter.

[4]The implications of the form of connection of the pressure coil vis-à-vis the current coil are discussed later.

Theory

If i_1 and i_2 are the instantaneous currents in the current and voltage coil, respectively, then, as a dynamometer instrument, the developed torque at any instant will be

$$T_{inst} \propto i_1 \times i_2$$

or $$T_{inst} = K\ i_1 \times v$$

since i_2 is proportional to v, and K a constant.

With $$v = V_m \sin \omega t$$

and $$i_1 = I_m \sin(\omega t - \phi), \text{ assuming a lagging p f load}$$

the average torque, responsible for deflection of the moving system, will be

$$T_{av} = KV_m I_m \int_0^T \frac{1}{T}\Big[\sin\omega t \times \sin(\omega t - \phi)\Big]dt,$$

over a time period, T

or, writing $$\omega t = \theta \text{ and } T = 2\pi$$

$$T_{av} = KV_m I_m \int_0^{2\pi} \frac{1}{2\pi}\Big[\sin\theta \times \sin(\theta - \phi)\Big]d\theta$$

$$= K\frac{1}{2}V_m I_m \cos\phi$$

$$= K\, V\, I \cos\phi,$$

V and I being the rms values of the voltage across the load and load current, respectively.

Therefore, finally $T_{av} = P$, the average power in the load.

The fact that the deflection of the wattmeter is proportional to P, the scale of the instrument is nearly uniformly divided as depicted in Fig.7.4.

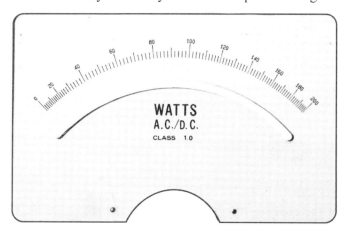

Fig.7.4 : Scale of a dynamometer wattmeter

Thus, a wattmeter is ideally suited to measure the power in an AC circuit *directly* when used with appropriate current and voltage coils ranges. In a 1-phase wattmeter, the current coil may be rated at 5, 10 or 20 A; the voltage coil 0-300 V. For larger values of currents and PD across the load, instrument transformers of suitable ratings may be employed as discussed later. The various terminals of the wattmeter are marked as indicated in Fig.7.3. Note the 'linking' of terminal Ⓜ of the current coil with "COM" of the pressure coil.

Torque expression of a single-phase dynamometer wattmeter related to mutual inductance

Referring to the schematic diagram of a wattmeter [see Fig.7.5],

let

total self-inductance of the moving coil	:	L_1
total self inductance of fixed coils	:	L_2
total resistance of the moving coil	:	R_1
total resistance of the fixed coils	:	R_2
instantaneous current through the moving coil (proportional to applied voltage, V)	:	i_1
instantaneous current through the fixed coils (usually the load current or a fraction of it)	:	i_2
mutual inductance between the two coils (dependent on relative position of the coils)	:	M

Then the instantaneous voltage and current relationships in the two sets of coils are

$$e_1 = R_1 i_1 + L_1 \frac{di_1}{dt} + M \frac{di_2}{dt}$$

$$e_2 = R_2 i_2 + L_2 \frac{di_2}{dt} + M \frac{di_1}{dt}$$

where e_1 and e_2 are the EMFs/PDs in the two circuits.

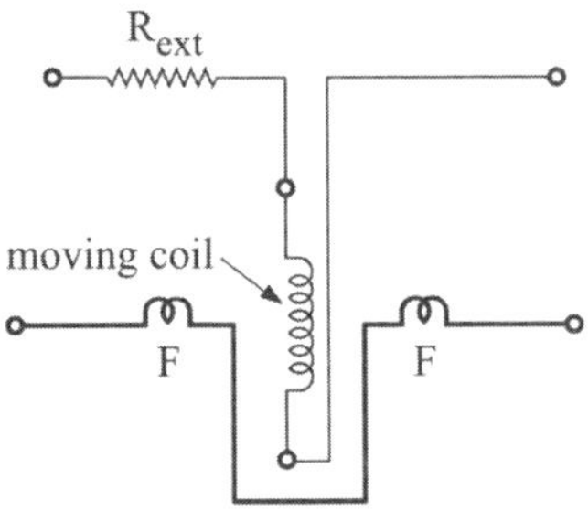

F, F : fixed coils

Fig.7.5 : Schematic of fixed and moving coil of a wattmeter

From the law of stored energy, mutual (stored) energy between the sets of coil is

$$e_r = M i_1 i_2$$

Now let the moving coil is deflected by a small angle δθ, resulting in (position dependent) change in M to be δM.

Then the corresponding small increase in the stored energy will be

$$\delta e_r = i_1 i_2 \delta M,$$

assuming i_1 and i_2 to be maintained constant.

Increase in the stored energy in the *moving* coil to maintain i_1 constant

$$\delta e_{r_1} = i_1 i_2 \delta M$$

Similarly, increase in the stored energy in the *fixed* coil to maintain i_2 constant

$$\delta e_{r_2} = i_1 i_2 \delta M$$

Also, if the torque due to control spring(s) is T, the work done by the spring(s) by movement through $\delta\theta$ will be $T \delta\theta$

From the principle of conservation of energy, total additional increase/input energy to the coils (to maintain i_1 and i_2 constant) must equal the change in *mutual* energy PLUS the mechanical work done, that is

$$\delta e_{r_1} + \delta e_{r_2} = \delta e_r + T \delta\theta$$

or $\qquad\qquad 2 i_1 i_2 \delta M = i_1 i_2 \delta M + T \delta\theta$

or $\qquad\qquad T \delta\theta = i_1 i_2 \delta M$

giving $\qquad\qquad T = i_1 i_2 \dfrac{\delta M}{\delta\theta}^1$

Now in terms of the voltage applied across the pressure coil and current through the current coil to act as a wattmeter, at any instant

$$i_1 = \frac{e_1}{R} = \frac{E_{1_m}}{R} \sin\omega t, \quad \text{where } R = R_1 + R_{ext}$$

and $\qquad\qquad i_2 = I_{2_m} \sin(\omega t - \phi),$

ϕ being the power factor angle of the load.

Substituting in the torque expression,

$$T = \frac{E_{1_m} I_{2_m}}{R} \left[\sin\omega t \times \sin(\omega t - \phi) \right] \frac{dM}{d\theta}$$

and, integrating over the complete cycle, the average torque

$$\boxed{T_{av} = \left[\frac{E_1 I_2}{R} \cos\phi \right] \times \frac{dM}{d\theta}}$$

If $dM/d\theta$, that is, the change of mutual inductance of the coils, is constant for the entire range, then

---

[1]The same expression was derived earlier, in general, for two coils having self and mutual inductances. [See Chapter V: Measuring Instruments].

$$T_{av} = K\, E_1\, I_2 \cos\phi$$

in which K is identified as the "wattmeter constant" that can be determined by design or calibration.

Errors Introduced by Wattmeters in Power Measurement

Although seemingly simple and straight-forward, the use of a wattmeter for the measurement of power is characterised by various possible errors due to

- voltage or pressure coil inductance
- pressure coil capacitance
- mutual inductance of the coils
- eddy currents induced in the metallic parts owing to time-varying magnetic field(s)
- temperature rise or (even) change of ambient temperature
- surrounding stray magnetic fields
- connection of pressure coil with respect to current coil and the load.

Clearly, some of the above errors will not manifest if the wattmeter is used in controlled environment, or, for example, used to measure power in a DC circuit.

Error due to voltage coil inductance

Having relatively large number of turns compared to the current coil, the voltage coil of a wattmeter usually has an inductance that cannot be neglected. The net result is that the pressure coil current is proportional to applied voltage, *but not in phase with it*; rather lagging it by a small angle.

Thus,

let r = resistance of the pressure coil (winding)

l = inductance of the coil

R_s = resistance connected externally in series with the coil to limit the current

V = voltage across the pressure coil

Then, the *impedance* of the pressure coil circuit is

$$Z = \sqrt{(r + R_s)^2 + (j\omega l)^2} \quad (\omega \text{ being the supply frequency}),$$

and the current through the coil,

$$i_p = \frac{V}{Z} = \frac{V}{\sqrt{(r + R_s)^2 + (j\omega l)^2}}$$

The voltage-coil current, i_p, will thus be lagging the PD by an angle

$$\beta = \tan^{-1}\left[\frac{\omega l}{r + R_s}\right]$$

and the phasor diagram of the (supply) voltage, load current and pressure coil current will be as shown in Fig.7.6. Ideally, the angle β should be zero; this means that either the reactance of the pressure coil (or l) should be zero, or R_s be 'infinite'[1].

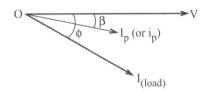

Fig.7.6 : Phasor diagram showing pressure coil current [not to scale]

Owing to finite β, the wattmeter deflection is proportional to

$I\,i_p\cos(\phi - \beta)$ and *not* to $I\,i_p\cos\phi$ as it should be, that is, to

$I\dfrac{V}{Z_p}\cos(\phi - \beta)$ where Z_p is the pressure-coil circuit impedance.

Now, from the impedance triangle pertaining to the pressure coil

$$Z_p = \frac{r + R_s}{\cos\beta}$$

and, therefore, the deflection is proportional to

$$I \times \frac{V}{r + R_s}\cos\beta \, \cos(\phi - \beta)$$

If the inductance of the voltage coil were zero, the wattmeter deflection will be proportional to

$$I \times \frac{V}{r + R_s}\cos\phi$$

and the wattmeter will read correctly at all frequencies and power factors. The ratio of the 'true' wattmeter reading to the actual reading is

$$\frac{\dfrac{IV}{r + R_s}\cos\phi}{\dfrac{IV}{r + R_s}\cos\beta \, \cos(\phi - \beta)} = \frac{\cos\phi}{\cos\beta \, \cos(\phi - \beta)}$$

and hence the true reading is given by

---

[1]In practice, the resistance R_s is made as large as possible in comparison with ωl, without reducing the coil current i_p and hence the deflection torque. At the same time, all efforts are made to limit the coil inductance.

$$\text{true reading} \atop \text{(or correct power)} = \frac{\cos\phi}{\cos\beta\ \cos(\phi-\beta)} \times \text{actual reading} \atop \text{(or apparent power)}$$

The ratio $\left[\cos\phi/\cos\beta\ \cos(\phi-\beta)\right]$ is known as "correction factor" of the wattmeter to account for non-zero inductance of the pressure coil.

Effect of load power factor

Considering the expression for the correction factor, the wattmeter will read high on (higher) lagging power factors of the load, since the effect of the voltage-coil inductance will be to bring the coil current more nearly in phase with the load current than if the inductance were zero and coil current in phase with the applied voltage.

If the power factor of the load were very low (that is, ϕ on the higher side), the wattmeter will read a higher value of the load power than what it should be, resulting in a serious error and calling for the use of the correction factor, unless special precautions are taken to compensate the effect of the inductance. The effect of angle of lag of the pressure-coil current, β, with respect to load power factor is depicted qualitatively in Fig.7.7 where, for very small β, the correction factor is nearly unity for all 'practical' load power factor angles.

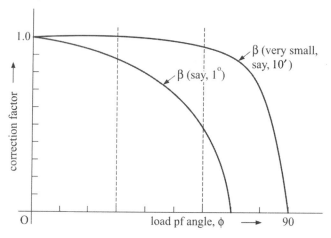

Fig.7.7 : Effect of β on the correction factor v/s load p f angle, ϕ

The compensation for the effect of the voltage-coil inductance is carried out in practice by connecting a small, precision capacitor across a part of the external resister, is series with the coil as considered later.

Error due to voltage-coil capacitance

The voltage coil may possess some capacitance owing to the high voltage across it and the effect of inter-turn capacitance of the winding; the 'total' coil capacitance may not be important in most cases, esp. in single-phase wattmeters. However, in wattmeters meant to be used in 3-phase circuits, the line-line voltage may lead to small, non-zero capacitance in the coil.

The effect of the coil capacitance in terms of its current and phasor relationship to the applied voltage (see Fig.7.8) may be similar to that due to the coil inductance except that the voltage-coil current will now *lead* the applied voltage by a small angle instead of lagging behind it. This means that the wattmeter will read *low* for lagging power factor loads, since the deflection will be proportional to

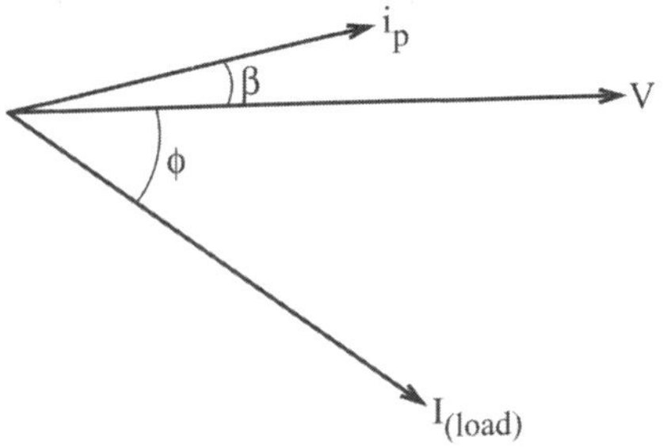

Fig.7.8 : Phasor diagram corresponding to capacitance effect of pressure coil

$$I \, i_p \cos(\phi + \beta)$$

instead of $I \, i_p \cos \phi$[1].

Effect of varying mutual inductance

The expression for deflection torque of a wattmeter contains the terms $dM/d\theta$. The modern instruments are designed such that M is (linearly) proportional to deflection within the working range of θ, and hence $dM/d\theta$ would be a 'constant' as shown in Fig.7.9. There is, therefore, no significant error introduced in the wattmeter on account of mutual inductance of the two coils, at nearly all loads and power factors.

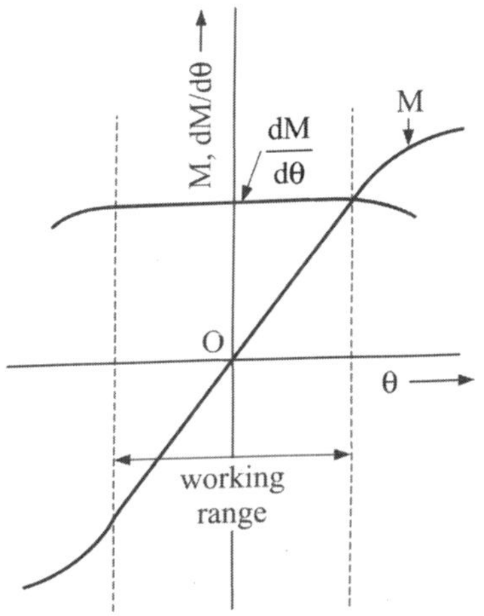

Fig. 7.9 : Variation of mutual inductance with deflection

Eddy-current errors

When used in AC circuits, the alternating flux inside the instrument will tend to induce eddy EMFs and currents in all metallic parts of the wattmeter, including the conductors. For this reason, large-section conductors are used for current coils and most 'inactive' parts are made of

---

[1]Clearly, if the capacitive reactance of the coil were to be equal to the inductive reactance, esp. for (slight) variation of frequency, the two will neutralise each other and the wattmeter will be error-free by itself.

non-metallic material to eliminate eddy currents and the effect of magnetic field(s) produced by them on the "working flux". Where possible, stranded conductors may be employed for winding the coils to minimise 'skin effect' and a change of resistance. Some unavoidable metal parts may be laminated.

Temparature effects

Changes in temperature could be

(a) internal, due to "I^2R" loss, resulting in temperature rise, esp. in the pressure coil and an increase in the series resistance;

(b) external, on account of wide fluctuations in the ambient temperature of increasing order, esp. encountered in tropical climates.

In both cases, the resistance, particularly that of the voltage-coil circuit, may be affected appreciably resulting in the change of magnitude and phase-angle relationship of the pressure-coil current.

Stray magnetic fields and shielding

Since the "working flux" of a typical dynamometer instrument is rather low, the wattmeter if used surrounded by stray magnetic fields (for example, in the vicinity of cables carrying hundreds of ampere)[1], may be seriously affected by their influence on the flux due to coils and hence in turn affect the deflection torque.

Screening

In practice, wattmeters are generally screened from, or shielded against, the stray magnetic fields by providing suitable screens on the inside of the casing (itself made of wood). The screen is either a sheet of high-permeability magnetic material (and this can be expensive) or non-magnetic material such as aluminium[2].

Methods of connection of pressure coil with respect to the current coil

There are two possible ways in which wattmeter connections can be made in a 1-phase circuit as shown in Fig.7.10. In practice, the current coil is always

---

[1]A phenomenon that may be encountered in on-site (or in the test lab) testing of large electric machines such as motors, generators and transformers.

[2]In the former, the stray field is "trapped" in the 'walls' of the magnetic material, being deflected almost along the surface due to its high relative permeability. In the latter, eddy currents induced in aluminium sheet produce their own magnetic field to neutralise the stray field as per Lenz's law. The design of the shield is thus carried out according to the principles involved.

connected in series with the load. A comparison of their various aspects is given below the figures.

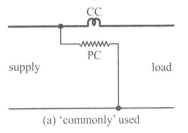

(a) 'commonly' used

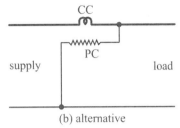

(b) alternative

The voltage coil is connected across the supply.

Whilst the current coil carries the true load current, the potential drop across the pressure coil is the sum of the voltage across the load PLUS the potential drop across the current coil. Thus in this case, the wattmeter measures the power in the load as well as the I^2R_C loss in the current coil (R_C being the coil resistance). Clearly, if the current is small, the potential drop across the current coil, as also the I^2R_C loss in it, can be very small and this method is preferable without causing much error.

The voltage coil is connected directly across the load.

In this case, the current coil now carries the load as well as the current through the pressure coil, whereas the voltage across the pressure coil is the true voltage across the load. Therefore, in this scheme, the wattmeter measures the power of the load PLUS the (ohmic) power lost in the pressure coil. Hence, if the load current is quite high, in comparison to the current taken by the pressure coil, this method can be used without introducing much error.

Fig.7.10 : Methods of connecting wattmeter in a 1-phase circuit

It is clear that neither method measures the power in the load uniquely or correctly.

Compensation for power loss in voltage coil

In the connection scheme shown in Fig.7.10(b), the effect of 'extra' current flowing through the current coil can be neutralised by providing a "compensating coil", or winding, as shown in Fig.7.11.

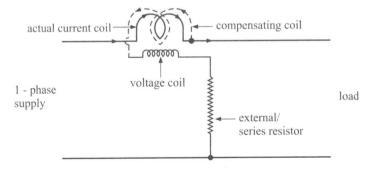

Fig.7.11 : Compensating coil in a wattmeter

The coil is connected in series with the voltage coil and along with the current coil such that it carries a (small) current equal and opposite to that in the main current coil in the 'forward' direction. Thus, the magnetic effect of the latter is nearly completely neutralised by the magnetic field produced by the compensating coil current in the opposite sense. This means that if there were no load current, and yet the voltage coil were energized, the deflection in the instrument will be zero owing to currents in the current and compensating coils being in the opposite direction.

Therefore, if a wattmeter is provided with a compensating coil as above, it is possible to use the connection scheme of Fig.7.10(b) for all values of the load current.

THE USE OF INSTRUMENT TRANSFORMERS WITH WATTMETERS

When the system voltage and current (and hence the power) are large (for example, voltage and current being more than about 600 V and 50 A, respectively), it becomes imperative to use appropriate potential and current transformers in association with a wattmeter to be able to measure the load power[1]. A circuit showing the use of a CT and PT for measurement of power with a wattmeter is shown in Fig.7.12.

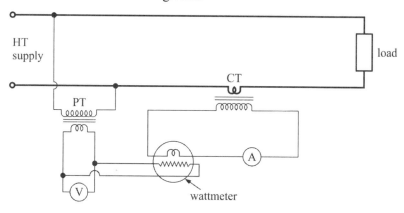

Fig.7.12 : Circuit showing connections of PT and CT with a wattmeter for measurement of power

In the circuit, the CT burden consists of current coil of the wattmeter and an ammeter if needed to measure the load current. Similarly, the PT burden comprises the wattmeter pressure coil and a voltmeter to measure the system voltage. Clearly, use of any additional instruments in the CT and PT

---

[1]Apart from facilitating measurement of power, the same PT and CT may be utilised to measure the primary voltage and current in modern, large power systems without, of course, exceeding the limits imposed by the burden and other design aspects.

secondary circuit must be limited as per requirement. When instrument transformers are used for power measurement as above, they usually introduce errors of measurement due to their ratio and phase angle errors, in addition to the errors an account of the wattmeter itself, and necessary corrections must be applied to obtain true power in the load.

To Derive the Correction Factor for using CT and PT

Refer to the phasor diagram shown in Fig.7.13 for the voltages and currents pertaining to the load, and those in the wattmeter pressure and current coils for a lagging power factor.

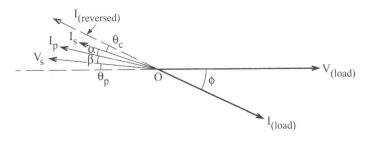

Fig.7.13 : Phasor diagram for measurement of power using instrument transformers and a wattmeter

Let

V = voltage across the load

I = load current

ϕ = phase or power-factor angle of the load

I_S = current in the secondary of the CT; also the current through current coil of the wattmeter

I_p = current in the pressure coil of the wattmeter

V_S = voltage across secondary of the PT; also the PD across pressure coil of the wattmeter

θ_c = phase angle (error) of the CT

θ_p = phase angle (error) of the PT

β = angle of lag of the wattmeter pressure coil on account of the coil inductance

α = phase angle between the currents in the current and pressure coils of the wattmeter, that is, the angle between I_p and I_S

Then

"True" power in the load, $P = V I \cos \phi$

or

$$P \propto \cos\phi \text{ for given V and I}$$

Reading of the wattmeter $\propto V_S I_S \cos(\alpha + \beta)$

where V_S and I_S are derived using *actual* transformation ratios of the PT and CT and $(\alpha + \beta)$ is the phase angle between V_S and I_S.

Therefore, correction (ratio) in terms of the phase angle of the load, ϕ, and phase angle of voltage and currents in the wattmeter,

$$K = \frac{\cos\phi}{\cos(\alpha+\beta)} = \frac{\cos\phi}{\cos\alpha\cos\beta},$$

approximately, neglecting the sine terms since α and β are very small.

Expressing α in terms of phase angles of the PT and CT

$$\cos\alpha = \cos(\phi - \theta_p - \theta_c - \beta)$$

and

$$K^1 = \frac{\cos\phi}{\cos\beta\cos\left(\phi - \theta_p - \theta_c - \beta\right)}$$

Hence, the power consumed in the load

P = K × *actual* ratio of PT × *actual* ratio of CT × wattmeter reading

If the ratio and phase angle errors of the PT and CT were considered negligible, the power in the load will simply be given by

P = nominal ratio of the PT × nominal ratio of CT × wattmeter reading

MEASUREMENT OF POWER IN A 1-PHASE CIRCUIT WITHOUT USING A WATTMETER

A useful (and interesting!) variation on the measurement of power in a *single-phase* circuit is the methods using instruments other than a wattmeter. Two methods are in use.

A. Three Voltmeter Method

The connections and corresponding phasor diagram for this method are given in Fig.7.14. Three voltmeters of appropriate range are connected together with the load as shown. The load current can be monitored by connecting an ammeter in series with the load, but is not essential for the theory behind the method.

---

[1]Although derived for (usually) a lagging power factor load, a correction factor can similarly be derived for a leading power factor load (in a rare case) by reference to an appropriate phasor diagram.

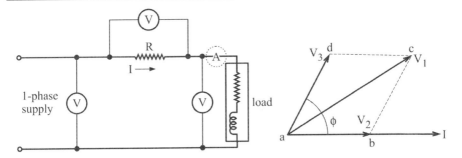

Fig.7.14 : Three-voltmeter method of power-measurement
[R is a non-inductive resistance in series with the load]

Let

V_1 = voltage of the supply

V_2 = voltage across the (standard, known) resistor, R

V_3 = voltage across the load

From the parallelogram a b c d of the phasors,

$$V_1^2 = V_2^2 + V_3^2 + 2\, V_2 V_3 \cos\phi \qquad \text{[an identity]}$$

or
$$V_1^2 = V_2^2 + V_3^2 + 2\, R\, I\, V_3 \cos\phi \qquad \text{since } V_2 = R\, I$$

But $V_3 I \cos\phi = P$, the power consumed in the load

Then
$$V_1^2 = V_2^2 + V_3^2 + 2\, P\, R$$

and
$$P = \frac{V_1^2 - V_2^2 - V_3^2}{2\, R} \ W$$

Also, the load power factor is given by

$$\cos\phi = \frac{V_1^2 - V_2^2 - V_3^2}{2\, V_2 V_3}$$

As expected, both the expressions are dimensionally correct.

It is clear that the method is based on the assumption that the current in the resistor R is the same as the load current; that is, no current is drawn by the voltmeters, and that the resistor is entirely non-inductive. The method is best suited to obtain power consumed by appliances taking a small current at a correspondingly high voltage. Also, for good accuracy, the resistor R should be chosen to make V_2 and V_3 of the same order as nearly as possible.

B. Three Ammeter Method

This method is similar to the above, but uses three *ammeters* connected as shown in Fig.7.15. The phasor diagram pertaining to the method is also given. The ammeter A_1 measures a current that is phasor sum of the load current and that taken by the non inductive resistor, R^1.

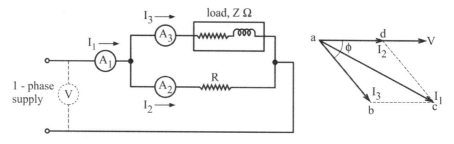

Fig.7.15 : 3-ammeter method of power measurement

The three currents drawn by the circuit are measured by the three ammeters, and displaced in phase as shown in the phasor diagram.

Again, from the parallelogram a b c d,

$$I_1^2 = I_2^2 + I_3^2 + 2\, I_2\, I_3\, \cos\phi$$

or, writing
$$I_2 = \frac{V}{R},$$

$$I_1^2 = I_2^2 + I_3^2 + 2 \left(\frac{V}{R}\right) I_3 \cos\phi$$

But $\quad V\, I_3 \cos\phi = P$, the power in the load,

and therefore

$$I_1^2 = I_2^2 + I_3^2 + 2 \left(\frac{P}{R}\right)$$

whence
$$P = \frac{\left(I_1^2 - I_2^2 - I_3^2\right) \times R}{2}\ W$$

and the load power factor, $\cos\phi = \dfrac{I_1^2 - I_2^2 - I_3^2}{2\, I_2 I_3}$

This method is more suited to appliances taking high currents at relatively low voltages; for example, several amperes at a few tens of volts. For better accuracy, the resistor R should be chosen so as to make I_2 and I_3 of the same order.

---

[1]From the connection scheme, and based on three currents as against three voltages, the method would appear to be the '*dual*' of the 3-voltmeter method.

Example

In a test by the 3-voltmeter method the following readings were obtained:

voltage across the mains	: 180 V, @ 50 Hz
voltage across the non-inductive resistor of 6 Ω	: 88 V
voltage across the load	: 106 V

Calculate the self inductance and effective resistance of the load and power consumed in it.

Power consumed by the load

$$P = \frac{(180)^2 - (106)^2 - (88)^2}{2 \times 6} = 1118.33 \text{ W}$$

Current through the load,

$$I = \frac{88}{6} = 14.67 \text{ A}$$

∴ load impedance, $Z_L = \dfrac{106}{14.67} = 7.23 \ \Omega$

From $P = V I \cos \phi,$

load power factor, $\cos \phi = \dfrac{1118.33}{106 \times 14.67} = 0.719$

and $\phi = \cos^{-1} 0.719 = 44°$

Since the power factor angle also applies to the load impedance,

load resistance, $R_L = Z_L \cos \phi$

$$= 7.23 \times 0.719 = 5.2 \ \Omega$$

and load reactance, $X_L = Z_L \sin \phi$

$$= 7.23 \times \sin 44 = 5.02 \ \Omega$$

At 50 Hz frequency,

the load inductance, $L = \dfrac{5.02}{2\pi \times 50} = 0.016 \text{ H}$

MEASUREMENT OF POWER IN THREE-PHASE SYSTEMS

Three-wattmeter Method

This method is based on using one wattmeter each in the three phases of the load, reading power per phase independently. The sum of the readings of the three wattmeters will give the total power. The schematic of the method for the star and delta connected loads is given in Fig.7.16.

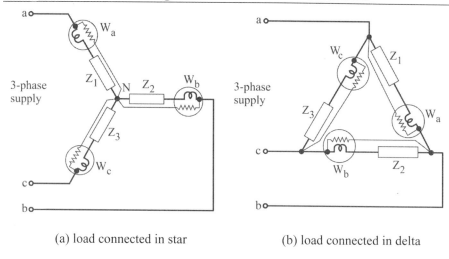

(a) load connected in star   (b) load connected in delta

Fig.7.16 : Three-wattmeter method for measuring three-phase power

Here, Z_1, Z_2 and Z_3 represent the load impedances in each of the phases.

Since the power is independently measured, the method has the advantage in that the load as well as the supply need not be balanced. The neutral point is to be available in the case of star-connected load to connect the wattmeters. Also, closed 'delta' must be openable when the load is delta-connected to insert the wattmeter current coil, and this forms a major disadvantage in this case. The advantage of the method is that it gives a clear indication of power in each phase. However, no direct information is available about the power factor unless it is deduced from the knowledge of PD across the load(s) and current through the same.

The Two-wattmeter Method

This method, being of much practical importance, is the commonest and universally used for all power measurements in *three-phase* circuits/systems. The chief advantage of the method is that it can be used for even unbalanced loads, *connected in star or delta*, but the supply itself must be balanced. Also, in the case of *balanced* loads, the method provides the *numerical* value of the load power factor in each phase, derived from the readings of the wattmeters as shown later. However, the nature of the power factor, whether lagging or leading cannot be ascertained. The connections for measurement of power in a star-connected load are shown in Fig.7.17 which also shows the various *instantaneous* and RMS voltages and currents.

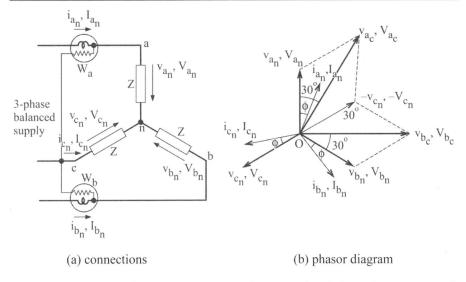

(a) connections   (b) phasor diagram

Fig.7.17 : Two-wattmeter method for measuring power in a balanced star-connected load

The method is based on the scheme of connecting current coils of the wattmeters in *any two* of the phases; a and b in the present case. The pressure coils are connected across those two phase terminals and the *remaining one* (phase c here) as shown in the figure.

Theory

To prove that the two wattmeters *do* measure the total (3-phase) power in the circuit.

In terms of magnitudes of instantaneous quantities, the total power,

$$p = v_{a_n} i_{a_n} + v_{b_n} i_{b_n} + v_{c_n} i_{c_n}$$

At the junction, n, by Kirchhoff's law

$$i_{a_n} + i_{b_n} + i_{c_n} = 0$$

or

$$i_{c_n} = -i_{a_n} - i_{b_n}$$

Substituting in the expression for power,

$$p = v_{a_n} i_{a_n} + v_{b_n} i_{b_n} + v_{c_n} \left(-i_{a_n} - i_{b_n} \right)$$

$$= \left(v_{a_n} - v_{c_n} \right) i_{a_n} + \left(v_{b_n} - v_{c_n} \right) i_{b_n}$$

However, $\left(v_{a_n} - v_{c_n} \right)$ and $\left(v_{b_n} - v_{c_n} \right)$ are the instantaneous voltages across the pressure coils of wattmeters W_a and W_b, respectively.

Hence,

p = reading of wattmeter, W_a + reading of wattmeter, W_b

= *total* (average) power in the circuit, irrespective of the circuit being balanced or not; that is, in general, $i_{a_n} \neq i_{b_n} \neq i_{c_n}$ and $v_{a_n}, v_{b_n}, v_{c_n}$ can be different[1].

Readings of the wattmeters considering RMS quantities and the total power

Referring to the phasor diagram of Fig.7.17(b), for a balanced load, the line voltages V_{a_c} and V_{b_c} considering the rms values, are

$$V_{a_c} = V_{a_n} - V_{c_n} = V_{a_n} + V_{c_n} \text{ (reversed)}$$

$$V_{b_c} = V_{b_n} - V_{c_n} = V_{b_n} + V_{c_n} \text{ (reversed)}$$

These are the voltages applied across the pressure coils of the wattmeters whilst the respective current coils carry currents I_{a_n} and I_{b_n}. As seen, the *phase differences* between the voltages (across pressure coils) and currents (through the current coils) are

for wattmeter $W_a : (30° - \phi)$;   and for $W_b : (30° + \phi)$

Hence, the readings of the two wattmeters will be

$$P_a = V_{a_c} I_{a_n} \cos(30° - \phi)$$

$$P_b = V_{b_c} I_{b_n} \cos(30° + \phi)$$

Since in a balanced supply and load

$$V_{a_c} = V_{b_c} = V \qquad \text{[L–L voltage, numerically]}$$

and

$$I_{a_n} = I_{b_n} = I$$

[line, or phase current in this case, numerically]

$$P_a = V I \cos(30° - \phi)$$

$$= V I (\cos 30° \cos \phi + \sin 30° \sin \phi)$$

---

[1]The method points to an important premise: the power in an n-wire system, fed from a *balanced* supply, can be measured using only (n – 1) wattmeters, irrespective of the load being balanced or unbalanced. Specifically, it follows from (the) Blondel's theorem, based on a paper delivered in 1893 at the International Electric Congress in Chicago, which states that in a system of N wires (or electrical conductors), (N–1) electrical meter elements (wattmeters or energy meters) when properly connected, will suffice to measure the total electrical power or energy taken, the connections being such that each wire carries a current coil and the corresponding voltage coil being connected between that wire and another wire to which all the voltage coils are connected and which itself carries no current coil.

$$P_b = V I \cos(30° + \phi)$$

$$= V I (\cos 30° \cos \phi - \sin 30° \sin \phi)$$

and $\qquad P_a + P_b = 2 V I \cos 30° \cos \phi$

$$= \sqrt{3} \, V I \cos \phi$$

$$= total \text{ power of the 3-phase load}$$

Here, in general, V and I represent *line* voltage and current in the phases.

Now, taking a difference of P_a and P_b

$$P_a - P_b = 2 V I \sin 30° \sin \phi = V I \sin \phi$$

or $\qquad \dfrac{P_a - P_b}{P_a + P_n} = \dfrac{\tan \phi}{\sqrt{3}}$

from which

$$\phi = \tan^{-1}\left(\sqrt{3}\,\frac{P_a - P_b}{P_a + P_b}\right)^1$$

Implications of load power factor and wattmeters readings

A. In general, since the wattmeter readings depend on $(30° - \phi)$ and $(30° + \phi)$ as the phase differences between the *line* voltages and currents, these will be different depending on the value of ϕ and, following a *given sequence of connections* in the circuit, the total power is seen to be given by arithmetic sum of the two readings; both P_a and P_b being positive.

B. For a *balanced* load, having unity power factor or $\phi = 0$, the two wattmeter readings will be *equal*, representing a special and indicative case of power measurement.

C. When the load is such, for example predominantly inductive with very poor power factor, such that $\phi > 60°$ and $(30° + \phi)$ is greater than 90°, one of the wattmeters will tend to read negative (once again, assuming a proper sense of connections). This situation must then be corrected by reversing the connections of *either* the current *or* the pressure coil of the particular wattmeter. Under the circumstance, this wattmeter reading is

---

[1]Note, however, that the "common" power factor so derived is for a *balanced* load as assumed here, the current in each phase being shown to be lagging by the same power factor angle, ϕ, [see Fig.7.17(b)].

reckoned negative, and the total power is the *difference* of the two readings, or the *algebraic* sum[1].

D. When ϕ is exactly equal to $60°$ in the balanced load (or $\cos \phi = 0.5$), one of the wattmeters will simply read "zero" whilst the other will read 'total' power.

E. Another interesting (and rather extreme) case is of "zero power factor" load, (that is, $\phi = 90°$) for example a *purely* inductive or capacitive circuit. In this case, both wattmeter readings will be equal, *but of opposite signs*, such that the net power is zero.

One-wattmeter Method

This is a special case of the application of the 2-wattmeter method, using only one wattmeter, and can be used only when the load and indeed the supply are balanced.

The connection scheme for this method is shown in Fig.7.18. The current coil of the wattmeter is connected in one of the phases/lines for a star-connected load, with one end of the pressure coil also connected to the same line. The other end is connected alternately to first one, and then the other, of the remaining two lines by means of a special two-way switch, S^2.

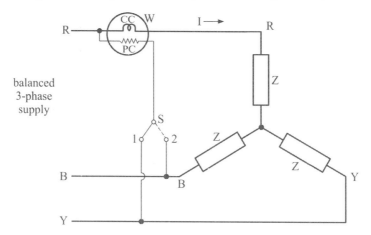

Fig.7.18 : Power measurement in a 3-phase balanced circuit using one wattmeter

Then, similar to the 2-wattmeter method, the wattmeter, with switch S in position 1, will measure

---

[1]This can happen, typically, when making measurement of iron loss in a machine or transformer which may have a low to very low power factor of the order of 0.3 to 0.5.

[2]It is important that the switch is "break"-before-"make" type.

$$P_1 = V I \cos(30° + \phi), \text{ say,}$$

where V is the *line* voltage across the pressure coil, and, when the switch is thrown to position 2, will measure

$$P_2 = V I \cos(30° - \phi),$$

and the total power will be given by algebraic sum of the two readings.

The power factor angle will be given by a similar expression as derived for the 2-wattmeter method.

The method is generally suitable for small loads.

Measurement of Power in a Three-phase Balanced Circuit Using Two CTs[1]

This method is particularly suited when the load is delta connected, that is, no neutral were available. Only one wattmeter is required. The connection scheme is shown in Fig.7.19. The method employs two *identical* CTs which have their primaries connected in series with two of the phases. The secondaries are cross-connected, that is in a differential mode, with two of the common terminals connected to the wattmeter current coil as shown. The CTs are chosen, preferably with 1:1 nominal ratio. The pressure coil of the wattmeter is connected across the two lines which carry the CTs primaries. The remaining, third phase is directly connected to the supply.

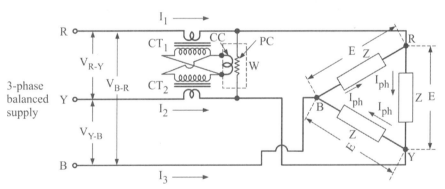

Fig.7.19 : Measurement of power in a 3-phase delta connected balanced load using two CTs

The phasor diagram and theory

The phasor diagram of voltages and currents related to the measurement is given in Fig.7.20.

---

[1]As described by H.M. Barlow, circa 1928.

An alternative to the method comprises two 'identical' PTs, instead of CTs, connected appropriately in the circuit.

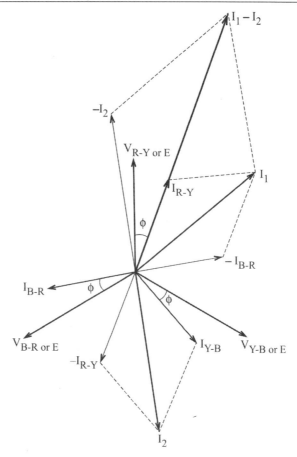

Fig. 7.20 : Phasor diagram for measurement of power using two CTs

Referring to the phasor diagram, with the supply and load being balanced, let the currents

$I_{R-Y} = I_{Y-B} = I_{B-R} = I_{ph}$, the current in each phase of the load

lagging the respectively phase voltages V_{R-Y}, V_{Y-B} and V_{B-R} by an angle ϕ; these voltages being also the line voltages of the supply, since the load is delta connected, and across each phase (= E). From the circuit diagram, I_1 is the phasor sum of I_{R-Y} and I_{B-R}. Similarly, I_2 is the phasor sum of I_{Y-B} and I_{R-Y}.

The wattmeter current coil carries the current $I_1 - I_2$ whilst the pressure coil, connected across phases R and Y is supplied with (line) voltage V_{R-Y}.

As seen from the phasor diagram, $\left| I_1 - I_2 \right| = 3 \times \left| I_{ph} \right|$ and lags the voltage by the angle ϕ.

∴   the wattmeter reading $= V_{R\text{-}Y} \times (I_1 - I_2)$

$$= 3 \times V_{R\text{-}Y} \times I_{ph}$$

$$= \text{total power consumed in the load.}$$

If the CTs of other than a nominal ratio of 1:1 were used, the total power will be

P = wattmeter reading × nominal ratio of the CTs[1]

WATTMETERS FOR 'SPECIAL' PURPOSE(S)

Unity Power Factor and Low Power Factor Wattmeters

When designed and constructed properly such that

(a)   the current coil carries the actual load current (or a known proportion of it; for example, in association with a CT), and

(b)   the current in the voltage coil is proportional to, *and very nearly in phase with*, the applied voltage (or the potential drop across the load)

the wattmeter will measure the 'true' power in the load, within the specified accuracy. A test (or self check) for the wattmeter will be that it measures equally on DC and AC supply. Such wattmeters are typically known as "unity power factor" wattmeters, well suited for power measurement in (AC) systems where the load power factor is high; for example, about 0.8 or 0.9 if not exactly unity, such as in load tests on electrical machines.

Low power factor wattmeters

In contrast, as depicted in the phasor diagram of Fig.7.6, at low and very low power factors of the loads, serious errors may result owing to inadequately designed pressure coil. Some of the measures that can be incorporated in the design of wattmeters meant for measurement of power having low power factor are

(i)   The pressure coil resistance and inductance

In a wattmeter, the deflection torque being proportional to V I cos ϕ or $I_L I_P$ cos ϕ, can be very small when cos ϕ, the power factor, is low. Thus, to increase the deflection torque for a given load current, it may help to increase the voltage coil current in magnitude by using a low resistance pressure coil. At the same time, the coil should be designed to have a smaller inductance so as to control the overall wattmeter power factor, esp. on lagging loads that are mostly encountered.

---

[1]However, it is implied that the CTs are of very good accuracy so that their errors do not affect the measured power.

Additionally, as discussed, compensation for the pressure coil inductance may be provided by connecting a suitable capacitor, C, across a section of the series resistor, as shown in Fig.7.21, preferably to form a 'tuned' circuit.

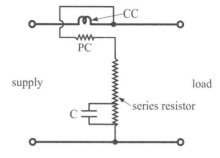

Fig.7.21 : Compensation for pressure coil inductance using a capacitor

(ii) Pressure-coil current

In addition, a low power factor wattmeter is invariably provided with a compensating coil in association with the current coil, designed and wound so as to nullify the effect of magnetic field produced by the pressure-coil current as explained earlier. [See Fig.7.11].

(iii) Control springs and torque

To match the smaller deflection torque in the wattmeters, 'finer' control springs are used, producing matching equal and opposite control torque. This, however, makes the instrument more delicate and must be handled with utmost care. In some cases, for measurement of small power at low power factor, a wattmeter of torsional-head type, eliminating control springs altogether may be more suitable.

Electronic Wattmeters

A dynamometer wattmeter is designed, calibrated and is suitable for measurement of power at power frequencies. For power measurements at higher frequencies, esp. for measuring low power, outside the range of dynamometer wattmeters, an electronic wattmeter may be employed. These wattmeters can be analogue or digital type.

Analogue type

In analogue electronic wattmeters, the principle of operation depends on the collection of electrical signals proportional to the 'load' current and voltage across it, by using appropriate transducers or devices, amplify the signals if too small and then input these to an electronic multiplier. The output, in the form of product of the two (maintaining the original phase difference between the load current and voltage) is the power which can be read on a calibrated instrument. Up to some frequency range, an example of the electronic multiplier is a Hall generator when operated under 'controlled' conditions[1].

Digital type

With modern, advanced digital technology, a wattmeter based on digitisation technique(s) can offer a wide range of application and accuracy.

After obtaining the signals proportional to load voltage and current using electromagnetic devices as in the case of analogue instruments, and after signal conditioning (usually amplification), a digital electronic wattmeter samples the signals thousands of times per second using Analogue-to-Digital (or A/D) converters of appropriate design, the sampling frequency being dependent on the supply/load frequency. The converters' outputs are then fed to a multiplier unit where the average of the (instantaneous) voltage multiplied by the (instantaneous) current is the true power which can be displayed on an LCD indicating 'window' (or on an analogue meter, incorporating a D/A converter). The instrument may be designed so that the true power *divided* by the apparent volt ampere (VA), another product from an electronic multiplier, can provide the load power factor. A 'computer circuit' linked to the instrument may use the sampled values to calculate, and indicate, RMS voltage, RMS current and VA, in addition to the power and power factor. More sophisticated models may retain the information in the 'computer memory' over an extended period of time and can transmit it to field equipment or a control location; for example, a control room or SCADA records centre.

When used in this manner, the digital electronic wattmeter can be a more versatile instrument, though at a higher initial cost.

VAR Meter

Measurement of reactive power in a 3-phase balanced load

In AC systems/working, measurement of reactive power forms an important requirement. When the circuit or load is balanced, this can be accomplished by using a "VAR meter". This instrument is essentially a wattmeter, but by connecting the current and pressure coil in a special manner can measure reactive (rather the active, but interpreted as reactive owing to the connections) power consumed in the load. Consider the three phase balanced load shown in Fig.7.22 and the corresponding phasor diagram.

---

[1]See Appendix VIII for details and use of such a device for various applications, including power measurement.

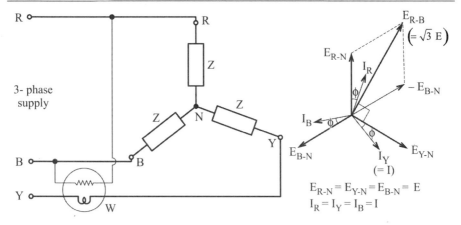

Fig.7.22 : Measurement of reactive power in a 3-phase balanced circuit and the phasor diagram

As shown, the current coil of the instrument is connected in the Y phase whilst the pressure coil is across the remaining two phases, that is, across phases R and B. The load current in each phase lags the phase voltage by an angle ϕ, such that the active power in each phase is $E I \cos \phi$, and total (active) power $3 E I \cos \phi$, where E is the voltage per phase and I the load current per phase. The total *reactive* power of the load will thus be $3 E I \sin \phi$.

Referring to the phasor diagram, the voltage across the instrument pressure coil is $E_{R\text{-}B}$ which *leads* the load current in phase Y, carried by the current coil, by an angle $(90° + \phi)$. Then, according to the principle of operation of a dynamometer instrument, the 'quantity' measured by the instrument is

$$Q = \sqrt{3}\ E \times I \times \cos(90° + \phi)$$
$$= -\sqrt{3}\ E I\ \sin\phi$$

But $E I \sin \phi$ is the reactive power of the load per phase

$\therefore$ reactive power/phase $= \dfrac{|Q|}{\sqrt{3}}$

and the total reactive power of the circuit

$$= 3 \times \dfrac{|Q|}{\sqrt{3}}$$

or $\qquad\qquad\qquad = \sqrt{3} \times |Q|$

that is, the reading of the wattmeter multiplied by $\sqrt{3}$.

The instrument when used for the purpose can be calibrated to read the total reactive power as above, thereby working as a VAR meter.

Double- and Three-element Wattmeters

These wattmeters have been developed to facilitate 'direct' measurement of power in three-phase systems, without the necessity to use more than one wattmeter, as in the 2-wattmeter method for 3-phase power measurement.

Double-element wattmeter

This is designed as, and consists of, two separate wattmeter movements housed in one case with the two moving coils (assuming a dynamometer-type design) mounted on the *same* spindle, with each moving coil itself being placed in between two fixed coils, of course, so that the total deflecting torque acting on the moving system will be the 'sum' of the torques produced by the two component wattmeter working systems. The reading of the instrument thus gives the total power in the circuit directly, the addition of individual-movement power (similar to that in the two wattmeter method) being carried out by the instrument itself. The instrument is thus very much suited to measure power in a 3-phase, 3-wire system. Some features of such a wattmeter are depicted in Fig.7.23.

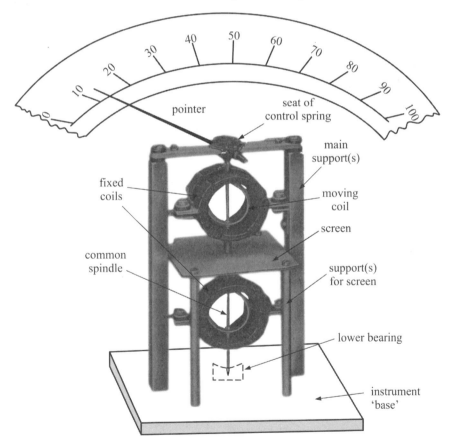

Fig.7.23 : A double-element wattmeter

An important requirement in such a wattmeter is that there shall be no magnetic interference of fields provided by one set of coils on that of the other. For example, the magnetic field of the lower-element fixed coils must not influence the moving coil of the upper-element and vice-versa, and thus lead to un-estimable errors. In order to eliminate such a possibility, and to ensure that the total torque shall be truly the sum of two torques produced independently by the two sets of coils, an appropriate magnetic shield or screen, consisting of high-permeability sheet material (such as soft iron or, preferably, mu-metal), is interposed between the two elements as indicated in Fig.7.23.

With the two coil sets housed in the same case, the connections of the wattmeter in a 3-phase, 3-wire system, to a delta-connected load (balanced or unbalanced) are made as shown in Fig.7.24.

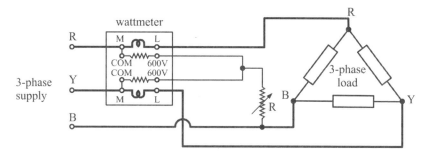

Fig.7.24 : Measurement of power in a 3-phase, 3-wire circuit using a double-element wattmeter

Basically, the connections are the same as those for the measurement of power by the two-wattmeter method. A small variation is that the ends of the two pressure coils are connected to the third phase (the one not carrying the current coil) jointly through an adjustable resistor, R, instead of directly. The purpose of this resistor is to compensate for any *difference* in the magnitudes of the pressure coil currents, as a result of possible mutual magnetic interference from one set of coils on the other, owing to different self inductance of the coils. The adjustment can be carried out during calibration by comparing the wattmeter reading from the total power obtained by two-wattmeter method.

Three-element wattmeter

This is similar to the double-element wattmeter and comprises THREE "single-phase" elements housed in a single case, with three sets of fixed and moving coils, the latter being mounted on a single moving system, carrying a pointer, control springs, damping vanes and balance weights. The total power is then measured by virtue of the resultant torque, being the sum of torques produced individually by the three component elements.

The wattmeter is of particular use in a 3-phase, *4-wire* system to measure the total power. The connections are made similar to the three-wattmeter method used in a 3-phase star-connected load, with the connection of the neutral forming the fourth wire. As in the case of a double-element wattmeter, it is essential to eliminate mutual magnetic interference among the sets of coils. For this, two nickel-iron sheets are interposed between the adjacent fixed coils to provide the magnetic screening.

In practice, whilst double-element wattmeters are in common use for measurement of power in 3-phase circuits, connected in delta or star (where neutral is not used at all), three-element wattmeter are rarely used on account of complexity of construction and 'handling'.

CALIBRATION OF WATTMETERS

Using DC Supply

Here, the power is simply given by the product of load current and the voltage across it, and is steady with time.

Comparison with a standard wattmeter

In this method, the wattmeter under test is connected in conjunction with a standard wattmeter of nearly the same specification, to a common load. The current coils of the wattmeters are connected in series whilst the pressure coils are in parallel across a common supply. The readings of the two wattmeters are compared at different loads to determine the error (or calibration graph) of the wattmeter under test.

Phantom (or fictitious) loading

The above method of direct loading has the disadvantage of wasting considerable power (usually as I^2R or heat loss) in the loading device, esp. if the wattmeter is to be tested at full rating[1]. This can be avoided by using the scheme known as phantom loading as shown in Fig.7.25.

---

[1]For example, if the wattmeter is rated at 300 V and 10 A, a power equal to 3000 W will be wasted when the wattmeter is tested at rated load; if the time of testing is also taken into account this would amount to great loss of energy.

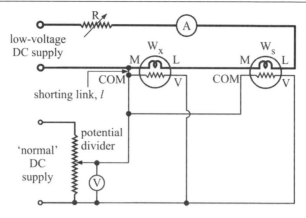

Fig.7.25 : Phantom method of wattmeter testing

W_x and W_s are the wattmeter under test and a standard wattmeter, respectively, whose current coils are still connected in series, *but fed from a separate, low-voltage DC supply*. The fictitious 'load' can be adjusted by the regulating resistor, R, and measured by an ammeter connected in series. The two pressure coils, connected in parallel, are supplied from another DC supply, the voltage being adjusted to the 'normal' value with the help of the potential divider as shown. The voltmeter reads the common voltage across either pressure coil. An equalising link, *l*, is necessary to join "M" and "COM" terminals of the current and pressure coil to maintain them at the same potential and to make the wattmeters to operate[1]. The calibration curve of the wattmeter can then be obtained by comparing the two readings as before[2].

Calibration using a DC potentiometer

The method using a (precision) DC potentiometer is superior to the previous one from the point of accuracy. The schematic connections for the method are shown in Fig.7.26.

---

[1]Without this, whilst the current and voltage across the respective coils may be set at appropriate values, no deflection will take place in the wattmeters.

[2]In the absence of any direct loading, the power loss is limited to the I^2R losses in the current and pressure coils which taken together will be only a small fraction of the total rated load.

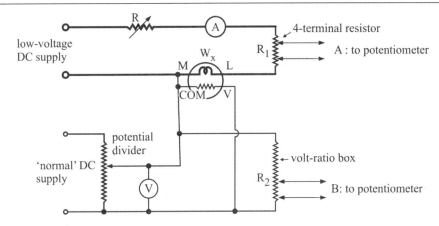

Fig.7.26 : Calibration of the wattmeter using a DC potentiometer

In the method, the wattmeter under test is used on its own and no standard wattmeter is required. The current coil is still fed from a low-voltage DC supply, regulated by the variable resistor, R. A standard, 4-terminal resistor, R_1, of low value (say, 0.1 Ω) is connected in series with the current coil, capable of carrying up to the rated current. The voltage coil of the wattmeter is connected to another DC supply, the voltage across the coil being adjustable by the potential divider. A suitable volt-ratio box, R_2, is connected across the pressure coil, a small voltage drop from which (say, about 1.5 V) is input to the potentiometer.

The potential drops A and B across the resistors R_1 and R_2 are connected in turn to, and measured by means of, the DC potentiometer, corresponding to a given setting of current and voltage for the wattmeter. The potentiometer readings provide the circuit current and voltage in terms of the measured potential drops and values/settings of resistors R_1 and R_2, giving true values of the 'power', compared to that read by the wattmeter.

AC Calibration

Calibration with a standard wattmeter

This method is essentially the same as the one used for DC testing using direct or phantom loading. The DC supplies are replaced by appropriate AC supplies; the low-voltage AC supply for the current coils being obtained using a step-down transformer whilst for the pressure coil a variac or dimmerstat is used. It is important that both the supplies are obtained from the same source to eliminate errors due to differring frequency and waveform. If the circuit is used exactly as shown in Fig.7.25, the test can be performed *only at one, given power factor*, being due to the phase difference between various current and voltage values.

Test at different power factors

In most cases, it is desirable to be able to vary power factor of the 'load' when carrying out the calibration; esp. when testing unity and low power-factor wattmeters. In the direct-loading method, this is achieved by using loads comprising variable resistors, inductors and/or capacitors to facilitate various lagging/leading power factors. However, if the fictitious method is to be employed (for its obvious advantages), it must be possible to alter the phase of the voltage applied across the voltage coil with respect to the current through the current coil. This is effected in practice by means of two common methods:

A. Using a Phase-shifting Transformer or Phase Shifter

In this arrangement, the current coils of the two wattmeters, W_x the one being tested or calibrated and W_S the standard one, are connected in series and fed from a step-down transformer, the value of the load current being adjusted by the regulating resistor, R. The two pressure coils, connected in parallel, are supplied from the rotor of the phase shifter, the magnitude of the applied voltage being adjustable with the help of a variac, whilst the phase angle between the applied voltage and 'load' current can be varied by adjusting the angular position of the rotor. Here, too, the current to the current coils and supply to the stator winding of the phase shifter are from the same source, once again to ensure the same supply frequency and waveform. The circuit connections are shown in Fig.7.27.

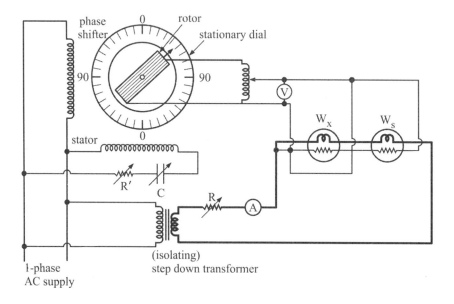

Fig.7.27 : AC calibration of a wattmeter using a phase shifter

B. Using a Coupled Motor-alternators Set

This scheme is based on the use of two 1-phase alternators, of similar electrical design, their rotors being coupled together by mounting on a common shaft and driven by a variable-speed motor; for example, a separately-excited DC motor, provided both with armature as well as field control. One of the alternators has a fixed stator, that is, bolted on to a foundation as usual, but the other is "cradle-mounted". This allows the stator of this alternator to be swung or capable of being rotated in relation to its rotor through any angle, in one or the 'reverse' direction that can be large enough. The angular difference or position is measureable on a stationary dial, as depicted in Fig.7.28. By adjusting the angular position of the second-alternator stator with respect to a reference axis of the fixed stator, the phase angle between the electrical outputs of the two stators can be varied.

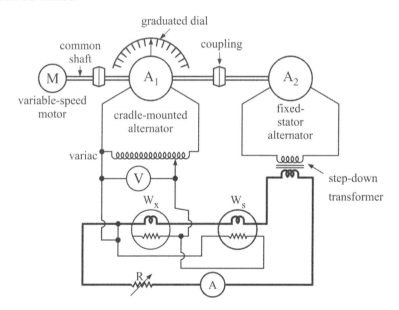

Fig.7.28 : Wattmeter testing using AC supplies from an alternators set

As in the previous method, the two current coils, in series, are fed from one of the alternators (for example, the one with the fixed stator; A_2 in the figure) through a suitable step-down transformer and a current regulating resistor, R, to adjust the 'load' current as shown. The two pressure coils, in parallel, are connected to the variable voltage supply (using a variac) obtained from the other alternator, A_1.

Although an expensive arrangement, the method has the advantage of common frequency (owing to common speed of rotation) and waveform of the two supplies; whilst any phase angle, representing lagging or leading power factor condition, can be effected by positioning the stator of one alternator with respect to the other.

Calibration using an AC potentiometer

This method is similar to the DC potentiometer testing of a wattmeter. However, as the name suggests, it employs an AC potentiometer[1] to measure 'true' current and voltage through the current coil and across the pressure coil, respectively, to provide 'true' power in the circuit. Again, the wattmeter is tested on its own; no comparison with a standard wattmeter being necessary. The circuit connections are shown in Fig.7.29 and the arrangement is self-explanatory.

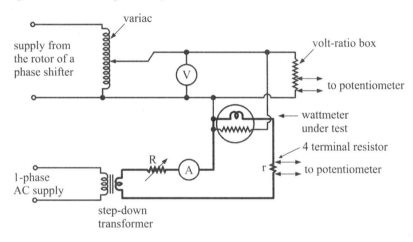

Fig.7.29 : Wattmeter calibration using an AC potentiometer

The two supplies to the current and pressure coils are arranged as shown. The supply to the current coil is from the same source as that to the stator of the phase shifter, the output from the rotor of which is connected to the pressure coil of the wattmeter. The phase difference between the current coil and pressure coil voltage is adjusted by changing angular position of the rotor of the phase shifter. The current coil circuit carries a small, standard resistance the output across which can be measured by the potentiometer in the first instance. A volt-ratio box is connected across the pressure coil to measure the voltage, V, in the circuit. The true power measured by means of the potentiometer is then compared with the reading of the wattmeter at different current settings (whilst maintaining the voltage across the pressure coil constant) to obtain the calibration curve.

---

[1]See Chapter IX for a detailed discussion of various potentiometers and their applications.

WORKED EXAMPLES

1. The mutual inductance between fixed and moving coils of a dynamometer wattmeter is expressed as $M = 0.01178 \times (\cos\theta/I_1)$ where I_1 is the current through the fixed coils and θ the deflection. Estimate the deflecting torque of the wattmeter carrying a current of 0.05 A at 0.8 p f through the moving coil when the deflection is (i) $60°$, (ii) $45°$

Given
$$M = 0.01178 \times \left(\frac{\cos\theta}{I_1}\right)$$

$\therefore$
$$\frac{dM}{d\theta} = -\left(\frac{0.01178}{I_1}\right)\sin\theta$$

Mean deflection torque of the wattmeter is given by
$$T_d = I_1 I_2 \cos\phi\left(\frac{dM}{d\theta}\right)$$

or
$$T_d = I_1 I_2 \cos\phi\left(\frac{-0.01178}{I_1}\sin\theta\right)$$

$$= -0.01178\, I_2 \cos\phi \sin\theta$$

or
$$|T_d| = 0.01178 \times 0.05 \times 0.8 \times \sin\theta$$

$$= 4.712 \times 10^{-4} \times \sin\theta \ \text{Nm}$$

(i)   at $\theta = 60°$, $\sin\theta = 0.866$

$\therefore$
$$T_d = 4.712 \times 10^{-4} \times 0.866 = 4.08 \times 10^{-4} \ \text{Nm}$$

(ii)   at $\theta = 45°$, $\sin\theta = 0.707$

$\therefore$
$$T_d = 4.712 \times 10^{-4} \times 0.707 = 3.332 \times 10^{-4} \ \text{Nm}$$

2. The voltage coil of a dynamometer wattmeter has an inductive reactance which is 0.5% of its resistance at normal frequency. Determine the correction factor for power measurement when the load power factor is (i) 0.8 lag, (ii) 0.5 lag.

Let the resistance of the voltage coil be R_p Ω. Then the reactance of the coil, X_p, is 0.5% of R_p or $0.005\,R_p$.

Let θ = phase angle between applied voltage V and voltage or pressure coil current, I_p

Then $\quad\quad\quad\quad \tan\theta = \dfrac{X_p}{R_p} = 0.005$

and $\quad\quad\quad\quad \theta = \tan^{-1} 0.005 = 0.2865°$

also, $\quad\quad\quad \cos\theta \simeq 1.0$

Let ϕ = phase angle between the applied voltage V and *load* current I, and load power factor = $\cos\phi$

Now the correction factor for potential coil inductance is

$$K = \dfrac{\cos\phi}{\cos\theta \ \cos(\phi - \theta)}$$

(i) at $\quad \cos\phi = 0.8$ (lag) or $\phi = 36.87°$

$$K = \dfrac{0.8}{1.0 \times \cos(36.87 - 0.2865)} = 0.996 \ \text{(very approx)}$$

(ii) at $\quad \cos\phi = 0.5$ (lag) or $\phi = 60°$

$$K = \dfrac{0.5}{1.0 \times \cos(60 - 0.2865)} = 0.992$$

3. Two wattmeters connected to measure power in a 3-phase, 440 V, 50 Hz, delta-connected balanced load system gave readings of 5 kW and 1kW, the latter being obtained by reversing the connection of current coil of the wattmeter. Determine what value of capacitance to be connected in each phase so as to make only one wattmeter read total power.

The wattmeters readings are

$$W_1 = 5000 \ \text{W}, \quad\quad W_2 = -1000 \ \text{W}$$

$\therefore$ total power, $\quad P = W_1 + W_2 = 4000 \ \text{W}$

and the pf angle, $\quad \phi = \tan^{-1}\left[\dfrac{\sqrt{3}(W_1 - W_2)}{W_1 + W_2}\right]$

$$= \tan^{-1}\left[\dfrac{\sqrt{3} \times 6000}{4000}\right]$$

$$= \tan^{-1}\left(\dfrac{3\sqrt{3}}{2}\right) = 69°$$

and pf $\quad\quad\quad = \cos\phi = 0.358$

In a 3-phase (balanced) system, power

$$P = \sqrt{3} V_L I_L \cos \phi$$
$$= \sqrt{3} \times 440 \times I_L \times 0.358 = 4000$$

from which

$$I_L = \frac{4000}{\sqrt{3} \times 440 \times 0.358} = 14.66 \text{ A}$$

Since the system/load is delta connected,

$$\text{the current/phase} = \frac{14.66}{\sqrt{3}} = 8.46 \text{ A,}$$

and load impedance/phase

$$= \frac{440}{8.46} = 52 \ \Omega$$

$$\text{Resistance/phase} \quad = \frac{4000}{3 \times (8.46)^2} = 18.62 \ \Omega$$

$$\therefore \text{ reactance/phase} \quad = \sqrt{52^2 - 18.62^2} = 48.54 \ \Omega$$

In order that one of the wattmeters should read zero, the p f of each phase should be 0.5.

$$\therefore \qquad\qquad \cos\phi = 0.5 \ \text{ or } \ \phi = 60°$$

and $\qquad\qquad \tan\phi = 1.732 \ = \dfrac{X}{R}$

$\therefore$ "new" reactance per phase, $X = R \tan\phi$

$$= 18.62 \times 1.732$$
$$= 32.25 \ \Omega$$

$\therefore$ the capacitive reactance per phase required for compensation

$$= 48.54 - 32.25$$
$$= 16.29 \ \Omega$$

and the capacitance per phase

$$C = \frac{1}{2\pi \times 50 \times 16.29} = 195.5 \ \mu\text{F}$$

4. A 250 V, 10 A dynamometer wattmeter has resistances of 0.5 Ω and 12,500 Ω of the current and potential coils, respectively. Calculate the % error due to each of the two methods of connection of the wattmeter with unity p f loads at 250 V and currents of (a) 4 A, (b) 12 A. Neglect error due to the pressure coil inductance.

The two alternative modes of connection of the wattmeter are as shown below:

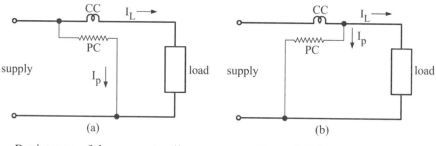

(a)   (b)

Resistance of the current coil, $\qquad$ $R_{cc} = 0.5\ \Omega$

Resistance of the pressure coil, $\qquad$ $R_{pc} = 12{,}500\ \Omega$

A. At load current of 4 A and unity p f,

(a) reading of the wattmeter

$$= V\ I\ \cos\phi$$

$$= 250 \times 4 \times 1 = 1000\ W$$

power loss in the current coil $= I_L^2\ R_{cc} = 4^2 \times 0.5 = 8\ W$

and error $\qquad = \dfrac{8}{1000} \times 100 \quad \text{or} \quad 0.8\%$

(b) For this connection,

power loss in pressure coil $= \dfrac{V^2}{R_p}$

$$= \dfrac{250^2}{12{,}500} = 5\ W$$

$\therefore$ error in this case $\quad = 0.5\%$

B. At load current of 12 A and unity p f,

(a) wattmeter reading $\quad = 250 \times 12 \times 1 = 3000\ W$

Power loss in the current coil $= 12^2 \times 0.5 = 72\ W$

and error $\qquad = \dfrac{72}{3000} \times 100 = 2.4\%$

(b) Power loss in the pressure coil $= \dfrac{250^2}{12{,}500} = 5\ W$

and error $\qquad = \dfrac{5}{3000} \times 100 = 0.166\%$

[This shows that for large load currents it is advantageous and more accurate to follow connection scheme (b)]

5. The load in a 1-phase AC circuit comprises a 1-phase transformer, the primary of which in series with an ammeter is connected to a 1-phase 230 V supply. A resistance of 200 Ω, in series with another ammeter is connected across the same supply. The total current of the circuit is measured by another ammeter, being 13 A. If the current through the resistor is 2 A and that into the transformer is 12 A, calculate (a) the input power, and (b) its power factor.

The circuit is as shown.

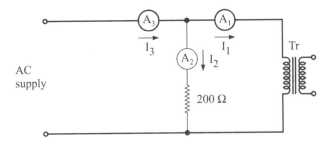

The observed currents are:

$$I_1 = 12 \text{ A}$$
$$I_2 = 2 \text{ A}$$
$$I_3 = 13 \text{ A}$$

This configuration corresponds to the case of power measurement in a 1-phase circuit by the 3-ammeter method. As derived in the text, the power consumed by the 'load' is

$$P = \frac{I_3^2 - I_1^2 - I_2^2}{2} \times R \quad W$$

$$= \frac{13.0^2 - 12.0^2 - 2.0^2}{2} \times 200$$

$$= 2100 \text{ W}$$

The power factor of the load is given by

$$\cos\phi = \frac{I_3^2 - I_1^2 - I_2^2}{2 \, I_1 I_2}$$

$$= \frac{13.0^2 - 12.0^2 - 2.0^2}{2 \times 12.0 \times 2.0} = 0.4375$$

6. An electrodynamic wattmeter is used to measure the power in a single-phase load. The load voltage is 120 V and the load current 8 A at a lagging pf of 0.1. The wattmeter voltage circuit has a resistance of 4000 Ω and an inductance of 31.9 mH, and is connected directly across the load. Estimate the percentage error in the wattmeter reading.

The connection of the load and wattmeter is as shown.

'Load' power $= V I \cos \phi$

$= 120 \times 8 \times 0.1$

$= 96$ W

and load pf angle, $\phi = \cos^{-1} 0.1 = 84.26°$

Power loss in the voltage circuit $= \dfrac{120^2}{4000} = 3.6$ W

Thus the wattmeter would read

$$96 + 3.6 \quad \text{or} \quad 99.6 \text{ W}$$

Now the voltage circuit reactance

$$= 2\pi \times 50 \times 31.9 \times 10^{-3} = 10 \ \Omega$$

∴ voltage circuit phase angle,

$$\alpha = \tan^{-1}\left(\dfrac{10}{4000}\right)$$

$$= 0.143°$$

Hence the *error factor* due to voltage coil inductance

$$= \dfrac{\cos(\phi - \alpha)}{\cos \phi}$$

$$= \dfrac{\cos(84.26° - 0.143°)}{\cos 84.26°}$$

$$= \dfrac{0.1025}{0.1} = 1.025$$

Thus, the wattmeter reading is

$$99.6 \times 1.025 \quad \text{or} \quad 102.1 \text{ W}$$

∴ percentage error in the wattmeter $= \dfrac{102.1 - 96}{96} \times 100 = 6.35\%$ (high)

7. A dynamometer wattmeter is used to measure the power factor of a cable, the frequency of the supply being 50 Hz. The cable capacitance is 0.6 μF. The wattmeter potential circuit is connected across the supply and has a resistance of 2700 Ω and an inductance of 0.058 H. The current coil has a resistance of 9.6 Ω; its inductance being negligible. The wattmeter deflection was made zero by adding an inductance of 0.041 H in series with the potential circuit. Draw a phasor diagram to illustrate these conditions and calculate the power factor of the cable.

The initial and later condition of the potential coil circuit is shown in figures below, together with the corresponding phasor diagrams.

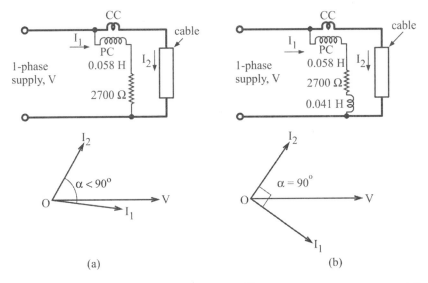

(a)   (b)

For the wattmeter to read zero in the 2nd case, the currents I_1 and I_2 through the pressure and current coil, respectively, should be in *phase quadrature*.

Now the *total* reactance of the pressure coil circuit

$$= 2\,\pi \times 50 \times (0.058 + 0.041)$$

$$= 31.086\ \Omega$$

and its impedance angle (or phase angle with respect to the supply voltage)

$$\alpha_1 = \tan^{-1}\left(\frac{31.086}{2700}\right) = 0.659^{\circ}$$

∴ the phase angle of the current through the cable/current coil

$$= 90.0 - 0.659$$

or $\qquad \alpha_2 = 89.341°$

Hence, the power factor of the cable

$$\cos \alpha_2 = \cos 89.341°$$

$$= 0.0115$$

8. Three identical wattmeters W_R, W_Y and W_B are connected to measure power in a balanced load supplied from a 3-phase balanced source. The current coils of the wattmeters carry their respective line currents, and the voltage circuits are connected between the respective supply terminals and common junction, S. When S is 'free', each wattmeter reads 2500 W. When S is connected to the R terminal the reading of W_Y is unchanged. Determine the corresponding reading of W_B and the p f of the load. Assume the phase sequence to be R-Y-B and ignore instrument losses.

The two connection schemes are shown below.

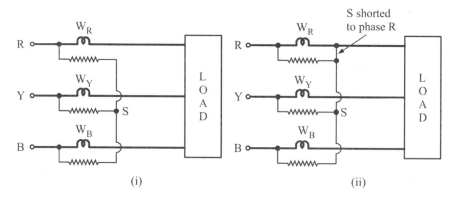

(i)   (ii)

With reference to the first case, the total power into the load is

$$P = 3 \times 2500 \quad \text{or} \quad 7500 \text{ W}$$

In the second case, the condition is that of a two-wattmeter method

The wattmeter reading of W_Y is still 2500 W

∴ reading of wattmeter,

$$W_B = 7500 - 2500$$

$$= 5000 \text{ W}$$

The p f angle is given by

$$\phi = \tan^{-1}\left(\sqrt{3} \times \frac{W_Y - W_B}{W_Y + W_B}\right), \text{ as per the phase sequence}$$

$$= \tan^{-1}\left(\sqrt{3} \times \frac{2500 - 5000}{7500}\right) \simeq -30°$$

∴ power factor $= \cos\phi = 0.866 \text{ (leading)}$

9. The three line currents of a 3-phase, balanced, 400 V, 50 Hz supply with phase sequence R-Y-B are:

$$\dot{I}_R = 3 \underline{/30°} \text{ A}$$

$$\dot{I}_Y = 4 \underline{/300°} \text{ A}$$

and $$\dot{I}_B = 5 \underline{/157°} \text{ A}$$

The voltage $\dot{V}_{Y\text{-}B}$ is taken as the reference phasor, being $400 \underline{/0°}$ V.

Determine the readings of each of the wattmeters when connected as figures (a) and (b), shown below:

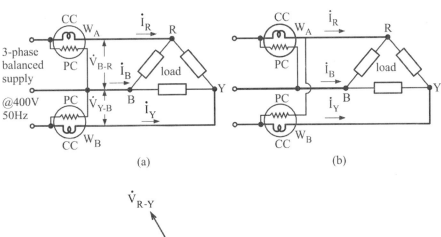

(a)   (b)

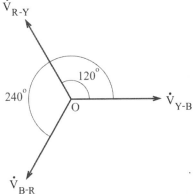

The three *line* voltages are:

$$\dot{V}_{Y\text{-}B} = 400 \underline{/0^\circ}, \text{ as given}$$

$$\dot{V}_{R\text{-}Y} = 400 \underline{/120^\circ}$$

$$\dot{V}_{B\text{-}R} = 400 \underline{/240^\circ}$$

as shown in the phasor diagram.

The reading of each of the wattmeters will be given by

$$W = \text{(rms voltage across the pressure coil)} \times$$
$$\text{(rms current through the current coil)} \times$$
$$\text{(cosine of the phase angle between the}$$
$$\text{above two)}$$

(A) Hence the readings of wattmeters in the first case will be

$$W_A = V_{R\text{-}B} \times I_R \times \cos[(240^\circ - 180^\circ) - 30^\circ]$$
$$= 400 \times 3 \times \cos 30^\circ$$
$$= 1039 \text{ W}$$
$$W_B = V_{Y\text{-}B} \times I_Y \cos(0^\circ - 300^\circ)$$
$$= 400 \times 4 \times \cos(-300^\circ)$$
$$= 800 \text{ W}$$

(B) In the second case, the PD across the pressure coil of wattmeter W_B is $V_{Y\text{-}R}$. The reading of wattmeter W_A is unaffected, still being 1039 W.

However

$$W_B = V_{Y\text{-}R} \times I_Y \times \cos[(120^\circ - 180^\circ) - 300^\circ]$$
$$= 400 \times 4 \times \cos(-360^\circ)$$
$$= 1600 \text{ W}$$

EXERCISES

1. In a particular measurement of power in a 3-phase balanced circuit using two wattmeters, the readings obtained were 5000 W and 1000 W. Calculate the total power in the circuit and power factor if (a) both meters read 'direct', (b) connection of pressure coil of one of the wattmeters were reversed.

    [6000 W, 0.656; 4000 W, 0.359

2. The input to a 500 V, 3-phase induction motor is measured at no load by the two-wattmeter method, to be 30 kW and the 'load' power factor as 0.4. Determine the readings of the two wattmeters.

    [34.85 kW, – 4.85 kW

3. In a two-wattmeter method of power measurement in a 3-phase system, the readings of the two wattmeters are 5000 W and 1500 W, respectively, the latter reading being obtained after reversing the connections to the current coil. Calculate the power and power factor of the load.

    [3500 W, 0.474

4. The three-voltmeter method was used to measure power in a 1-phase circuit and the following observations were made:

voltage of the supply	:	230 V
voltage across the load	:	150 V
PD across a standard 10 Ω resistor	:	90 V

    Calculate the power supplied to the load and its power factor.

    [1115 W, 0.826

5. A wattmeter reads 5540 W when its current coil is connected in "R" phase and the pressure coil across the "R" phase and the neutral of a 3-phase, balanced, star-connected load at 400 V and 30 A. What will be the reading of the wattmeter if the current coil is still connected in the "R" phase, but the pressure coil is now across the "B" and "Y" phases, the phase sequence being R-Y-B? What does the wattmeter reading represent?

    [7.2 "kW"; $\sqrt{3} \times$ reactive power/phase

6. The 2-wattmeter method is used to measure power in a 3-phase, star-connected load with wattmeter W_A reading 6000 W and W_B reading (– 1000 W), its reading having been obtained by reversing its pressure coil connections.
    (a) Determine the load power factor.
    (b) If the 3-phase supply comprises 400 V (L–L) at 50 Hz, what would be the value of a capacitance that may be connected in each phase so as to make the entire power measured by wattmeter W_A?

    [(a) 0.381; (b) C = 985 μF in each phase

QUIZ QUESTIONS

1. Analogous to AC, power in DC circuit can be said to have unity power factor.

   ☐ true   ☐ false

2. Power in a single-phase AC circuit is given by

   ☐ $V \times I$ ☐ $V \times I \times \cos\phi$

   ☐ $V \times I \times \sin\phi$ ☐ $V \times I \times \tan\phi$

   with standard notation

3. A dynamometer wattmeter can be used to measure

   ☐ only "DC" power ☐ only "AC" power

   ☐ either ☐ neither

4. The scale of a typical dynamometer wattmeter can be

   ☐ cramped at the beginning ☐ uniform

   ☐ camped at the middle ☐ none of these

5. The type of damping generally used in a dynamometer wattmeter is

   ☐ eddy-current damping ☐ air-friction damping

   ☐ fluid-friction damping ☐ gravity damping

6. In a dynamometer wattmeter the current coil is usually the fixed coils and pressure coil is the moving coil.

   ☐ true   ☐ false

7. The measurement of power in a 3-phase, 4-wire system requires

   ☐ 2 wattmeters ☐ 3 wattmeters

   ☐ 4 wattmeters ☐ any number of wattmeters

8. The power in a 3-phase circuit is measured using two wattmeters. If reading of one of the wattmeters is negative and different from the other, the power factor is

   ☐ Unity ☐ 0.5 (lagging)

   ☐ 0.5 (leading) ☐ less than 0.5 (lagging)

9. The other two well-known methods of measuring power in a 1-phase circuit are

   (a) _____

   (b) _____

10. The supply in the two-wattmeter method of power measurement has to be balanced.

    ☐ true   ☐ false

11. Two-wattmeter method will measure power correctly whether the load is balanced or unbalanced.

    ☐ true   ☐ false

12. Two-wattmeter method will measure power in a star-connected load only.

    ☐ true   ☐ false

13. In an "n-wire" system, the number of wattmeters required to measure the power correctly are

    ☐ n   ☐ n − 1

    ☐ n + 1   ☐ none of these

14. The two main sources of error in a dynamometer wattmeter are

    (a) _____

    (b) _____

15. If in a 3-phase balanced load the current coil of a wattmeter is connected in "R" phase and pressure coil across phases "B" and "Y" the wattmeter reading will be proportional to _____ in the circuit

16. The two usual methods of testing of a dynamometer wattmeter are

    (a) _____

    (b) _____

17. A preferred method of testing of a wattmeter is _____

18. A wattmeter may also be manipulated to measure energy consumed in a load.

    ☐ true   ☐ false

VIII : Measurement of Energy

VIII

MEASUREMENT OF ENERGY

RECALL

The Concept of Energy

In electrostatics, the work done by an 'external agent' in taking a unit positive charge against an electric field, by a distance in time t, was defined as potential difference (or PD) between the two points. This definition applies equally well to points on a conductor carrying a current.

In the figure given below D represents any electrical device (or circuit elements) such as a lamp, a motor or a heater, essentially a 'seat' of some form of "energy".

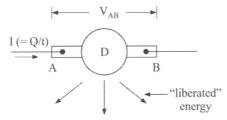

Assume a current of I flowing through the device from terminal A to B. If the current flows for t sec, the charge Q which it would carry from A to B will be

$$Q = I \times t$$

Suppose that the device D 'liberates' a total amount of energy E in the time t, in the form of light (from the lamp), mechanical work (from the motor) or heat (from the heater) then E is the amount of equivalent *electrical energy* given up by the charge Q in passing through the device from A to B. From the definition of potential difference (PD), therefore,

$$E = Q \, V_{AB}$$

Using the expression for Q $(= I\,t)$, the energy equation can be written

$$E = (I\,t) \times V_{AB}$$

or
$$= (V_{AB} \times I) \times t$$

In MKS system of units if V_{AB} is in volt, I in ampere and t in second, E is given in joule.

Writing $P = V_{AB} \times I$ as power, rating or capacity of the device, the *basic* equation of electrical energy is given by

$$E = P \times t$$

showing that electrical energy is simply the product of electric power in watt delivered-for/consumed in t seconds, as is commonly known or understood.[1]

---

[1]While the power P in a device or circuit may largely be a constant quantity, the element of time, t, may require different treatment as shown later.

ELECTRICAL ENERGY AS A QUANTITY

When dealing with the measurement of electrical energy as a quantity, the scheme may essentially be same as the measurement of power, except that the instrument must also take into account the length or duration of *time* for which the power is supplied into a given system or load. It follows that any power measuring device when 'coupled' with a time recording device or mechanism will indicate (or rather register) the energy supplied to the load; such a comprehensive device being called an "energy meter". Most (such) energy meters differ from wattmeters, or power measuring instruments, in the manner in which the time is accounted for.

DC and AC Energy

In the case of DC load, the electric power is simply

$$P = V \times I$$

and thus the energy is

$$E = (V \times I) \times t$$

In an AC system, the power and energy are

$$P = V \times I \times \cos\phi$$

and

$$E = (V \times I \times \cos\phi) \times t,$$

respectively.

If the power is constant in either case, the energy will simply be proportional to time.

Usually in a load, whilst the supply voltage may be essentially constant, the load current may vary[1] over a (long) period of time (or even the power factor may change with time in the case of AC load) and true energy consumed over the time period T will strictly be given by

$$E = \int_0^T p \, dt$$

where p represents the circuit power at *any instant*.

Thus, an energy measuring instrument or meter must incorporate a power sensing or measuring components, 'linked' to a suitable time integrating mechanism, and yet indicate directly the (total) energy consumed in the load.

Unit of electrical energy

The electrical energy is almost invariably measured in watt-hour or kilowatt-hour, abbreviated Wh (or kWh), when the power is expressed in watt (W) or kilowatt (kW), respectively, and the time of interest is reckoned in hours.

Thus,   1 kWh = 1 kW × 1 hr,

or an appropriate combination of "power" and "time"; for example, an electric bulb of 100 W rating will consume 1 kWh when lit for 10 hours.

Types of Energy Meter

There are broadly two types of energy meter, mainly according to the power supply in which they are used. Thus, there are

  A. meters for use in DC circuits

  B. meters for AC loads.

METERS FOR DC CIRCUITS

With the universal use of AC at present for all purposes, the meters/devices for DC are of rather academic interest and historic importance only, although with their some distinct advantages, there is no reason why these cannot be used in AC systems also, incorporating a suitable rectifier between the AC supply and the meter and following appropriate calibration, as shown in Fig.8.1. Here, the AC load current is converted into a proportionate DC voltage using a suitable CT and bridge rectifier whilst the supply voltage is fed to the energy meter (device) via another similar arrangement.

---

[1]Typical examples are the domestic loads at 1-phase or 3-phase supply and industrial loads comprising various types of load, including motors of different types, ratings and hence of various power factors.

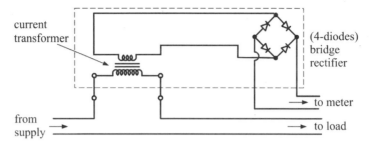

Fig.8.1 : Scheme for using a DC energy meter in AC supply

The energy meters for exclusive use in DC circuits are mainly of two types.

A. Electrolytic Meters

These are essentially *ampere-hour* meters as their readings are based on the mass of metal deposited, or the quantity of gas liberated, from an electrolytic solution owing to the flow of the direct current.

This means that the readings, or record, are merely indicative of number of ampere-hours, or coulombs, passed through the meter in a given interval (or hours)[1].

Although the scheme and construction being cumbersome in a real, practical sense, these meters are 'cheap' and simple in operation and have the advantage of good accuracy even at very small loads. They are also unaffected by stray magnetic fields since they do not depend on any magnetic effect of the current; and since there are no moving parts, frictional errors are altogether absent.

B. Mercury Motor Meters

The principle of operation of this type of ampere-hour or energy meter is illustrated in Fig.8.2.

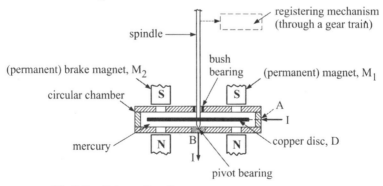

Fig.8.2 : Schematic of a mercury motor energy meter

---

[1]At a 'constant' supply voltage, however, the meter can be calibrated to record the energy consumed in joule or watt-hours.

Construction and action

The (direct) current I is fed into the meter at the point A on the circumference of a shallow circular chamber filled with mercury. A copper disc, D, mounted on the meter spindle is free to rotate within the chamber, in good contact with the mercury. The flow of current is from mercury into the edge of the disc, radially towards the centre, and then out through the lower, pivot bearing at point B as shown. The path of the current is under the influence of a permanent magnet M_1, located at a radial end. The interaction of the current with this field produces a driving or motor torque, making the disc to rotate at a certain speed; this torque being proportional to the (load) current. Another permanent magnet M_2 located nearly diametrically opposite to M_1, produces a braking torque, resulting from the interaction of the eddy currents produced in the (rotating) disc with the magnetic field due to M_2. A steady speed of rotation of the disc is achieved when the driving torque is numerically equal to the braking torque. The number of revolutions of the disc in a given time, or $\int_0^T N\,dt$, will thus be proportional to the total quantity of electricity, $\int_0^T I\,dt$, in ampere hours[1]. The spindle drives a registering mechanism, comprising a gear train and dials to read the energy consumed. Since the rotation of the disc is the main feature, the meter is affected by various friction and drag errors that may have to be compensated. Also, the speed of rotation versus the braking torque, may have to be suitably adjusted for a given load.

METERS FOR AC SYSTEMS

There are mainly two categories of energy meter for use in AC systems:

1. 'standard' electromechanical energy meters; also called the induction-type watt-hour (energy) meters

2. now increasingly in use the electronic or solid-state energy meters.

Induction-type Energy Meters

Till recently, these meters, also known as AC motor-type meters, have been in use universally for the measurement of a variety of energy in single- and poly-phase AC circuits, manufactured by hundreds of companies around the world with very close competition.

---

[1]Once again, at a constant supply voltage, the registered quantity can directly indicate the energy consumed in watt-hours.

Constructional features

This type of meter represents a unique example of a complex electromagnetic device. The essential constructional features of a single-phase, induction-type energy meter are illustrated in Fig.8.3[1].

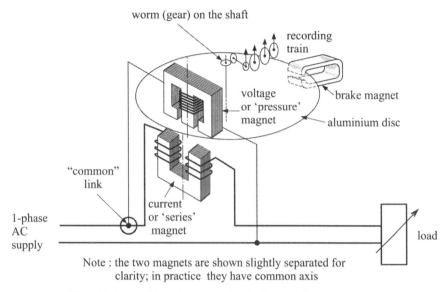

Fig.8.3 : Isometric schematic of a 1-phase AC energy meter

The meter mainly comprises

(a)   Pair of Electromagnets

The key part of the energy meter is a pair of (AC) electromagnets, mounted on the main frame of the meter *on to one side,* one above the other. The upper magnet, known as voltage magnet, carries the voltage (or pressure) coil on the central limb of an "E"-shaped laminated core and is wound with large number of turns to be connected across the supply. The lower, or the current magnet, is fitted directly beneath the voltage magnet in the same plane and carries a coil (known as current or series coil) on the adjacent limbs of a "U"-shaped core, also laminated. This coil is wound with fewer number of turns of wire of suitable cross-section or diameter so as to carry the normal, rated load current (or a fraction of it), and is connected in series with the load.

The vertical axis of the central limb of the voltage magnet is aligned with the mid central line of the current magnet as shown. When energised, the two magnets are responsible for producing the desired driving torque as discussed later.

---

[1]Clearly, the meter has some common similarity to mercury motor energy meter meant for DC.

(b)   The Aluminium Disc

An aluminium disc, about 8 cm in diameter and about 0.5 mm thick, is mounted on the same frame so as to be free to rotate in a horizontal plane, one end of the disc being in the airgap between the two electromagnets. The disc is made of high-purity aluminium, stippled to ensure absolute flatness and to prevent deformation due to stress release. The upper surface of the disc at the peripheral edge is marked with 100 equally-spaced graduations and the edge itself has nearly 250 regularly cut serrations for any stroboscopic measurement or testing of speed of rotation. As discussed later, two small (about 1.5 mm dia) anti-creep holes are drilled diametrically apart, about 1 cm from the edge to prevent creeping at no load.

(c)   The Brake Magnet

Nearly diametrically opposite to the location of the electromagnets is mounted a permanent magnet, called the "brake magnet", with an airgap through which passes the aluminium disc. The purpose of the brake magnet is to provide the necessary braking torque.

The magnet comprises a highly stable and strong ALNICO magnet piece of rectangular shape, possessing very high retentivity and energy content for long life and unaffected by usual ambient temperature variations. The magnet piece is fitted to a mild steel yoke or assembly, in turn fitted to the meter frame. For calibrating the meter at rated load, the magnet assembly is provided with a mechanism to adjust the radial orientation of the magnet pole with respect to the centre of the rotor disc or its axis of rotation by turning the pinion mating the assembly plate; the adjustment of the magnet can, in fact, be carried out both laterally in horizontal plane as well as in and out from the axis of rotation of the disc.

The design and fixture of the magnet assembly is such that the disc is easily removable without having to dismantle the magnet. The details of a typical brake magnet assembly are shown in Fig.8.4.

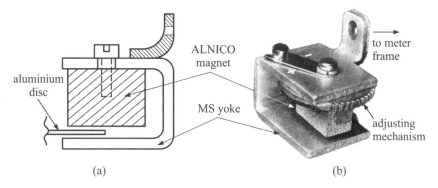

Fig.8.4 : Brake magnet assembly

(d)   Shaft (or spindle) and Bearings

The disc is fixed by die-casting process to a tubular (brass) light-weight shaft or spindle at right angles, such that the shaft is free to rotate about a vertical axis. The upper end of the shaft is fitted into a specially designed receptacle screwed (with a lock nut) to the meter frame. The receptacle is provided with a thin hard steel pin passing into the upper hollow end of the shaft, similar to a tiny bush bearing, allowing the shaft 'free' suspension and rotation.

The lower end of the shaft is fitted with a tiny polished, hard steel ball, resting on a highly polished jewel or synthetic sapphire bearing which carries the weight or thrust of the moving system. The sapphire which is cut with its axis correctly oriented so as to match the axis of rotation of the disc is mounted and housed in an accurately machined holder. The holder is threaded on the outside for being fitted to the lower part of the meter frame and is capable of vertical adjustment for proper 'alignment' of the disc between the airgaps of the two electromagnets as also the brake magnet. The bearing holder can be locked in its final position with the help of a lock nut.

(e)   Registering Mechanism

To record or for direct reading of the energy consumption, a special "windows type" recording train or clock mechanism of the cyclometer type is fixed on the meter frame, with a worm gear fitted on the upper side of the shaft and a gear assembly, driving a number of meshed gears to record the consumption normally from one tenth of a unit of energy (or kWh) to 9999 units, that is, having 5-digit wheels reading in units of 1000, 100, 10, 1 and 0.1[1]. The digit wheels or gears and 'transfer' pinions are made of stable metal so as not to be adversely affected by ambient conditions. Fig.8.5 shows a typical registering mechanism.

Fig.8.5 : A meter registering mechanism

(f)   Compensation Mechanisms

The meter is invariably provided with various compensation mechanisms, usually attached to the two electromagnets and also supported on the meter frame to compensate for "friction" and "phase angle" errors as discussed later. These mechanisms are in the form of adjustable devices, the position of which relative to the magnets can be altered/adjusted by loosening and tightening of appropriate screws.

---

[1]With the 'reading' on the recording device returning to 0000 at the end, and starting all over again.

(g)   Terminal Block

The lower end of the meter frame is fitted with a terminal block where the ends of the two (magnets') coils are terminated. The block also facilitates external connections to the supply and load as shown in Fig.8.6. The block is provided with a cover plate for safety and having a tamper-proof sealing arrangement.

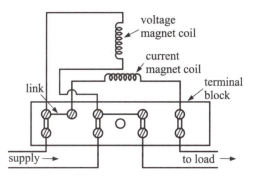

Fig.8.6 : Schematic of terminal block of an energy meter

(h)   The Case

The entire meter assembly is housed in a metal case, having special fitting screws (to be tightened from the back) with holes to thread a seal to prevent tampering of the meter, and a glass cover on the upper side to allow meter readings to be monitored or recorded.

A picture of the inside of a typical single-phase energy meter indicating various key parts is shown in Fig.8.7.

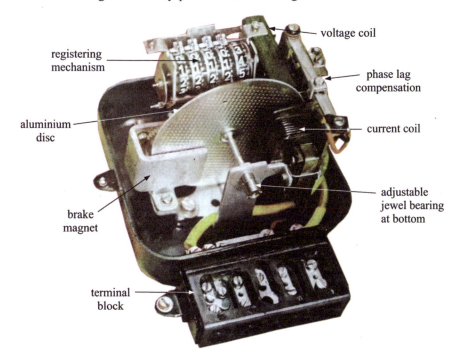

Fig.8.7 : A picture showing constructional details of an actual energy meter

Theory of Operation

Production of driving torque

Magnetic Fields Due to Electromagnets

Voltage magnet and coil

This magnet carries a coil of several hundred turns of thin wire, is connected across the supply and thus carries a current proportional to the supply voltage. The nearly closed magnetic circuit of this coil produces a very high ratio of inductance to resistance, thus causing the coil current, and hence the flux in the core due to it, to lag the supply voltage by nearly $90°$ – an important requirement for error-free operation of the meter as discussed later. The major portion of the flux passes across the narrow gap between the central and side limbs, but some amount also passes through the plane of the disc as shown in Fig.8.8(a). The flux actually passing into the aluminium disc is the useful flux, responsible for production of driving torque.

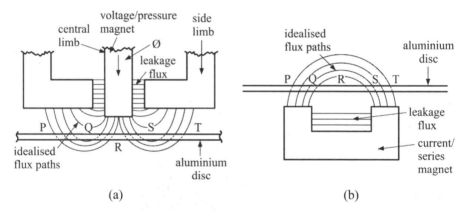

Fig.8.8 : Magnetic fluxes due to the two electromagnets

Current magnet

The current magnet consisting of a coil wound with a small number of turns and inserted on the two side limbs of the core carries the load current, either directly or from the secondary of a CT. The magnet flux due to this magnet crosses into the disc as shown in Fig.8.8(b) and is essentially *in phase* with the current. There is also some unavoidable leakage flux across the two limbs owing to the "open" construction of the core.

It is the *simultaneous* presence of the two fluxes in the disc at common regions marked Q and S in Fig.8.8(a) and (b), and their interaction with each other involving induced eddy currents in the disc that gives rise to the required driving torque. This, in fact, forms an important requirement of all induction-type instruments, in general.

Single electromagnet and aluminium disc: zero net torque

Consider a flux given by $\emptyset = \emptyset_m \sin\theta$, $\theta = \omega t$. Suppose it produces eddy currents in the disc such that $i = I_m \sin(\theta - \alpha)$, the induced currents lagging the applied AC flux by an angle α. The two quantities interact such that an instantaneous torque is produced, given by

$$T_{inst} \propto \emptyset\, i$$

and the mean torque

$$T_m \propto \frac{1}{\pi} \int_0^\pi \emptyset\, i\, d\theta$$

or

$$T_m = K\frac{1}{\pi} \int_0^\pi \emptyset_m\, I_m\, \sin\theta\, \sin(\theta - \alpha)\, d\theta$$

where K is a constant.

or

$$T_m = \frac{K\,\emptyset_m\, I_m}{2\pi}\left[\theta\cos\alpha - \frac{\sin(2\theta - \alpha)}{2}\right]_0^\pi$$

$$= \frac{K\,\emptyset_m\, I_m}{2\pi}\left[\pi\cos\alpha\right]$$

or

$$T_m \propto \emptyset_{rms}\, I \cos\alpha$$

where $\emptyset_{rms}$ and I are the "rms" values of the flux and current, respectively.

Now if there were an aluminium disc[1] under the influence of a single electromagnet, excited by AC supply as shown in Fig.8.9(a), the phasor relationship between the various electrical quantities will be as shown in Fig.8.9(b).

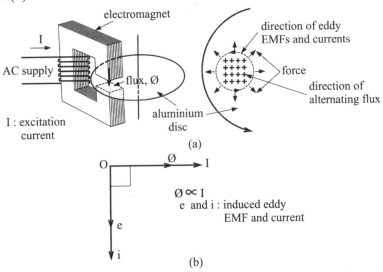

Fig.8.9 : Single electromagnet and induced eddy currents in an aluminium disc

---

[1]The aluminium disc could, in effect or in theory, be replaced by any metallic disc.

Clearly, the phase angle between the flux and the induced eddy currents, that is α, is 90° and hence the mean developed torque is zero. The disc would thus remain stationary and the effect of eddy current will only be to heat up the disc.

As brought out later, it is imperative to provide *at least two* alternating fluxes simultaneously and two distinct eddy current paths in order to produce a useful driving torque in the disc. The process involved for the purpose is known as "phase splitting" of fluxes and incorporates

(a) electric or (b) electromagnetic means.

These are

(a) Using a "second" (electro)magnet that is *displaced in space,* in which the winding is, for example, shunted by a suitable resistor so that the currents in the two windings differ in phase from each other.

(b) Using a single magnet still, but with a slit in the 'pole' of the magnet around one part of which is fitted a copper shading band. The eddy currents in the short-circuited copper ring and flux due to these then cause the flux in this part of the magnet to be phase displaced from the flux in the un-shaded portion.

The two schemes are illustrated in Fig.8.10.

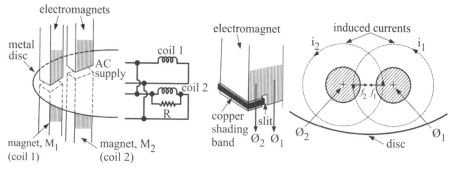

(a) electrical phase splitting   (b) electromagnetic phase splitting

Fig.8.10 : Electrical and electromagnetic phase splitting to produce phase-displaced fluxes

The phasor diagram corresponding to shading-ring scheme is shown in Fig.8.11, bringing out the essence of production of non-zero deflecting torque. With proper design, it is possible to achieve desirable phase displacement and obtain maximum driving torque.

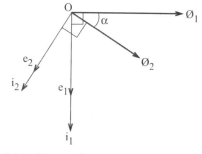

Fig.8.11 : Phasor diagram for copper shading band scheme

General theory of torque production with phase splitting

Refer to Fig.8.12 which shows the arrangement of two alternating fluxes acting on a thin aluminium disc, D, free to rotate about a vertical axis through its centre. P_1 and P_2 are two laminated iron poles producing fluxes $\emptyset_1$ and $\emptyset_2$, the fluxes impinging vertically downward on the disc. Let e_1 and i_1 are the induced eddy EMF and current in the disc due to $\emptyset_1$ and e_2 and i_2 those due to $\emptyset_2$, respectively, as shown in the figure. The directions of i_1 and i_2 are as given by the Lenz's law, assuming the fluxes $\emptyset_1$ and $\emptyset_2$ being as directed and both *increasing* at the instant of consideration.

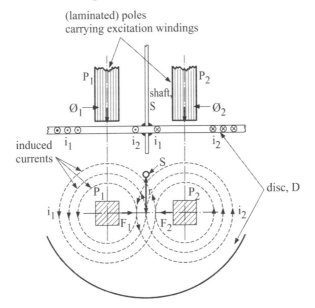

Fig.8.12 : Two alternating fluxes acting on an aluminium disc

Let
$$\emptyset_1 = \emptyset_{1m} \sin \omega t$$
$$\emptyset_2 = \emptyset_{2m} \sin (\omega t - \alpha)$$

where α is the phase difference between $\emptyset_1$ and $\emptyset_2$ as a result of "phase splitting".

Then
$$e_1 = \frac{d\emptyset_1}{dt} = \omega \, \emptyset_{1m} \cos \omega t$$

and
$$i_1 = \frac{e_1}{R} = \frac{\omega \, \emptyset_{1m}}{R} \cos \omega t$$

where R represents the resistance of the path of eddy currents in the disc, assumed resistive, that is, neglecting inductance.

Similarly,

$$i_2 = \frac{\omega \, \varnothing_{2m}}{R} \cos(\omega t - \alpha)$$

assuming 'symmetrical' location of the poles P_1 and P_2 about the axis of disc rotation and thus identical path(s) for eddy current(s) i_2, and same resistance R.

The interaction of eddy currents with fluxes of appropriate poles results in production of forces according to the law $F = B \, i \, l$.

Thus, the force due to P_1 or $\varnothing_1$, interacting with currents i_2 is given by[1]

$$F_1 = K \, \varnothing_1 \, i_2, \text{ where K is a constant,}$$

directed as shown.

Similarly,

$$F_2 = K \, \varnothing_2 \, i_1, \text{ assuming same constant, K.}$$

If the forces were acting at an effective distance r from the centre of the disc or axis of rotation as shown, the corresponding instantaneous torques will be

$$T_1 = F_1 \, r \quad \text{and} \quad T_2 = F_2 \, r$$

and the net, resultant *instantaneous* torque on the disc to cause its rotation (in *either* direction) will be

$$T = r \times [K \, \varnothing_1 \, i_2 \sim K \, \varnothing_2 \, i_1]$$

Substituting for the instantaneous values of $\varnothing_1$, $\varnothing_2$, i_1, i_2,

$$T = \frac{K \, r \, \omega}{R} \left[\varnothing_{1m} \sin \omega t \, \varnothing_{2m} \cos(\omega t - \alpha) \sim \varnothing_{2m} \sin(\omega t - \alpha) \, \varnothing_{1m} \cos \omega t \right]$$

$$\text{or } T = \frac{K \, r \, \omega}{R} \, \varnothing_{1m} \varnothing_{2m} \left[\sin \omega t \, \cos(\omega t - \alpha) - \cos \omega t \, \sin(\omega t - \alpha) \right]$$

$$\qquad\qquad\qquad\qquad\qquad\qquad - \text{ assuming I term} > \text{II term}$$

$$= \frac{K \, r \, \omega}{R} \, \varnothing_{1m} \varnothing_{2m} \sin \alpha, \text{ by expanding the trigonometric terms on the RHS and simplifying}$$

$$= K' \, \varnothing_{1m} \, \varnothing_{2m} \sin \alpha, \text{ assuming R, r and } \omega \text{ to be constants,}$$

---

[1]It is the interaction as mentioned that would produce any non-zero, useful torque; the interaction of $\varnothing_1$ with i_1 would yield zero useful torque and result only in heating due to $i_1^2 R$ loss.

and will be a maximum when $\alpha = 90^{o}$[1].

Also, since the torque expression is independent of ωt or t, the torque has the same value at all instants of time.

It is clear that for a large torque, the eddy currents path resistance, R, must be as small as possible. Therefore, the disc material should be of low resistivity; for example, copper or aluminium. Owing to lesser density and cost, aluminium is invariably preferred, resulting in quite high torque/weight ratio.

Expression for meter torque in practice

In an actual energy meter, the two fluxes are produced by the two electromagnets – voltage and current - by virtue of the currents in the windings, being proportional to the supply voltage and the load current, respectively. Thus, the net driving torque, and the speed of rotation of the disc, should be proportional to the power in the load and the number of revolutions in a given time must correspond to the *energy* consumed. Note that, in an energised meter, the disc revolves continuously at a certain speed and hence EMFs will be induced in the disc *dynamically* (according to the B *l* v rule) as it cuts through the (net) flux of the two electromagnets. This is additional to the *statically* indeed EMFs due to the alternating flux of the magnets impinging on the disc as discussed before. The full-load speed of rotation in most meters in use is, however, only about 40 or 50 rpm and, thus, the dynamically induced EMF is much smaller in magnitude compared to the statically induced EMF $\left(\text{according to d}\emptyset/\text{dt action}\right)$ at normal supply frequency of, say, 50 Hz, and (the former) can be disregarded.

Theory

In the given single phase circuit comprising the supply and load, let the applied voltage be given by

$$e = E_m \sin \omega t$$

---

[1]It follows that if α were zero, that is, the two fluxes were derived from one single pole, as considered earlier, the net torque will be zero.

Further, if the reactance X of the induced currents path were taken into account, then R must be replaced by Z where $Z = \sqrt{R^2 + X^2}$, and the induced currents will lag behind the respective EMFs by an angle $\beta \left(= \tan^{-1} \dfrac{X}{R} \right)$. The torque expression will then be

$$T = \frac{K \, r \, \omega}{Z} \emptyset_{1m} \, \emptyset_{2m} \sin \alpha \, \cos \beta$$

In normal practice, $\beta \simeq 10°$, or very small, and may be ignored.

and the load current $i = I_m \sin(\omega t - \phi)$

at any instant, where ϕ is the angle of lag of the current with respect to e.

   The flux in the current (or series) magnet will then be

$$\emptyset_{se} = K\, I_m \sin(\omega t - \phi),\ K \text{ a constant,}$$

being essentially in phase with it.

   From the expression of $e = \dfrac{d\emptyset}{dt}$, the flux in the voltage coil will be

$$\emptyset_{sh} = K' \int e\ dt = -K'\,\frac{E_m}{\omega}\cos\omega t\,,\ K' \text{ another constant}$$

The induced EMF due to the flux $\emptyset_{se}$ is given by

$$e_{se} = -\left(\frac{d\emptyset_{se}}{dt}\right) = -K\,I_m\,\omega\cos(\omega t - \phi)$$

and the corresponding induced eddy currents

$$i_{se} = \left(\frac{e_{se}}{Z}\right) = -\frac{K\,I_m\,\omega}{Z}\cos(\omega t - \phi - \beta)$$

assuming a general case of Z being the impedance of the eddy currents path and β its phase angle. Similarly, the induced EMF due to flux in the voltage magnet and corresponding eddy currents are given by

$$e_{sh} = -\left(\frac{d\emptyset_{sh}}{dt}\right) = -K'\,E_m\,\sin\omega t$$

and

$$i_{sh} = \left(\frac{e_{sh}}{Z}\right) = -\frac{K'\,E_m}{Z}\sin(\omega t - \beta)$$

assuming the same impedance for the eddy current path(s).

   The *instantaneous* torque in the disc is then produced by the interaction of $\emptyset_{sh}$ with i_{se} and $\emptyset_{se}$ with i_{sh}, the net torque being

$$T_{inst} = \frac{KK'}{Z}E_m I_m\Big[\cos\omega t\ \cos(\omega t - \phi - \beta) + \sin(\omega t - \phi)\ \sin(\omega t - \beta)\Big]$$

and the mean torque

$$T_m = \frac{KK'}{Z}E_m I_m\,\frac{1}{2}\Big[\cos(\phi + \beta) + \cos(\phi - \beta)\Big]$$

$$= \frac{KK'}{Z}E_m I_m\,\cos\beta\ \cos\phi$$

$$= \left[\frac{KK'}{Z}\cos\beta\right]E\,I\cos\phi$$

where E and I are the rms values of the supply voltage and load current, respectively.

If Z and β are assumed constant,

$$T_m \propto E\,I\cos\phi,$$

that is, power in the circuit, and is independent of the supply frequency. Further, since the torque does not depend on a term involving time, it does not fluctuate cyclically, resulting in a *steady* speed of rotation of the disc.

The phasor diagram pertaining to the above case is shown in Fig.8.13. Note the position of the phasors representing fluxes $\varnothing_{se}$ and $\varnothing_{sh}$ and the angle 'α' between them. As deduced earlier, α should be nearly $90°$ for maximum driving torque.

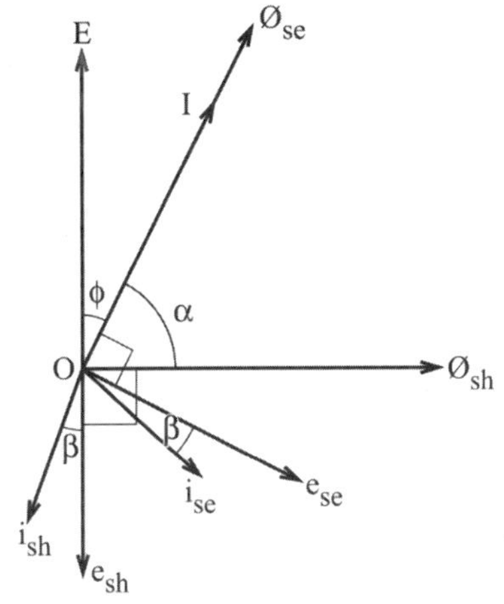

Fig.8.13 : Phasor diagram of induction-type energy meter

Braking torque

The part of a meter showing location of the brake magnet is reproduced in Fig.8.14.

Let the speed of rotation of the disc corresponding to a given load is N rpm or ω rad/s. If the gap flux density is B, then, neglecting fringing, the EMF induced by the "flux-cutting rule" (e = B*l*v), in an effective conductor length in the disc, will be given by

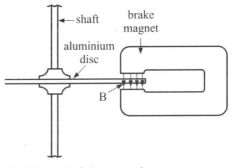

Fig.8.14 : Brake magnet in an energy meter

$$e = K_1\,\varnothing\,\omega \quad\text{or}\quad K_2\,\varnothing\,N$$

where K_2 is a constant to account for the 'conductor' length, area of pole faces (of the magnet), angular speed ω and the effective distance of 'conductor' from the axis of rotation. If r represents the resistance of the currents in the disc caused by the induced EMF, e, then

$$i = \frac{e}{r} = \frac{K_2\,\varnothing N}{r}$$

Now the braking torque will be produced by the interaction of i with Ø, or the braking torque,

$$T_B = K_3\,\varnothing\,i\,R, \quad K_3 \text{ being another constant,}$$

at an 'effective' radius R from the axis of rotation.

Substituting for i

$$T_B = K \, \emptyset^2 \, N \left(\frac{R}{r} \right), \quad \text{where } K = K_2 K_3$$

For a *permanent* magnet $\emptyset$ is constant. If R and r are also assumed constant in a practical meter

$$T_B \propto N$$

Hence, a steady speed of rotation of the disc is attained when the deflecting or operating torque due to power in the load is equal to the braking torque produced by the brake magnet[1].

Meter Constant

A quantity of practical importance related to an energy meter, in association with the speed of rotation, is the "meter constant", defined (and specified on the meter "name plate") as the number of revolutions made by the disc corresponding to 1 unit of electricity or one kWh. The actual value of the constant for a given meter would naturally depend on its design and may vary from 500 to 3000 rev/kWh, usually in inverse proportion to the current rating at the same rated voltage.

Typical Errors in an Energy Meter and Their Compensation

There are FOUR important sources of error in an energy meter that are inherent. Special provisions are made to compensate for these errors.

Frictional error and 'light-load' compensation

A continuous frictional error largely occurs at the lower, jewel bearing which is usually designed as highly-polished cup-shaped jewel seat at the centre of which rests the steel ball fitted at the lower end of the tubular shaft carrying the disc. This would theoretically mean a 'point' contact between the meter shaft and the bearing. However, even with best efforts and every care taken to design and manufacture the bearings, both the lower jewel as well as the top pin-type, there can be appreciable friction so as to seriously affect the rotation of the disc, particularly at light load (that is, a small fraction of the full-load current) when the 'normal' torque as computed above may be very low[2]. In order to overcome the bearing friction and ensure accurate registration, it is imperative to provide in the meter a small

---

[1]Since the braking torque is proportional to square of the flux in the gap of the magnet it emphasises the necessity of quality of the magnet; the gap flux must not alter with time or with other parameters such as the ambient temperature.

[2]The situation may become progressively worse in course of time – say, in a few years – when the rotation of the disc may become quite sticky, the meter slowing down and recording low consumption. Whilst this may be in the interest of the consumer, the utility or supplier may suffer loss of revenue!

additional torque as nearly as possible equal to the frictional torque, practically independent of the load on the meter and which acts in the direction of rotation.

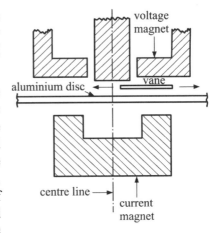

Fig.8.15 : Arrangement for friction compensation

The compensation is usually in the form of providing a small, thin metallic 'shading' vane, situated between the center limb of the voltage magnet and the disc and slightly to one side of the limb centre-line as shown in Fig.8.15. The interaction between the portions of the flux which are shaded and un-shaded by the vane and the (eddy) currents induced by the fluxes in the disc results in a small driving torque to compensate for the bearing friction. To adjust the magnitude of the torque so-developed, the position of the vane to the left or right can be altered by turning a screw provided for the purpose. The final position of the vane can be locked by means of another pair of screws. The meter is tested for 5% of the full load, rated current, at rated voltage and unity power factor to check for compensation of the frictional error[1].

Comment

Despite the compensation for friction having been carried out at the recommended low load current, there may be some "very light" loads that may go un-registered owing to the load current being too small to overcome the friction. Consider, for example, a meter rated at 1-phase, 230 V, 5 A, 50 Hz. The compensation is carried out at 5% of 5 A or 250 mA. With the load consisting of so-called 'night bulbs', rated at 10 W @ 230 V, each bulb would draw 43.4 mA. Therefore, electric consumption of even 5 such bulbs lit at the same time at night *with no other load on*, will go un-registered by the meter.

The second instance is that of multitude of devices operating at 1.5 to 6 V; for example, cell-phone chargers, musical devices, small

---

[1]A modern, innovative development to eliminate bearing friction at lower end nearly completely is the increasing use of "magnetic suspension" bearing. This comprises two small, flat permanent magnets, about one cm in diameter, one magnet being attached to the shaft and the other to the meter frame incorporating a specially designed assembly which allows the adjustment of the relative position of magnets. The force of repulsion between the magnets with the same polarity opposite each other results in the desirable contact-free bearing action.

'transistor' radios and so on. These operate typically at current levels of a few mA, DC. On the AC side, the current may be less than a mA and will certainly not be registered by the meter. The situation can be alarming when millions of consumers use such devices at the same time, leading to appreciable loss of revenue to the utility!

Full-load rotational error and compensation

The meter is designed to perform satisfactorily when the driving torque at any load is equal to the braking torque due to the brake magnet. That this is so is ensured whilst the meter is carrying full/rated current, at rated supply voltage and unity power factor when the developed or driving torque is at its maximum. Then, if the two torques are not balanced, the meter will tend to run fast or slow depending on relative values of the torques. With the driving torque as obtained above, the braking torque is adjusted/made equal to the driving torque by

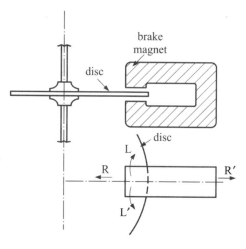

Fig.8.16 : Compensation for (full-load) rotational speed

adjusting the position of the brake magnet, either in the radial direction $R-R'$ (that is, moving the magnet towards or away from the shaft), or laterally by rotating it about its vertical axis $(L-L')$ as shown in Fig.8.16.

Similar to the friction compensation, the adjustment is carried out by means of a screw and locking the magnet in position after final adjustment[1]. [See Fig.8.4(b)].

Power-factor error and "lag" adjustment or compensation

It is an essential requirement for accurate registration of energy in the meter that the flux due to the voltage magnet should lag the applied voltage by exactly $90°(E)$ as had been assumed in working out the theory of operation of an induction-type meter in general. However, in practice, the angle may be less than $90°$, mainly owing to the pressure-coil resistance. As a result, the meter which has been adjusted for the position of the brake magnet under full-load, unity power-factor condition to register correctly at "healthy"

---

[1]Typically, only one adjustment is carried out, viz. at full load and unity power factor; it being assumed that the adjustment (or equality of the two torques) will hold at other values of power, too.

power factors approaching unity, will be greatly in error at other power factors, particularly at lower *lagging* power factor loads.

Provision must, therefore, be made to correct the above "angle of lag". Fig.8.17 shows the method of compensation for this requirement in the form of a shading band or loop (of a few turns of copper wire) placed surrounding the lower end of central limb of the voltage magnet, the two ends of the loop being taken out and connected to a resistive wire bent as shown in the figure. The effective length of this bent loop can be adjusted by means of the shorting clamp, fitted with a screw. Thus, depending on the clamp position, the total length of the shading loop and hence the induced current through it can be altered. The flux produced by this current when added to the main flux results in the total flux which is almost exactly in phase quadrature with the applied voltage as depicted by the phasor diagram in Fig.8.17. With this compensation, the meter is made to register correctly for loads varying in power factors from 0.5 lag to unity[1].

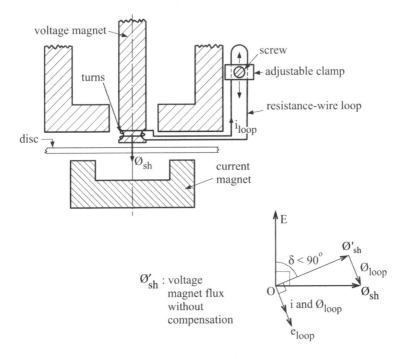

Fig.8.17 : Phase lag compensation

---

[1]See schedule of tests discussed later for the detail of test for this compensation.

The Case of Phase Angle between Supply Voltage and Flux due to it being Less than 90°

For a general power factor angle, ϕ, of the load, less than unity and lagging or leading, a correction in the expression for the driving torque may be required when the flux due to the voltage magnet is out of phase by an angle less than 90°, notwithstanding the compensation for the shortcoming. This condition is expressed by the phaser diagram of Fig.8.18.

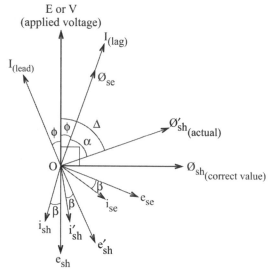

Fig.8.18 : Phasor diagram for the case of $\Delta < 90°$

A. Assume first that the flux due to voltage magnet is lagging exactly by 90°, that is, $\Delta = 90°$.

Then as derived earlier, the driving torque,

$$T_d = K\,\varnothing_{sh}\,\varnothing_{se}\,\cos\beta\,\cos\phi$$

where β is the angle by which eddy currents lag the corresponding induced EMFs in the two coils.

Since $\varnothing_{sh} \propto V$ and $\varnothing_{se} \propto I$,

$$T_d = K'\,V\,I\,\cos\beta\,\cos\phi,$$

and with $\beta \simeq 0$ and $\cos\beta \simeq 1$,

T_d will simply be $K'\,V\,I\,\cos\phi$

or $T_d \propto V\,I\,\cos\phi$, as derived before.

If α is the phase angle between I and $\varnothing_{sh}$, or between $\varnothing_{se}$ and $\varnothing_{sh}$

$$\alpha = 90 - \phi \text{ and } \cos\phi = \sin\alpha$$

Then $T_d = K'\,V\,I\,\cos\beta\,\sin\alpha$

B. Now consider the case when $\Delta < 90°$ and $\alpha = \Delta - \phi$

Then $\qquad\qquad T_d' = K'\,V\,I\,\cos\beta\,\sin(\Delta - \phi)$

and the error, on account of the flux due to voltage magnet being not exactly lagging by 90°, will be

$$= \frac{T_d' - T_d}{T_d} \times 100\%$$

or

$$= \frac{\sin(\Delta - \phi) - \cos\phi}{\cos\phi} \times 100\%$$

Example

A single-phase energy meter when tested at normal rating of 240 V, 10 A is 1% slow at unity power factor. Estimate the error at rated volt-ampere when the load factor is

(i) 0.8 (lag),   (ii) 0.8 (lead).

The meter is 1 % slow at 1.0 pf

That is

$$\frac{\sin(\Delta - \phi) - \cos\phi}{\cos\phi} \times 100 = -1$$

Putting $\phi = 0$ and $\cos\phi = 1.0$

$$\text{error} = \frac{\sin\Delta - 1}{1} = -0.01$$

from which

$$\Delta = \sin^{-1}(0.99) \text{ or } 81.89° \text{ (as against } 90°)$$

(i) At 0.8 (lag) power factor, that is $\phi = 36.86°$

$$\text{the error} = \frac{\sin(81.89° - 36.86°) - 0.8}{0.8} \text{ or } -11.56\%$$

showing that the meter is considerably *slow*.

(ii) At 0.8 (lead) power factor, that is $\phi = -36.86°$

$$\text{the error} = \frac{\sin(81.89° + 36.86°) - 0.8}{0.8} \text{ or } +9.59\%$$

the meter being appreciably fast.

If calculated at other power factors; for example at 0.5 lag or lead, the errors could be worse, being − 25.43% at 0.5 lag pf and + 23.43% at 0.5 lead pf.

The mechanisms for various compensations in an actual meter are depicted in Fig.8.19.

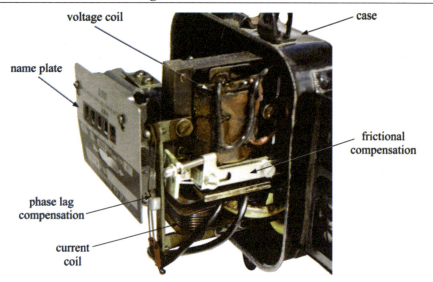

Fig.8.19 : Compensation mechanisms in a typical meter

"Creep" and its elimination

In some meters a slow, but continuous rotation of the disc is observed when only the voltage magnet is excited, the (load) current through the current/series magnet being zero. If unchecked, the meter will register 'consumption' over a period of time which can be appreciable. This slow and unwarranted disc rotation is called *creep* and may usually be due to

(a)   incorrect or over compensation of the meter for frictional load;

(b)   appreciably higher supply voltage, much above the normal, fed to the voltage magnet coil.

The 'standard' method to eliminate creep in modern meters is to drill two small holes in the disc at opposite ends, about 2 mm in diameter and about 1cm inside from the outer edge. The disc will come to rest when one of the holes comes near the edge of the *leading* pole of the voltage electromagnet. The action may be understood by reference to the flux paths shown in Fig.8.20.

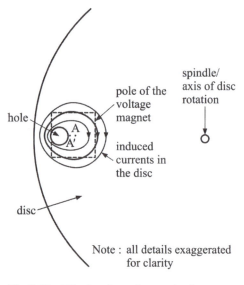

Fig.8.20 : Elimination of creep in the meter

When one of the holes is under the edge of the pole of the voltage magnet, the usual circular eddy currents in the disc (around the hole) are distorted as shown, becoming somewhat elliptical (or 'stressed'), causing the effective centre of the equivalent magnetic pole due to eddy currents to shift towards A′. A very small torque is thus produced (by current and flux interaction) tending to align A, A′ along the line/axis A–A′, A being the 'centre' of the magnet pole.

Thus, the disc may creep until the hole reaches a position as defined above. Clearly, the rotation of the disc due to creep is limited to a maximum of 180°.

In some meters, a tiny piece of iron wire is attached to the edge of the disc. The force of attraction of the brake magnet upon this iron wire would then prevent continuous rotation of the disc under no load conditions.

Other meter errors

In addition to the errors due to incorrect settings, and steps taken to set them right as described above, there are a few more errors that might affect the operation of an energy meter. Some of these are considered below.

Error due to Variation of Supply Voltage

Apart from the change of driving torque according to $T_d \propto V \, I \cos \phi$, a variation of supply voltage is accompanied by two effects:

  (i)   a slight change of braking torque for although in an actual meter this torque is largely proportional to the EMF and currents induced in the disc due to flux of the permanent magnet, it is partly also due to cutting of the alternating flux(es) of the voltage and current magnets. For this reason, the flux due to the permanent magnet is kept as high as possible to swamp the effect of the alternating fluxes of the main electromagnets

  (ii) the voltage variation may have some effect on the phase compensation which is normally carried out at rated supply voltage.

Usually, the supply is maintained at the specified, rated value and thus any occasional variation for small intervals may not seriously affect the meter operation.

Error due to Overload

The term 'overload' here signifies a current through the current coil, which even though not large enough to cause damage to the coil, is much in excess of the marked current rating for the meter. A direct consequence of very high current in the current coil is that the meter tends to *run slow* owing to saturation of the current magnet core, and a slight increase in the braking torque produced by the flux of this magnet. Accordingly, the normal operating flux of the current magnet is kept as small as possible, corresponding to the near linear part of the B-H curve, much below the saturation level.

Error due to Change of Frequency

The expression for the driving torque of an induction-type meter shows that the torque is essentially independent of the supply frequency. This is true for the main operation; however, there are a few possible effects of frequency that might modify the behavior of the meter to some extent:

For example,

(i)   alteration of the ratio of reactance (changed value owing to change of frequency) to resistance in the voltage magnet coil, to affect the *phase* of the coil current;

(ii) alteration of the reactance of the phase-compensation loop that may further affect the phase of the voltage flux;

(iii) changes in the reactance of the induced-current paths in the disc, resulting in changed magnitudes and phase of these currents.

However, the variation of supply frequency is usually kept within close limits (± 3% of the rated value)[1] in a power system, as stated before, for more important requirements such as to maintain stability in modern, complex inter-connected power systems, and hence the effect on meter operation might not be serious.

Error due to Change of Temperature

The main effect of temperature variation (usually the ambient temperature) on the action of a meter is to alter the resistance of the induced-currents paths in the aluminium disc that might affect both the driving and braking torques. An increase of temperature may also increase the resistance of the voltage magnet coil which consists of large number of turns using 'thin' wire and thus alter the ratio of coil inductance to resistance. The other effect is possible changes in the magnetic properties of the brake magnet; however, the use of modern, high-quality materials (ALNICO[2] being one of them) for these magnets eliminates the possibility of change of magnetism with temperature.

---

[1]More stringently allowed to vary only between 49.5 and 50.5 Hz in the case of huge inter-connected systems when the normal frequency is 50 Hz.

[2]ALNICO is a costly permanent-magnet material and hence only the minimum possible size for the brake magnet is used; usually of rectangular shape that can be easily manufactured, fitted to the mild steel yoke. [See Fig.8.4].

Error due to (incorrect) Mounting

In order that the meter may operate without any lateral drag and therefore altered frictional characteristics, the meter must be mounted on the distribution panel in an essentially vertical position.

TESTING OF ENERGY METERS[1]

Owing to long-term repercussions of errors compounded or 'integrated' over the period of use, all energy meters undergo thorough testing at works before put to service. These usually include tests at various loads and power factors.

Schedule of Checks and Adjustments

1. Check for Creep

   The voltage magnets are fed from a stabilised supply at rated PLUS 10% voltage whilst the current magnets are un-excited. The disc rotation is observed; the disc coming to rest as soon as one of the holes reaches the edge of the leading pole of the voltage magnet. No adjustment is usually warranted.

2. Long Duration Test

   The current fed to the meter is adjusted to rated/full load current at *unity* power factor. The time of a given number of revolutions, say 40, of the disc is compared with that of the "rotational sub-standard" (RSS) energy meter. The position of brake magnet is adjusted in the optimum direction depending on whether the meter is running too slow or too fast. [See Fig.8.16].

3. Light Load Adjustment

   In this test the current is adjusted to 5% of the rated value, at unity power factor and the speed of rotation of the meters is observed; alternatively, the time for a given number of revolutions is measured, or vice-versa, that is, the number of revolutions in a given time.

   The adjustment for correction of speed is carried out by means of position of the vane under the voltage magnet as discussed previously.

4. Lag Adjustment

   This adjustment is carried out by exciting the current magnet at rated current, but at 0.5-lag power factor. The speed of rotation of the disc is

---

[1]In the testing lab of a meter manufacturing industry, a number of meters, usually up to 20, are tested at a time, all hung in a row (in a vertical position) on a test bench, with their current magnets connected in series and voltage magnets in parallel. The testing scheme and equipment, discussed later, is thus common to all the meters and the operator checks and makes necessary adjustments in succession, moving from one meter to the next.

adjusted by altering the position of the shorting clamp across the wires of the loop as shown in Fig.8.17.

The cycle of steps 2, 3 and 4 is repeated till all the meters check out. A meter not capable of adjustment after repeated cycle of checks is discarded.

Connection Schemes for Testing

Since all AC (induction-type) energy meters are essentially tested at unity as well as 0.5 lag power factor at various load current and rated supply, it is imperative that suitable means must be provided to not only vary the load current, but also to alter the phase angle between the supply to the load and the current through it.

Apparatus Required

The various apparatus required to meet all the test requirements are:

(a) a sub-standard dynamometer wattmeter

(b) sub-standard (dynamometer-type) voltmeter and ammeter

(c) a Rotating Sub-Standard (or RSS), that is, an energy meter of sub-standard accuracy

   – all the above must be of appropriate ranges, matching the specification of the meter(s) under test

(d) a stop watch or any other suitable timing device for accurate measurement of time

(e) a device capable of altering (and measuring) the phase difference between the voltage and load currents[1]; this can be in the form of

   (i)   a motor and two alternators set, one of the two alternators being cradle mounted

   (ii) a phase shifter or phase-shifting transformer as described previously. [See Chapter VII: Measurement of Power].

Method using direct loading

In this method, the circuit diagram for which is shown in Fig.8.21, the connected load is an actual load at a chosen power factor and read by the wattmeter, W. The disc of the meter starts rotating as soon as the supply is switched on. From an instant, usually with reference to the red mark corresponding to "zero" graduation on the disc as it passes, say, the leading

---

[1]In its simple form, a unity p f load can be a variable resistor, or a bank of incandescent lamps of different wattage. The lagging or leading p f load can be arranged, respectively, by a combination of variable resistance and inductance, or variable resistance and capacitance, connected in series.

edge of the brake magnet, the time on the stop watch is noted for a given number of revolutions (which can be very large in number for a 'long duration' test) and the selected load. From the meter constant, specified on the name plate, the power corresponding to the recorded number of revolutions is worked out and compared with the true power measured by the wattmeter.

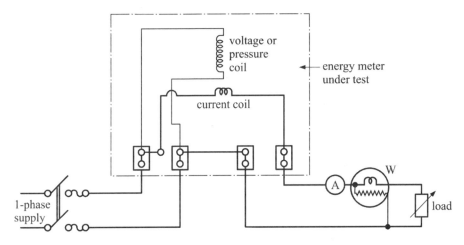

Fig.8.21 : Circuit diagram for testing of meter by direct loading

Then

$$\text{the \% error} = \frac{\text{'power' recorded by the meter} - \text{wattmeter reading}}{\text{power in the meter}} \times 100\%$$

that is, $\% \text{ error} = \dfrac{\text{apparent power} - \text{actual power}}{\text{apparent power}} \times 100\%$

[Here the term "power" may also be interpreted as energy].

However, the method is cumbersome and wasteful owing to considerable heat loss in the resistors or other loading devices.

Methods using "phantom" loading

When the rating of the meter to be tested is high; for example, rated full-load current being 20-50 A, the direct-loading method, although straightforward, has the serious disadvantage of involving considerable waste of *energy* in terms of I^2R loss(es) in the resistors and shortening of life of lamps when these constitute the load. Under the circumstances, recourse is made to the methods using "phantom" or fictitious loading where the voltage and current coils are fed through two *independent* sources of adjustable supplies. Not only the current through the current magnet coil can be set to any value up to the full load without the use of any R-L or R-C circuit(s), the phase of the current with respect to the supply voltage can be adjusted more flexibly to any desired value, lagging or leading, using one of the two common methods mentioned earlier.

The circuit diagrams and connections for this scheme of testing are essentially the same as described for testing of wattmeters in Chapter VII, except that an RSS replaces the sub-standard wattmeter for the purpose of comparison where required. Then the speed of rotation of the discs of the two meters, during the entire schedule of tests, can be compared using the stop watch and noting the number of revolutions as before, and the error worked out.

Example

A single-phase, 230 V, 20 A energy meter has a meter constant of 480 rev/kWh. If during a full load test, the disc makes 40 revolutions in 66 seconds, calculate the meter error.

The full-load 'power' is $230 \times 20 = 4600$ W or 4.6 kW

In one hour the meter should make 4.6×480 revotutions

The corresponding speed of the disc is

$$= \frac{4.6 \times 480}{60} \text{ or } 36.8 \text{ rpm}$$

∴ correct time for 40 revolutions should be

$$\frac{40}{36.8} \times 60 \text{ or } 65.2$$

With the actual time taken being 66 s the time taken is $(66 - 65.2)$ or 0.8 s too long.

Hence the meter error is $\dfrac{0.8}{65.2} \times 100$ or 1.2% *slow*

[Alternatively, the apparent and actual power can be deduced from the observations and error computed in the usual manner; in both cases, the meter constant plays the key role].

Three-phase Energy Measurement

The measurement of energy in a 3-phase system can be accomplished similar to the power, that is, using three single-phase (induction type) meters in a 3-phase, 4-wire system and two meters in a 3-phase, 3-wire system. The connections are also made as in the case of power measurement using three or two wattmeters, respectively.

Three-phase Meter

An alternative to the use of two or three single-phase meters is to use a three-phase (induction type) meter which is more convenient for the consumers connected to a 3-phase supply. Thus, a 3-phase meter for use in a 3-wire system[1] consists of two sets of electromagnets, *but only one brake magnet.*

The meter incorporates two aluminium discs, capable of rotating independently through the airgaps of the voltage and current magnet pairs, are mounted on a *single* spindle or shaft and geared to a common registration or recording device. Each element has its own friction and lag compensation mechanism. When excited, the total torque on the moving system will be the sum of the torques from the individual elements (the summation being carried out 'mechanically'), similar to the net deflecting torque in the two-element dynamometer-type wattmeter. If the individual elements produce equal torques for a given load, the registration will be the total energy in the three-phase system. To ensure this, the two pressure coils are connected in parallel across a 1-phase supply with the current coils in series opposition. If the torques produced by the individual elements are equal, *the disc would not rotate.* If there is rotation, the "light-load" adjustment of one of the meters is adjusted till the disc is stationary. Other cheeks and adjustments are carried out as in the case of the single-phase energy meter, *separately on the top and bottom element.*

A 3-phase energy meter as above may be subject to errors due to interaction between the two elements, with the flux due to one element interfering with the currents induced in the other disc. This may be eliminated by providing a suitable shield, or screen, between the two elements similar to a 3-phase wattmeter. The meter may also tend to have relatively large friction errors due to the greater weight of the rotating part (the single spindle or shaft carrying two discs). However, this can be minimised by elaborate design of the jewel bearing and carrying out repeat checks of the light-load adjustments.

It is always advantageous to employ a 3-phase meter in a 3-phase system for if such a meter operates at load power factors below 0.5, the torque of one element reverses, but the net torque due to both the elements is still proportional to the power, and in the forward direction, recorded on the single register[2].

ELECTRONIC OR SOLID-STATE ENERGY METERS

The old, conventional electro-mechanical or induction-type energy meters dealt with above have been in use universally over several decades and, of

---

[1]Or 4-wire system, with the neutral wire not connected into the meter.

[2]If under the circumstances, two 1-phase meters were used, one will register in the reverse direction. Thus, if this results in a net reverse registration of that meter over a period of time, there is a possibility of confusion when considering total consumption pertaining to the customer.

course, still continue to be used in good number. However, these are increasingly being replaced by electronic or solid-state energy meters, mainly characterised by the absence of the rotating disc and associated errors like friction.

There are a number of advantages that are claimed in favour of these meters: The meters

- employ modern static electronic components and elements with better quality and stability, and are devoid of any rotational part;
- are more accurate and reliable and have much less power consumption, there being no electromagnets[1];
- can record very low loads;
- can be remotely located, at a 'safe' place;
- have much longer life, usually free from maintenance requirement;
- are tamper-proof in many ways;
- in addition to the energy consumed in a given duration, they can also indicate/record maximum demand, power factor and reactive power etc. during the period;
- the design may also include electronic clock mechanism to compute the cost, rather than just the quantity, of the electricity consumed over a given period with the pricing varying by the time of day, day of week, and/or seasonally;
- can be designed as smart meters, so as to "choose" the use of electric supply from a utility that is economic, or at low tariff, and in the interest of the customer; for example, during "peak", "off peak" or usual period(s);
- can also be designed as "pre-paid" meters to guarantee the due revenue to a utility or vendor.

Their only 'disadvantage' would seem to be the initial cost, being much more than a conventional meter of the same rating, but the meter may turn out to be cost effective in the long run. The meters may also suffer "sudden" component failure and start registering erratically in which case a replacement may be unavoidable.

---

[1]Although they, too, depend on the use of electromagnetic transducers such as PTs and CTs, where required, which may have inherent power loss.

Specification of a Typical Electronic Meter

Single- or three-phase

Rated voltage	:	230 V (400 V L-L for 3-phase)
Rated (or base) current	:	20 A (per phase for 3-phase load)
Rated frequency	:	50 Hz
Manufactured to	:	IS 13779, 1EC 1036
Class (or accuracy)	:	1.0 [as compared to 2.0, usually for conventional meters]
Meter constant	:	1600 IMP/kWh
Minimum starting current	:	0.4% of base current (cf. 5% for which conventional meters are tested)
Power loss in voltage circuit	:	< 1.5 W

Technology

Most solid-state meters use a PT and a CT to measure the supply voltage and load current, respectively, so that the meter can be located remotely from the main supply terminals – an advantage even in domestic applications. The outputs from the instrument transformer secondaries are thus main meter inputs or quantities.

Design

The design of the meter incorporates the following main modules or parts:

- Power supply

  The AC supply drawn from the source is suitably stepped down and rectified to provide smooth, stabilised DC power supply at various voltages used for the electronic circuitry.

- Metering module

  This is provided with the voltage and current inputs and has a voltage reference and samplers, followed by an "Analogue-to-Digital Converter" (or ADC) to digitise all the inputs. These inputs are then processed using a digital signal processer to derive the various quantities of interest such as power, energy etc.

- Processing and communication section

  The main function of this section is the communication using various protocols and interface with other add-on modules connected as slaves to it. This can be in the form of a microcontroller.

- LCD display

    This module is designed to display in LCD form the quantity of interest, computed and stored in the microcontroller[1].

    The basic principle of operation of a solid-state or electronic energy meter using a Hall generator as a multiplier is shown in Fig.8.22 whilst a block diagram of a 3-phase meter is illustrated in Fig.8.23.

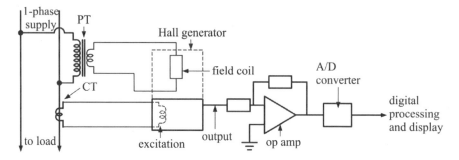

Fig.8.22 : Principle of operation of a 1-phase electronic energy meter

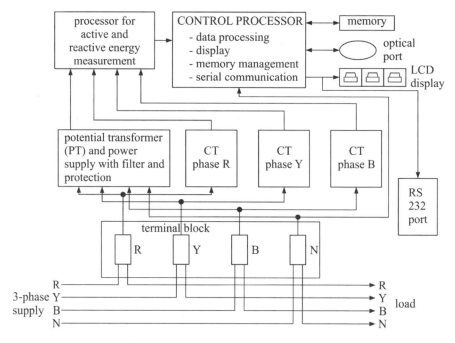

Fig.8.23 : Block diagram of a 3-phase electronic energy meter

---

[1]The 'frequency' and duration of integration, depending on the circuit load, is indicated by a flashing LED.

Sources of Error

Whilst considered generally superior to the induction-type or conventional energy meters, a serious source of long-term error in the solid-state meter is the drift in the preamplifier, followed by the precision of the voltage reference. Both of these may also vary with temperature which may be appreciable if the meter is installed outdoor.

INDUSTRIAL METERING

In domestic, low-voltage distribution systems it is usually adequate to register only the useful energy, that is, kilowatt-hours, with the load being mostly at unity p f, using a 1-phase or 3-phase meter as above. However, in industrial applications, it becomes essential or even mandatory to consider and record the reactive energy consumption (kVARh), too. The total energy consumption is then given by kilovolt-ampere hour (or kVAh), being the appropriate resultant of kWh and kVARh.

This requirement becomes imperative on account of ever-present 'problem' of varying power factor which may be less than unity during various intervals of energy recording, owing to unavoidable use of a variety of drives (and motors) and other loads with inherent poor power factor.

The industrial customers, invariably consuming bulk electrical power, are almost always charged with a tariff that should account for active as well as reactive energy consumption and, additionally, a fixed or varying charge towards maximum demand encountered by the customer during a given period – a day, month or year. This necessitates a metering system that is far more comprehensive compared to the use of a single meter, usually in domestic sector.

Measurement of kVARh

The 'active' energy, or kWh, meter depends for its working on the input of the (supply) voltage and an in-phase (that is, the one multiplied by cos ϕ; ϕ being the power factor angle) component of the load current[1]. Clearly, the corresponding kVARh metering would necessitate the same voltage, but a current component that is in *phase quadrature* with the voltage. The measurement of the reactive energy can therefore be effected by using a watt-hour meter described earlier in which either the voltage flux or the current flux is given a phase displacement of $90°$. Accordingly, an induction-type watt-hour meter with voltage flux *in phase* with the voltage (rather in quadrature as in the usual energy meter) and current flux in phase with the current will register VARh or kVARh.

---

[1]Or quantities proportional to these using instrument transformers.

For the purpose, in a usual two element 3-phase energy meter, the input voltages are arranged to be in phase quadrature with the normal L-L voltages by using special schemes of connections. One such scheme, along with the appropriate phasor diagram of voltages is shown in Fig.8.24.

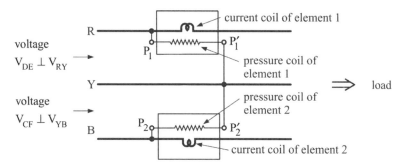

(a) two-element, 3-phase energy meter

voltages to pressure coils $P_1 - P_1'$ and $P_2 - P_2'$ are to be in phase quadrature with V_{RY} and V_{YB} as derived below

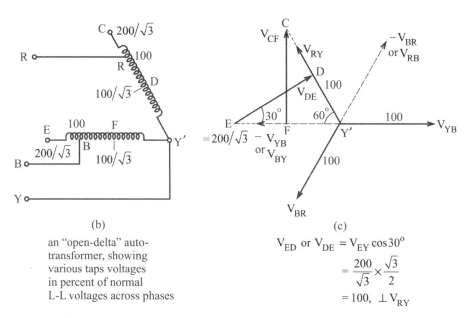

(b)

an "open-delta" auto-transformer, showing various taps voltages in percent of normal L-L voltages across phases

(c)

V_{ED} or $V_{DE} = V_{EY} \cos 30°$

$= \dfrac{200}{\sqrt{3}} \times \dfrac{\sqrt{3}}{2}$

$= 100, \perp V_{RY}$

Fig.8.24 : Schematic of connections to an induction energy meter to measure reactive energy

As shown in Fig.8.24(a), the current coils of the two-element 'wattmeter', are connected in phases R and B. The pressure coils $P_1 - P_1'$ and $P_2 - P_2'$ are fed from the pairs of tappings DE (voltage V_{DE}) and CF (voltage V_{CF}) of the open-delta-connected (special) auto transformer, shown in Fig.8.24(b), such that the voltages V_{DE} and V_{CF} are with a phase difference of 90° with respect to 'normal' line voltages V_{RY} and V_{YB}, respectively, *but equal in magnitude to original line voltages* as derived in the phasor diagram of Fig.8.24(c). Thus, as reasoned out earlier the meter registers the VARh of the three phase load/circuit

Single element method

The schematic of a simple method to measure reactive energy for *balanced* loads is shown in Fig.8.25, as discussed earlier for measurement of reactive power. [See Chapter VII]. The current coil of the (1-phase) energy meter is connected in the R phase whilst the pressure coil is connected across the remaining phases B and Y.

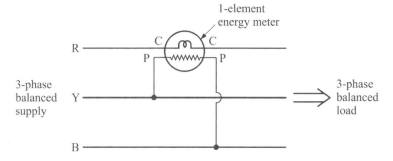

Fig.8.25 : Measurement of reactive energy in a balanced load

The recording of the meter is

$$VARh = \sqrt{3}\ V_L I_L\ \sin\phi$$

with the usual notation. With the connections to the voltage and current coil of the energy meter arranged as above, the meter can be calibrated to register VARh directly by allowing for the factor $\sqrt{3}$ in the gear train.

Measurement of Total Energy, VAh or kVAh

A normal (active) energy meter used to record kVAh

Referring to phasor diagram of voltage, current and fluxes for a 1-phase kWh meter, shown in Fig.8.26, let the "voltage" flux ϕ_{sh} to lag the voltage by $90°$, with the eddy-current angle, β, to

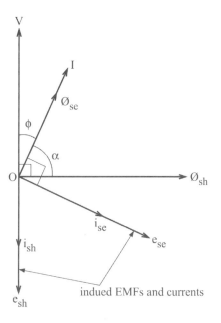

Fig.8.26 : Phasor diagram for a 'normal' energy meter

be zero. If the load p f angle is ϕ, the deflection torque in the meter will be given by

$$T_D = K V I \cos \phi$$

If the phase difference between $\emptyset_{sh}$ and load current I is α, then

$$\alpha = 90 - \phi$$

and $T_D = K V I \sin \alpha$

Now let $\emptyset_{sh}$ to lag V by $90 + \phi^1$, so that $\alpha = 90°$

Then $T_D = K V I \sin 90°$

$$= K V I$$

and the meter would record total energy consumed in a given time.

However, if the load power factor deviates from ϕ during the course of operation, an error will be encountered since the meter is modified according to $\phi + \alpha = 90 + \phi$. Thus, for continuously varying power factor the method may be hardly applicable as discussed next.

Total energy for changing power factor

The relationship between the active and reactive components of the load current and the total current *lagging* the applied voltage by an angle ϕ can be depicted by the sides of a right-angled triangle as shown in Fig.8.27(a).

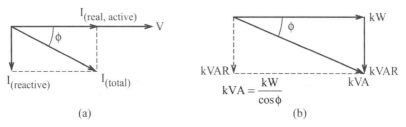

(a)   (b)

Fig.8.27 : Current components and power at a constant p f

In a circuit in such a case, the active, reactive and total *power*, at a *given time* and *constant* power factor, will also be given by a right-angled triangle as shown in Fig.8.27(b), and the relationship will apply to energy consumed in the same given time, so that

$$kVAh = \sqrt{(kWh)^2 + (kVARh)^2}$$

---

[1]This can, for example, be achieved by over phase compensation by using loops of abnormal thickness on the shunt magnet, or using other suitable means.

In practice, however, the load power factor may seldom remain constant over a length of time. Hence, in a system with changing power factor the total kVAh will be the arithmetic sum of the kVAh values corresponding to constant power-factor periods as shown in Fig.8.28 where the segments OA, AB, BC represent the values of kVAh for three constant power-factor periods. Each

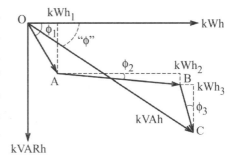

Fig.8.28 : kVAh values for three different power factors

of these segments is obtained by using the above relationship, and the diagram such as in Fig.8.27(b), for power factor angles ϕ_1, ϕ_2 and ϕ_3, the process being extendable to more and more angles. The total kVAh is the 'line' OABC such that

$$kVAh = OA + AB + BC$$

where $$OA = \frac{kWh_1}{\cos \phi_1}, \; AB = \frac{kWh_2}{\cos \phi_2}, \; BC = \frac{kWh_3}{\cos \phi_3},$$

which is more than the length of the straight line OC, representing the "average" value of kVAh corresponding to (an assumed) constant power factor "ϕ" over the entire duration of interest.

Strictly speaking, the measurement of kVAh should be the *integral* of the kilovolt-amperes with time,

or $$kVAh = \int_0^t kVA \; dt$$

and should be carried out taking into account the product of (usually constant) voltage across the supply/load and varying total current, the latter being dependent on the changing load power factor. At a working level, the above requirement can be fulfilled by suitably combining the 'actions' of a watt-hour meter and a reactive volt-ampere-hour meter to drive a meter recording kVAh continuously.

Tri-vector Meters

These are special meters, housed in a single case and produce a simultaneous record and display of all the three integrated quantities, viz. the kWh, kVARh and kVAh as indicated schematically in Fig.8.29.

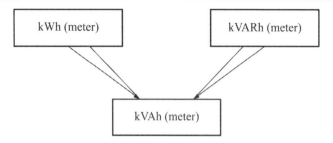

Fig.8.29 : Principle of operation of a tri-vector meter

One type of three-phase kVAh meter incorporates one induction-type kWh meter and one kVARh meter, the shafts or extensions from the two (meters) drive a 'sphere' by means of special friction discs, the 'mechanical' arrangement being such that the resultant rotation of the sphere drives another disc by friction, the shaft attached to this disc driving a mechanism to record kVAh at any instant. The schematic of the mechanism in a very simplified form, the actual operation being rather complex, is depicted in Fig.8.30(a) whilst the mathematical relationship is defined in Fig.8.30(b).

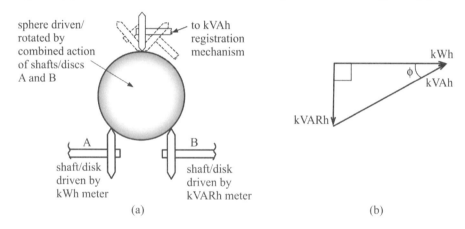

Fig.8.30 : A device using a sphere and 'friction' discs to register kVAh

Thus, the kVAh corresponding to a given power factor *at any instant* is measured and registered according to the expression defined earlier, the speed of rotation of the sphere being automatically proportional to square-root of the sum of squares of the two meter spindle, or shaft, speeds.

Landis and Gyr tri-vector meter

This is another specially designed and built meter, historically developed by the British firm Landis and Gyr, that can register true value of kVAh correct to within ± 1 per cent. This meter also consists of a kWh meter and a kVARh meter in one case, incorporating a special "summator" mounted to link the movement of the two meters. The summator comprises a unique, rather complicated system of gearing, driven by the two meters in a manner as discussed above and resulting in driving a mechanism to register the kVAh.

The general principle behind the movement of gearing is

- at 'low' power factor angles, say 0 to 10°, the load/system kWh is very nearly equal to kVAh, within very small error, and the registration is predominantly by the kWh meter movement;

- at 'high' power factor angles, say 80° to 90°, the kVARh is very nearly equal to kVAh, and thus the registration is predominantly by the kVARh meter movement;

- for intermediate values of power factor angles, the summation, and hence the registration, will obviously be by the combined, resultant action of movements of both the meters. The result is that the kVAh, measured at various power factors, is almost constant, the registered value being dependent on the values of kWh and kVARh, and a given power factor.

The 'action' of the summator in terms of movements of the two meters with respect to variation of power factor will be somewhat like that depicted in Fig.8.31.

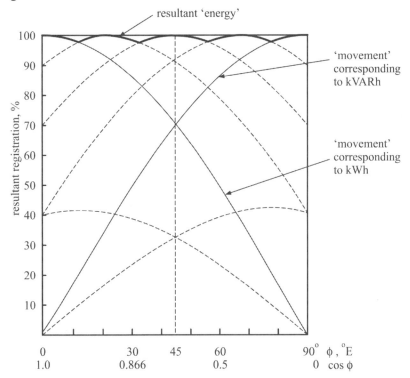

Fig.8.31 : Relative movement of "summator" at various power factors in Landis and Gyr tri-vector meter

Maximum Demand Indicators

In most industrial metering, an essential requirement is to measure/record the maximum value of demand of electric power in terms of kVA or MVA, usually the latter during, say, a day or any given interval of time. This is of interest to both the consumer and the utility as very often tariffs are also based on the value of maximum demand.

Measuring instruments for the purpose, called "maximum demand indicators", are specially-designed meters, meant to record the maximum power or "rate of energy consumption" by a consumer during a prescribed time interval. The consumer will then be charged based on both: the maximum demand during that period as well as the total number of units consumed, kWh and kVARh or kVAh, over the hours pertaining to that period. The usual period of interest is a day or 24 hours, sub-divided into intervals of ½ hour each.

The meters are designed to measure and record maximum demand during each of the (half-hour) interval, rather than to respond to momentary heavy demands, and the maximum of the 24-hour duration forms the basis for computing charges related to maximum demand.

Such indicators may be in the form of a separate, stand-alone instrument; however, these are usually in the form of an attachment to a normal energy or kWh meter.

Merz-Price maximum demand indicator

This is a widely used instrument comprising a special attachment to an already connected watt-hour meter to provide basic movement to the unit. The meter is designed to indicate maximum *consumption* during a half-hour[1] interval, over a quarter of a year (or any other duration), between consecutive re-setting of the instrument[2].

The essential features of this type of indicator are schematically shown in Fig.8.32.

---

[1]Usually so, but not limited to, and can be any other chosen interval.

[2]If the consumption during any "half" hour is K units, the maximum demand in kW, say, will simply be K/0.5 kW, and *maximum of all such demands* over one quarter year being similarly arrived at.

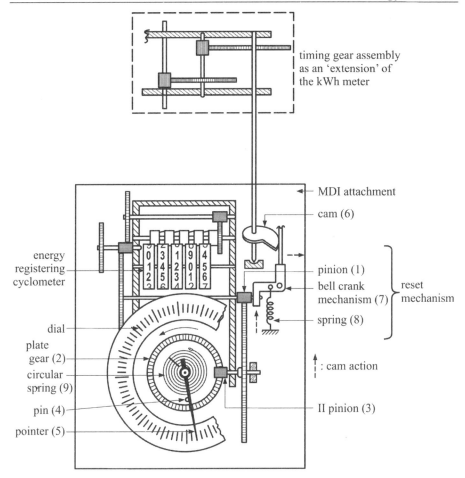

timing gear assembly
as an 'extension' of
the kWh meter

MDI attachment

cam (6)

energy
registering
cyclometer

pinion (1)

bell crank
mechanism (7)

reset
mechanism

spring (8)

dial

plate
gear (2)

circular
spring (9)

: cam action

pin (4)

II pinion (3)

pointer (5)

Fig.8.32 : Schematic of a Merz-Price maximum demand indicator

Working

The working of the indicator can be explained as follows:

- at the beginning of the half-hour cycle, the pinion **1** from the energy meter, starts anticlockwise rotation of the 'plate' gear **2** through another pinion **3**;

- pin **4** attached to the plate gear in turn drives forward the pointer **5**, over the graduated dial (usually calibrated to read kVA, or MVA) for the (pre-set) interval of 1/2 hour, the position of the pointer being indicative of the energy consumed, or the demand (in kVA or MVA as the case may be), during the interval;

- at the end of this time interval, the cam **6** momentarily disengages the pinion **1** by the action of the bell crank mechanism **7**, normally keeping the pinion engaged to the plate gear by the downward force of spring **8**;

- this allows the pin **4** with its driving mechanism, or plate gear, to return to zero position under the action of the circular spring **9**;

- the pointer **5**, however, *does not return to zero*, but is left stationary by a special friction action/device and continues to indicate the number of units consumed in the previous half-hour period (or the kVA or MVA corresponding to that);

- during the next half-hour period initiated by engaging of pinions **1** and **3** (the latter 'linked' with the plate gear), the pin **4** is again driven forward, moving anti-clockwise, but the pointer is moved forward *only if the energy consumed in this period happens to exceed that consumed in all preceding periods*;

- thus, the "maximum demand" over the interval of interest – 24 hours, a week or one-quarter of the year – is obtained by the *last* position of the pointer;

- the pointier itself can then be reset to zero on the dial to commence the cycle all over for the next interval of interest; for example, the next quarter of the year.

An important feature of this indicator is that a reasonably heavy demand should exist during the half-hour period so as to activate the pointer.

Comment

Latest Developments: Smart Meters

With continual development, an improved category of *electronic* meters, known as Smart Meters, are now available in many countries, a special feature of these being their full remote control using high-frequency carrier transmission. The meters can thus be 'read' from centralised control rooms of the utility or power-supply company. They can provide "bill alerts", generate timely and *accurate* billing, measure usage time, are tamper-proof and can even facilitate dis-connention of supply without human intervention.

The meters can store six months of data, in the event of delayed billing and payments, and serve for miscellaneous requirements; for example, to measure and maintain a record of consumption in the premises for various types of load to help in planning of added generation capacity. A further advancement may allow the consumers the choice of overall economy of "buying" electricity variously from different suppliers, esp. by staggering the consumption at various hours of the day.

WORKED EXAMPLES

1. A single-phase energy meter having a constant of 100 revolutions/kWh makes 360 revolutions when connected to a load of 42 A at 230 V and p f of 0.4 (lag) for 1 hour.

   Determine the percentage error.

   Power supplied to the load,

   $$W_L = 230 \times 42 \times 0.4$$

   $$= 3864 \text{ W} \quad \text{or } 3.864 \text{ kW}$$

   Energy supplied in one hour or true energy

   $$= 3.864 \times 1 \quad \text{or} \quad 3.864 \text{ kWh}$$

   Indicated energy

   $$= \frac{\text{no. of revolutions made}}{\text{meter constant}}$$

   $$= \frac{360}{100} \text{ or } 3.6 \text{ kWh}$$

   ∴ $$\% \text{ error} = \frac{\text{indicated energy} - \text{true energy}}{\text{indicated energy}} \times 100$$

   $$= \frac{3.6 - 3.864}{3.6} \times 100$$

   $$= -7.33 \%$$

   [since the indicated energy is less than the actual energy, the meter is running slow; this is also indicated by the negative sign].

2. An energy meter is calibrated on a 250 V supply for a steady current of 15 A at unity p f for 5 hours. If the meter readings before and after the test are 8234.21 kWh and 8253.13 kWh, respectively, calculate the percentage error. If the meter spindle turns through 290 revolutions during 5 minutes when a current of 20 A is passed at 250 V and 0.87 p f, calculate the meter constant.

   Actual energy consumed at given supply and load

   $$= 250 \times 15 \times 1.0 \times 5 \times 10^{-3}$$

   $$= 18.75 \text{ kWh}$$

Consumption shown by the meter

$$= 8253.13 - 8234.21$$

$$= 18.92 \text{ kWh}$$

$$\therefore \qquad \% \text{ error} = \frac{18.92 - 18.75}{18.92} \times 100$$

$$= + 0.898 \%$$

[showing that the meter is running fast]

In the second case, energy recorded

$$= 250 \times 20 \times 0.87 \times \left(\frac{5}{60}\right) \times 10^{-3}$$

$$= 362.5 \times 10^{-3} \text{ kWh}$$

$$\therefore \qquad \text{meter constant} = \frac{290}{0.3625} \quad \text{or} \quad 800 \text{ rev/kWh}$$

3. A correctly adjusted single-phase, 240 V, 50 Hz induction watt-hour meter has a constant of 600 rev/kWh. Determine the speed of the meter disc for a current of 10 A at a p f of 0.8 (lag).

If the meter lag adjustment is altered so that the phase angle between the applied voltage and voltage coil flux is 86° (instead of 90°), calculate the error introduced in the meter readings at (a) unity p f, (b) 0.5 p f (lag).

Actual energy consumed in one minute

$$= 240 \times 10 \times 0.8 \times \left(\frac{1}{60}\right) \times 10^{-3}$$

$$= 0.032 \text{ kWh}$$

Meter constant is 600 rev/kWh

∴ number of revolutions made in 1 minute corresponding to 0.032 kWh

$$= 0.032 \times 600 \quad \text{or} \quad 19.2$$

∴ speed of the meter disc is 19.2 rpm

In the second case, when the phase angle, Δ, is 90°, the meter speed is

$$N \propto V \text{ I } \cos \phi \text{ or } K \text{ V I } \cos \phi, \text{ K a constant}$$

When Δ is not equal to 90°,

$$N \propto V \text{ I } \sin (\Delta - \phi) \quad \text{or} \quad K \text{ V I } \sin (\Delta - \phi)$$

∴ error introduced due to incorrect lag adjustment

$$= K \text{ V I} \frac{\sin(\Delta - \phi) - \cos \phi}{K \text{ V I } \cos \phi} \times 100\%$$

When $\Delta = 86°$ and $\phi = 0$ (unity pf),

$$\text{error} = \frac{\sin(86° - 0°) - 1}{1} \times 100$$

$$= -0.24\%$$

When $\Delta = 86°$ and $\phi = 60°$ (0.5 pf lag),

$$\text{error} = \frac{\sin(86° - 60°) - \cos 60°}{\cos 60°} \times 100$$

$$= -12.3\%$$

[Note the appreciable error at low power factor]

4. The meter constant of a 3-phase, 3-element energy meter is 0.12 rev/kWh. If the meter is used with a PT of ratio 22 kV/110 V and a CT of ratio 500/5 A, find the error expressed as a percentage of the current reading from the following test figures on the meter side:

line voltage : 110 V, current : 5.25 A, pf : 1.0

time to complete 40 revolutions : 61 s

Actual energy consumed during the test period

$$= \sqrt{3} \times \text{PT ratio} \times \text{CT ratio} \times V_s \times I_s \times \cos\phi \times t$$

$$= \sqrt{3} \times \frac{22{,}000}{110} \times \frac{500}{5} \times 110 \times 5.25 \times 1.0 \times \frac{61}{3600} \times 10^{-3}$$

(t in hour)

$$= 339 \text{ kWh}$$

Energy recorded by the meter during the same period

$$= \frac{\text{no. of revolutions made}}{\text{meter constant}}$$

$$= \frac{40}{0.12}$$

$$= 333.3 \text{ kWh}$$

$\therefore$ percent error $= \dfrac{333.3 - 339}{333.3} \times 100$

$$= -1.71\%$$

[that is, the meter is slow by 1.71 %]

5. An industrial customer has a Merz-Price maximum demand indicator fitted with an integrating device. Over a month, the following readings are recorded:

kVARh = 125,000;   kWh = 150,000

maximum kVAR demand = 180;   maximum kW demand = 200

If the tariff for the consumer is Rs. 20 per kVA + Rs. 0.15 per kVAh, calculate the monthly bill.

What is the average p f of the consumer's load and the load factor?

$$\text{Total kVAh in a month} = \sqrt{(\text{kWh})^2 + (\text{kVARh})^2}$$

$$= \sqrt{(125,000)^2 + (150,000)^2}$$

$$= 195,256$$

$$\text{Maximum kVA} = \sqrt{(\text{kW})^2 + (\text{kVAR})^2}$$

$$= \sqrt{(200)^2 + (180)^2}$$

$$= 269$$

$$\therefore \text{ monthly bill} = 269 \times 20 + 0.15 \times 195,256$$

$$= 5,381 + 29,288$$

$$= \text{Rs. } 34,669$$

Average power factor,

$$\cos\phi_{av} = \frac{\text{kWh}}{\text{kVAh}}$$

$$= \frac{125,000}{195,256}$$

$$= 0.64$$

$$\text{Load factor} = \frac{\text{no. of units consumed}}{\text{maximum demand} \times \text{no. of hours}}$$

$$= \frac{125,000}{200 \times 30 \times 24}$$

$$= 0.868$$

[Also, average consumption cost per unit

$$= \frac{34,669}{125,000}$$

$$= \text{Re. } 0.277 \text{ per kWh}]$$

EXERCISES

1. An energy meter is designed to make 100 revolutions to record 1 kWh. Calculate the number of revolutions made by it when connected to a load of 40 A, at 230 V and 0.4 pf for one hour. If it actually makes 360 revolutions, find the percentage error.

$$[368; -2.22 \%$$

2. In an energy meter, in a test run of 30-minute duration with a constant current of 5 A, the meter was found to register 0.51 kWh. If the meter is to be used in a 200 V circuit, find its error and state if it is running fast or slow.

$$[1.96 \%; \quad \text{fast}$$

3. The number of revolutions per kWh for a 230 V, 10 A watt-hour meter is 900. On test at half load the time for 20 revolutions of the disc is found to be 69 s. Determine the meter error at half load.

$$[0.812 \%$$

4. A 50 A, 230 V energy meter on full-load test makes 61 revolutions in 37 s. If the normal disc speed is 520 revolutions per kWh, find the percentage error.

$$[-11.77 \%$$

QUIZ QUESTIONS

1. The induction-type meter generally used to measure electrical energy is known as
    - □ kWh meter
    - □ wattmeter
    - □ ampere-hour meter
    - □ avometer

2. A typical, conventional energy meter works on the principle of
    _____

3. An induction-type energy meter comprises
    - □ a pair of electromagnets
    - □ an aluminium disc
    - □ a brake magnet
    - □ all of these

4. Special bearings are used in energy meters at lower end to reduce friction. These are
    - □ jewel bearings
    - □ ball bearings
    - □ bush bearings
    - □ none of these

5. Name four important parts of an induction energy meter
    - a. _____   b. _____
    - c. _____   d. _____

6. The aluminium disc in an energy meter can be replaced by
    - □ a brass disc
    - □ a copper disc
    - □ an iron disc
    - □ a plastic disc

7. The current coil of a typical energy meter comprises a small number of turns of thick wire.
    - □ true   □ false

8. The potential coil of an energy meter consists of a large number of turns of thin wire.
    - □ true   □ false

9. For correct operation, the flux due to voltage coil in an induction energy meter must lag the voltage by
    - □ $0°$
    - □ $45°$
    - □ $90°$
    - □ any angle

10. The name plate of a 1-phase energy meter specifies
    - □ operating voltage
    - □ limiting load current
    - □ meter constant
    - □ all of these

11. A 50 A, 230 V energy meter with a meter constant of 520 rev/kWh makes 61 revolutions in 37 seconds on full load, unity p f. What is its percentage error?

☐   + 1%   ☐   – 1%

☐   – 0.76%   ☐   none of these

12 "Creep" in an energy meter is a phenomenon in which the disc rotates even when load current is zero.

☐ true   ☐ false

13. The creep in an induction-type meter is due to

☐   over compensation for friction   ☐   over voltage

☐   undue vibrations   ☐   all of these

14. The friction compensation in the energy meter is carried out by

☐   adjusting the brake magnet

☐   adjusting shunt magnet

☐   adjusting the series magnet

☐   shorting loops in the voltage magnet core

15. The speed of rotation of the disc on load in the meter can be corrected by

☐   adjusting the position of the brake magnet

☐   lowering the disc in the gap

☐   adjusting the series magnet

☐   none of these

16. Energy meters are provided with special means to make them_____

17. The creep in the energy meter can be arrested by _____

18. A 3-phase, 2-elements energy meter can work in a

☐   3-phase, 3-wire circuit   ☐   3-phase, 4-wire system

☐   3-phase, n-wire system   ☐   3-phase, 5-wire system

19. A 3-phase energy meter can measure correctly in a 3-phase balanced or un-balanced network.

☐ true   ☐ false

20. The two methods of testing an energy meter are

a._____   b._____

21. The electronic energy meters are fast replacing conventional meters because these are _____

22. A maximum demand indicator is a kind of energy meter.

    ☐ true   ☐ false

23. A tri-vector meter can measure

    a. _____   b. _____

    c. _____   d. _____

24. A typical maximum demand indicator still in use is known as _____

IX : Potentiometers

IX

POTENTIOMETERS

RECALL

Measurement of PD in a Circuit

A straightforward means to measure PD in an electric circuit is to use a suitable voltmeter of appropriate range which can read the PD directly on the scale. However, in reading the PD

  (i) the voltmeter draws a small current from the circuit for its working, the current desired to be as small as possible, theoretically zero;

  (ii) if the PD is small, say a few mV, it is difficult to obtain a voltmeter (or a milli-voltmeter) of sufficient accuracy that may not draw appreciable current and not 'load' the circuit, to provide accurate value of the PD[1].

In general, the accuracy of such methods using an appropriate meter, a "pointer"-type instrument, may be up to 1% owing to the presence of moving parts etc. When greater accuracy is required, elaborate measuring schemes may be necessary.

Another requirement may be to measure the "open-circuit" EMF of a cell or a battery which cannot be measured accurately by means of a voltmeter owing to the current drawn from the cell and thus altering its EMF value. Thus, a simple voltmeter method is not suitable for measurement of small to very small PDs or EMFs.

In such cases, recourse is invariably made to the use of a well-designed "potentiometer", which forms one of the most useful and widely used electric circuits, to provide high degree of accuracy.

---

[1]Although an electronic voltmeter with an extremely high input impedance (say, using a cathode-follower circuit) may be used for the purpose.

BASIC (DC) POTENTIOMETER

Constructional Features

A potentiometer basically comprises a wire AB of uniform cross section and about 1 metre long across which is connected a battery via a key and a variable resistance to adjust the current through the wire as shown in Fig.9.1. These form the main part of the potentiometer. When the key is closed, a suitable current flows continuously through the wire and is known as the main potentiometer-wire current. An ammeter may be connected in series to read this current.

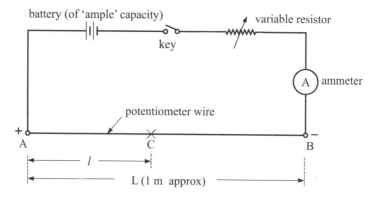

Fig.9.1 : A basic DC Potentiometer

Essential requirements

For satisfactory operation and results, it is imperative that

- the wire is made of high-quality resistive material with extremely low temperature coefficient so that its resistance does not vary during the test due to the heat produced by the flow of current[1];
- the wire is of uniform cross-section all through its length so that its resistance per unit length (for example, per cm) and hence the potential drop/cm is the same from A to B, when carrying a current;
- the battery should be of 'ample' capacity so that the current supplied by it does not vary during the test.

Principle of Operation

For a *steady* current I through the potentiometer wire, the potential drop across AB will be IR where R is the total resistance of the wire. If the wire length is L cm, the PD per cm will be IR/L and for any length l cm, measured from point A, say, the PD will be given by (IR/L) × l V. With (IR/L) being a constant, it follows that the PD at any point on the wire is proportional to the distance of the point (for example point C in the figure) from A, or length l of the wire, being zero at A and 'full value' at B. Thus, PD at C

---

[1]The wire used invariably consists of MANGANIN, an alloy of copper, manganese, iron and nickel, discovered by Weston in 1889, and has a temperature co-efficient of the order of 0.0005 per cent per °C.

$$PD_{(c)} = K \times l \ V$$

with K having been deduced from the knowledge of I, R and L as above.

Measuring an EMF

Suppose it is required to measure the (*open circuit*) EMF of a cell. For this, the cell is connected in the potentiometer circuit as shown is Fig.9.2, forming the "test" circuit. It is essential to connect the positive terminal of the cell to terminal A of the main circuit which is also at positive potential by virtue of its connection to the supply battery. A centre-zero galvanometer, G, and a key are connected in series with the cell as shown.

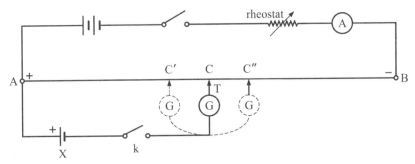

Fig.9.2 : Measuring the EMF of a cell

The variable point on the potentiometer wire, also known as the slide wire, is shown by C; in practice, this is effected by a "jokey" that can be slid along the wire and pressed at a particular point to make a good contact with the wire.

Now suppose that the galvanometer terminal T being at sliding point C and galvanometer key, k, pressed, the galvanometer shows no deflection (or is in a "null" state). This would imply that the EMF of the cell, X, is exactly balanced by the slide wire potential drop up to the point C. If the length AC of the wire is l, and the potentiometer constant K deduced as before, the cell EMF will simply be given by

$$cell_{(EMF)} = K \times l \ V$$

Clearly, for any other position of T, say at C' or C", and k pressed, the galvanometer will show a deflection to left or right depending on the cell EMF being greater or smaller than the PD along the wire at C' or C". This also means that various trials may have to be made before galvanometer 'null' is achieved, or that the point C is identified.

As revealed above, the most important aspect of using a potentiometer for measuring PDs is that it uses a null, or no-deflection, method to obtain results. No current is drawn from the cell at the null and so true open-circuit EMF of the cell can be measured. Likewise, no current flows through the galvanometer and accuracy related to it is limited by its sensitivity. The results obtained from potentiometer measurement are thus most accurate and values correct to about 1 part in 100,000 (or 0.001%) can be achieved by using a well-designed potentiometer.

Precautions using potentiometers and factors affecting accuracy of measurements

To achieve good results, the following steps must be observed.

   (i)   It is most important to observe appropriate polarity markings of all the apparatus, esp. the connection of 'positive' terminal (almost invariably the left end) of the slide wire. If not so, not only the test will be jeopardised, serious damage may occur to the galvanometer or the cell.

  (ii) The precision with which the balance point of a potentiometer can be arrived at depends on the *sensitivity* of the galvanometer. Paradoxically, a very sensitive galvanometer is prone to damage when the setting of the variable contact (the jokey) is far off the balance. In practice, therefore, a galvanometer is protected during off-balance conditions by keeping it in "low-sensitivity" stage. Two methods of effecting this are illustrated in Fig.9.3. In figure (a), the series resistance is kept at highest value for lowest sensitivity and gradually reduced to zero when the sensitivity of the galvanometer would be maximum. Referring to figure (b), the shunt resistance is kept at zero (the galvanometer fully shunted) and gradually increased to obtain highest galvanometer sensitivity.

(a) using series resistance   (b) using shunt resistance

Fig.9.3 : Methods of controlling galvanometer sensitivity

(iii)   Since the PD to be measured is obtained in terms of potential drop per cm, multiplied by the length of the slide wire up to the point of balance, the accuracy of measurement would be affected by the accuracy with which the length can be measured. This would point to the use of as long a potentiometer wire as feasible; however, this may not always be possible in practice[1].

---

[1]Modern potentiometers have got round this dilemma by using 'concentrated' coil resistances of high precision in series with a small length of slide wire as described later.

(iv)   The main circuit of the potentiometer should be 'stable' during the test. This implies a constant slide wire current (derived from a battery of 'good' AH capacity) and a constant wire resistance, R.

(v)   It follows that there is an upper limit of the PD that can be measured using a potentiometer which should be slightly less than the total potential drop across the slide wire (given by $l = L$). For measuring higher PDs than the limiting values, special potential (drop) devices such as a "volt-ratio" box will have to be employed as considered later.

Measurement of an un-known PD by the method of comparison

Suppose that there are two cells or batteries of which one is of un-known EMF that is to be measured whilst the EMF of the other is known[1]. The cell of which the EMF is known is first connected in the potentiometer circuit and balance obtained as usual. Let the balance in this case is obtained at slide wire length l_1. The cell with un-known EMF is then connected and balance obtained as before, at length l_2, say.

Then, EMF of the known cell

$$E_1 = K \times l_1, \qquad \text{K being the slide-wire constant,}$$

and that of the un-known

$$E_2 = K \times l_2$$

Dividing

$$\frac{E_1}{E_2} = \frac{l_1}{l_2}$$

and hence

$$E_2 = \left(\frac{l_2}{l_1}\right) \times E_1$$

Clearly, notwithstanding the other sources of error, the accuracy of measurement thus depends on the accuracy with which l_1 and l_2 are measured.

Error(s) due to Thermo-electric EMFs

Thermo-electric EMFs may find their way at joints or junctions formed by conductors and terminals of dissimilar metals, if at different temperatures, resulting in errors of obtaining a balance. To eliminate such errors, the entire potentiometer set-up, and parts thereof, must be maintained at the same temperature. However, this may not be possible for 'ordinary' slide wire potentiometer made of 1 m or longer wire length.

---

[1]This could in most cases be a "standard" cell, employed in nearly all modern potentiometer experiments, as discussed later.

An alternative may be to reverse the main circuit connections; that is, connect the main battery first across A-B as usual and obtain the balance and then repeat the test by reversing the battery connections with positive terminal now connected to B instead to A [see Fig.9.2]. In the latter case, care being taken that the positive terminal of the cell under test is now connected to end B of the wire. An average value corresponding to the two balance lengths may then be taken as the 'true' EMF of the cell.

PRACTICAL POTENTIOMETERS[1]

Potentiometers in modern-day use are of two types:

A.   DC potentiometers

B.   AC potentiometers

according to their use for measurement in a DC or AC circuit as the name implies.

Potentiometers for DC Circuits

Dial-type potentiometers

A major draw-back of the slide-wire potentiometer comprising a long resistance wire and a linear, graduated scale beneath it, is its limited accuracy, in terms of the *least count* (that is, the ability to measure smallest value of the PD), and cumbersome configuration requiring a good deal of space; for example a table of appropriate size.

The above limitation is eliminated in "dial-type" potentiometers, now invariably in common use in engineering laboratories. This alternative design of a potentiometer, housed in a compact box, essentially comprises a pair of dials: one incorporating a number of discrete values of equal resistances in the form of (self-)inductance-free coils, wound using manganin wire; the other, in series with these coils, being the actual slide wire with a continuously moving contact having a total resistance equal to one of the above discrete resistances. The other basic features and principle of operation is similar to the ordinary single, slide-wire type potentiometer of the basic design.

Crompton DC Potentiometer

A two-dials type potentiometer, used extensively in most tests for DC measurements, is the Crompton potentiometer, developed by R.E. Crompton

---

[1]It is interesting that even though a potentiometer (word) ends with the suffix "meter", it is *not* a meter in true sense of the word although it does comprise a galvanometer as an essential accessory for null detection. Strictly, it is only a specially connected circuit or a device.

in the early 1900s. The schematic diagram of such a potentiometer is given in Fig.9.4.

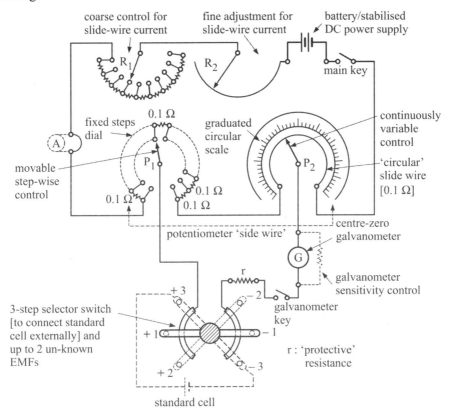

Fig.9.4 : Schematic of the DC Crompton potentiometer

As shown, the potentiometer 'slide wire' consists of two parts:

(a) one, with up to fourteen (sometimes 18 or more : only a few being shown in the figure) 'fixed' coils, each having a resistance of, (say) 0.1 ohm, arranged in a circular fashion and connected in a rotational manner by means of the contact stud P_1; and

(b) another, a 'graduated' slide wire (graduations marked and seen on a circular scale on the top surface of the instrument), having a *total* resistance of 0.1 ohm and in series with the fixed coils, the contact position of which is continuously adjustable by stud P_2; varying the length (and hence the resistance) of the slide wire in association with the fixed coils, the *total* resistance of the circuit can be changed effectively, from zero to the limiting value.

The studs P_1 and P_2 are connected to a 3-way selector, or change-over, switch in series with a galvanometer and its key to facilitate external connections to a standard cell[1] or un-known EMFs/PDs in turn, by choosing contacts +1, −1; +2, −2 etc. All appropriate terminals are marked with polarities positive (+) and negative (−) for ease of making proper connections and to avoid the possibility of any damage to the potentiometer (and esp. the galvanometer) due to 'wrong' connections. As stated before, the supply battery must be of ample capacity, sometimes replaced by an electronically stabilised DC supply of same voltage and connected correctly with regard to polarity. If required, an ammeter may be connected at the appropriate terminals to record the main, slide-wire current (also known as 'standard' current); else these terminals may simply be shorted. A small resistance r is connected in series with the galvanometer to protect the standard cell when the former is shorted under minimum sensitivity condition.

Method of Use

Standardisation

The potentiometer is first "standardised" which means it is made "direct reading" by adjustment of the main or slide-wire current from the battery with the help of variable resistors R_1 and R_2. The step-wise procedure is

- connect the main battery or the stabilised supply at terminals as marked;
- connect the Weston (standard) cell at appropriate terminals of the change-over switch[2];
- set the galvanometer in the LEAST sensitivity position (the shunt resistance fully cut out);
- adjust the setting of the "slide wire" dials to read 1.0186 V (the standard cell EMF) by using *both* fixed-resistance and slide-wire parts in turn;
- adjust R_1 and R_2 to approximately mid-way position and close the (main) key;
- press the galvanometer key momentarily and keep adjusting R_1 and R_2 to get NULL in the galvanometer;
- repeat the preceding two steps by gradually increasing the galvanometer sensitivity to its maximum value.

The potentiometer is now standardised.

Do NOT alter the settings of resisters R_1 and R_2. Note the reading of the ammeter, if connected, to record the standard slide-wire current.

---

[1]Almost invariably, the standard cell used in conjunction with the potentiometer would be a Weston cell, having an open circuit EMF of 1.0186 V.

[2]In some instruments, the standard cell is in-built as an accessory and can simply be switched in. However, for reliable operation and good accuracy, it is preferable to use an independent, well-maintained cell.

Measuring the Un-known EMF or PD[1]

- the cell of which the EMF is to be measured is connected across terminals +1 and –1 (or +2 and –2) of the selector switch and contacts made;
- with the galvanometer in the least sensitive position, adjust the settings of the two dials (studs P_1 and P_2) successively to obtain null in the galvanometer;
- by increasing the galvanometer sensitivity in gradual steps and observing null whilst adjusting the dial settings, the value of the un-known EMF (or PD in a circuit) can be obtained *directly* from the readings on the dials[2].

Typical Applications of the Crompton Potentiometer

Calibration of ammeter, voltmeter and wattmeter

Ammeter

The calibration of an ammeter using the potentiometer requires to 'generate' a PD proportional to the current passing through the ammeter, as shown in Fig.9.5. For this, the ammeter is connected in series with a (4-terminal) standard resistance of low value and a variable resistor, or 'load', to adjust the current at the desired value. The circuit is fed from a suitable, adjustable DC supply as shown. After setting the ammeter current to a given value, the PD across the standard resistance is connected to, and measured on, the potentiometer, supposed to have been stadardised as described. The 'correct' value of the current is then deduced by dividing the measured PD by the (standard) resistance value, the latter being so chosen as to give a PD of about 1 V for the highest ammeter current.

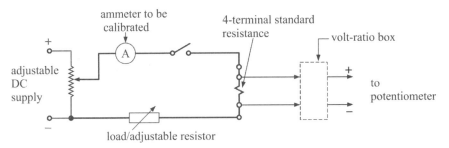

Fig.9.5 : Calibration of an ammeter using the potentiometer

---

[1]At times it might help if the un-known voltage may be measured first by means of a suitable voltmeter, before proceeding with the measurement using the potentiometer.

[2]The Crompton potentiometer described in detail above is to demonstrate the construction and use of a *typical* two-dials type of DC potentiometer. Other superior versions, with more elaborate construction and better accuracy are now available.

A volt-ratio box may be incorporated in the potentiometer circuit, shown dotted in the figure, if the PD across the standard resistance exceeds the limiting value of the potentiometer when the ammeter current is too large. The test is repeated for different chosen values of ammeter current and compared to obtain the calibration curve. Observe that the same test set-up may be used for simply *measuring* the current in a circuit instead of calibration of the ammeter.

Voltmeter

The circuit for the purpose of voltmeter calibration is shown in Fig.9.6. The readings of the voltmeter are compared with those derived from PD measured on the potentiometer, multiplied by the ratio of the volt-ratio box. In order that 'convenient' PDs may be obtained for measurement on the potentiometer, a suitable volt-ratio box is usually employed. The internal circuit connections of a typical volt-ratio box, to which a maximum of 300 V can be input to give about 2 V output that can be connected to potentiometer test terminals, is shown in Fig.9.6(b). In effect, it is a special type of a potential divider having fixed tappings to provide various voltage inputs, with corresponding small PD for connection to the potentiometer.

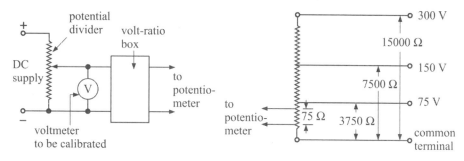

(a) Circuit for voltmeter calibration   (b) Construction of a typical volt-ratio box

Fig.9.6 : Calibration requirements for a voltmeter

Thus, if the leads to the potentiometer are taken from two tapping points, the resistance between them being, say, 75 Ω, then for measuring a voltage of the order of 150 V, connected between the terminals "common" and "150 V", respectively, the resistance between these tappings is chosen to be 7500 Ω. For this voltage, the PD across terminals connected to the potentiometer will then be (75/7500) ×150 or 1.5 V – an easily 'manageable' PD that can be directly measured[1].

---

[1]Instead of being marked "75 V", "150 V" etc., many volt-ratio boxes have the marking in terms of multiplication factors: for example, "x 50", "x 100", "x 150" etc.

Since, at balance, no current flows in the galvanometer circuit, that is, no current is drawn from the volt-ratio box (or that there is no "loading"), such a device provides exact sub-division of the applied voltage, considering that the resistances between the tappings are correctly adjusted, and constructed using a material such as manganin. Volt-ratio boxes are available for a variety of input voltage range, from 200 V to 1000 V, having (total) internal resistance of 10,000 Ω to 50,000 Ω and an accuracy of voltage division of ± 0.1 per cent.

Wattmeter

The calibration of a wattmeter using a standardised potentiometer combines the features of calibration of an ammeter and voltmeter as illustrated in Fig.9.7.

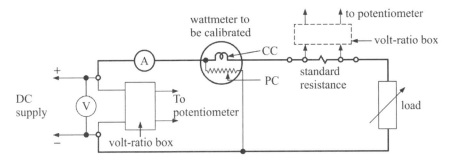

Fig.9.7 : Circuit for calibration of a wattmeter

The current coil of the wattmeter to be calibrated is connected in series with a suitable (4-terminal) standard resistor and an ammeter (to indicate the chosen load current) to an adjustable load that should remain constant at the given setting during the test. The DC supply is connected across the load, or rather the pressure/voltage coil of the wattmeter, as shown; it also forms the input to a suitable volt-ratio box, the PD output from which is connected to the potentiometer terminals, say +2, –2.

The two PDs measured by the potentiometer in turn – one across the standard resistance and the other across the volt-ratio box as above – provide the load current and supply voltage, respectively. Thus, 'true' power in the load

$$P_{true} = \frac{PD \text{ across the resistor}}{(standard) \text{ resistance value}}$$

$$\times \left(PD \text{ across the volt-ratio box} \times \text{multiplying factor} \right)$$

which can be compared with the wattmeter reading for a given load current to obtain the desired calibration curve.

Measurement of Resistance

A dial-type potentiometer with good accuracy and resolution can also be used to measure an un-known resistance, esp. the low resistances which are difficult to measure accurately by other means. These resistances are usually of 4-terminal type, the series terminals being connected to carry a given current and shunt terminals to provide potential drop for measurement purpose. The key advantage behind the use of a potentiometer for such application is the same as before: no current being drawn from the circuit at balance, the method thus affording maximum accuracy. The connections to the potentiometer for such a measurement are shown in Fig.9.8. The method depends on the use of a *known* resistance of nearly the same value, connected in series with the un-known.

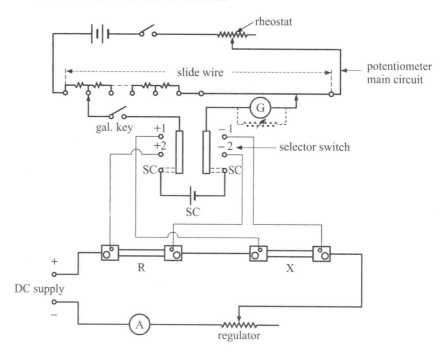

Fig.9.8 : Measurement of a 4-terminal resistance using potentiometer

R and X are the known and un-known resistances, respectively, connected in series so as to carry the same current. Their potential terminals are connected to the potentiometer via pairs of terminals such as +1, −1 and +2, −2 of a selector switch as shown. The PDs across the two are measured in turn on the potentiometer which are related to the two resistances by the ratio:

$$= \frac{\text{PD across R}}{\text{PD across X}}$$

from which X can be obtained in terms of R.

Example

A Crompton potentiometer has an 18-step dial, each representing 0.1 V. The dial resistance per step is 10 Ω. The slide wire part is circular, having 11 turns (arranged helically, one above the other) and a total resistance of 11 Ω. Each turn has 100 divisions and interpolation up to $1/4^{th}$ of a division. The main circuit battery voltage is 6 V.

Calculate (a) working current, (b) setting of the rheostat (in the main circuit), (c) measuring range, and (d) resolution of the potentiometer.

Voltage drop across each step 0.1 V

Resistance/step = 10 Ω

(a) $\therefore$ current through the wire or working current

$$= \left(\frac{0.1}{10}\right) = 0.01 \text{ A or 10 mA}$$

EMF of the battery in the main circuit = 6 V

(b) Now total resistance in the main circuit

= total dial resistance + slide wire resistance + resistance of the rheostat

$= 18 \times 10 + 11 + R = 191 + R \ \Omega$,

where R is the desired rheostat resistance.

But total main circuit resistance is given by battery voltage/working current

or $$\frac{6}{0.01} \ \Omega$$

$\therefore$ $$191 + R = \frac{6}{0.01}, \text{ giving } R = 409 \ \Omega$$

(c) The measuring range of the potentiometer is derived as

(voltage drop across the dial + voltage drop across the slide wire) for the working current

$$= (18 \times 0.1) + (11 \times 0.01) \text{ or } 1.91 \text{ V}$$

(d) Resolution (or least count)

Resistance of 11 turns of the slide wire	=	11 Ω
Resistance/turn	=	1 Ω
Slide wire (or working) current	=	0.01 A
Voltage drop across each turn	=	$1 \times 0.01 = 0.01$ V
Each turn has/corresponds to 100 divisions		

$$\therefore \text{ voltage drop/division} = \left(\frac{0.01}{100}\right) \text{ or } 0.1\,\text{mV}$$

$$\text{Interpolation can be had to} = \frac{1}{4} \text{ division}$$

$$\therefore \text{ resolution} = \left(\frac{10^{-4}}{4}\right) \text{ or } 0.025\,\text{mV}$$

This shows that the given potentiometer is capable of measuring 0.025 mV to 1.91 V.

Potentiometers for AC Measurements

Considering the accuracy afforded by DC potentiometer(s), it was inevitable that their use was extended to measure un-known potential drops in AC circuits/devices. Thus, the basic components/parts of an AC potentiometer are the same as for a DC potentiometer, viz.,

- a slide wire circuit, using discrete coils as well as an actual slide wire in series;
- a source of adjustable and dependable *AC* supply;
- control resistor(s) to adjust slide wire current and an ammeter to record the same;
- a null detector which may be an appropriate galvanometer or any other device, suitable for use with alternating current;
- miscellaneous devices like selector switch, keys etc.

In addition, however, there would be some special requirements for a given potentiometer as discussed later.

Comparison with DC potentiometer and measurements

(i) An important difference between a DC and AC potentiometer is that whereas in the former only the *magnitude* of the un-known EMF or PD is to be measured, and "balanced" in the potentiometer, in the AC potentiometer the *magnitude and the phases* of PDs in relation to a given/chosen reference must be balanced appropriately. Consider, for example, a simple R, L series circuit and the various phasors associated with it as shown in Fig.9.9.

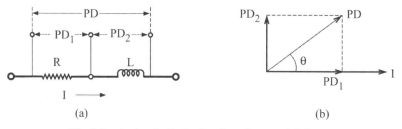

(a)   (b)

Fig.9.9 : A simple R, L circuit and potential drops

If the current, I, common to R and L in series is taken as a reference phasor [see Fig.9.9(b)], then the PD across the resistor, PD_1, will be in phase with the current whilst that across the inductance, PD_2, will be in phase quadrature. The resultant potential drop, PD, will be leading the current by an angle θ, given by $\tan^{-1}(PD_2/PD_1)$.

Thus, if the potential drops PD_1, PD_2 and PD were to be measured on the (AC) potentiometer, the results *must* reflect both the magnitudes of these potential drops AND their phasor relationships with the phasor I^1.

(ii) Next, to avoid the possibility of errors due to (supply) waveform and frequency and their variation during the measurement and to maintain these to be identified through the entire test set-up (including that forming the source of the PD to be measured), it is essential that *both* the circuits are fed from the SAME supply, as illustrated in Fig.9.10[2].

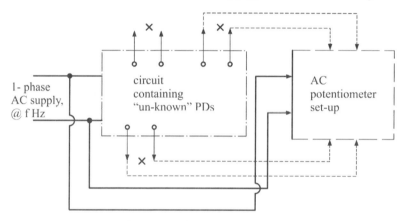

✗ : various PDs to be measured

Fig.9.10 : AC supply to the potentiometer

(iii) Also, there is no "standard" for AC EMF/source, similar to the Weston cell. Thus, whereas a DC potentiometer can easily be standardised using the Weston cell, it is not so straightforward in the case of AC potentiometers which use special means for standardisation. This may typically consist of first using a DC supply, followed by an appropriate AC supply, the key device for the process of standardisation being a common instrument used for both; for example, a dynamometer type, sub-standard ammeter, known as a "transfer" instrument.

---

[1]For the academic reasoning, any other phasor (e.g. any one of the three above) could be the *reference* phasor, the requirement of measurement being the same.

[2]As will be seen later, elaborate means are employed to ensure this requirement in different types of AC potentiometer.

Types of AC Potentiometer

Ac potentiometers are broadly classified as

  (a)   polar, and

  (b)   co-ordinate

type, depending on the manner in which the measurement of the un-known PD is carried out to yield its magnitude *and* phase difference.

  In the polar form, the PD is measured in terms of its (absolute) *magnitude* AND the *angle of phase* with respect to a reference phasor, usually the current passing through the main circuit as shown in Fig.9.11(a), the angle itself capable of being measured from 0 to 360°. Accordingly, the measured PD is expressed as

$$PD = |PD| \, \underline{/\theta}$$

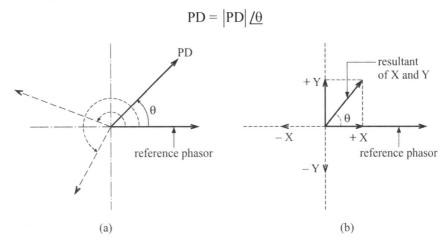

  (a)   (b)

Fig.9.11 : Basis of polar and co-ordinate type AC potentiometers

  In the co-ordinate type, as the name implies, the un-known PD is measured in terms of its *components* along in-phase and (phase) quadrature ordinates of the Argand diagram as shown in Fig.9.11(b). The components can be measured as ±X and ±Y on the potentiometer.

  Then, in this case, the magnitude of the un-known PD is given by

$$|PD| = \sqrt{(\pm X)^2 + (\pm Y)^2}$$

whereas the phase 'difference' angle expressed as

$$\theta = \tan^{-1} \frac{(\pm Y)}{(\pm X)}$$

Drysdale-Tinsley polar-type potentiometer

Developed by the British scientist G.V. Drysdale, this potentiometer belongs to the first category and consists of the usual DC type (potentiometer) with the slide-wire circuit in two parts: the discrete coil resistances and a proper slide wire. The schematic of the internal circuitry of the potentiometer is shown in Fig.9.12. Instead of being just connected to a DC supply to energise the main circuit, the potentiometer is provided with a two-way changeover switch (a DPDT) to allow its connection to a DC supply in one position and then switch over to AC supply for actual AC measurements.

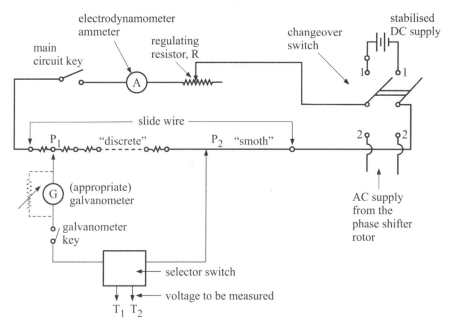

Fig.9.12 : Schematic of Drysdale potentiometer

AC Supply from the Rotor of a Phase Shifter

An essential accessory of the potentiometer is a phase shifter, or phase shifting transformer, described in Chapter VII. The primary or the stator of the phase shifter is fed from the same supply that feeds the test circuit. Thus, the current through the circuit automatically becomes the reference phasor and the PD measured on the potentiometer yields the required phase angle by virtue of the main circuit being supplied from the rotor winding, the spatial position of the rotor being indicative of the angle from $0°$ to $360°$.

The other essential requirement is the inclusion of a precision type electrodynamometer ammeter of sub-standard accuracy connected in the main circuit as shown. This is used for the purpose of standardisation of the potentiometer, first with DC and then AC supply and thus acting as a "transfer" instrument.

Operation of the Potentiometer

Standardisation

Using DC supply

- The main/slide wire circuit is energised by direct current by connecting the change-over switch to DC supply.
- A DC (D'Arsonval) galvanometer is connected at G and a Weston cell across terminals marked $T_1 T_2$.
- The sliding contacts P_1 and P_2 are set to read the standard cell EMF (1.0186 V).
- The potentiometer is balanced as usual by adjusting the main circuit current by means of the regulating resistor, R, and with progressive increase of the galvanometer sensitivity.
- The current in the ammeter is noted to be known as the "standard" current.

Using AC supply

- Replace the D'Arsonval galvanometer by the vibration galvanometer; this may, however, be shorted as it is not required at this stage.
- Remove the standard cell.
- Move the change-over switch to terminals 2, 2, allowing the slide wire to be energised by the AC supply from the phase-shifter rotor.
- The regulating resistance, R, is now adjusted such that the dynamometer ammeter reads the same standard current as during DC standardisation. The setting of R must not be disturbed.

   The potentiometer is now standardised on AC.

Measurement of Un-known PD/EMF

- Activate the vibration galvanometer by removing the shunt, but adjust its sensitivity to a minimum.
- Connect the PD or EMF to be measured across terminals T_1, T_2.
- Adjust successively the sliding contacts P_1 and P_2 *and* the (angular) position of the phase shifter rotor until balance is achieved, as indicated by the vibration galvanometer (or other indicator) at its most sensitive setting.
- At balance, the magnitude of the un-known EMF is given by the reading of the potentiometer dial in terms of position of contacts P_1 and P_2, whilst the phase angle with respect to the reference phasor is read off the position of the phase-shifter rotor.

Gall-Tinsley co-ordinate potentiometer

This potentiometer, developed by D.C.Gall and H.Tinsley, measures an AC voltage in terms of its two components that are in phase quadrature with each other. Hence the name.

It essentially comprises two separate potentiometer circuits, housed in the same case. One is called the "in-phase" potentiometer and measures that component of the un-known voltage during the test which is *in phase with the supply*. The other is the "quadrature" potentiometer which measures the other component of the un-known voltage, in phase quadrature with the other.

This is achieved by supplying the two slide-wire currents, *which are themselves in phase quadrature* with each other, by employing special means of supplying the two currents. Provision is made to measure the components 'in phase' or phase opposition. Thus, the first component may be measured as $\pm V_X$ and the second as $\pm V_Y$. The magnitude of the un-known EMF or PD and its phasor relationship with respect to the reference phasor (usually the current through the in-phase slide wire) can then be derived as brought out earlier, viz.

$$|V| = \sqrt{(\pm V_X)^2 + (\pm V_Y)^2}$$

and

$$\theta = \tan^{-1}\left(\frac{\pm V_Y}{\pm V_X}\right)$$

The 'limiting' values of $|V|$ that can be measured are specified on the instrument. The schematic of a typical potentiometer, showing both slide wires, is given in Fig.9.13.

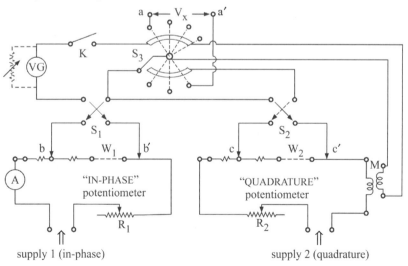

Fig.9.13 : Circuit diagram of Gall-Tinsley coordinate potentiometer

Constructional Details

In the diagram

W_1, W_2	:	In-phase and quadrature slide wires of the usual construction (that is, discrete coils and continuously variable slide wire)
A	:	Precision ammeter of electrodynamometer type
VG	:	Vibration galvanometer
M	:	Precision-type mutual inductance with negligible resistance
S_1, S_2	:	Reversing or sign-changing switches (to change signs of V_X and/or V_Y)
S_3	:	Special selector switch to connect un-known voltages (in the state shown in the figure, it allows the output of M to be connected into the in-phase potentiometer slide wire)
R_1, R_2	:	Controlling resistors to adjust current(s) in the slide wires

Supplies to the Potentiometer

The supplies for the two potentiometer slide wires are obtained from two transformers, fed from a 1-phase supply as shown in Fig.9.14.

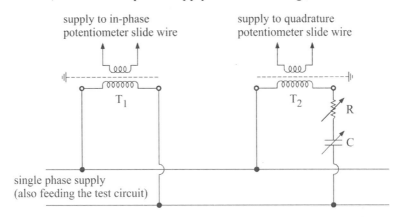

Fig.9.14 : Supplies for the co-ordinate potentiometer

T_1 and T_2 are two specially designed step-down transformers to supply the in-phase and quadrature slide wire circuits of the potentiometer, respectively. The secondary voltage of each transformer is usually 6 V. They also serve as "isolation" transformer, to isolate the potentiometer from the line (or the primary supply) and are usually provided with earthed screens between the two windings as additional shielding. The primary winding of the transformer T_2 carries a precision-type variable resistor, R, and a 'loss-free', precision variable capacitor, C, as shown; both can be adjusted so as to make the secondary voltage of T_2 exactly in phase quadrature with the output of T_1 as required.

Operation of the Potentiometer

Standardisation

A. On DC supply

Similar to the previous cases, this potentiometer, too, is standardised before making any measurement(s). The first step is to do so using a DC supply.

- Connect a battery, or stabilised DC supply, of 2 to 6 V across terminals of the *in-phase* slide wire.
- Connect a standard (Weston) cell across terminals a - a′ and a DC (D'Arsonval) galvanometer in place of the vibration galvanometer.
- Set the sliding contacts b - b′ on the slide wire to read the standard cell EMF, 1.0186 V.
- Adjust the variable resistor, R_1, until balance is obtained.
- Note and record the ammeter reading; this is the standard current for the potentiometer slide wires *and is usually 50 mA.*

B. On AC supply

(i) *In-phase potentiometer*

- Disconnect the standard cell and DC supply.
- Replace the DC galvanometer by the vibration galvanometer.
- Connect the AC supply from transformer T_1 to the in-phase slide wire.
- Adjust the resistor R_1 so that the (dynamometer) ammeter reads the standard current as noted during DC standardisation.

(ii) *Quadrature potentiometer*

To standardise the quadrature potentiometer, it is essential to make *its* slide-wire current *exactly* equal to the standard current AND in phase quadrature with the current in the in-phase slide wire.

For this

- Connect supply from transformer T_2 to the terminals of the quadrature potentiometer, while disconnecting the supply to the in-phase potentiometer.
- Connect the output of the mutual inductor secondary to the slide wire sliding contacts b - b′ of the *in-phase* potentiometer as "unknown" EMF. Set the contacts b - b′ to read the value of this output[1].

---

[1]Assuming the quadrature potentiometer slide wire current to be 50 mA as desired (being the "standard" current) and the mutual inductance of M being 0.0318 H (say), the induced EMF across the secondary of M will be $e = 2\pi \times 50 \times 0.0318 \times 0.05$ or 0.5 V, corresponding to its primary current as 50 mA and supply frequency of 50 Hz. The contacts b - b′ are thus set to read 0.5 V.

- Balance is now obtained by adjusting the resister R_2 and with the vibration galvanometer at increasing sensitivity. This ensures that the current in "quadrature" slide wire is also 50 mA, but in phase-quadrature with the in-phase slide wire.

The quadrature slide wire is thus also standardised. The settings of R_1 and R_2 must not be disturbed.

Comment

The preceding step implies that if

(i)   the secondary voltage from T_2 is in phase-quadrature with that of T_1, achieved by virtue of R and C in the primary circuit of T_2; and

(ii) the current in the quadrature slide wire were to be exactly equal to the standard current (being 50 mA as assumed), which is also the primary current of M,

the secondary, or induced EMF in M will be *in phase* (or phase opposition) with the slide wire current of the in-phase potentiometer, being the time derivative of the primary current and hence can be balanced on the in-phase potentiometer, using the switch S_1, if necessary.

Measurement of the Un-known Voltage/PD

- The un-known voltage is switched into the potentiometer circuit(s) by means of the selector switch, S_3, using any of the terminal pairs.

- In this position of S_3, the un-known potential drop is connected to the two slide wire contacts b-b' and c-c' and the portions of wires across them are *in series* as shown in Fig.9.15 (follow the arrows)[1].

- Balance is obtained by successive/alternate adjustment of *both* pairs of sliding contacts b-b' and c-c' in conjunction with the vibration galvanometer, together with the reversing switches S_1 and S_2, if required.

- At balance, the reading/setting of the in-phase slide wire, together with the position of switch, S_1, gives the magnitude and sign of the in-phase component of the un-known voltage. Similarly, the magnitude and sign of the quadrature component is obtained from the reading/setting of contact c-c' on the quadrature slide wire along with the position of switch S_2.

---

[1]Observe that if switches S_1 and S_2 are reversed as shown dotted, the direction of *both* slide-wire currents will reverse, without any change in the 'external' circuit, for example the galvanometer. However, if only one of them is reversed, currents in the two slide wires will be in phase opposition.

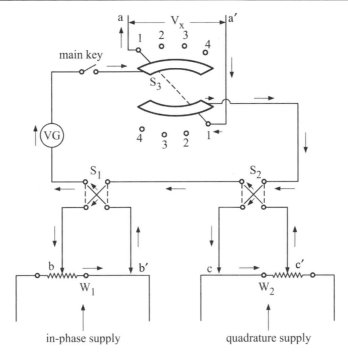

Fig.9.15 : Potentiometer slide-wires circuit to measure the un-known EMF

- The un-known voltage is then given by

$$\dot{V} = \pm V_{in-phase} \pm j V_{quad}$$

in complex form as discussed earlier.

Possible Errors Associated with the Potentiometer

The co-ordinate potentiometer just described is prone to various errors:

(a) slight difference in the reading of the dynamometer ammeter on AC as compared to DC, thus resulting in the standard current being different from that on DC;

(b) an error in the nominal value of mutual inductance, M, which is part of calculating the secondary induced EMF for an assumed/given in-phase potentiometer standard current may cause an error in the standardisation of the quadrature potentiometer;

(c) any error in knowing the supply frequency which again is a part of calculation of secondary EMF of M may lead to the same error as at (b);

(d) a difference in the construction of the two slide wires, assumed identical, may results in the error of measuring components of the un-known voltage;

(e) the presence of harmonics, if any, may lead to an error during standardisation, the latter being based on the DC and rms value during

DC and AC standardisation; also the standardisation of the quadrature slide wire assumes a sinusoidal current in the circuit;

(f)   the quality of R and C used for phase splitting, may affect the phasor relation of the slide-wires current (assumed to be in quadrature) and hence make the standardisation inaccurate/erroneous.

Practical Applications of AC Potentiometers

As the AC potentiometer is a unique and most universally used device for alternating current measurements, particularly at low voltages (or even higher ones, using a volt-ratio box), there are numerous applications where it can be used very effectively. Some of the most common ones are

   (i)   calibration of an ammeter or a voltmeter

  (ii)   calibration of a wattmeter

 (iii)   measurement of power

 (iv)   measurement of self inductance of a coil

  (v)   magnetic measurements[1]

Calibration of an Ammeter or a Voltmeter

This is carried out in the same manner as for DC instruments using a DC potentiometer, described earlier.

Calibration of a Wattmeter

The circuit used for the purpose is shown in Fig.9.16.

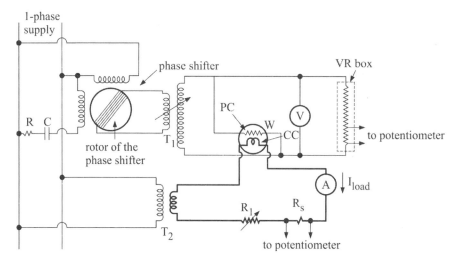

Fig.9.16 : Circuit for calibration of a wattmeter using a polar-type potentiometer

In the diagram, W is the wattmeter to be calibrated.

---

[1]This is dealt with later in Chapter XII.

The procedure

- the pressure coil of the wattmeter is connected to the secondary of an adjustable and isolating transformer T_1, the primary of which is fed from the rotor of a phase shifter;

- the current coil is fed from another isolating (usually a step-down) transformer T_2, its primary being supplied from the *same* 1-phase supply that energises the stator or primary of the phase shifter;

- a regulating resistor R_1 and a standard resistance R_S are connected in series with the current coil as shown;

- the load current, read on the ammeter A, is adjusted with the help of R_1 whilst the voltage to the pressure coil of the voltmeter is adjusted using T_1, along with its phase angle with respect to the load current as given out by the angular position of the phase-shifter rotor;

- the potential drops across the VR box and the standard resistances, R_S, are measured in turn on the potentiometer to give 'true' values of the circuit/load voltage and current;

- the 'true' power is then computed from the above, as in the case of DC-potentiometer test, and compared with the wattmeter reading.

Measurement of power

The circuit requirements for this are shown in Fig.9.17 which is self-explanatory. Using a polar potentiometer, the load current is obtained

$$\dot{I} = |\dot{I}| \angle \theta_1$$

and the voltage across the load as

$$\dot{V} = |\dot{V}| \angle \theta_2$$

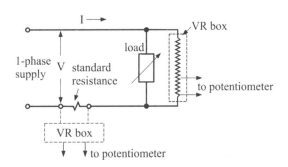

Fig.9.17 : Circuit for measurement of power

with $\dot{I}$ and $\dot{V}$ being expressed in terms of the PDs measured on the potentiometer and the value of standard resistance and multiplying factor of the volt-ratio box, respectively.

Then the power is given by

$$P = |\dot{V}| \times |\dot{I}| \cos \theta$$

where θ equals $\theta_1 \sim \theta_2$.

Measurement of Self Inductance of a Coil

The circuit for this test is shown in Fig.9.18.

The following steps form the procedure

- The coil having a resistance R in series with its self inductance L is connected across a 1-phase supply
- A 4-terminal standard resistance, R_S, is connected in series with the coil
- The two potential drops marked $\dot{V}_1$ and $\dot{V}_2$ are measured on the potentiometer.

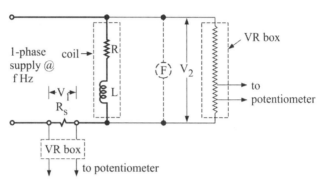

Fig.9.18 : Circuit for measurement of self-inductance of a coil

Let these be

$$\dot{V}_1 = |\dot{V}_1| \, \underline{/\theta_1} \text{ and } \dot{V}_2 = |\dot{V}_2| \, \underline{/\theta_2}$$

Then, circuit current

$$\dot{i} = \frac{\dot{V}_1}{R_S} = |\dot{i}| \, \underline{/\theta_1}$$

and the coil impedance is

$$\dot{Z} = R + j X_L = \frac{\dot{V}_2}{\dot{i}} = \frac{|\dot{V}_2| \, \underline{/\theta_2}}{|\dot{i}| \, \underline{/\theta_1}}$$

$$= \frac{|\dot{V}_2| \, R_S}{|\dot{V}_1|} \, \underline{/\theta_2 - \theta_1}$$

Then, coil resistance, $R = \dot{Z} \cos \theta$

and its reactance, $X_L = \dot{Z} \sin \theta$, where $\theta = \theta_2 - \theta_1$

from which its inductance can be deduced, given by

$$L = \frac{X_L}{2 \pi f}$$

f being the supply frequency and must be known, or measured during the text using a 'reliable' frequency meter.

Example

In a test to determine impedance of a coil using a co-ordinate potentiometer, voltage across a 1.0 Ω standard resistance is

+ 0.952 V on in-phase, and

– 0.34 V on quadrature slide wire.

Voltage across a 10 : 1 potential divider across the coil is

+ 1.35 V on the in-phase, and

+ 1.28 V on the quadrature slide wire.

Calculate the resistance and reactance of the coil.

As measured, the voltage across the coil is

$$\dot{V} = 10 \times (1.35 + j1.28) \quad \text{or} \quad (13.5 + j12.8) \text{ V}$$

and current through the coil is

$$\dot{I} = (0.952 - j0.34) \text{ A}$$

∴ the coil impedance

$$\dot{Z} = \frac{13.5 + j\,12.8}{0.952 - j\,0.34}$$

$$= \frac{18.6\,\underline{/43.5^\circ}}{1.011\,\underline{/-19.6^\circ}}$$

$$= 18.4\,\underline{/63.1^\circ}$$

$$= 8.324 + j\,16.4$$

∴ coil resistance $= 8.324 \ \Omega$

and coil reactance $= 16.4 \ \Omega$

WORKED EXAMPLES

1. A conventional slide-wire DC potentiometer is supplied from a battery of 4 V, having negligible internal resistance. The slide wire is laid over a scale and has the full working length of 200 cm. The resistance of the wire is 100 Ω. A cell of known EMF of 1.018 V is used to 'standardise' the potentiometer and gives a balance at 101.8 cm.

    (i)   Find the working current of the slide wire and setting of the regulating rheostat.

    (ii) If the slide-wire scale has divisions marked in mm and each division can be interpolated to $1/5^{th}$, determine the resolution of the potentiometer.

    The configuration of the potentiometer is as shown.

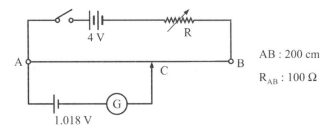

    The standard cell of EMF 1.018 V is balanced at a distance AC of 101.8 cm

    ∴ PD per unit length of the potentiometer wire

$$= \frac{1.018}{101.8} = 0.01 \text{ V/cm or } 10 \text{ mV/cm}$$

    Resistance of the slide wire per unit length

$$= \frac{100}{200} \text{ or } 0.5 \text{ Ω/cm}$$

    ∴ the slide wire current, $I = \dfrac{PD/cm}{resistance/cm}$

$$= \frac{0.01}{0.5}$$

$$= 0.02 \text{ A   or   } 20 \text{ mA}$$

    Let R be the regulating resistance in the main circuit.

    Then $I = \dfrac{4}{R+100} = 0.02 \,(A)$

    whence $R = 100 \text{ Ω}$

Resolution:

Least count of the slide wire

$$= \frac{1}{5} \quad \text{or} \quad 0.2 \text{ mm}$$

PD/cm = 10 mV

$$\therefore \text{PD per 0.2 mm} = \left(\frac{0.01}{10 \times 5}\right) = 0.0002 \text{ V/cm}$$

or the resolution = 0.2 mV/mm

2. A simple potentiometer circuit is set up as in the figure of example 1 above, the uniform wire, AB, being 1 m long and having a resistance of 2 Ω. The internal resistance of the 4 V battery is negligible. If the regulating resistance were given a value of 2.4 Ω, what would be the length AC for zero galvanometer deflection?

If R were made 1.0 Ω and the 1.5 V cell (in place of the standard cell) and galvanometer were replaced by a voltmeter of resistance 20 Ω, what would be the reading of the voltmeter if the contact C were placed at the mid-point of AB?

Slide-wire current $= \dfrac{4}{2 + 2.4} = 0.91 \text{ A}$

PD across AB $= 2 \times 0.91 \quad \text{or} \quad 1.82 \text{ V}$

and this is equivalent to 100 cm.

Now $\quad \dfrac{1.82}{1.5} = \dfrac{100}{L_{AC}}$ at balance, $L_{AC} =$ length of wire AC

$\therefore \quad\quad L_{AC} = \dfrac{150}{1.82} = 82.4 \text{ cm}$

The circuit corresponding to second part is as shown

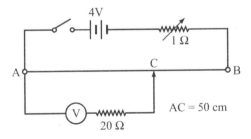

Regulating resistance $\qquad = 1\,\Omega$

Slide-wire current $\qquad = \dfrac{4}{(2+1)} = 1.333$ A

PD across length AC $\qquad = 1.333 \times 1$ or 1.333 V

Current through the voltmeter $\qquad = \dfrac{1.333}{20+1} = 0.0635$ A

and the voltmeter reading $\qquad = 0.0635 \times 20 = 1.27$ V

3. A standard cell of 1.0185 V balances at 50 cm on a simple DC potentiometer comprising a slide wire of length 100 cm.

Calculate

(a) the EMF of a cell that balances at 72 cm

(b) the percentage error in a voltmeter that balances at 64.5 cm while reading 1.33 V

(c) the percentage error in an ammeter that reads 0.43 A and balances at 43.2 cm with a PD across a 2 Ω standard resistance in series with the meter

With the balance at 50 cm for the standard cell of 1.0185 V,

$$PD/cm = \frac{1.0185}{50} = 0.02037 \text{ V/cm}$$

(a) With the (un-known EMF) cell, the balance is obtained at 72 cm

$\therefore$ EMF of the cell $\qquad = 0.02037 \times 72$

$\qquad\qquad = 1.46664$ V

(b) Indicated reading of the voltmeter

$$V_{ind.} = 1.33 \text{ V}$$

'Correct' value using the potentiometer

$$V_{corr.} = 64.5 \times 0.02037$$

$$= 1.3138 \text{ V}$$

$\therefore \qquad$ % error $= \dfrac{V_{ind.} - V_{corr.}}{V_{ind.}} \times 100$

$$= \frac{1.33 - 1.3138}{1.33} \times 100$$

$$= 1.218 \%$$

(c) Indicated reading of the ammeter

$$I_{ind.} = 0.43 \text{ A}$$

With balance at 43.2 cm, PD across the 2 Ω resistance

$$= 43.2 \times 0.02037 \quad \text{or} \quad 0.879984 \text{ V}$$

∴ 'correct' current through the ammeter

$$I_{corr.} = \frac{0.879984}{2} \quad \text{or} \quad 0.439992 \text{ A}$$

∴ $$\% \text{ error} = \frac{0.43 - 0.439992}{0.43} \times 100$$

$$= -2.324 \%$$

4. The circuit in the figure shown below is used to measure the EMF of a thermocouple, T. AB is a uniform wire of length 100 cm and resistance of 2 Ω. With K_1 closed and K_2 open, the balance is obtained at a length of 90 cm. With K_2 closed and K_1 open, the balance length is 45 cm. What is the EMF of the thermocouple? What is the value of R if the internal resistance of the driver cell is negligible?

The circuit

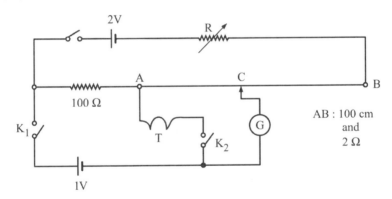

(a) With key K_1 closed, the balance for 1 V cell is obtained at 90 cm from point A, or a resistance of 1.8 Ω

∴ total resistance up to C

$$= 100 + 1.8 \quad \text{or} \quad 101.8 \ \Omega$$

and the PD across AB

$$= \frac{2}{101.8} \times 1 \quad \left[\text{since cell EMF} = 1\text{V}; \ R_{AB} = 2\Omega\right]$$

$$= 0.0196 \text{ V} \quad \text{or} \quad 19.6 \text{ mV}$$

With key K_1 open and K_2 closed, the resistance 100 Ω is not in circuit and the thermocouple is balanced on slide wire AB, such that AC is equal to 45 cm

$$\therefore \qquad \frac{AC}{AB} = \frac{E_T}{19.6} = \frac{45}{100}$$

where E_T is the thermocouple EMF

Hence,

$$E_T = \frac{19.6 \times 45}{100} \text{ or } 8.82 \text{ mV}$$

(b) Considering the series circuit comprising the battery (of 2 V), the series resistances R and 100 Ω and the slide wire AB, the slide-wire current, corresponding to the first balance is given by

$$I_{\text{slide-wire}} = \frac{1}{101.8} = 0.00982 \text{ A} \quad \text{or} \quad 9.82 \text{ mA}$$

This is also given by

$$9.82 = \frac{2 \times 1000}{102 + R}$$

or $2000 = 1001.64 + 9.82 \text{ R}$

whence $R = 101.66 \ \Omega$

5. A Crompton DC potentiometer has a dial comprising 18 steps, each having a fixed resistance of 10 Ω. There is also a slide wire connected in series with the steps having a resistance of 10 Ω and divided into 100 divisions. At standardisation, the working current of the potentiometer is 10 mA. Each division of the slide wire is capable of interpolation to $1/5^{th}$ of a division. Calculate the resolution of the potentiometer.

With the working current being 10 mA, PD across the slide wire is

$$10 \times 10 \quad \text{or} \quad 100 \text{ mV}$$

PD across each division $= \dfrac{100}{100} \quad$ or $\quad$ 1 mV

Since each division can be interpolated to $1/5^{th}$, the resolution is

$$\frac{1}{5} \times 1 \quad \text{or} \quad 0.2 \text{ mV}$$

6. A simple slide wire potentiometer is 100 cm long and has a resistance of $100\,\Omega$. The service battery has an EMF of $4\,V$ and negligible internal resistance. A standard cell of 1.018 V and internal resistance of $1\,\Omega$ is connected in series with a galvanometer having a resistance of $20\,\Omega$ to make the potentiometer direct reading. If the balance is obtained at 88.5 cm on the slide wire, calculate the value of regulating resistance in the main circuit of the potentiometer.

The potentiometer circuit is as shown below

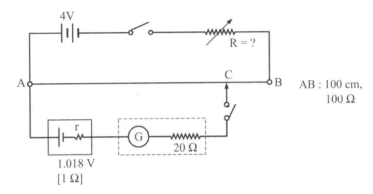

Let the regulating resistance be R Ω

Then current through the slide wire

$$I = \frac{4}{R+100}\,A$$

The potential drop across AC at balance, AC being 88.5 cm (or equivalent to 88.5 Ω) will be

$$PD_{AC} = \frac{4}{R+100} \times 88.5\ V$$

This must equal the EMF of the standard cell or 1.018 V, since no current flows through the cell and galvanometer at balance

or $\frac{4}{R+100} \times 88.5 = 1.018$

whence $R = 247.74\ \Omega$

7. A small choke coil is tested using a co-ordinate potentiometer. The coil is connected in series with a standard resistance of $1\,\Omega$ to measure the current. The PD across the resistance is found to be $(0.9 - j\,0.85)\,V$, whilst that across the coil as a whole is measured to be $(1.5 + j\,0.45)\,V$. Determine the power loss in the coil.

The PD across the standard resistance of $1\ \Omega$ gives the current through the coil directly

Thus

$$I_{coil} = 0.9 - j\ 0.85\ A, \text{ and lagging}$$

or $\qquad\qquad = 1.238 \underline{/-43.36°}\ A$

The PD across the coil is

$$PD_{coil} = 1.5 + j\ 0.45\ V$$

or $\qquad\qquad = 1.566 \underline{/+16.699°}\ V$

$\therefore$ power loss in the coil is given by

$$P = 1.566 \times 1.238 \times \cos(16.699° + 43.36°)$$

$$= 0.967\ W$$

8. In the measurement of power of a load by a polar-type AC potentiometer, the following observations were recorded:

Voltage across a $0.2\ \Omega$ standard resistance in series with the load is $1.46\ \underline{/32°}\ V$

Voltage across a 200:1 volt-ratio box across the line is $1.37\ \underline{/56°}\ V$

Determine the current, voltage, power and power factor associated with the load.

Voltage across the $0.2\ \Omega$ standard resistance $= 1.46\ \underline{/32°}\ V$

$\therefore$ current through the load

$$I_{load} = \frac{1.46\ \underline{/32°}}{0.2}$$

$$= 7.3\ \underline{/32°}\ A$$

PD across the volt-ratio box is $= 1.37\ \underline{/56°}\ V$

$\therefore$ PD across the load is

$$PD_{load} = 200 \times 1.37\ \underline{/56°} \quad \text{or} \quad = 274\ \underline{/56°}\ V$$

Phase difference between the supply voltage and the line (or load) current is

$$\phi = 56° - 32° = 24°$$

$\therefore$ power supplied to the circuit

$$= V\ I\ \cos\phi$$

$$= 274 \times 7.3 \times \cos 24°$$

$$= 1827.274\ W$$

From this must be subtracted the power lost in the standard resistor to obtain the power into the load.

or power *into* the load

$$P_{load} = 1827.274 - (7.3)^2 \times 0.2$$

$$= 1816.616 \text{ W}$$

Now PD across the load

$$= 274\underline{/56°} - 1.46\underline{/32°}$$

$$= (153.22 + j\,227.156) - (1.238 + j\,0.7737)$$

$$= 151.982 + j\,226.383$$

$$= 272.668\underline{/56.125°} \text{ V}$$

∴   phase angle of the load

$$\phi_{load} = 56.125° - 32°$$

$$= 24.125°$$

and power factor of the load

$$= \cos 24.125°$$

or $$= 0.9126 \text{ (lag)}$$

EXERCISES

1. The EMF of a battery A is balanced at a length of 75 cm on a DC potentiometer wire. The EMF of a 'standard' cell, 1.02 V, is balanced at a length of 50 cm. What is the EMF of A? Calculate the new balance length if battery A has an internal resistance of 2 Ω and a resistor of 8 Ω is shunted across its terminals.

[1.53 V; 60 cm

2. The "driver" cell of a potentiometer has an EMF of 2 V and negligible internal resistance. The potentiometer wire has a resistance of 3 Ω. Calculate the resistance needed in series with the wire if a PD of 5 mV is required across the potentiometer wire. The wire is 100 cm long and a balance length of 60 cm is obtained for a thermocouple EMF, E. What is the value of E?

[1197 Ω, 3 mV

3. In the circuit shown below, the EMF E_S of a 'standard' cell is 1.02 V and this is balanced by the PD across a resistance of 2040 Ω in series with a potentiometer wire AB. If AB is 100 cm long and has a resistance of 4 Ω, calculate the length AC on it which balances the EMF 1.2 mV of the thermocouple XY.

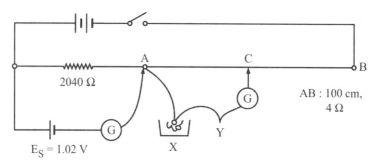

[60 cm

4. In a simple 100 cm long potentiometer, balance is obtained at 60 cm when a standard cell of 1.0186 V is used. Calculate

(a)   the EMF of a cell which balances at 75 cm;

(b)   the current flowing through a standard resistance of 2 Ω, the PD across which is balanced at 66 cm;

(c)   the voltage across the high of a 100/1 volt-ratio box, the "low" side of which is balanced at 84 cm;

(d)   the percentage error in an ammeter that indicates 0.28 A when the PD for the same current through a 2.5 Ω resistor is balanced at 40 cm;

(e) the percentage error in a voltmeter reading 1.25 V when the PD across it is balanced at 72 cm.

[1.27325 V, 0.56023 A, 142.6 V, +3%, +2.216%

5. A co-ordinate potentiometer is used to measure power in an AC circuit. The voltage across a 0.1 Ω standard resistance, in series with the load, is $(0.35 - j\,0.10)$ V. The voltage across a 300:1 potential divider connected across the supply is $(0.8 + j\,0.15)$ V. Determine the power consumed by the load and its power factor. Neglect power loss in the standard resistor.

[795W, 0.8944

6. Calculate the inductance of a coil from the following measurements on an AC polar-type potentiometer:

PD across a 0.3 Ω standard resistor connected in series with the coil

$$= 0.612\,\underline{/12°6'}\ \text{V}$$

PD across the test via a 100:1 volt-ratio box $= 0.781\,\underline{/50°48'}\ \text{V}$

The supply frequency is 50 Hz.

[76.2 mH

QUIZ QUESTIONS

1. A potentiometer can measure
   - □ only current
   - □ only voltage
   - □ power
   - □ all in a circuit

2. A DC potentiometer is preferable to a DC voltmeter of small range for measuring small potential drops because it
   - □ is more expensive
   - □ is more accurate
   - □ looks impressive
   - □ none of these

3. Basically a potentiometer works on the principle of comparing two PDs.
   - □ true   □ false

4. For higher accuracy, the slide wire of a potentiometer should be
   - □ as long as possible
   - □ as short as possible
   - □ as thin as possible
   - □ as thick as possible

5. A potentiometer provides accurate measurement of test EMF because at null condition no current is drawn from the cell.
   - □ true   □ false

6. A potentiometer is basically a
   - □ deflection type instrument
   - □ null-type instrument
   - □ digital instrument
   - □ deflection-cum-null-type instrument

7. A potentiometer is usually standardised first in order to make it
   - □ accurate in operation
   - □ precise to measure un-known PD
   - □ accurate and direct reading
   - □ all of these

8. The open circuit EMF of the Weston standard cell used for standardisation of potentiometer is
   - □ 1.0 V
   - □ 1.0186 V
   - □ 1.1 V
   - □ 1.186 V

9. The Crompton DC potentiometer used in the lab is
   - □ simple slide-wire type
   - □ step resistors type
   - □ step resistors AND slide-wire type
   - □ none of these

10. Higher voltages can be measured with the help of a typical potentiometer by using
    □   a volt-ratio box   □   a rheostat potential divider
    □   a potential transformer   □   a step-down transformer

11. In a test using a DC potentiometer, the following readings were obtained:

    (a) PD across the un-known low resister = 0.4221 V

    (b) PD across a 0.1Ω standard resistor = 1.0235 V

    The value of the un-known resistor is
    □   0.4221 Ω   □   0.041208 Ω
    □   1.0235 Ω   □   0.1 Ω

12. In a test on a simple slide-wire potentiometer, the PD across a 0.1 Ω standard resistor is balanced at 75 cm whilst the standard cell of EMF 1.45 V is balanced at 50 cm.

    The current in the circuit is
    □   7.5 A   □   14.5 A
    □   21.75 A   □   2.175 A

13. Like in DC, in an AC potentiometer, it is enough to measure
    □   only PD across an element
    □   phase angle
    □   power consumed
    □   both PD AND phase angle across the element

14. The AC potentiometer that measures two components of the un-known voltage is called
    □   polar type   □   co-ordinate type
    □   slide-wire type   □   quadrature type

15. For accurate measurement, it is essential that slide wire current(s) in an AC potentiometer is derived from the same source that supplies the test circuit.
    □ true   □ false

16. The standardisation of a typical AC potentiometer is carried out by using
    □   an AC 'standard'
    □   a DC standard cell and a galvanometer
    □   a DC standard cell and a transfer instrument
    □   an AC standard and "AC" galvanometer

17. The device used for phase angle adjustment in a polar potentiometer is called _____.

18. The readings of a polar-type potentiometer when measuring reactance of a coil are:

    $V = 27.8 \underline{/29.7^\circ}$ V; $I = 12 \underline{/13.8^\circ}$ A

    The reactance of the coil is

    □   2.317 Ω   □   0.634 Ω

    □   2.78 Ω   □   − 2.22 Ω

X : Measurement of Resistance

X

MEASUREMENT OF RESISTANCE

RECALL

A resistance forms an important and ever present part of an electric circuit, whether AC or DC. The term derives from the property of an element "to resist", in the present case the flow of current in the circuit. The knowledge of the 'correct' value of the circuit resistance is therefore vital in most applications and devices for their optimum design and performance.

Ohm's Law

This forms the basis of all studies related to resistance. The law known after the German physicist George Ohm, enunciated by him in 1827, states that "the current through a conductor between two points is directly proportional to the potential difference or voltage across the two points, and inversely proportional to the *resistance* between them." Accordingly, the resistance R of a conductor, a resistor or any device in an electric circuit characterised by a resistive property, real or equivalent, is defined as the ratio (V/I), that is, R given by

$$R = \frac{V}{I}$$

where V is the PD across the conductor etc. and I the current flowing through it as shown in the simple, adjoining figure. If V and I are in volt and ampere, respectively, R is obtained in ohm[1], denoted by the

Greek symbol, Ω. Thus, by measuring V and I in a suitable test set-up, R can be derived using the above relationship. It is, however, important that the value of R does not vary during the test due to, say, heating caused by the flow of current.

The other, indirect method of *estimating* the resistance is to measure the physical parameters that define the resistance, and using the formula

---

[1]The unit of resistance in MKS system, named after the physicist.

$$R = \rho \, \frac{L}{A}$$

where L is the length of the conductor, A its area of cross-section and ρ the resistivity. Thus, by measuring L and A and with the knowledge of ρ, R can be obtained; in ohm[1] when L is in metre, A in sq. metre and ρ in ohm-metre. However, the accuracy with which R is obtained would clearly depend on the accuracy of measurement of L and A, and the knowledge of *actual* value of ρ of the material of the conductor. It is, therefore, much preferable to measure the resistance directly, using a suitable method.

---

[1]The other divisions or multiples of ohm commonly used are :

micro-ohm, or $\mu\Omega$, is 10^{-6} ohm

milli-ohm, or $m\Omega$, is 10^{-3} ohm

kilo-ohm, or $k\Omega$, is 10^{3} ohm

meg-ohm, or $M\Omega$, is 10^{6} ohm.

CLASSIFICATION OF RESISTANCES

Broadly speaking, and particularly from the point of measurement, resistances can be classified in three categories:

A. Low Resistances

All resistances having values from, say, micro-ohm to *one ohm* are classified as low resistances. Typical examples of resistances in this category are those of armature and series windings of DC machines, shunts used for extension of ammeter range, cable lengths, leads used in measurement schemes, contacts (at terminals or sockets) etc.

B. Medium Resistances

This category includes all resistances above 1 Ω and up to about 100,000 Ω. Typical examples in practice are the resistances of most appliances (e.g. in domestic use) and devices, including shunt windings of DC machines, rheostats used in labs, potential dividers, potentiometer coils etc.

C. High Resistances

Resistances of values above 100 $k\Omega$ and up to hundreds of meg-ohm (theoretically infinity) fall in this category. A common example is resistances of insulators, used variously in power systems and industrial and domestic appliances.

Note, however, that a classification as above cannot be assumed to be rigid, but only as a guide to a method of measurement to be adopted, in general. Considerable overlap or flexibility may be possible and a given method may be applicable for measurement of a 'low' or 'medium' resistance, the ultimate aim being the accuracy of the results obtained.

MEASUREMENT OF LOW RESISTANCE

When using a given method to measure a low resistance, say of the order of a few milliohm, the common "problems" that might be encountered are

  (i) errors introduced due to contact and lead resistances, not easily estimable and which may be comparable to the resistance to be measured;

  (ii) thermoelectric EMFs that may arise at joints or contacts, made using dissimilar metals and may affect the PD involving the un-known resistance;

  (iii) subjective errors as the quantities such as the PDs and currents being measured may be too small;

  (iv) errors associated with the instruments such as voltmeter, ammeter and resistance box(es) etc.;

  (v) ambient temperature variations during the test (or 'self' heating and temperature rise in the resistance itself) which may result in un-steady measurements.

Depending on the *actual* value of the un-known resistance, one of the following methods may be used for measurement of a low resistance:

  (a) voltmeter and ammeter method

  (b) substitution method

  (c) potentiometer method

  (d) a suitable bridge method, standard or a special one

Of these, the first two methods may be more useful for measuring resistances falling in 'medium' resistance category; for example, $1\,\Omega$ to few tens of ohms. However, with or without some corrections, the method may also be applied to measure low resistances closer to $1\,\Omega$.

Voltmeter and Ammeter Method

This method being the simplest of all, but an approximate one to some extent, requires a simple circuit shown in Fig.10.1. If used with good care, the method can yield results with better than 1 percent accuracy.

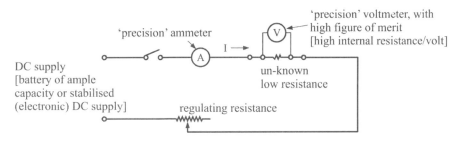

Fig.10.1 : Voltmeter and ammeter method of measurement

The method is self-explanatory: for various adjusted values of the steady circuit current, the PD across the resistance is measured and the resistance value derived. The 'average' of about six readings can be assumed to be the value of the un-known resistance.

Errors due to 'internal' resistance of the voltmeter

If the voltmeter resistance is not 'sufficiently' large as compared to the resistance being measured, the former is bound to 'load' the circuit, or shunt the resistance, thereby resulting in error of measurement of PD. In other words, the current read on the ammeter will not be 'true' current through the resistance, but sum of the current through the resistance and that drawn by the voltmeter; this being an appreciable fraction. Consider, for example, the following:

A resistance has the actual value of 2 Ω. It is tested using the voltmeter-ammeter method. The voltmeter used has an internal resistance of 10 Ω. The current drawn from the supply is 200 mA. What would be the reading of the voltmeter?

The configuration of the problem is as shown in the adjoining figure.

The "un-known" resistance is in parallel with the voltmeter resistance.

The parallel equivalent of the two is

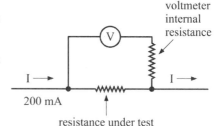

$$R = \frac{2 \times 10}{2 + 10} = 1.667 \ \Omega$$

Voltage drop across the combination

$$PD = 1.667 \times I = 1.667 \times 0.2 \quad \text{or} \quad 0.334 \text{ V, assuming}$$

the 'through' current to remain unchanged.

This would also be the reading of the voltmeter.

The correct value should have been

$$2.0 \times 0.2 \quad \text{or} \quad 0.4 \text{ V}$$

Hence, the error in the voltmeter reading

$$= \frac{\text{apparent reading} - \text{actual reading}}{\text{apparent reading}} \times 100\%$$

$$= \frac{0.334 - 0.4}{0.334} \times 100 \quad \text{or} \quad -19.76 \ \%$$

This means that the voltmeter reads "low" on account of its internal resistance shunting the resistance being measured. The resistance value derived from the measured PD will be 0.334/0.2 or 1.67 Ω as against the actual value of 2 Ω; the error of measurement thus being – 19.76%, same as for the PD. Also, this example assumes that the ammeter is error-free.

Clearly, a voltmeter having a much higher internal resistance (resistance of the coil, plus a resistance in series) would result in much less error[1].

Correction for Shunting Effect of the Voltmeter

In general

Let the actual value of the resistance be R, resistance of the voltmeter, R_v and measured value of the resistance be R_m. The current through the ammeter be I. Assume that both ammeter and voltmeter read correctly.

Then, the resistance of R and R_v in parallel is

$$\frac{R \times R_v}{R + R_v}$$

and volt drop across $\quad R = \dfrac{R \times R_v}{R + R_v} \times I$

$$= \text{voltmeter reading}$$

and the measured resistance given by

$$R_m = \left(\frac{R \times R_v}{R + R_v} \times I \right) \Big/ I$$

$$= \frac{R \times R_v}{R + R_v}$$

from which $\quad R = \dfrac{R_m \times R_v}{R_v - R_m}$

Thus, by knowing R_v and R_m from the ratio of voltmeter and ammeter readings, R can be deduced.

Substitution Method

The basis of this method is to compare the un-known resistance with a variable standard resistance (for example, that in a decade resistance box capable of varying resistance value from, say, 0.001 ohm to 100 ohm or

---

[1]For example, if the voltmeter resistance were 100 Ω, the error in the measured value of the resistance will only be about – 2 %; and only – 0.4 % for the voltmeter resistance of 500 Ω.

more) by *substituting* the two resistances, in turn, in a simple series circuit as shown in Fig.10.2.

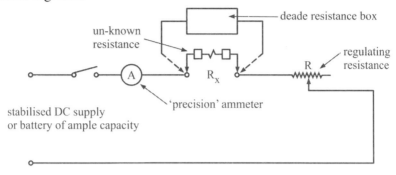

Fig.10.2 : Circuit for substitution method of resistance measurement

Initially, the un-known resistance is inserted in the circuit and the regulating resistance adjusted to read a suitable current (which may depend on the 'safe' current-carrying capacity of the resistance) on the ammeter. The un-known resistance is then replaced by the standard resistance, such as a decade resistance box, the value of which is adjusted, without altering the regulating resistance setting such that the ammeter reads the same current as before. The value of the un-known resistance is then equal to the value of the standard resistance.

The method is better than the voltmeter-ammeter method in that there is no error due to the current through the ammeter being different from that through the resistance to be measured, caused by the current drawn by the voltmeter. However, it is important that the DC supply is constant throughout so that the current does not vary during the substitution/measurement.

Potentiometer Method

In this method, the un-known resistance is compared with a standard resistance of nearly the *same order* of magnitude to achieve best results. Since a potentiometer is capable of measuring very low values of potential drops, the method can yield quite accurate results even for very low resistances. The principle of measurement is based on measuring PD across the un-known and standard resistance in turn[1]. Then

$$\text{un-known resistance} = \frac{\text{PD across the un-known}}{\text{PD across the standard}} \times \text{standard resistance}$$

[1]See Chapter IX for details.

Bridge Methods

The simplest form of a bridge that can be used with a reasonable accuracy if used with care is the well-known Wheatstone bridge, invented in 1843 by the English physicist Sir Charles Wheatstone, comprising four resistance "arms" as these are known and connected as illustrated in Fig.10.3[1]. The circuit, most widely used in the 'standard' configuration or its variation(s), forms an important basis for measurement of not just the resistance(s), but other circuit parameters like inductance and capacitance as described in the next chapter. When the bridge is "balanced" using appropriate procedure as discussed later, the un-known resistance, R, in the figure can simply be obtained as

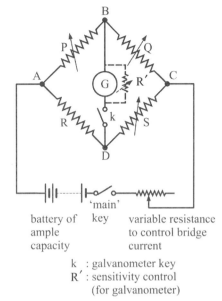

battery of ample capacity

key

variable resistance to control bridge current

k : galvanometer key
R′ : sensitivity control
     (for galvanometer)

Fig.10.3 : The "basic" Wheatstone bridge

$$R = \frac{P}{Q} \times S$$

where P, Q, and S are known, non-inductive precision resistances, usually variable.

Kelvin double bridge

This method, due to Lord Kelvin, is one of the best and most commonly used device for precise measurement of low to very-low resistances. It is essentially a modification or development of the simple 4-arm Wheatstone bridge, and the chief advantage of the bridge is that errors due to contact and leads resistances which can be comparable to the un-known resistance are almost completely eliminated. The schematic of the bridge is shown in Fig.10.4.

In the figure, R_X is the un-known low resistance, connected in series with a standard resistance R_S of the same order of magnitude, and both preferably of the 4-terminal type. The current terminals of the two resistances are

---

[1]It is interesting to know why this form of circuit comprising FOUR arms as shown is called a "bridge". One explanation may be that at balance, when no current passes through the galvanometer, the current in the adjoining arms simply 'rides' over the terminal connecting the galvanometer, similar to a bridge over a river.

shorted by a low-resistance link r as shown. A battery, an ammeter and a regulating resistance complete the 'series' circuit. P, Q, p and q are four known non-inductive resistances; one of the two pairs – P and Q or p and q – being variable. P and Q are usually known as outer whilst p and q are called the inner resistances or branches.

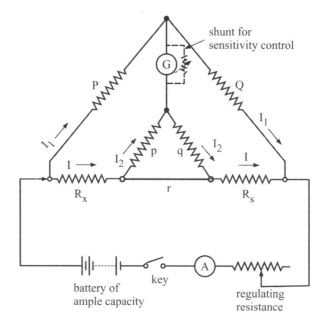

Fig.10.4 : Circuit showing connections of Kelvin double bridge

As shown, the four resistances are connected to form *two* sets of ratio arms, one including the un-known resistance and the other the standard resistance[1]. A sensitive galvanometer is connected across the junctions of P, Q and p, q.

Operation and Theory

A suitable current, usually comparable to the rated current of the un-known resistance, is passed through the circuit and this must be kept constant throughout the test. The resistances P and Q, or p and q, are varied whilst maintaining the ratio P/Q equal to p/q until the galvanometer shows null at its most sensitive setting. The bridge is then said to be balanced, as usual.

To derive the expression for R_X in terms of other known resistances, it is helpful to transform the "Δ" consisting of resistances p, q and r in Fig.10.4 into an equivalent "Y" as shown in Fig.10.5.

---

[1]Hence, perhaps, the name: Kelvin *double* bridge.

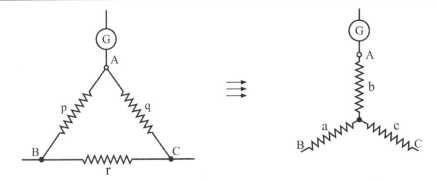

Fig.10.5 : Delta of resistances p, q and r and equivalent star network

The values of the resistances comprising equivalent star are given by

$$a = \frac{p\,r}{p+q+r}, \quad b = \frac{p\,q}{p+q+r}, \quad c = \frac{q\,r}{p+q+r}$$

using the standard Δ-Y transformation relationships.

The circuit of Kelvin double bridge in Fig.10.4 then takes the form shown in Fig.10.6. At balance, there is no current passing though the galvanometer and so the resistance b is ineffective. The bridge thus takes the appearance of the simple Wheatstone bridge with the four arms consisting of P, Q, (R_S + c) and (R_X + a). Hence, under balanced condition,

$$\frac{P}{Q} = \frac{R_X + a}{R_S + c}$$

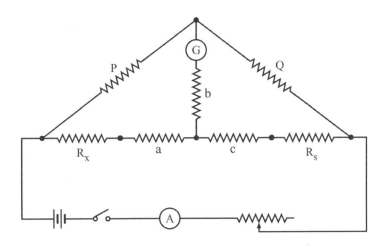

Fig.10.6 : Kelvin double bridge as modified by Δ-Y conversion

from which

$$R_X = \frac{P}{Q} \times (R_S + c) - a$$

Substituting for the values of c and a

$$R_X = \frac{P}{Q} \times R_S + \frac{P}{Q} \times \frac{q\,r}{p+q+r} - \frac{p\,r}{p+q+r}$$

$$= \frac{P}{Q} \times R_S + \frac{q\,r}{p+q+r}\left[\frac{P}{Q} - \frac{p}{q}\right]$$

The second term of this expression can be made very small in comparison to the first by

(i)   making the link resistance r negligible (by using, for example, a link of 'substantial' area of cross-section and very small length);

(ii)   making the ratio P/Q very nearly equal to p/q (which is possible in practice, irrespective of the value of link resistance, r).

Under the circumstances, disregarding the second term, the un-known resistance is simply given by

$$\boxed{R_X = \frac{P}{Q} \times R_S}$$

Errors due to thermo-electric EMFs

Apart from the errors due to contact resistance, there can be errors arising from thermo-electric EMFs owing to the use of dissimilar metal wires and terminal materials. As per standard practice, these can be eliminated by reversing the direction of current and performing two tests. The mean of the two readings should then be taken as the correct value of R_X.

MEASUREMENT OF MEDIUM RESISTANCE

A positive aspect pertaining to measurement of a medium resistance which may range from a few ohms to several hundred ohms, is that the contact and leads resistances (being only a few milliohm or even less) do not cause any error as for low resistances and can be disregarded. The measured values are thus generally true and are affected only by the errors associated with the measuring instruments and apparatus, and subjective errors.

The three commonly used methods for measuring medium resistance are[1]

(a) Voltmeter and ammeter method

(b) Substitution method

(c) Wheatstone bridge

Of these, the first two have already been dealt with whilst describing measurement of low resistances and are generally suited much better for medium resistances with relatively less errors and good accuracy.

Wheatstone Bridge

Although only briefly explained in relation to measurement of low resistances, the method is discussed in more detail here as it is most widely adopted for medium resistances with far better accuracy compared to either of the other two methods. The general, or rather 'standard', composition of a typical Wheatstone bridge is as shown in Fig.10.7.

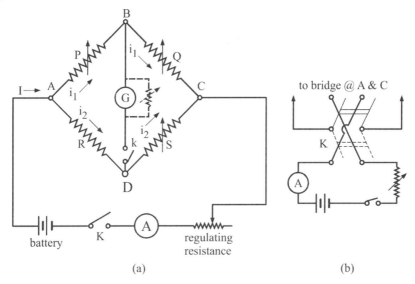

Fig.10.7 : Connections of a Wheatstone bridge

The bridge comprises four resistances of which one, R, being the un-known whilst the other three, P, Q, and S are known, precision, non-inductive, variable resistances, usually in the form of decade resistance boxes of appropriate range. A sensitive, *centre-zero* galvanometer, with the

---

[1]For quick measurement, and for most 'practical purposes, a well-designed ohm-meter can also be used to give a fairly acceptable value of the resistance. A variety of such ohm-meters are now available, affording a reasonable accuracy of measurement. The principle of operation of one type is discussed later in this chapter. However, the method is only 'indicative', considering the accuracy of the instrument itself.

provision to vary its sensitivity, is connected across one diametrically opposite pair of nodes such as B and D, in series with a push-button key. The other pair of nodes, A and C in the figure, is fed from a battery of ample capacity, or a stabilised DC supply of appropriate voltage, carrying in series a key K and a regulating resistance. Although not necessary for the measurement, an ammeter may also be connected in series with the regulating resistance to maintain the current constant at a low, safe value during the course of the test.

In order to eliminate errors arising from any thermo-electric EMF(s), the 'one-way' key may usually be replaced by a reversing key as shown in Fig.10.7(b).

Operation of the bridge and simple theory

With fixed values of resistance assigned to P and Q in arms AB and BC – usually known as "ratio arms" - such that $P/Q = 1$, 0.1, 0.01 etc. in succession, the resistance S is adjusted appropriately (as discussed later by way of an example) till the bridge is balanced as indicated by null in the galvanometer when in its most sensitive setting.

At this stage

(i)   the terminals B and D are at the same potential;

(ii)   the same currents pass through resistances P and Q, and R and S (i_1 and i_2 in the figure);

(iii)   the PD across AB is the same as across AD, and that across BC same as that across CD.

Thus

$$i_1 P = i_2 R$$
$$i_1 Q = i_2 S$$

Dividing the two to eliminate i_1 and i_2

$$\frac{P}{Q} = \frac{R}{S}$$

or

$$\boxed{R = \frac{P}{Q} \times S}$$

giving the un-known resistance R in terms of the *ratio* of resistances P and Q, multiplied by the standard resistance S[1].

Clearly, the accuracy of measurement depends on

(i)   the accuracy with which P, Q and S are known;

---

[1]Once again, it may be helpful if the value of resistance R is obtained approximately by means of, say, an ohm-meter or a quick test using voltmeter-ammeter method.

(ii) the sensitively of the galvanometer;

(iii) the constancy of the bridge current from the battery during the test.

As indicated, the test is repeated by reversing the current in the circuit. The mean value of the two readings should then be taken as the true value of the un-known resistance.

An Example of Measurement in Practice

Suppose that the resistance to be measured has a value of 78.64 ohm; its approximate value being about 80 ohm as measured by means of an ohm-meter. The resistance is connected (as R) in the bridge network as shown in Fig.10.7. The ratio arms P and Q comprise two 3-step decade resistance boxes, to be able to vary their values from 10 ohm to 1000 ohm in steps of 10, 100, 1000 Ω. The resistance S consists of another, 4-dial (or steps) decade resistance box, containing multiple of units, tens, hundreds and thousands of ohm. It is necessary to make all connections tightly at various terminals, using short leads of reasonably thick cross-section. The galvanometer must be shunted suitably to vary its sensitivity.

The procedure can be explained in terms of the following steps:

- Set both P and Q equal to 10 ohm each[1]
- Set S = 1 ohm; galvanometer in its *least* sensitive setting
- Energise the bridge by switching on the battery circuit
- Press the galvanometer key, k, momentarily; disconnecting immediately if the galvanometer deflection is excessive
- Note the direction of deflection of the galvanometer, that is, to the left or right
- Set S = 1000 ohm and press key k again momentarily
- Note again the direction of deflection of the galvanometer
- If the two deflections are in *opposite* directions, it would show that the un-known resistance (R) has a value somewhere between 1 and 1000 ohm
- Keep increasing the setting of S to, say, 10, 20, . . . , 50 ohm up to 80 ohm and observe the galvanometer deflection which will be in opposite directions for S to be 70 and 80 ohm
- This would show that the resistance R lies between 70 and 80 ohm

---

[1]It is assumed that the decade boxes are 100 percent accurate in terms of resistance values through entire range; in practice, this may not be so!

- Repeat by adjusting S to be 78 and 79 ohm; the opposite deflection of the galvanometer, with its sensitivity increased, if required, would indicate that the value of R lies almost between 78 and 79 ohm

To obtain the value of R to first and second decimal place

- Change the setting of Q to 100 ohm whilst P still = 10 ohm (this would make P/Q = 0.1)
- Repeat the test by increasing S to 780, . . 786 and 787 ohm and observe the galvanometer deflections
- For S = 786 and 787, the deflection of the galvanometer will be in opposite direction, showing that R has a value between 78.6 and 78.7 ohm, taking into account the ratio P/Q = 0.1
- Set Q = 1000 ohm whilst P still = 10 ohm, so that P/Q = 0.01
- Repeat the test by setting the value of S up to 7863 and 7865, the galvanometer showing deflections in opposite directions (with near maximum sensitivity)
- Finally, for S = 7864, it will be observed that complete null is obtained with the galvanometer *in its most sensitive state*
- Thus, R will be given by

$$R = 7864 \times 0.01 \text{ or } 78.64 \text{ ohm}$$

The whole procedure can then be repeated by reversing the battery connection and current in the circuit. If the effect of thermo-electric EMF, if any, is negligible, final balance will be achieved with P = 10, Q = 1000 and S = 7864 ohm, as before.

With the settings of P, Q and S as above, if an alteration of 1 ohm in S disturbs the balance of the bridge, then the sensitivity of the bridge can be reckoned to be 1 part in 7864. Clearly, by using a "5-dial" decade box for S whilst decade boxes for P and Q chosen such that P/Q can be set at 0.001, it should be possible to obtain the value of the un-known resistance up to 3rd decimal place. However, the sensitivity of the galvanometer must match with this elaboration[1].

Carey-Foster Slide-wire Bridge

A bridge that can be considered a modification of the standard Wheatstone bridge with higher precision is the Carey-Foster slide-wire bridge, specially suited for comparison of two nearly equal resistances, one of which can be the un-known resistance. The schematic of this bridge is shown in Fig.10.8.

---

[1]It has been found in practice that by observing all precautions, an accuracy of a few parts in 10,000 can usually be achieved with a simple Wheatstone bridge.

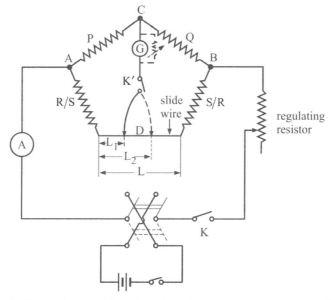

Fig.10.8 : Carey-Foster slide-wire bridge for measuring medium resistance

In addition to the four resistance branches as in a Wheatstone bridge, one of them being the un-known resistance, the bridge includes a slide wire (similar to that in a 'basic' potentiometer) of length L, joining the resistances R and S.

Operation of the bridge

- To begin with, resistances P and Q are first adjusted so that the ratios R/Q and R/S are approximately equal. This can easily be achieved by using suitable decade resistance boxes of good accuracy.

- Exact balance of the bridge is first obtained by adjusting the position of the sliding contact D on the slide wire, with the galvanometer in the full-sensitivity state towards the end.

  Let L_1 be the distance of D from the left-hand end for this balance.

- A second balance is then obtained by interchanging the resistances R and S in the circuit (the balance shown by the dotted arrow on the slide wire).

  Let the distance of the contact D be L_2 for this balance.

  [A special switch may be used to interchange R and S without disturbing the circuit].

- Then, similar to a Wheatstone bridge, for the first balance

$$\frac{P}{Q} = \frac{R + L_1 r}{S + (L - L_1) r}$$

where r represents the resistance of the slide wire per unit length.

- For the second balance

$$\frac{P}{Q} = \frac{S + L_2 r}{R + (L - L_2) r}$$

For a simplified derivation of the final expression, add 1 to both sides in the two expressions for P/Q. Then

$$\frac{P}{Q} + 1 = \frac{R + L_1 r + S + (L - L_1) r}{S + (L - L_1) r} = \frac{R + S + L r}{S + (L - L_1) r}$$

also

$$\frac{P}{Q} + 1 = \frac{S + L_2 r + R + (L - L_2) r}{R + (L - L_2) r} = \frac{S + R + L r}{R + (L - L_2) r}$$

Equating the two expression for P/Q + 1

$$S + (L - L_1) r = R + (L - L_2) r$$

or

$$S - R = (L_1 - L_2) r$$

or

$$R = S + (L_2 - L_1) r^1$$

Thus, the un-known resistance is obtained in terms of the known, standard resistance S *and* the difference of the two lengths of the slide wire, corresponding to two balances. Note that the resistances P and Q do not figure in the expression for R.

Once again, the test may be repeated by reversing the bridge current using the reversing key to eliminate thermo-electric EMF errors, and mean value of R obtained.

Clearly, the accuracy of the method depends on

(i) the current in the circuit being constant during the test(s);

(ii) the accuracy with which r is known and the lengths L_1 and L_2 are measured;

(iii) the sensitivity of the galvanometer, in determining the balance points (or the lengths L_1 and L_2).

Practical Applications of Resistance Measurement by Wheatstone Bridge

The fact that for a set of four given values of resistance in the four arms of a Wheatstone bridge, an out-of-balance voltage would appear in the galvanometer circuit when the bridge is *not* balanced can be used for a

---

[1]Alternatively, the above mathematically satisfactory derivations can be interpreted as the two "P" values considered to be given by $R + L_1 r$ and $S + L_2 r$, respectively, from the two tests and equated to yield the same value for R. Similarly, consider the two values of "Q" as $S + (L - L_1) r$ and $R + (L - L_2) r$ and equate.

number of applications to measure un-known *physical quantities* or *effects*. In the normal usage, the bridge *is* balanced to determine the single value of the un-known resistance in terms of the other three known resistances. However, if the usual un-known resistance forms a resistive 'effect', equivalent to or representing a physical quantity, say temperature at a point in a machine, the same can be incorporated in one of the arms of a Wheatstone bridge and the said effect may be 'measured' in terms of the out-of-balance potential drop that may be calibrated to read the physical quantity on a suitable meter or indicator.

Alternatively, the bridge may be balanced each time for the un-known resistance representing the varying physical quantity and the different values of the resistance so measured may be related to the quantity of interest. The general requirement thus is

physical quantity $\Rightarrow$ equivalent resistance.

Based on the above principle, Wheatstone bridges may be constituted (some of these being quite elaborate in form) to effect the following measurements as illustrative examples.

Using strain gauges[1]

Making use of the property that resistance of a strain gauge (wire) varies almost linearly with the stress (or strain) to which they are subjected, these can be employed in a bridge to measure

- strain(s) on surfaces and parts of a structure;
- pressure in a system;
- torque developed in a rotating machine;
- displacement;
- weight of a loaded truck or lorry on a "weigh bridge".

Using resistance temperature detectors or RTDs

- temperatures at various parts of, say, an electric machine.

Using appropriate transducers

- humidity
- moisture

Most of these applications are described in detail in a later chapter[2].

---

[1]These are special devices using very thin copper-nickel alloy or nickel-chrome wires, encapsulated in PVC covers, measuring typically about 1.2 cm long, 4-5 mm wide and about 0.1 mm thick.

[2]See Chapter XIII: "Measurement of Non-electrical Quantities".

Ohm-meter

This is a simple instrument, sometimes a part of a "multi-meter" (used to measure voltage, current and resistance in a circuit by using a selector switch and appropriate circuitry inside), designed for quick measurement of resistance of components or parts of a circuit. The other important use is to check "continuity" of a circuit or conductor.

The essential circuit of a simple ohm-meter is shown in Fig.10.9 to demonstrate the principle of operation.

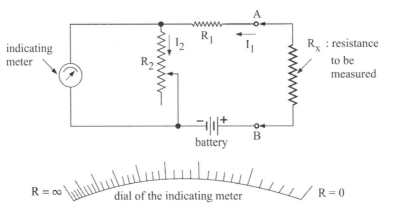

Fig.10.9 : Circuit of a simple ohm-meter

The ohm-meter shown is the simple series type, with R_1 as a fixed series resistance to limit the current through the battery (to suit the full deflection of the indicating instrument – usually the PMMC type). R_2 is an adjustable resistance connected across the meter, mainly for "zero adjustment". A and B are external terminals to connect the resistance (R_X) to be measured. When R_X is 'zero', that is, terminals A and B are simply shorted (a condition to check the continuity of a circuit or conductor), maximum current will flow through the indicating instrument. R_2 is then adjusted to give full deflection in the meter and to read "zero" on the meter scale. When terminals A and B are open which corresponds to R_X being "infinity", there is no current through the meter and it reads *its* zero. This is indicated as $R_X = \infty$ on the dial. Thus, the dial of an ohm-meter is graduated in a *reverse* order and 'measures' the un-known resistance from *right to left*.

For other values of R_X between 0 and ∞, the deflection will be between these limits, depending on the current allowed to flow through the meter. The dial of the instrument is calibrated accordingly.

In practice, there may be many variations of the meter; for example to cover the range of the resistance to be measured in steps such as x10, x100 etc.

It is important to check the condition of the battery (usually only a cell) periodically to ensure the satisfactory working of the meter. This is sometimes indicated when it is not possible to adjust the zero using the shunt potentiometer, R_2.

MEASUREMENT OF HIGH RESISTANCE[1]

The requirements for the measurement of high resistance, typically in megohms, may be 'diametrically' opposite to those associated with the measurement of low or very-low resistances. Thus, the methods of measurement described so far may not be suitable for measurement of a high resistance. The main 'problem' that is encountered is on account of the existence of leakage, or surface, currents which are invariably present and which may often be comparable with the current *through* the resistance itself, required for the purpose of measurement. This calls for special methods to be devised to take such leakage currents into account; or make them negligible in comparison to the 'main' current.

Loss of Charge Method

This is the simplest method, easily applicable and widely used for the measurement of resistances which do not have a noticeable capacitive effect or possibility of an error caused by leakage/surface or absorption current(s). The method essentially consists of using a 'pure' capacitor of capacitance C being charged first from a battery of sufficient capacity and then discharged through the resistance, R_X, to be measured as shown in Fig.10.10. A voltmeter, usually of electrostatic type is connected in parallel with the capacitor and the resistance as shown. The voltmeter is first used to measure the voltage across the capacitor as it gets (fully) charged and later the potential difference across the resistance at different time as the former discharges with time through the resistance. The electrostatic voltmeter is used on account of its extremely high (input) resistance and hence practically no current drawn during the test.

---

[1]The common examples of high resistance are those representing the insulation resistances of cables (of small lengths as specimen or 'total' length) and a variety of insulators used in HV or EHV power systems. Due to the presence of high voltages, the resistance is also invariably accompanied by a capacitance.

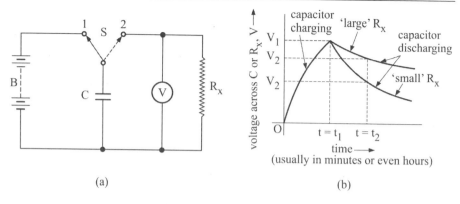

(a)   (b)

Fig.10.10 : Circuit for loss-of-charge method and the charge-discharge 'cycle'

In the circuit, S is a two-way switch which is thrown to position 1 to charge the capacitor to a voltage V_1 [see Fig.10.10(b)][1]. At that instant the switch is *quickly* moved to position 2 when the discharge of the capacitor through resistance R_X begins. During the interval of discharge, if V is the terminal voltage of the capacitor (also the PD across the resistor) at any time t, Q being the corresponding charge and C its capacitance, the current i at the instant is given by

$$i = -\frac{dQ}{dt} = -C\frac{dV}{dt}$$

(negative sign representing the discharge process)

But i is also equal to V/R_X, R_X being in series with C.

$\therefore$
$$-C\frac{dV}{dt} = \frac{V}{R_X}$$

or
$$C\frac{dV}{dt} + \frac{V}{R_X} = 0$$

Solving this equation for V gives

$$V = V_1 \in^{-t/CR_X}$$

where V_1 is the initial voltage across C, at $t = t_1$ when the switch is moved to position 2.

Taking log to the base $\in$

$$\log_\in V = \log_\in V_1 - \frac{t}{CR_X} \qquad (\text{since } \log_\in \in = 1)$$

---

[1]If the capacitor is fully discharged at the beginning of the test, there may be a sudden rush of charging current.

or $R_X = \dfrac{t}{C \log_\in \left(\dfrac{V_1}{V}\right)}$ or $= \dfrac{0.4343\, t}{C \log_{10}\left(\dfrac{V_1}{V}\right)}$,   converting log to base 10;

$$0.4343 = \frac{1}{\log_\in 10}$$

If C is in farad and t in second, R_X will be given in ohm.

Thus, if the voltmeter reading after t seconds [$= t_2 - t_1$, as in Fig.10.10(b)] is V_2, meaning that the capacitor has discharged from V_1 to V_2, the un-known resistance will be given by

$$R_X = \frac{0.4343\, t}{C \log_{10}\left(\dfrac{V_1}{V_2}\right)}$$

and the accuracy being dependent on the accuracy with which C is known and that of measuring time and PD across the resistance[1].

To account for 'finite' resistance(s) of the capacitor and voltmeter

Let the *combined* resistance of the capacitor and voltmeter in parallel be R_1 and assumed to be determined by test. Initially, R_X is connected across the capacitor in discharge mode, but effectively also has R_1 connected in parallel with it.

Then, the capacitor having been charged to a value V_1 at time t_1 is made to discharge through the two resistances in parallel. If R' is the combined resistance of R_1 and R_X in parallel and the PD across it falls to V_2 in time t, then

$$V_2 = V_1 \in^{-t/CR'}$$

From this, the value of R' can be derived as before.

The test is then repeated with R_X disconnected, the capacitor discharging through R_1 which is deduced from the value of PD in the measured time. The un-known resistance R_X is then obtained from the expression

$$R' = \frac{R_X \times R_1}{R_X + R_1}$$

---

[1]The assumption of C being a pure capacitance may not hold in practice and its loss component (an equivalent resistor) may have to be taken into account if it is comparable to the resistance being measured. Also, the duration of test, if R_X is very large, may extend to several *hours* and the effect of errors on the results due to change of temperature during the test may also have to be considered.

> ### Comment
>
> If the (high) resistance to be measured is the insulation resistance of an (imperfect) *capacitor* or *cable*, a single test would suffice. In Fig.10.10, C will still represent the capacitance part, and must be known, and R_X its insulation resistance. The cable (or the capacitor) would simply discharge through R_X and can be computed knowing the PD in a known, measured time interval, as before.

Price's Guard-wire Method

This is one method particularly aimed at measuring insulation resistance of a *cable* where the errors due to surface leakage current are eliminated by the use of a device called a "guard wire". The principle behind the method is illustrated by the schematic shown in Fig.10.11.

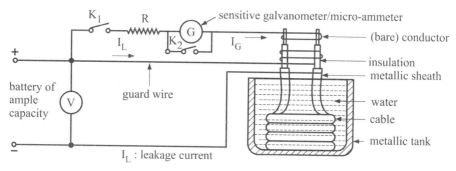

Fig.10.11 : Schematic of the Price's guard-wire method

To begin with, the length of the cable whose insulation resistance is to be measured is immersed in water in a coiled form, contained in a tank, for about 24 hours before making any measurements. This enables the water to soak through any microscopic 'defects' (such as perforations) that might exist in the insulation and also for the insulation to attain the same (uniform) temperature as that of water in the tank[1]. This, in a way, 'stabilises' physical condition of the cable.

The connections of the cable ends are made to a battery, of ample capacity or an alternative source of stabilised DC supply of about 500 V, in the following two ways:

    (a)   the outer covering, or the metallic sheath, is removed for a length of about 30 cm and a bare wire wound round the two-ends, this being

---

[1]For 'good' results, the water temperature should be maintained constant, ideally at the operating temperature of the cable, throughout the test by employing appropriate means.

connected directly to the positive terminal of the supply, so that any current which leaks across the insulation surface is taken direct to the supply via the outer sheath and does not pass through the galvanometer circuit;

(b) the insulation of the conductor itself is trimmed likewise to expose the bare conductor which is again wound with a wire and connected to positive terminal of the supply, in series with a (highly) sensitive galvanometer or a micro-ammeter (shunted by a key), a protective resistance, R, and main supply key as shown.

The DC supply, used in the circuit first charges the cable (acting as a capacitor) and later 'maintains' the current through the insulation. A precision voltmeter, usually an electrostatic type is used to measure the voltage across the insulation.

Operation

- The galvanometer key K_2 is closed and the cable is charged by closing the main key K_1. This may take some time; closing galvanometer key ensures its safety from sudden inrush of the charging current which may be quite high.

- Once the cable is nearly fully charged, the galvanometer is inserted in the circuit by opening key K_2 when a current I_G, usually a few micro-ampere, would flow from bare conductor part of the cable through sheath to the negative terminal of the supply.

The insulation resistance of the cable is then simply given by

$$R_{cable} = \frac{V}{I_G},$$

usually in meg-ohm.

PORTABLE INSULATION RESISTANCE TESTING INSTRUMENTS

Megger

This is the commonest of the various portable insulation-resistance testers and has been in extensive use for decades. As the name suggests, the instrument dial is calibrated to read the insulation resistance in meg-ohm. The schematic of a typical megger illustrating the principle of operation and movement of the pointer under two extreme cases of the un-known resistance, R_X, being 0 or ∞ is shown in Fig.10.12.

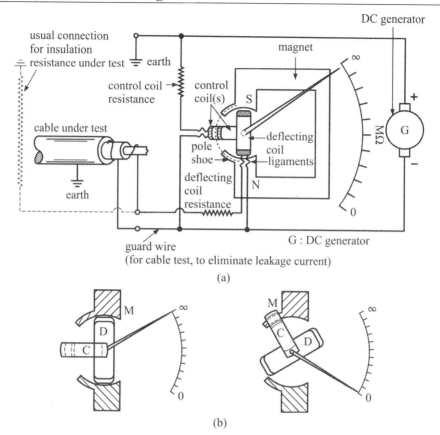

C: Control coil(s);   D: Deflection coil;   M: Permanent magnet

Fig.10.12 :   Schematic of a megger to show principle of operation

(a) internal constructional details

(b) deflection of the pointer in two extreme cases

Constructional details

The moving system of the instrument consists of two coils, called control coil and deflection coil, respectively, rigidly mounted at nearly right angles, the terminals of the coils being connected in parallel across a small generator, that of the deflecting coil being in series with the resistance to be measured, one end of which is earthed and thus linked to the earthed positive terminal of the generator as shown. The polarities are arranged such that the torque produced by them under the action of a magnetic field are in opposition. Both the coils together, particularly the deflecting coil, are designed to move inside the airgap of a permanent magnet having extended pole shoes as shown. The deflecting coil is connected in series with

(i)   a fixed (deflecting-circuit) resistance and

(ii)   the resistance under test.

The control coil, actually constructed in two parts, also carries a resistance, all connected in series. The two parts are arranged with the number of turns, and radii of action, such that, for external magnetic fields (of uniform intensity), their torque cancel one another, thus giving an astatic combination to eliminate effect of external/stray magnetic fields.

The generator housed in the same box/case is a permanent-magnet DC generator, capable of generating 100, 250, 500, 1000 or 2500 V depending on the design and requirement[1]. The generator is coupled to a handle, external to the case, and is hand-driven through gearing and a centrifugally controlled clutch mechanism. The latter is designed to slip at a pre-determined speed, irrespective of the speed at which the handle is (manually) rotated. This ensures steady voltage for use in the instrument.

Operation

- When the un-known resistance is 'finite', that is, a value between zero and infinity, the instrument pointer takes a steady position on the dial under the combined influence of the deflecting and the control coils such that the torque due to deflecting coil = torque due to control coil, the two acting in opposition.

- In one extreme case when R_X is infinity (an "open" circuit), the deflecting coil is open-circuited, carrying no current and acting as a "dummy". Only the control coil carries a current, producing a magnetic field such that the coil would align with the permanent-magnet field. The pointer thus indicates "∞" as shown in Fig.10.12(b).

- In the other extreme case when R_X is zero, corresponding to a "dead" short, maximum current, but restricted by the deflection-circuit resistance, would flow in the deflecting coil. This would produce a strong magnetic field, and large deflecting torque, tending its axis to align with the field of the permanent magnet.

- At the same time, the control torque increases as the control coil moves to embrace more field by getting into the extended (permanent-magnet) pole shoe as shown even when the magnitude of current through the coil remains the same, being limited by the series resistance.

- Under the action of the two coils, or rather the torques produced by them, the pointer takes a steady position indicating "0" on the dial.

- For any "intermediate" value of the un-known resistance and the two equal and opposite torques, the pointer reads the value on the scale, calibrated in meg-ohm.

---

[1]This actually forms the basis of specification and rating of a given megger for use in different circuits and anticipated value of the insulation resistance.

WORKED EXAMPLES

1. In an "ammeter and voltmeter" method to measure a 'medium' resistance, a voltmeter of internal resistance of $500\,\Omega$ and a milliammeter of $1\,\Omega$ resistance are used. If the voltmeter reads $20\,V$ and the milliammeter $100\,mA$, calculate the resistance value if the voltmeter is connected across

    (i)   the un-known resistance and the millimeter connected in series;

    (ii) the un-known resistance with the milliammeter connected on the supply side.

(i) The connection is as shown.

Measured value of *total* resistance across AB

$$= \frac{20}{0.1} \text{ or } 200\,\Omega$$

$$= R_x + 1.0$$

$$\therefore R_x = 200 - 1 = 199\,\Omega$$

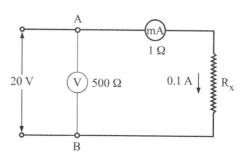

(ii) In this case the resistance is connected as shown.

The "measured" value is $\frac{20}{0.1}$ or $200\,\Omega$ as before.

But the total resistance across AB is

$$R = \frac{R_x \times 500}{500 + R_x} = 200\,\Omega$$

whence

$$3R_x = 1000$$

or $R_x = 333.33\,\Omega$

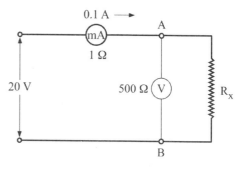

2. In the measurement of a 'medium' resistance by the substitution method, a standard resistance of $100\,k\Omega$ is used. The galvanometer with an internal resistance of $2000\,\Omega$ gives a deflection of 46 divisions with the un-known resistance in series and 40 divisions with the standard resistance. Determine the value of the un-known resistance.

The circuit is as shown

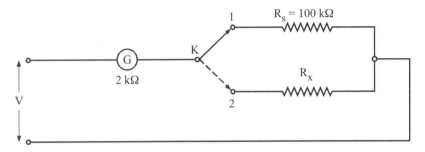

With the key K in position 1,

total circuit resistance $= 100 + 2$ or $102 \text{ k}\Omega$.

For the applied supply voltage of V,

the circuit current is $I = \dfrac{V}{102}$ mA

and is proportional to 40 divisions in the galvanometer.

With the key in position 2,

total circuit resistance will be $(R_X + 2) \text{ k}\Omega$ (assuming R_X in $\text{k}\Omega$) and

the circuit current

$$= \frac{V}{R_X + 2} \text{ mA}$$

$$\propto 46 \text{ divisions}$$

∴ $\qquad \dfrac{V}{102} : \dfrac{V}{R_X + 2} \quad :: 40 : 46$

or $\qquad \dfrac{R_X + 2}{102} = \dfrac{40}{46}$

whence $\qquad R_X = 86.69 \text{ k}\Omega$

3. A 4-arm Wheatstone bridge comprises the following branch resistances:
AB: 200 Ω; BC: 20 Ω; CD: 8 Ω and DA: 100 Ω.

The bridge is configured in the usual manner with the galvanometer having an internal resistance 20 Ω being connected across BD. The bridge is supplied from a steady DC source of 20 V, connected across AC. If the bridge as connected above is NOT balanced, find the current through the galvanometer. What should be the resistance connected in the arm DA for the bridge to be balanced?

The bridge is connected as shown in Fig. A. Since P/Q ≠ R/S, the bridge is not balanced; the terminals B and D are not at the same potential and thus a current would flow through the galvanometer.

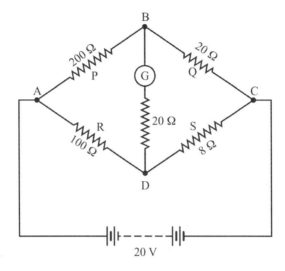

Fig. A

Apply Thevenin's theorem to obtain the "open circuit" voltage and Thevenin equivalent resistance across BD. The equivalent circuit for this is shown in Fig.B.

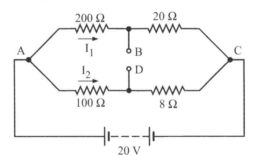

Fig. B

(a)   To derive the open circuit PD across BD, V_{TH}

Referring to Fig. B,

$$I_1 = \frac{20}{220} = 0.091 \text{ A}$$

$$I_2 = \frac{20}{108} = 0.185 \text{ A}$$

Potential drop B to A,

$$V_{B \to A} = (-) \, 200 \times I_1 = -18.2 \text{ V}$$

Potential drop D to A,

$$V_{D \to A} = (-) \; 100 \times I_2 = -18.5 \text{ V}$$

$$\therefore \qquad V_{B\text{-}D} = 18.5 - 18.2 = 0.3 \text{ V} = V_{TH}$$

(b) To derive Thevenin equivalent resistance, R_{TH}

With the galvanometer removed and the source of EMF suppressed, the circuit for the purpose is shown in Fig. C, with successive reduction.

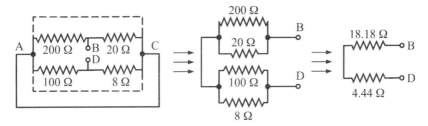

Fig. C : AC → shorted at the upper and lower end

From the figure, resistance across BD or Thevenin equivalent resistance,

$$R_{TH} = 18.18 + 4.44 = 22.62 \; \Omega$$

The Thevenin equivalent circuit is thus as shown in Fig. D.

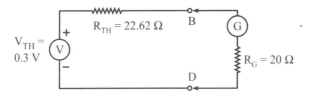

Fig. D

Total resistance in the circuit

$$= 22.62 + 20.0$$

or $\qquad R_{total} = 42.62 \; \Omega$

$\therefore$ out of balance current through the galvanometer

$$= \frac{V_{TH}}{R_{total}}$$

$$= \frac{0.3}{42.62} = 7 \text{ mA (approx)}$$

For balancing the bridge

$$\frac{P}{Q} = \frac{R}{S}$$

or

$$\frac{200}{20} = \frac{R}{8}$$

whence

$$R = 80 \ \Omega$$

4. A 4-terminal resistor of approximately 50 $\mu\Omega$ resistance was measured using the Kelvin's double bridge having the following composition:

    standard resistor   : 100.03 $\mu\Omega$

    inner ratio arms   : 100.31 and 200.0 Ω

    outer ratio arms   : 100.24 and 200.0 Ω

    connecting (link)

    (resistance between the

    4-terminal resistors)   : 700 $\mu\Omega$

Determine the un-known resistance to the nearest 0.01 $\mu\Omega$

The circuit connections of the bridge are as shown.

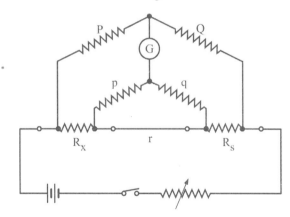

Here

$$P = 100.24 \ \Omega$$
$$Q = 200.0 \ \Omega$$
$$p = 100.31 \ \Omega$$
$$q = 200.0 \ \Omega$$
$$R_S = 100.03 \ \mu\Omega$$
$$r = 700 \ \mu\Omega$$

Since $\dfrac{P}{Q} \neq \dfrac{P}{q}$, full expression for R_X must be used, that is

$$R_X = \frac{P}{Q} \times R_S + \frac{q\,r}{p+q+r}\left[\frac{P}{Q} - \frac{p}{q}\right]$$

Now
$$\frac{P}{Q} \times R_S = \frac{100.24}{200} \times 100.03 \times 10^{-6} \quad \text{or} \quad 50.135 \ \mu\Omega$$

$$\frac{q\,r}{p+q+r} = \frac{200 \times 700 \times 10^{-6}}{100.31 + 200} \quad \text{or} \ 466.1849 \ \mu\Omega$$
[neglecting r in comparsion to p and q]

and
$$\frac{P}{Q} - \frac{p}{q} = \frac{100.24}{200} - \frac{100.31}{200} \quad \text{or} \ -0.00035$$

$\therefore \qquad R_X = 50.135 + (466.1849) \times (-0.00035) \ \text{or} \ 49.9718 \ \mu\Omega$

5. In a "loss-of-charge" method, a capacitor of capacitance 2.5 μF is charged to a voltage of 500 V. When connected across an electrostatic voltmeter, the voltmeter reads 300 V in 15 minutes. The test is repeated with a high resistance R in parallel with the voltmeter and the capacitor. If the voltmeter reading drops from 500 V to 300 V in 10 minutes, what is the value of the resistor R?

(a) With (only) the charged capacitor connected across the voltmeter, PD at the end of 15 minutes is 300 V.

Let R_1 = resultant of voltmeter resistance and leakage resistance of the capacitor (in parallel)

Then
$$R_1 = \frac{0.4343 \times t\,(\text{in s})}{C\,(\text{in farad}) \times \log_{10}\left(\dfrac{E}{V}\right)}$$

E being the initial voltage (= 500 V here)

or
$$R_1 = \frac{0.4343 \times 15 \times 60}{\left(2.5 \times 10^{-6}\right) \times \log_{10}\left(\dfrac{500}{300}\right)}$$

$$= 704.75 \ \text{M}\Omega$$

(b) When the un-known resistance R_X is connected across the voltmeter and capacitor in parallel, the voltage, V, drops to 300 V in 10 minutes[1].

Let R_2 = resultant of voltmeter resistance, leakage resistance of the capacitor and R_X, all in parallel.

Then
$$R_2 = \frac{0.4343 \times 10 \times 60}{\left(2.5 \times 10^{-6}\right) \times \log_{10}\left(\dfrac{500}{300}\right)}$$

$$= 469.834 \text{ M}\Omega$$

Clearly, R_2 is the parallel equivalent of R_1 and R_X

or
$$\frac{1}{R_2} = \frac{1}{R_1} + \frac{1}{R_X}$$

or
$$\frac{1}{R_X} = \frac{R_1 - R_2}{R_1 R_2}$$

$$= \frac{704.75 - 469.834}{704.75 \times 469.834}$$

whence $R_X = 1409.5 \text{ M}\Omega$

---

[1] The drop of voltage is faster since the resultant resistance of three resistances in parallel is relatively less.

EXERCISES

1. In a Wheatstone bridge the ratio arms P and Q are 100 Ω and 10,000 Ω, respectively. The standard resistance S = 8962 Ω at balance. The supply battery has an EMF of 1.5 V and negligible internal resistance. The galvanometer has a resistance of 500 Ω. Determine the value of the "un-known" resistance, R, and the out-of-balance current if S is increased by 1 Ω.

   [89.63 Ω, 0.745×10^{-8} A

2. In the measurement of an un-known resistance by Kelvin double bridge, the following observations were made:

   Outer ratio arms = inner ratio arms = 1:100

   Standard resistance = 20 Ω

   Determine the value of the un-known resistance.

   [0.2 Ω

3. A Kelvin double bridge is balanced with the following constants:

   Outer ratio arms   :   100 Ω and 1000 Ω

   Inner ratio arms   :   99.92 Ω and 1000.6 Ω

   Resistance of the link   :   0.1 Ω

   Standard resistance   :   0.00377 Ω

   Calculate the value of the un-known resistance.

   [389.72 μΩ

4. In a "loss of charge" method, the insulation resistance of a short length of cable is measured. During the test, the voltage falls from 100 V to 80 V in 20 s; the capacitor having a capacitance of 0.0005 μF. Calculate the insulation resistance of the cable.

   [112149.8 MΩ

5. The following observations were made in a "loss-of-charge" method for measuring insulation resistance of a cable:

   Capacitance used   : 0.2 μF

   Capacitor charged to a voltage across the capacitor to 400 V.

   Time for drop of voltage across the capacitor to 250 V   : 90 s

   Determine the value of the resistance.

   [957.45 MΩ

6. In an experimental determination of a high resistance R by a discharging-condenser method, the following readings were noted:

    The charged condenser of capacitance 12.5 μF was connected across an electrostatic voltmeter and R in parallel, and the voltage recorded at different times:

Seconds	0	100	200	300	400	500	600
Volt	150	121	97	83	65	57	43

    A further set of observations was made with R disconnected as given below:

Seconds	0	200	400	600
Volt	150	143	133	121

    Plot the variations of voltage as above and determine the value of R.

    [57.5 MΩ

QUIZ QUESTIONS

1. Resistances are generally classified as

    (a) _____   (b)_____

    (c)_____

2. Low resistances are typically fabricated with

    □ 2 terminals   □ 4 terminals

    □ 6 terminals   □ none of these

3. Most serious errors of measurement associated with low resistances is

    □ effect of ambient temperature   □ material of the resistor

    □ contact resistance   □ all of these

4. A typical 'medium'-resistance values can be

    □ less than 1 Ω □ 1 Ω to 100,000 Ω

    □ greater than 100,000 Ω □ none of these

5. The simple "voltmeter-ammeter" method is best suited to measure a

    □ very low resistance   □ low resistance

    □ medium resistance   □ high resistance

6. In a 'voltmeter-ammeter' method, a voltmeter with 5 Ω internal resistance is connected across a resistor having an actual value of 1 Ω. If the net current drawn by the combination is 100 mA, the measured value of the resistance is

    □ 1 Ω □ 0.833 Ω

    □ 1.2 Ω □ none of these

7. In the substitution method of resistance measurement, the PD across the un-known resistance is same as that across standard resistance.

    □ true   □ false

8. A 'standard' Wheatstone bridge is most suited for the measured of a

    □ low resistance   □ medium resistance

    □ high resistance   □ none of these

9. The Wheatstone bridge works on the principle of

    □ measuring PD across the un-known resistor

    □ measuring and equating currents in the branches

    □ null detection

    □ variable supply voltage

10. A Wheatstone bridge comprises ratio arms of 1000 Ω and 100 Ω. The 'standard' arm consists of a 4-decade box of 1 Ω, 10 Ω, 100 Ω and 1000 Ω steps range. The bridge is suited to measure maximum and minimum resistance of

☐ 111100 Ω, 10 Ω   ☐ 1110 Ω, 1 Ω

☐ 11110 Ω, 1 Ω   ☐ 11100 Ω, 1 Ω

11. A Wheatstone bridge may not be used for precision measurement on account of errors due to

☐ resistance of connecting leads   ☐ thermo-electric EMFs

☐ contact resistance(s)   ☐ all of these

12. A Kelvin double bridge is ideally suited to measure low resistance of 4-terminals type.

☐ true   ☐ false

13. A Kelvin double bridge is possibly so named because it has double ratio arms.

☐ true   ☐ false

14. A Kelvin double bridge may reduce to an ordinary 4-arms bridge when

☐ the ratio of outer arms resistances equals that of inner arms resistances

☐ the two ratios are in excess of 1

☐ the two ratios are unequal by 1

☐ none of these

15. A typical Carey Foster bridge comprises

☐ 3 ratio arms and a slide wire   ☐ 4 ratio arms and a slide wire

☐ 2 ratio arms and one slide wire ☐ 1 ratio arm and the slide wire

16. In the "loss of charge method", the voltmeter generally used is

☐ PMMC type   ☐ dynamometer type

☐ electrostatic type   ☐ moving iron type

17. In the loss of charge method of measuring high resistance, the basic principle of operation is discharge of capacitor through a 'resistance'.

☐ true   ☐ false

18. The rate of reduction of voltage across the resistance in the loss of charge method is slower, the higher the resistance.

☐ true   ☐ false

19. A "guard" terminal is usually employed in measurement of high resistance to

    □ by pass the leakage current

    □ eliminate the effects of stray magnetic field

    □ provide safety to the operator

    □ none of these

20. The instrument called "Megger" is possibly so named because it is used to measure insulation resistances in meg-ohm.

    □ true   □ false

21. A megger is generally used to measure

    □ contact resistance of low resistors

    □ insulation resistance of cables

    □ resistance of leads in circuits

    □ armature resistance of DC machines

22. A typical megger draws its supply from

    □ 1-phase AC supply   □ a 12 V battery

    □ an inverter   □ an internal PM DC generator

XI : Alternating-current Bridge Methods

XI

ALTERNATING-CURRENT BRIDGE METHODS

RECALL

In most alternating-current circuits, the two quantities that are important to be measured in addition to the resistance[1], are *inductance* and *capacitance*. These can be measured by adopting simple methods, giving reasonably accurate results in most cases. However, when measurements with better accuracy are desired, particularly at frequencies other than or higher than the supply frequency, AC bridge methods are best suited and invariably used. A large number of bridges are available as discussed in this chapter, some of which are uniquely applicable for the measurement of an inductance or a capacitance.

The Genesis of an AC Bridge

Nearly all AC bridges are the modification of the basic or 'standard' Wheatstone bridge, described previously for the measurement of (DC) resistance and which comprised four resistances connected in a unique manner. One of the resistances would be the un-known; the other three being standard, non-inductive and variable resistances. At balance of the bridge, achieved by the adjustment of the known resistances, the un-known resistance, R_x, would be given by

$$R_x = \frac{P}{Q} S$$

where P, Q and S are the known resistances.

It was seen that the accuracy of measurement depended on the accuracy with which the resistances P, Q and S were known and the sensitivity of the galvanometer, and that with relative adjustment of P and Q, forming "ratio" arms of the bridge, results accurate up to two or more decimal places could be obtained. The same features and advantages, in general, also relate to AC bridges and bridge methods.

---

[1] This may simply be the "DC" resistance, measured using DC supply and will usually be equal to "AC" resistance at low frequencies; for very high frequency, the AC resistance may be greater than the DC resistance owing to skin effect.

A GENERAL 4-ARM AC BRIDGE

A 'standard' form of an AC bridge comprising four 'arms' is shown in Fig.11.1.

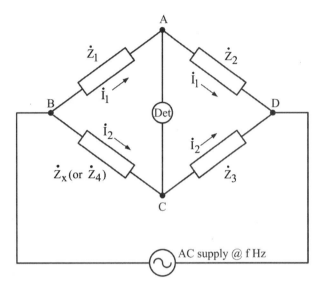

Fig.11.1 : A 4-arm AC bridge network

It differs from the usual 4-arm Wheatstone bridge in the following three main respects:

A. Bridge arms

The four arms of the bridge comprise four *impedances*, one of which being un-known (or Z_x), instead of simple resistances as in a Wheatstone bridge. The impedances themselves may be any of the three elements of an AC circuit, viz. resistance, inductance or capacitance; or a combination of any of these, connected in series or parallel, thus being complex network quantities.

B. Detector

The DC, D'Arsonval galvanometer is replaced by a suitable *detector* to detect null condition in the bridge and may usually be an "AC galvanometer", typically a "vibration galvanometer" or any other device depending (particularly) on the supply frequency[1].

---

[1]A *tuned* vibration galvanometer is ideally suited and frequently used when the supply is at power or other low frequency; other detectors are chosen for higher frequencies.

C. The supply

The supply to the bridge itself is an appropriate AC supply instead of a simple battery, derived from the usual 1-phase supply of power frequency and usually incorporating a step-down transformer. For higher frequency applications, an appropriate source, for example, a signal generator may be used as required.

Theory and Operation

The theory of the AC bridge can be worked out on the same lines as that of the Wheatstone bridge. At final balance, achieved by the adjustment of impedances $\dot{Z}_1$, $\dot{Z}_2$, $\dot{Z}_3$ with the detector being in most sensitive state, the PD across $\dot{Z}_1$ would equal that across $\dot{Z}_x$ and the PD across $\dot{Z}_2$ being equal to that across $\dot{Z}_3$; the points A and C being at the same potential [see Fig.11.1]. An important point to note is that the various potential drops MUST not only be equal, or balance, in magnitude, *but also in terms of phasor relationship* as well. To check this requirement, all AC bridges are accompanied by an appropriate phasor diagram, *drawn at the condition of balance*, showing all the potential differences in proper phasor relationship with one another. The 'reference' phasor in almost all cases is taken to be the current in one of the arms.

Thus, referring to Fig.11.1, the equations at balance are

$$\dot{I}_1 \, \dot{Z}_1 = \dot{I}_2 \, \dot{Z}_x \quad \text{and} \quad \dot{I}_1 \, \dot{Z}_2 = \dot{I}_2 \, \dot{Z}_3$$

Dividing to eliminate $\dot{I}_1$, $\dot{I}_2$,

$$\frac{\dot{Z}_1}{\dot{Z}_2} = \frac{\dot{Z}_x}{\dot{Z}_3}$$

from which the un-known impedance is obtained as

$$\dot{Z}_x = \frac{\dot{Z}_1}{\dot{Z}_2} \times \dot{Z}_3$$

To put the above expression in perspective and to understand the special features associated with an AC bridge, assume that

(a) $\dot{Z}_x$ comprises a coil of resistance R_x and self inductance L_x, connected in series the values of which are to be determined;

(b) $\dot{Z}_3$ is a *known* impedance, consisting of a resistance part R_3, in series with a self-inductance (part), L_3;

(c) $\dot{Z}_1$ and $\dot{Z}_2$ are just two *non-inductive* resistances of values R_1 and R_2.

(d) Then, at balance,

$$R_X + j \omega L_X = \frac{R_1}{R_2}(R_3 + j \omega L_3)$$

where $\omega = 2\pi f$ is the (angular) supply frequency which must usually be known.

Or $R_2(R_X + j \omega L_X) = R_1(R_3 + j \omega L_3)$

or $R_2 R_X + j \omega R_2 L_X = R_1 R_3 + j \omega R_1 L_3$

Equating real and imaginary parts

$$R_2 R_X = R_1 R_3$$

and $j \omega R_2 L_X = j \omega R_1 L_3$

giving

$$R_X = \frac{R_1}{R_2} \times R_3$$

$$L_X = \frac{R_1}{R_2} \times L_3$$

The following important inferences can be drawn from the above sample example:

(i) Making two of the four arms, $\dot{Z}_1$ and $\dot{Z}_2$ here, as known resistances which may be adjustable, reduces the bridge partly to a Wheatstone bridge in which $\dot{Z}_1$ and $\dot{Z}_2$, or rather R_1 and R_2, constitute the ratio arms (P and Q) accompanied by their obvious advantages of affording good accuracy of the results.

(ii) The un-known inductance is obtained simply in terms of ratio of the resistances R_1 and R_2 and inductance (part) of the known resistive-inductive impedance, *with the supply frequency no where appearing in the results and hence need not be known.*

(iii) This forms an important facet of the bridge, and of many more as discussed later, in that if the bridge is balanced for the fundamental frequency, it should also be balanced for any harmonic so that the waveform of the source supply need not be exactly sinusoidal or known[1].

---

[1]However, it must be realised that the effective resistance and reactance of a coil do vary with frequency, so that a bridge 'balanced' at the fundamental frequency may, in practice, not exactly be balanced for harmonics. Hence, it is better that the source supply be of sinusoidal waveform as far as possible. This would also help in achieving a "quick" balance when a detector, tuned to the single frequency, is employed.

(iv) *Two* balance equations are almost always obtained for an AC bridge at the balance condition(s), so that by equating real and imaginary parts the two un-known quantities, often a resistance and an inductance or capacitance, can be determined.

(v) It is usual, for the convenience of adjustment and to get the balance with minimum time incurred, to arrange the bridge to have only two variable components. This should reflect in the final balance equations, each one of which should contain one variable; the equations are then said to be independent and not coupled. In the above example, R_3 and L_3 can be identified as the variable components, the technique of balancing being to adjust L_3 until a minimum indication is obtained on the detector and then to adjust R_3 until a new, smaller (minimum) indication is obtained. Then revert to L_3 and so on until the detector indicates zero PD across AC with best sensitivity.

Most AC bridges are configured to follow the above technique, leading to ease of measurement.

The simple arrangement in the above example does not constitute a unique one and there can be a large number of possible combinations, dictated by the requirements and with the ultimate objective of achieving the final balance as quickly and accurately as possible. A particularly useful group of bridges is obtained if two of the four arms are pure, non-inductive resistances. Some of the possibilities of this type are listed in Table 11.1.

Table 11.1: AC bridge combinations with two resistance arms

Combination no.	$\dot{Z}_1$	$\dot{Z}_2$	$\dot{Z}_3$	$\dot{Z}_4$	Bridge to be identified as
1	R and L in series	R and L in series	R	R	Maxwell's (inductance)
2	R and L in parallel	R and L in parallel	R	R	
3	R and C in series	R and C in series	R	R	De Sauty's
4	R and C in series	R and C in parallel	R	R	Wien's
5	R and C in parallel	R and C in parallel	R	R	
6	R and L in series	R	R	R and C in parallel	Maxwell's (inductance-capacitance)
7	R and L in series	R	R	R and C in series	Hay's
8	L	L	R	R	
9	C	C	R	R	
10	L	R	R	C	

The table shows that almost all the bridges have two arms comprising 'pure' resistances, the first combination representing the case already dealt with. A few combinations or arrangements represent specific bridges, indentified by the names of scientists who developed them and which are in wide use, to be studied in detail later in the chapter. The last three listed in the table are of academic interest only as they represent *pure* inductance or capacitance in two of the four arms and are not obtainable in practice. The list is not exhaustive, but only to show that a large number of bridge circuits are possible.

There are also cases of more elaborate bridge circuits used for measurement of "lossy" capacitors or impure inductances, comprising more than four arms or branches. Further, bridges with two resistance arms may fall in the category of being with *ratio* arms or those with *product* arms, meaning whether the un-known quantities – resistance and inductance or capacitance – are obtained in terms of ratio of the two resistances or product of the two. Clearly, this would depend on the relative location of the resistances in the bridge with respect to other impedances: an example of the latter type is one of the combinations listed at 6, or 7.

It is also important to note that AC bridge methods, designed for the determination of inductance or capacitance, are essentially *comparison* methods. Thus, in the bridge considered above, the value of inductance L_x is obtained by comparison with the standard inductance L_3. This is also an example when the value of the un-known is obtained in terms of ratio of the known resistances. The impedances would in such cases be mostly similar, that is, both resistive-inductive, connected in adjacent branches or arms; if in *opposite* arms, these would be resistive-inductive v/s resistive-capacitive, and the un-known quantity would be obtained in terms of product of the two resistances in the other two arms. The other variation(s) may be the use of *mutual* inductance in one of the arms instead of a simple (pure) inductance.

Sources of Supply and Detectors in AC Bridges

Sources

The supply for AC bridges may normally be obtained from

- supply mains, in most cases using a suitable step-down transformer at the power frequency (50 or 60 Hz);
- a motor driven alternator, with excitation control, to provide an appropriate supply at varying (low) frequencies[1];
- an audio-frequency or radio-frequency oscillator, or "signal-generator"; this being capable of providing a supply of sinusoidal waveform and of desired magnitude at the chosen frequency.

---

[1] In this case, it is imperative that the waveform of the induced EMF be known, to be essentially sinusoidal.

In every case, apart from being able to adjust the supply voltage, it is necessary to ensure the frequency stability and freedom from harmonics, in general. The source may normally be coupled to the bridge by a screened transformer which may also act as an isolating transformer.

Detectors

For work at a single frequency, for example the power frequency, moving-coil vibration galvanometer, *tuned* to that frequency, may act as an ideal detector, and is extensively used, usually incorporating a lamp and scale arrangement to yield enhanced sensitivity. At higher frequencies, typically in audio-range; for example, commonly used 1 kHz, a telephone-type detector, or a head-phone, specially designed to be free from lower or higher frequency noise is frequently used. The null condition is indicated when the 1 kHz note disappears and the detector is absolutely silent[1].

For still higher frequencies, in the range of radio frequencies (several hundred kHz to MHz), a cathode-ray oscilloscope with adequate band-width and good sensitivity (in terms of mV or μV/cm) is best suited for the purpose. A distinct advantage with the CRO as detector is that the sensitivity can be enhanced by adjusting the gain of the vertical amplifier. The other very helpful aspect is that the indication is visual.

Vibration galvanometer

This forms an important requirement to detect the null condition in all AC bridge methods when fed from AC supply of power frequency.

Vibration Galvanometer of Moving Coil Type

A moving-coil vibration galvanometer is similar to the DC or D'Arsonval type in construction, primarily comprising a small coil, usually of single-turn – or just a loop – suspended between the poles of a permanent magnet. The main difference is that instead of an angular rotation, the coil is made to oscillate – or 'vibrate' – about its mean vertical position when carrying an *alternating* current. Hence the name!

The amplitude of vibrations/oscillations of the coil for a given current is maximum – also a reflection of the sensitivity of this type of galvanometer – when the "natural" frequency of vibration of the moving system is in resonance with the frequency of the AC supply. This is achieved by 'tuning' the system by appropriate means.

---

[1]This may, in many cases, lead to "subjective" error owing to sensitivity or hearing capability of the observer and may be mitigated to some extent by interposing a suitably tuned (audio) amplifier.

Duddell vibration galvanometer[1]

This is the frequently used vibration galvanometer, due to Duddell, which comprises a moving coil of a single loop, resulting in minimum mass and hence very small moment of inertia, of fine bronze (or platinum-silver) wire. The upper end of the loop passes over a tiny pulleys, being pulled tight by a spring attached to this pulley. The schematic arrangement is as illustrated in Fig.11.2. The loop of the wire is stretched over two small, prism-shaped ivory bridge pieces, one at the top and the other at the bottom as shown[2].

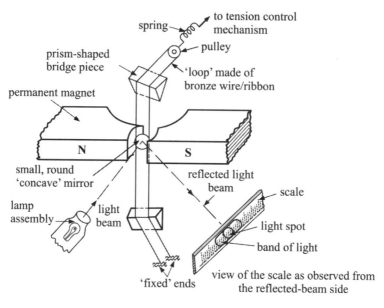

Fig.11.2 : Constructional details of Duddell moving-coil vibration galvanometer and lamp-and-scale arrangement

Operation

When a small alternating current passes through the loop, a couple is produced similar to that in a simple DC galvanometer, tending to deflect the loop clockwise (or anti-clock wise) about its vertical axis. With the direction of current reversed in the next half cycle, this couple also reverses, making the loop to deflect in the opposite direction, thus causing oscillation (or vibration) of the loop, as many times in a second as the frequency of the current.

---

[1]This is already described in Chapter IV: Visual Display and Analyses, but again discussed here in the present context.

[2]There are numerous variations of vibration galvanometers, depending on the way the coil is constructed and the mechanism of tuning used in the galvanometer. Duddell galvanometer is a typical case of these features.

Tuning

The strength of mechanical oscillations of the loop, or the amplitude, would naturally depend on the distance between the bridge pieces which control the "length" of the loop, as well as the tension provided by the spring. Thus, the "tuning" of the instrument, meaning to achieve maximum amplitude of oscillations corresponding to the AC supply frequency, can be achieved by

- adjusting the distance between the bridge pieces; this provides coarse tuning, and
- adjusting the tension of the spring beyond the pulley, for fine tuning.

The response of a typical tuned vibration galvanometer is shown in Fig.11.3 where it is seen that the deflection of the galvanometer reaches a sharp peak at the optimum frequency of the supply for a given current and galvanometer constant. Thus, the tuning process is strictly applicable, in theory, for a *given* (or chosen) frequency, usually the fundamental supply frequency. It is thus essential that the supply is sinusoidal in waveform, with no harmonics present[1].

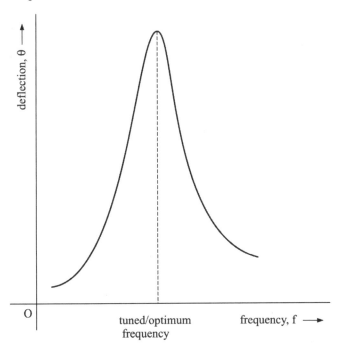

Fig.11.3 : Response curve of a tuned VG

---

[1]If this were not so, the working and sensitivity of the galvanometer will be seriously hampered; when used in an AC bridge this may lead to difficulty in achieving 'true' balance or null.

Lamp and scale arrangement

The galvanometer is invariably used along with a lamp-and-scale arrangement as indicated in the figure. A focused beam of light from a lamp assembly, placed about 1 metre away, is made to fall on a small (circular, concave) mirror fitted on the loop in the middle as shown. The focused reflected beam falls on a graduated scale and, since the loop carries an alternating current, would appear in the form of a band of light on the scale, being maximum in width for a *given* loop current if the galvanometer is properly tuned; this being an indication of the bridge to be out-of-balance. When there is no current flowing through the loop, the mirror will be stationary, resulting in the reflected beam to appear as circular, stationary spot on the scale – a condition that is strived for in an AC bridge at the instant of null.

In theory, this type of galvanometer can be designed for use in the frequency range of 100 Hz to over 1 kHz. The commonest use is, however, for power frequency of 50 Hz. A sensitivity of about 50 mm of light-beam width, on a scale about one metre distant, for a loop current of about one microampere can be achieved in practice.

Theory of operation of a vibration galvanometer

The equation of oscillatory motion of the moving system of the galvanometer can be expressed by the well-known second-order differential equation

$$a\frac{d^2\theta}{dt^2} + b\frac{d\theta}{dt} + c\,\theta = G\,i$$

where the various terms correspond to the galvanometer action are as follows:

(i)   The first term represents the *retarding* couple, dependent on the moment of inertia of the moving system and its angular acceleration. This couple is expressed as $a\dfrac{d^2\theta}{dt^2}$ in which "a" is the inertia constant or moment of inertia of the moving system (consisting of the loop, the mirror etc.) and $\dfrac{d^2\theta}{dt^2}$ the angular acceleration corresponding to the deflection θ from a reference datum.

(ii) The second term corresponds to another retarding couple arising from the *damping* effect of air friction and elastic hysteresis of the wire used for making the loop.

This couple is normally proportional to angular velocity of the moving system and can be expressed as $b\dfrac{d\theta}{dt}$ in which "b" is the (overall) "damping constant" and $\dfrac{d\theta}{dt}$ the angular velocity.

(iii) The third term relates to a couple in the form of a *control* torque, owing to the twist, torsion or elasticity of the suspension and is proportional to the deflection. This can be written as $c\,\theta$ where "c" can be indentified as "control" or "restoring" constant[1].

(iv) The term on the right-hand side of the expression represents the *deflection* torque, produced in the loop as a result of flow of current. Clearly, this must be equal to the sum of all "retarding" torques for equilibrium.

As in a DC galvanometer, G denotes the "galvanometer constant", given by

$$G = NBA$$

where N is the number of turns of the coil (simply 1 in the above case), B the airgap magnetic flux density due to the magnet and A the 'effective' area of the loop forming the "coil". Since the current through the loop is alternating current, i, in the equation of motion can be expressed as

$$i = I_m \cos \omega t$$

with I_m as the maximum value of the current and ω the angular frequency, equal to $2\pi f$.

The differential equation of the motion of the moving system can then be written:

$$a\dfrac{d^2\theta}{dt^2} + b\dfrac{d\theta}{dt} + c\,\theta = G\,I_m \cos \omega t$$

The solution of the above equation will yield

(a) a transitory part that may die rapidly with time and may be disregarded;

(b) an expression for θ, representing *sustained* oscillations or vibrations of angular frequency, ω.

Following the usual procedure (of solving a second-order differential equation), the angular deflection θ is given by

$$\theta = \dfrac{G\,I_m}{\sqrt{\left(c - a\,\omega^2\right)^2 + b^2\omega^2}}\left[\cos\left(\omega t - \beta\right)\right]$$

---

[1]It may be noted that both damping and control 'torques' in a vibration galvanometer may be very small, but present in practice to "balance" the operation.

where
$$\beta = \tan^{-1}\left[\frac{b\,\omega}{\left(c - a\,\omega^2\right)}\right]$$

The expression for θ shows that the moving system of the galvanometer will oscillate with an amplitude of

$$\hat{A} = \frac{G\,I_m}{\sqrt{\left(c - a\,\omega^2\right)^2 + b^2\omega^2}}$$

corresponding to a current $I_m\left(\text{or } I = I_m/\sqrt{2}, \text{ rms}\right)$,

and a frequency
$$f = \frac{\omega}{2\pi}.$$

Clearly, for 'good' performance, the value of $\hat{A}$ can be 'optimised' by controlling the values of the various constants a, b and c, for a given G and I_m.

RESISTORS, INDUCTORS AND CAPACITORS USED IN AC BRIDGES[1]

Resistors/Resistances

Fixed Resistances

The most important requirement of a resistance is extremely low temperature coefficient to maintain stability during the test and repeatability later.

In addition

(a) Low and very low resistances are mostly 4-terminals type, having two "current" and two "potential" terminals as discussed previously. The resistance usually consists of parallel strips of low temperature-coefficient material, soldered or brazed to copper or brass blocks at the two ends which are fitted with terminals for making "series" and "shunt" connections. The value of such a resistance may vary from a few milliohms to an ohm.

(b) Resistances of higher values are made of wires of silver-platinum or manganin, wound *non-inductively* on a special bobbin, in a manner shown in Fig.11.4, known as bifilar method of winding. In this type of winding,

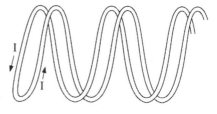

Fig.11.4 : A bifilar winding

---

[1]Also discussed in Chapters I and II.

also described in Chapter II, the wire is "doubled back" on itself before forming a coil. This gives the effect of two insulated wires, side by side in good contact, carrying currents in opposite directions nearly at the same instant. This results in the magnetic field due to the 'two' currents neutralising one another and hence resulting in negligible net inductive effect.

The wire is laid in one layer to ensure efficient cooling. The bobbin carrying the resistance coil is fitted inside a cylindrical case of ample volume, the coil ends being terminated on an ebonite top. To maintain uniform temperature, the inside of the case is filled with paraffin wax, or in some cases with insulating mineral oil. In practice, the resistance may be maintained at a constant temperature for some hours before being used for measurements.

Variable Resistances

These are usually in the form of (decade) resistance boxes containing a large number of resistance coils, each of a given resistance value, the 'total' resistance value being selected by rotary switches. The general, typical inside arrangement of such a box is illustrated in Fig.11.5 in which the value of resistance is varied by the rotation of a laminated copper brush or arm. Each of the coil itself is wound using bifilar technique and in single layer so as to eliminate any inductive effect. When used in high-voltage circuit/bridge, the box may be suitably screened which may be grounded to eliminate stray capacitance effect.

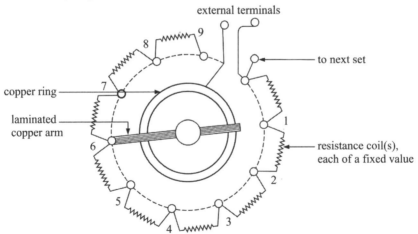

Fig.11.5 : Schematic of a dial-type decade resistance box

A decade box may usually comprise four or five dials, or sets of coils, in the steps of $0.001\,\Omega - 0.01\,\Omega$, $0.01\,\Omega - 0.1\,\Omega$, $0.1\,\Omega - 1.0\,\Omega$ etc., the upper limit of a box may be up to 10 kΩ to 100 kΩ, or sometimes even a few meg-ohm.

Inductors

Fixed Inductor or Inductance

These are usually constructed in the form of toroidal coils, designed with appropriate number of turns wound on formers of given size. The inductance of the coil can then be derived from its design constants. The wire size is chosen with care so as to minimise the resistance of the coil and consequent heating and temperature rise; and to make the self-resistance negligible compared to the inductance. For low values (for inductance of a few millihenry), the coils are essentially air-cored. For large values, the coils may be iron-cored, designed for low value of flux density in the core to avoid saturation effect. When operating in AC circuit, the design should also consider skin effect in the wire used for the winding, and its elimination.

Variable Inductors

Variable inductors or inductances, self as well as mutual, are widely required in AC bridge networks. These devices, called inductometers, are designed and constructed to satisfy the following requirements:

- they must have a very high inductance and negligible resistance (and hence large time constant), to qualify as "pure" inductance in bridge circuits;
- the variation of inductance should obey a linear or straight-line law for ease of setting or control, the dial being uniformly graduated;
- the inductance value from a given inductometer should cover as large a range as possible between maximum and minimum settings for the versatility of balancing a bridge where the latter largely depends on the adjustment of inductance in a particular branch;
- the inductance for a given position or setting must not vary with time or slight variation of frequency.

Mutual Inductors

Most variable inductors are designed and constructed for use as self or mutual inductor as they usually comprise two coils: one fixed and the other movable with respect to the former. When used as a self inductance, the fixed and moving coils are connected in *series* in an appropriate manner such that the (self) inductance will be given by

$$L = L_1 + L_2 + 2M$$

where L_1 and L_2 are the self inductances of fixed and moving coil, respectively, and M is the mutual inductance between the two, its value being dependent on the relative position of the moving coil with respect to the fixed coil. In order to minimise frequency errors and that due to skin effect, the coils are usually wound using stranded wires.

A Typical Inductometer

The construction of a typical inductometer, due to Ayrton and Perry, is shown schematically in Fig.11.6. The device consists of a moving coil which is mounted inside the fixed coil, fitted to a spindle to which is attached a pointer moving on a graduated scale. The bottom end of the spindle rests on a support bearing whilst the top end carries a knob or handle which can be rotated by hand. Rotation of the moving coil in clockwise or anticlockwise direction changes the inductance value, indicated on the scale which is calibrated to read the inductance directly. When used as self inductance, the ends of the two coils are connected internally in series.

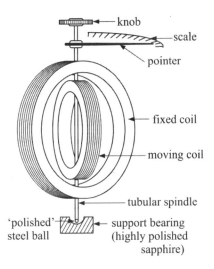

Fig.11.6 : Ayrton and Perry inductometer

Typically, the coils are wound as "short" air solenoids, with no magnetic material present in the winding and therefore resulting in near-linear variation of inductance even when carrying "appreciable" currents. When constructed for use in 'accurate' measurements, the coils may be wound on mahogany formers whose surfaces are spherical (or curved) to provide still linear variation.

Capacitors[1]

Fixed Capacitors

Parallel-plate capacitor(s)

Capacitors with fixed, known capacitance are essentially constructed in the 'standard' form using two or more parallel plates with air or any other suitable 'insulating' material as dielectric; air being preferable when better accuracy and freedom from "dielectric loss" is desired, owing to the fact that air is the only dielectric of which permittivity is definitely known. Two practical forms of "parallel-plate" capacitors are schematically shown in Fig.11.7. The second form, commonly employed in electronic circuitry, consists of long strips of tin foils of suitable width sandwiched with a paper strip of same width and rolled tightly to result in a small cylinder. This is

---

[1]A capacitor as a device for storing electric charge was almost accidentally invented by Van Musschenbrook of Leydan around 1746 and popularly became known as Leydan jar.

encapsulated suitably to impart strength and protection, with two leads brought out at either end for making connections.

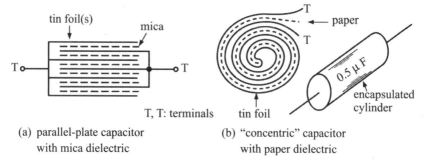

(a) parallel-plate capacitor
with mica dielectric

(b) "concentric" capacitor
with paper dielectric

Fig.11.7 : Typical forms of "parallel plate" capacitor

"Standard" (air) capacitor

For best results – accuracy and stability – fixed capacitors, also known as "standard" capacitors are of *concentric cylinder* type with "guard rings" and parallel-plate construction with guard plates and, of course, with air as dielectric as depicted in Fig.11.8. The guard ring/plate or screen, usually made of aluminium or copper acts as a 'protection' to the inner, 'live' plate, A, invariably circular in shape, the other

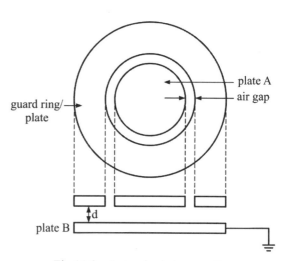

Fig.11.8 : A standard air capacitor

purpose of the ring being in affording exact calculation of the capacitance. The ring is of the same outside diameter as the opposite plate B and, when in use, is at the same potential as the inner plate A. This results in the (electrostatic) field being almost completely perpendicular to the plate even up to the edge of plate A. It is thus possible to calculate almost exactly the value of the capacitance from the dimensions of the capacitor.

Whilst affording best accuracy and stability with time, the "air" capacitors suffer from the following drawbacks:

- The relative permittivity being just one, the capacitance value could be quite small (usually in picofarad); to obtain higher capacitance, the size of plates has to be considerable, or the number of "units" have to be many more making the device rather cumbersome.

- The dielectric strength of the capacitor being low, it necessitates a larger gap if the PD across the plates is appreciable (for example, a few kilovolts) to avoid breakdown of air; in turn, a longer airgap would result in lower capacitance.

- There being no solid dielectric between the plates, elaborate support structure of insulating material will have to be provided to hold the plates in position to maintain a uniform airgap.

Capacitors with non-air Dielectric

For general laboratory measurements, capacitors with solid dielectric such as mica or paraffined paper, with the former being preferable owing to lower dielectric loss, are fabricated which may provide capacitance up to 1 μF for single-plate type and up to 5 μF or more for multi-plate construction or using decade capacitance boxes (with capacitors being connected in parallel). The capacitance value of such capacitors is ascertained by calibration rather than derived by calculations. It is also necessary to store (and use) these in controlled ambient environment to maintain accuracy and stability, esp. with regard to temperature.

Variable capacitors

These capacitors are usually of parallel-plate type with air as dielectric to result in negligible dielectric loss or power factor. A typical, *continuously* variable capacitor "unit" consists of two or more sets of plates of aluminium or brass (thin) plates or "wafers", approximately semi-circular in shape. One of the sets comprises moving plates, arranged to move inside the gap of other, fixed plates as shown in Fig.11.9(a). An actual assembly is shown in Fig.11.9(b) whereas a picture is shown in Fig.11.9(c).

The capacitance is varied by varying the areas of the moving plates interleaving or "common" with the fixed plates with the maximum value occurring when the moving plates are completely 'inside' the fixed plates. The plates are proportioned and shaped in a manner, not being exactly semicircular, such that the capacitance varies in almost exact proportion to the angle through which moving plates are turned, thus resulting in a near 'linear' variation. The plates should be sufficiently thick so as to avoid bending any time which may alter the capacitance and calibration, and all the corners should be rounded as shown.

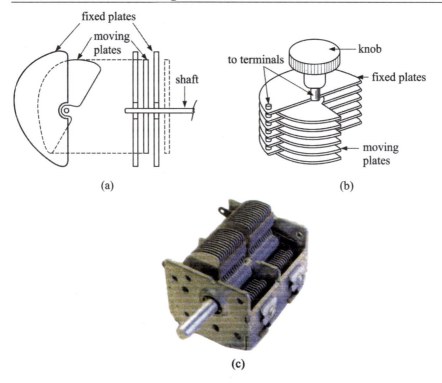

Fig.11.9 : Moving-plates, variable air capacitor

To increase the value of the capacitance, a number of such "basic" sets are mounted on a single shaft so that the capacitance of single sets add up to give the required higher value[1]. Any angular movement of the shaft rotates all the moving plates simultaneously inside the respective fixed plates.

When designed and constructed with care, such capacitors may be obtained with

- a power factor being less than 0.0001 at 1 kHz; ;
- a "temperature coefficient" of about 30-40 parts per million; and
- a residual "self inductance" between 0.04 and 0.07 μH.

for all settings. The values at other, lower or higher, frequencies vary only slightly.

Varying capacitance values *in steps* can be obtained in a circuit by means of fixed capacitors arranged in a decade box, similar to that used for decade resistance box(es).

---

[1]When used in this form, the capacitor is known as a "gang" capacitor [see Fig.11.9(b) and (c)], invariably used in a radio receiver in the RF stage to "tune in" different stations, their carrier frequencies being essentially spaced apart by nearly 10 kHz or more.

MEASUREMENT OF INDUCTANCE

The inductance, usually the self inductance (L) of a coil, can be measured employing one of the following methods:

  (i)   Ammeter and voltmeter method
  (ii) Three-ammeter or three-voltmeter method
  (iii) AC potentiometer method
  (iv) AC bridge methods

Ammeter and Voltmeter Method

This is the simplest of all methods, easily applicable for the measurement of inductance in the range of about 50 to 500 mH.

The method consists of passing a suitable current of normal, *known* frequency through the coil. The current and the potential drop across the coil is measured using an ammeter and voltmeter, respectively, of good accuracy, to yield impedance, $\dot{Z}$, of the coil. The DC resistance, R, of the coil – which may be the same as AC resistance if the frequency is not high and the coil is of air-core type – can be measured separately by suitable means. Then the (inductive) reactance of the coil will be given by

$$X_L = \sqrt{Z^2 - R^2}$$

and its inductance by

$$L = \frac{X_L}{2\,\pi\,f}\ H$$

where f is the supply frequency and must be accurately known.

Three-ammeter/Three-voltmeter Method

These methods have already been applied and discussed for the measurement of power in a single-phase circuit[1]. They are considered in brief here to show how the inductance of a coil can be 'measured' using either of the methods.

Three-ammeter Method

This method consists of connecting the coil under test in *parallel* with a standard known resistance, R, across a suitable supply of *known* frequency, f. Using three ammeters, the current through the coil, I_1, that through the standard resistance, I_2, and the total current, I, are measured. Then, from the phasor diagram of various currents and the supply voltage [see Fig.7.14, Chapter VII],

$$I^2 = I_1^2 + I_2^2 + 2\,I_1 I_2\,\cos\theta$$

where θ denotes the 'impedance' angle of the coil.

---

[1]See Chapter VII: "Measurement of Power."

From the above expression,

$$\cos\theta = \frac{I^2 - I_1^2 - I_2^2}{2\,I_1 I_2}$$

Also,

$$\cos\theta = \frac{r}{\sqrt{r^2 + (2\,\pi\,f\,L)^2}}$$

where r and L denote the resistance and (self) inductance of the coil.

Equating the two expressions for cos θ and simplifying,

$$L = \frac{1}{2\,\pi\,f}\sqrt{\frac{4\,r^2\,I_1^2\,I_2^2}{\left(I^2 - I_1^2 - I_2^2\right)^2} - r^2}\ \ H$$

Three-voltmeter Method

In this case the test circuit is made up of the coil connected in *series* with a known, standard resistance, across a supply of known frequency, f. The potential drops across the coil, V_1, that across the resistance, V_2, and supply voltage V, are measured using three voltmeters.

Then

$$V^2 = V_1^2 + V_2^2 + 2V_1 V_2 \cos\theta$$

[See phasor diagram of Fig. 7.13, Chapter VII]

giving

$$\cos\theta = \frac{V^2 - V_1^2 - V_2^2}{2V_1 V_2}$$

But

$$\cos\theta = \frac{r}{\sqrt{r^2 + (2\,\pi\,f\,L)^2}}$$

whence, equating and simplifying,

$$L = \frac{1}{2\,\pi\,f}\sqrt{\frac{4\,r^2\,V_1^2\,V_2^2}{\left(V^2 - V_1^2 - V_2^2\right)^2} - r^2}\ \ H$$

AC Potentiometer Method

This has already been dealt with under "application of AC potentiometers"[1].

---

[1] See Chapter IX: "Potentiometers."

AC Bridge Methods

These are the most extensively used methods for measuring self or mutual inductance of a coil, covering almost the entire range from micro-henry to henry.

Quality factor or "Q" of a coil

A term commonly associated with different types of coils (e.g. air- or iron-core), and which may sometimes influence the choice of a particular bridge for measurement purpose, is the "quality factor" or "Q" of the coil. This is defined as the ratio

$$Q = \frac{\omega L}{R}$$

where R and L represent the resistance and inductance of the coil, respectively, and ω the angular frequency of the AC supply.

It can be seen that Q of a coil is directly proportional to ω (or f) and L and indirectly to R. Thus, a coil of "high quality", having a high value of Q, is the one which possesses high inductance, operating at high frequency and has as low a resistance as possible. In theory, Q will tend to "infinity" if R tends to zero. These properties are of practical importance when the coils are to be "tuned" in a circuit[1]. In practice, R can never be zero and, apart from its "DC" value may depend on the skin effect, esp. at very high frequencies. A common means of designing a coil with high value of Q is to provide an iron core, the lateral position of which being adjustable, from one end relative to the other, by a screw action, thus helping in fine tuning of the coil. Clearly, a high Q value is a measure of the quality or efficacy of a coil, esp. in high-frequency applications.

Maxwell's bridge

This is the simplest bridge in which the R and L parameters of the coil under test are compared with a standard coil of known, but variable, resistance and inductance. The connections for the bridge are illustrated in Fig.11.10 which also shows the phasor diagram of the various quantities at balance[2].

---

[1]For example, in the RF stage of a radio receiver.

[2]Observe that the currents I_1, I_2 and voltages V_1, V_2, V etc. are *phasors*, to be denoted by a ($\cdot$) above the letters. However, this is omitted throughout the chapter for simplicity.

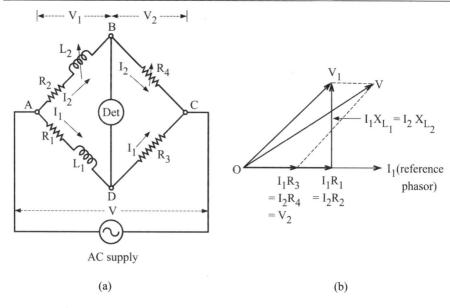

Fig.11.10 : Maxwell's bridge and phasor diagram

In the figure, R_1 and L_1 are the resistance and self inductance of the coil; L_2 is the variable, known inductance of negligible resistance in series with a resistance, R_2, which may be varied in steps of fixed values such as 10, 100, 1000 Ω etc., R_3 and R_4 are non-inductive resistances, the latter being variable. Det denotes the detector, as appropriate – usually a tuned vibration galvanometer.

At balance, the potential drops across AB and AD and those across BC and CD are equal.

Thus,

$$I_1 \times (R_1 + j\omega L_1) = I_2 \times (R_2 + j\omega L_2)$$

and

$$I_1 \times R_3 = I_2 \times R_4$$

or, by dividing,

$$\frac{R_1 + j\omega L_1}{R_3} = \frac{R_2 + j\omega L_2}{R_4}$$

Cross multiplying and equating real and imaginary parts,

$$R_1 R_4 = R_2 R_3, \qquad \text{giving } R_1 = \frac{R_2 R_3}{R_4}$$

and

$$\omega L_1 R_4 = \omega L_2 R_3, \qquad \text{giving } L_1 = \frac{L_2 R_3}{R_4}$$

It is seen that

- the expressions for R_1 and L_1 are dimensionally balanced (see the phasor diagram);

- the value of L_1 is obtained independent of the supply frequency which is an important feature of the bridge.

Maxwell's bridge with a capacitance

In this bridge which is a modification of the 'original' bridge, one of the arms consists of a standard, known capacitor (sometimes variable), connected in parallel with a resistance as shown in Fig.11.11.

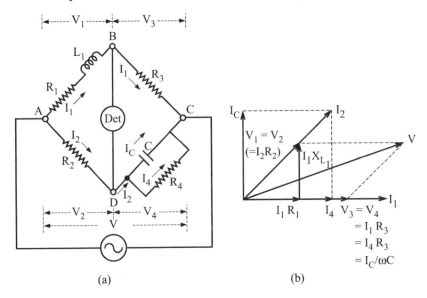

(a)   (b)

Fig.11.11 : Maxwell's bridge with a capacitor

As discussed earlier, the branch containing the capacitor occupies a position *opposite* to the coil under test. [See Table 11.1].

In the bridge, R_1 and L_1 represent the resistance and inductance, respectively, of the coil as before; R_2, R_3 and R_4 are known, non-inductive resistances and C a (loss-free) standard capacitor.

At balance, equating the relevant PDs

$$I_1(R_1 + j\omega L_1) = I_2 R_2$$

and
$$I_1 R_3 = I_2 \cfrac{1}{\left(\cfrac{1}{R_4} + j\omega C\right)} = \cfrac{I_2 R_4}{1 + j\omega R_4 C}$$

or, dividing,

$$\frac{R_1 + j\omega L_1}{R_3} = \frac{R_2(1 + j\omega R_4 C)}{R_4}$$

and, by cross multiplication,

$$R_1 R_4 + j\omega R_4 L_1 = R_2 R_3 + j\omega R_2 R_3 R_4 C$$

Equating real and imaginary parts

$$R_1 R_4 = R_2 R_3, \qquad\qquad \text{giving } R_1 = \frac{R_2 R_3}{R_4}$$

and $\qquad\qquad j\omega R_4 L_1 = j\omega R_2 R_3 R_4 C \qquad\qquad \text{giving } L_1 = R_2 R_3 C$

Here, too, the expression for L_1 is dimensionally correct and the value is, once again, obtained to be independent of frequency.

Operation

The bridge is balanced by varying C and R_4, giving them independent settings in turn. Physically, R_2 and R_3 may each be, say, 10, 100, 1000 or 10,000 ohm, to give a suitable value to the product $R_2 R_3$ which appears in both the expressions (for R_1 and L_1). C and R_4 may be decade capacitance and resistance boxes, respectively. For example, let R_2 = 10,000 ohm and R_3 = 100 ohm, then $R_2 R_3$ will be 10^6. If now the balance is obtained, the reading of C in *microfarad* will directly give the value of L_1 in henry.

The phasor diagram at balance, pertaining to the bridge, is shown in Fig.11.11(b). The bridge is one of the very general usefulness for the measurement of a wide range of inductance, at power and audio frequencies. However, a disadvantage is the requirement of a standard, variable capacitor of matching range and values which may be difficult and expensive to obtain.

The "Q" of the inductor at balance is given by

$$Q = \frac{\omega L_1}{R_1} = \omega C R_4$$

This is made use of in some "commercial" bridges where C (and ω) are fixed; R_4, being variable, is then calibrated in Q values. The bridge is, however, not suitable for high Q values since the required value of R_4 (for a 'reasonable' value of C) for balance may be very high, and not practicable.

Hay's bridge

This is a modification of the Maxwell's inductance-capacitance bridge in that the branch meant for obtaining balance comprises a standard capacitance and resistance in *series* rather in parallel as before. The connections of the bridge and the phasor diagram at balance, are shown in Fig.11.12.

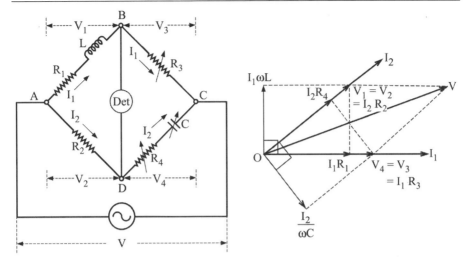

Fig.11.12 : Hay's bridge for measurement of inductance

The bridge is particularly suited for measuring relatively large inductances and hence to test coils with high "Q"; for example, the ones with iron cores.

Referring to the circuit, L is the inductance to be measured with R_1 its resistance, C is a standard, variable capacitor and R_2, R_3 and R_4 are non-inductive resistances as before. Balance is normally obtained by varying C, R_3 and R_4.

At balance,

$$V_1 = V_2 \quad \text{and} \quad V_3 = V_4$$

$\therefore$ $\qquad I_1 (R_1 + j \omega L) = I_2 R_2$

and $\qquad\qquad I_1 R_3 = I_2 \left(R_4 + \dfrac{1}{j\omega C} \right)$

Dividing to eliminate I_1 and I_2

$$\frac{R_1 + j\omega L}{R_3} = \frac{R_2}{R_4 - j/\omega C}$$

or $\qquad\qquad R_2 R_3 = R_1 R_4 - \dfrac{j R_1}{\omega C} + j\omega R_4 \, L + \dfrac{L}{C}$

Multiply all through by ωC. Then

$$\omega R_2 R_3 C = \omega R_1 R_4 C - j R_1 + j \omega^2 R_4 L C + \omega L$$

Equating real and imaginary parts

$$\omega\,R_2\,R_3\,C = \omega\,R_1\,R_4\,C + \omega\,L$$

and

$$R_1 = \omega^2\,R_4\,L\,C$$

On simplification

$$L = \frac{R_2 R_3 C}{1 + \omega^2 R_4^2 C^2}$$

and hence

$$R_1 = \frac{\omega^2 R_2 R_3 R_4 C^2}{1 + \omega^2 R_4^2 C^2}, \quad \text{substituting for L;}$$

both values being dependent on ω^2, making the accurate knowledge/measurement of supply frequency critical.

The Q (factor) of the coil is obtained from $\dfrac{\omega L}{R_1}$

or

$$Q = \frac{1}{\omega\,C\,R_4}$$

by substituting for R_1 and L.

Comment

It is interesting that if, as an alternative, the coil under test is assumed to comprise an inductance L in *parallel* with a resistance R_1, then writing the equations at balance as before

$$\frac{R_1 \times j\omega L}{R_1 + j\omega L}\left(R_4 - \frac{j}{\omega C}\right) = R_2 R_3$$

giving finally on simplification

$$R_1 = \frac{R_2 R_3}{R_4} \quad \text{and } L = R_2 R_3\, C$$

and the values are seen to be *independent of frequency.*

Conventionally, however, a coil may always be represented as a series combination of its resistance and self inductance.

Dimensional Balance

As a check for this bridge for its dimensional balance, consider the expression for the inductance,

$$L = \frac{R_2 R_3 C}{1 + \omega^2 R_4^2 C^2}$$

Dimensionally, this may reduce to

$$[L] = \left[\frac{1}{\omega^2 C} \right]$$

Now dimensions of L, C and ω, respectively, are

$$[L] = L\mu; \quad [C] = L^{-1} T^2 \mu^{-1} \quad \text{and} \quad [\omega] = T^{-1}$$

in EM system of units.

Substituting in the above expression for [L]

$$[L] = \frac{1}{T^{-2} L^{-1} T^2 \mu^{-1}}$$

$$= L\mu$$

$$= [L]$$

This shows that the expression for L derived from the bridge at balance is dimensionally correct. Similar results can be derived for the dimensions of R_1.

Anderson's bridge

This bridge is 'derived' from Maxwell's inductance-capacitance bridge in which an important aspect is the use of a fixed, standard capacitor instead of a variable one which may be difficult to obtain. This is also an example of deviation from the 'standard' four-arm bridge in that it comprises *five* branches, arranged to obtain a "quick" balance.

The construction of the bridge is shown in Fig.11.13. Once again, R_1, L represent the coil under test, C is the fixed, standard capacitor and R_2, R_3, R_4 and r are four non-inductive resistors. The bridge is first balanced using a DC supply and a DC galvanometer as detector. Resistances R_2, R_3 and R_4 are adjusted to get the balance. In this step, the inductance of the coil and capacitor, C, play no role and the bridge is balanced as a

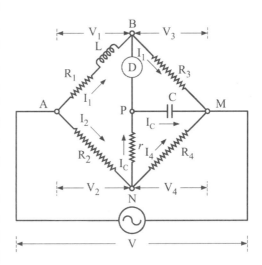

Fig.11.13 : Anderson's bridge

Wheatstone bridge comprising R_1, R_2, R_3 and R_4. The DC supply and galvanometer are then replaced by AC supply and appropriate detector; usually a headphone with the AC supply at 1 kHz. The bridge is now balanced by adjusting r, leaving R_2, R_3 and R_4 undisturbed.

Theory

To begin with, it may be helpful to convert the "star" comprising the resistances R_2, r and R_4 of the bridge circuit into an equivalent "delta" by using the appropriate transformation equations and rearranging the network.

The transformed bridge network is shown in Fig.11.14 where the resistances R_5, R_6 and R_7 are the "sides" of the equivalent delta, given by

$$R_5 = \frac{R_2 r + R_4 r + R_2 R_4}{R_4}$$

$$R_6 = \frac{R_2 r + R_4 r + R_2 R_4}{R_2}$$

$$R_7 = \frac{R_2 r + R_4 r + R_2 R_4}{r}$$

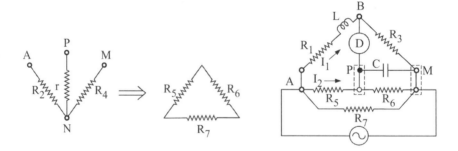

Fig.11.14 : The "star-delta" transformation and transformed Anderson's bridge

Corresponding to the "original" balanced condition of the bridge, the transformed bridge comprises A B M P since the resistance R_7 is 'ineffective', being directly connected across the supply and taking a continuous small current. Hence, the bridge is, in effect, the same as the Maxwell's inductance-capacitance bridge with C and R_6 in parallel, analysed earlier. Proceedings on the same lines, therefore, the inductance and resistance of the coil are given by

$$L = R_3 R_5 C \quad \text{and} \quad R_1 = \frac{R_3 R_5}{R_6}$$

Substituting for the values of R_5 and R_6 as derived in the transformation

$$R_1 = \frac{R_2 R_3}{R_4}$$

and

$$L = \frac{C\,R_3}{R_4}\left[r\left(R_2 + R_4\right) + R_2 R_4\right]$$

Here, too, the expression for inductance is independent of the supply frequency.

The phasor diagram applicable at balance is shown in Fig.11.15[1]. As seen from the expression for L, the method can also be used to measure the capacitance of the capacitor C if a calibrated self-inductance is available, the value of C being given by

$$C = \frac{L\,R_4}{R_3\left[r\left(R_2 + R_4\right) + R_2 R_4\right]}$$

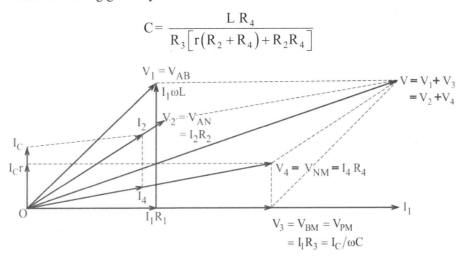

Fig.11.15 : Phasor diagram for Anderson's bridge at balance

Owen's bridge

This is yet another common, convenient method of measuring self-inductance and (AC) resistance of a coil in terms of a fixed, standard capacitance. The circuit diagram of the bridge is given in Fig.11.16. The coil under test, represented by R and L, is connected in series with a variable non-inductive resistor R_1 whilst the standard capacitor is C_S. C is another fixed capacitor, in series with a non-inductive resistor R_3. The balance of the bridge can be obtained by successive adjustment of the resistances R_1 and R_3.

---

[1]Being somewhat 'complicated' to "balance the various phasors", a step-by-step procedure to arrive at the phasor diagram is given in Appendix V.

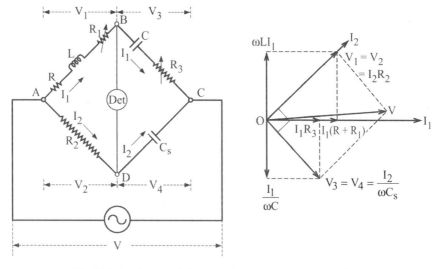

Fig.11.16 : Owen's bridge and phasor diagram at balance

At balance,

$$V_1 = V_2 \quad \text{and} \quad V_3 = V_4$$

By elimination of I_1 and I_2 and re-arranging

$$\frac{R + R_1 + j\omega L}{R_2} = \frac{R_3 - j/\omega C}{-j/\omega C_S}$$

or

$$\frac{-j(R + R_1)}{\omega C_S} + \frac{L}{C_S} = R_2 R_3 - \frac{jR_2}{\omega C}$$

Equating real and imaginary parts and simplifying,

$$L = R_2 R_3 C_S \quad \text{and} \quad R = R_2 \frac{C_S}{C} - R_1$$

It is seen that here also the expressions for R and L are independent of supply frequency and waveform[1]. Also, the bridge may be used for power loss measurement in the coil, being calculated from I^2R where I is the current through the coil and R the (AC) resistance as measured above[2].

Heaviside-Campbell bridge

This bridge is suited for the measurement of self inductance of a coil over a wide range, in terms of self- and mutual-inductance of another coil. No capacitance is used.

---

[1]If R_1 is "zero" (no additional resistance in series with the coil), the balance may be obtained by making C variable. Then, $R = (C_S/C) \times R_2$.

[2]Clearly, this is applicable in all the bridges where R and L of the coil are measured in terms of bridge components.

The connections of the bridge are shown in Fig.11.17. As usual, the coil under test is represented by a resistance R_1 and inductance L_1. The 'balancing' coil has a self inductance L_2 and variable, mutual inductance M, in series with its resistance R_2. The simplicity of the bridge is in the use of two non-inductive resistances R_3 and R_4 in the remaining two branches. The bridge is balanced by successive adjustment of R_3, R_4 and M. With the assumed currents flow at balance as shown

$$I_2(R_2 + j\omega L_2) + j\omega M I = I_1(R_1 + j\omega L_1)$$

and
$$I_2 R_4 = I_1 R_3$$

Also,
$$I = I_1 + I_2$$

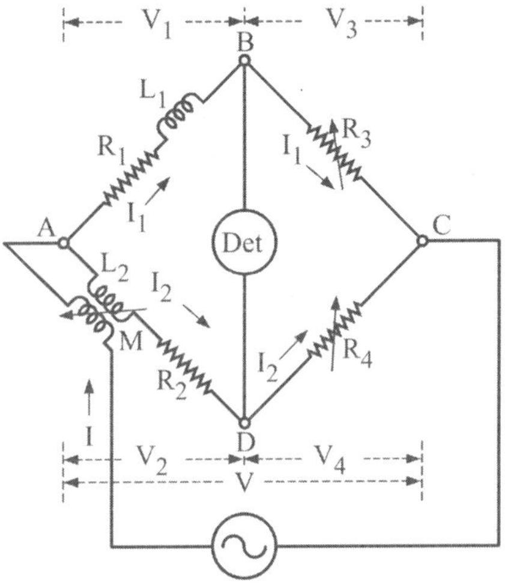

Fig.11.17 : Heaviside-Campbell bridge

Substituting for I in the first equation, rearranging it in terms of I_1 and I_2 and eliminating the two currents

$$\frac{R_2 + j\omega(L_2 + M)}{R_4} = \frac{R_1 + j\omega(L_1 - M)}{R_3}$$

or
$$R_3[R_2 + j\omega(L_2 + M)] = R_4[R_1 + j\omega(L_1 - M)]$$

Simplifying and equating real and imaginary terms

$$R_2 R_3 = R_1 R_4, \qquad \text{giving } R_1 = \frac{R_3}{R_4} R_2$$

and
$$R_3(L_2 + M) = R_4(L_1 - M)$$

from which
$$L_1 = \frac{R_3 L_2 + (R_3 + R_4) M^1}{R_4}$$

Once again, the values of R_1 and L_1 are obtained independent of supply frequency. Also, the resistances R_3 and R_4 constitute "ratio" arms of the bridge, such that if $R_3 = R_4$, then

$$R_1 = R_2 \quad \text{and} \quad L_1 = L_2 + 2 M$$

MEASUREMENT OF CAPACITANCE

Most capacitors in practice are imperfect to a degree, comprising a pure capacitance AND a resistance representing dielectric loss which may be appreciable at high frequencies. This is similar to an inductive coil where a resistance, usually assumed in series carrying the same current, represents the "I^2R" loss in the winding ; as well as skin effect if the current is of high frequency. It is, therefore, imperative to include the resistance part of the capacitor when making measurement using a given method. In general, the loss component can be accounted for in two ways.

Symbolic Representation of an Imperfect Capacitor

A. Loss Component as a Series Resistor

This constitutes a series R, C circuit as shown in Fig.11.18, together with the corresponding phasor diagram. When fed from an AC supply at a voltage V, a current I would flow, with PDs V_R and V_C across the resistance and capacitance (both assumed 'pure') parts, respectively, as shown.

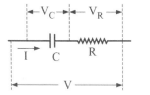 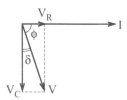

Fig.11.18 : An imperfect capacitor represented as C and R in series and phasor diagram

Then,
$$V^2 = V_R^2 + V_C^2$$

or
$$(Z I)^2 = (R I)^2 + \left(\frac{I}{j \omega C}\right)^2$$

where Z is the 'impedance' of the capacitor.

---

[1]Alternatively, if the components R_1 and L_1 of the coil are known (by another measurement), the *mutual* inductance M of the coil can be derived in terms of L_1, L_2, R_3 and R_4 as
$$M = \frac{R_4 L_1 - R_3 L_2}{R_3 + R_4},$$

From the phasor diagram

$$\tan\phi = \frac{V_C}{V_R} = \frac{\left(\dfrac{I}{\omega C}\right)}{I R} \qquad \text{or} \qquad \frac{1}{\omega C R}$$

Also,

$$\tan\delta = \frac{V_R}{V_C} = \frac{I R}{\dfrac{I}{\omega C}} = \omega C R$$

or

$$\delta = \tan^{-1}(\omega C R)$$

In this form, tan δ is known as the "loss factor" of the capacitor and δ the "loss angle". Clearly, this comes into play on account of R: lower the value of R, lower the loss or loss content in the capacitor. Ideally, if R is zero, there is no loss and the capacitor would be loss free or perfect. The loss factor, or the angle δ, is also representative of "loss-property" of the dielectric used in the capacitor, or its relative permittivity[1].

B. Loss Component as a Parallel Resistance

As an alternative, the loss component of a capacitor may be represented by a resistance in *parallel* with the pure capacitance part of it as shown in Fig.11.19, with the corresponding phasor diagram.

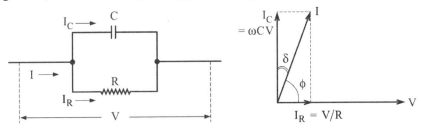

Fig.11.19 : An imperfect capacitor as a parallel combination of C and R and the phasor diagram

Here, the resultant *current* is given by

$$I^2 = I_R^{\,2} + I_C^{\,2}$$

when the circuit is fed from an AC supply of voltage V and frequency $f = \omega/2\pi$ Hz.

Or

$$\left(\frac{V}{Z}\right)^2 = \left(\frac{V}{R}\right)^2 + (\omega C V)^2$$

---

[1] With the relative permittivity of "air" being one, an "air-capacitor" will have the least (dielectric) loss and hence may be considered nearest to a prefect capacitor.

The loss factor is given by

$$\tan \delta = \frac{I_R}{I_C} = \frac{\dfrac{V}{R}}{\omega C V} \qquad \text{or} \qquad \frac{1}{\omega C R}$$

and the loss angle

$$\delta = \tan^{-1}\left(\frac{1}{\omega C R}\right)$$

In this case, *larger* the value of R, smaller will be the loss factor, or loss in the capacitor. In the limit, if R would tend to infinity, there would be no loss and the capacitor would be a perfect one[1].

Whilst making measurements, either of the two forms may be assumed.

Measurement of Capacitance Using Ammeter and Voltmeter Method

This is similar to the measurement of inductance of a coil. If an alternating voltage of pure sinusoidal waveform and known frequency is applied to a capacitor of capacitance C farad, a current of $I = \omega C V$ will flow where V is the applied AC voltage. The current, which may be rather small, may be measured by a sensitive ammeter whilst the voltage across the capacitor may be measured using a voltmeter having very high internal resistance, preferably an electrostatic voltmeter. The capacitance will then simply be given by

$$C = \frac{I}{\omega V} \; F$$

AC Potentiometer Method

In this method, the capacitor is connected in series with a known, standard resistance and fed from a supply as above. The PDs across the capacitor and the resistance are measured in turn using an AC potentiometer, incorporating a VR box across the capacitor, if necessary. Then the circuit current is

---

[1]From the phasor diagram

$$I_C = I \cos \delta = \omega \, C V, \text{ or } I = \left(\frac{\omega C V}{\cos \delta}\right).$$

Hence, power loss in the capacitor

$$P = V I \cos \phi = V I \cos (90 - \delta) \quad \text{or} \quad P = V I \sin \delta.$$

Substituting for I, $\qquad P = V \left(\dfrac{\omega C V}{\cos \delta}\right) \sin \delta \quad \text{or} \quad V^2 \omega C \tan \delta$

and thus proportional to $\tan \delta$ for given V, ω and C.

derived from the PD across the resistance and, knowing the voltage across the capacitor, the capacitance can be calculated as above[1].

Capacitance Measurement Using AC Bridges

De-Sauty's bridge

This bridge is the "capacitance equivalent" of the Maxwell's inductance bridge where the coil of un-known inductance is replaced by the capacitor under test, a standard capacitor forming one of the remaining branches.

Measuring a 'Pure' Capacitance

The bridge, and the corresponding phasor diagram at balance are shown in Fig.11.20 where C_1 is the capacitor under test and C_2 the standard capacitor. Both the capacitors are assumed to be loss-free, a typical example being an "air" capacitor. The bridge is balanced by varying R_1 or R_2 and thus it forms the simplest bridge of the type.

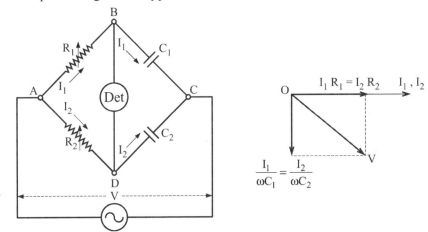

Fig.11.20 : De-Sauty's bridge for measuring loss-free capacitor and phasor diagram

At balance,

$$I_1 R_1 = I_2 R_2$$

and

$$-\frac{j}{\omega C_1} I_1 = -\frac{j}{\omega C_2} I_2$$

Thus,

$$\frac{R_1}{R_2} = \frac{C_2}{C_1}$$

---

[1]In both the cases, the capacitance is assumed to be loss free. A correction can be made accounting for the loss component, if necessary.

or
$$C_1 = \frac{R_2}{R_1} C_2$$

and the expression for C_1 is independents of the supply frequency.

In practice, it is observed that maximum sensitivity is achieved if C_2 is comparable to C_1.

Bridge for an 'Imperfect' Capacitor

If the capacitor under test is lossy, with the loss component represented by a resistance in *series*, the bridge would take the form shown in Fig.11.21 in which C and r represent the capacitor and its loss component. Then a 'standard' capacitor, C_S, is also to be used with *its* loss effect represented by a resistance r_S in series. An additional resistance, R_2, is connected in series with the standard capacitor for final balance.

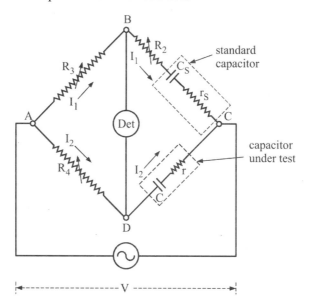

Fig.11.21 : De-Sauty's bridge for imperfect capacitor

At balance

$$\frac{r_S + R_2 - \dfrac{j}{\omega C_S}}{r - \dfrac{j}{\omega C}} = \frac{R_3}{R_4}$$

Simplifying and equating real and imaginary parts,

$$C = \frac{R_3}{R_4} C_S$$

and
$$r = \frac{R_4}{R_3}(r_S + R_2)$$

Loss Factor

The above expressions can also be used to determine the loss factor of the capacitor.

Thus

$$\tan \delta = \omega\, C\, r = \omega \frac{R_3}{R_4} C_S \times \frac{R_4}{R_3}(r_S + R_2)$$

$$= \omega\, C_S\, r_S + \omega\, C_S\, R_2$$

$$= \omega\, C_S\, R_2 + \tan \delta_S$$

where $\tan \delta_S$ is the loss factor of the standard capacitor.

Wien's bridge

This bridge is frequently employed to measure capacitance of an imperfect capacitor in which the loss component is represented by a resistor in *parallel* with the capacitor. A typical case is the specimen of a cable. The connections of the bridge are given in Fig.11.22 where C_1 is the capacitance of the imperfect capacitor shunted by a resistance R_1. C_2 is a 'standard' *air* capacitor and R_2, R_3, R_4 are variable, non-inductive resistances, used to obtain the balance. The phasor diagram at balance is also shown in the figure.

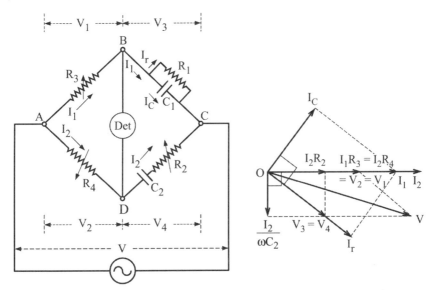

Fig.11.22 : Wien's bridge and phasor diagram at balance

At balance

$$I_1 R_3 = I_2 R_4$$

and

$$I_1\left(\frac{R_1}{1+j\omega R_1 C_1}\right) = I_2\left(R_2 - \frac{j}{\omega C_2}\right)$$

or

$$\frac{R_3\left(1+j\omega R_1 C_1\right)}{R_1} = \frac{R_4}{R_2 - \dfrac{j}{\omega C_2}}$$

Solving in the usual manner and by elimination,

$$R_1 = \frac{R_3\left(1+\omega^2 R_2^2 C_2^2\right)}{\omega^2 R_2 R_4 C_2^2}$$

and

$$C_1 = \frac{\left(R_4/R_3\right)C_2}{1+\omega^2 R_2^2 C_2^2}$$

Dimensional balance

This is another instance when a check on dimensional balance for C_1 should be carried out to verify the validity of the expression for the capacitance.

Thus, from the above expression the dimensions of the two sides can be written

$$\left[C_1\right] \equiv \left[\frac{C_2}{\omega^2 R_2^2 C_2^2}\right]$$

Now, in EM units

$$[\omega] = T^{-1}$$
$$[R] = LT^{-1}\mu$$
$$[C] = L^{-1}T^2\mu^{-1}$$

∴   dimensions of the RHS of the expression for C_1

$$= \frac{L^{-1}T^2\mu^{-1}}{T^{-2}L^2T^{-2}\mu^2 L^{-2}T^4\mu^{-2}}$$
$$= L^{-1} T^2 \mu^{-1}$$
$$= [C]$$

The expression for C_1 as above is, therefore, dimensionally correct.

A special case

The Wien's bridge can also be successfully used for measurement of (supply) frequency if R_1 and C_1 are known from a different, previous measurement. This is because both the expressions for R_1 and C_1 are coupled to the frequency, ω.

Thus, if R_3 and R_4 are fixed resistances having a ratio $\dfrac{R_3}{R_4} = 2$, similar to ratio arms of a Wheatstone bridge, $C_1 = C_2 = C$ and $R_1 = R_2 = R$, then both balance equations become

$$\omega\,C\,R = 1$$

The frequency is then simply obtained as

$$f = \frac{1}{2\pi C R}$$

The method can be used very satisfactorily up to low, audio frequencies. At higher frequencies, loss effect in the capacitors and stray capacitance effects make the method less satisfactory.

Schering bridge

The Schering bridge is the most widely used method for measurement of capacitance, esp. for measuring dielectric loss and power factor associated with the capacitor. There are two types of Schering bridge in common use.

A. 'Low'-voltage Schering Bridge

The circuit of the bridge for use with low supply voltage is shown in Fig.11.23.

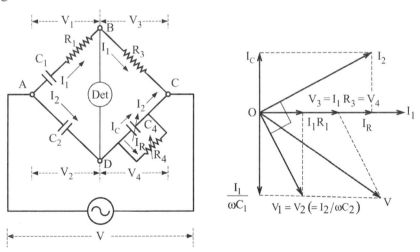

Fig.11.23 : Low-voltage Schering bridge and phasor diagram at balance

C_1 is the un-known capacitance and R_1 the resistance representing its loss component, in series combination. C_2 is a fixed, loss-free capacitor. R_3 and R_4 are non-inductive, standard resistors, of which R_4 is variable. C_4, in parallel with R_4, is another loss-free, variable capacitor. The bridge is balanced by adjusting R_4 and C_4.

At balance, equating various PDs,

$$I_1\left(R_1 + \frac{1}{j\omega C_1}\right) = I_2\left(\frac{1}{j\omega C_2}\right)$$

and

$$I_1 R_3 = I_2\left(\frac{R_4}{1 + j\omega R_4 C_4}\right)$$

or

$$\frac{\left(R_1 + \frac{1}{j\omega C_1}\right)}{R_3} = \frac{1 + j\omega R_4 C_4}{R_4 \times (j\omega C_2)}$$

Simplifying and equating real and imaginary parts

$$R_1 R_4 = \frac{R_3 R_4 C_4}{C_2}, \quad \text{giving} \quad R_1 = R_3 \times \frac{C_4}{C_2}$$

and

$$\frac{R_4}{C_1} = \frac{R_3}{C_2}, \quad \text{giving} \quad C_1 = \frac{R_4}{R_3} \times C_2$$

It is seen that the expressions for R_1 and C_1 are dimensionally correct and independent of frequency.

Loss factor

Since the loss component of the capacitor is represented by a series resistance, the loss factor is given by

$$\tan \delta = \omega R_1 C_1$$

Substituting for R_1 and C_1

$$\tan \delta = \omega\left(\frac{R_3 C_4}{C_2}\right)\left(\frac{C_2 R_4}{R_3}\right)$$

$$= \omega R_4 C_4$$

and the loss angle, $\delta = \tan^{-1}(\omega R_4 C_4)$

B. High-voltage Schering Bridge

This bridge is similar to the low-voltage Schering bridge, except that

(a) a single-phase, high-voltage supply of up to $100\,\text{kV}$ is used as the source; this being necessary to allow a measurable current to flow in the branches comprising (low-value) capacitances, offering an extremely high reactance/impedance, esp. at power frequency;

(b) representation of loss component of the capacitor by a very high resistance in *parallel* with the capacitor;

(c) use of appropriate screens to guard the low-voltage ('lower' two) branches and the detector (a high-sensitivity vibration galvanometer or head-phone, worn by the operator); and also to avoid errors due to inter-capacitance between the high-and low-voltage branches.

The schematic of the bridge is shown in Fig.11.24.

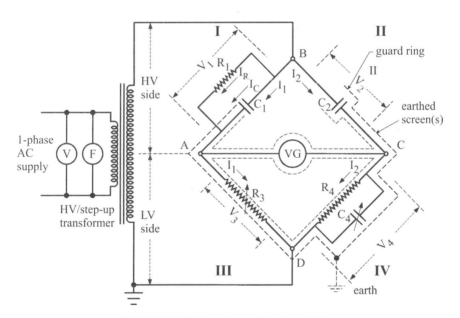

Fig.11.24 : High-voltage Schering bridge

In the bridge circuit, C_1 is the capacitor under test and R_1 a resistance across C_1, representing its dielectric loss. C_2 is a standard air capacitor, with a guard ring which is connected to the screen and grounded. R_3 and R_4 are non-inductive resistors, the former being variable. C_4 is a loss-free variable capacitor, connected across R_4. The bridge is balanced by successive adjustment of R_3 and C_4.

At balance,

$$V_1 = V_2 \quad \text{and} \quad V_3 = V_4$$

or
$$I_1\left(\frac{R_1}{1+j\omega R_1 C_1}\right) = I_2\left(\frac{1}{j\omega C_2}\right)$$

and
$$I_1 R_3 = I_2\left(\frac{R_4}{1+j\omega R_4 C_4}\right)$$

Dividing to eliminate I_1 and I_2

$$\frac{\left(\dfrac{R_1}{1+j\omega R_1 C_1}\right)}{R_3} = \frac{\left(\dfrac{1}{j\omega C_2}\right)(1+j\omega R_4 C_4)}{R_4}$$

or
$$j\omega R_1 R_4 C_2 = R_3(1+j\omega R_1 C_1)(1+j\omega R_4 C_4)$$

Simplifying and equating real and imaginary parts

$$R_1 = \frac{R_3\left(1+\omega^2 R_4^2 C_4^2\right)}{\omega^2 R_4^2 C_2 C_4}$$

and
$$C_1 = \frac{R_4}{R_3} \times \frac{C_2}{1+\omega^2 R_4^2 C_4^2}$$

Also, both R_1 and C_1 depend on the knowledge of the supply frequency[1].

Loss factor and loss angle

When the loss component of a capacitor is represented by a resistance in parallel with it, the loss factor, $\tan\delta$ is expressed as

$$\tan\delta = \frac{1}{\omega C R}$$

as derived before.

Therefore, for the capacitor under test

$$\tan\delta = \frac{1}{\omega C_1 R_1}$$

Substituting for C_1 and R_1

$$\tan\delta = \frac{1}{\omega} \times \frac{1}{\dfrac{R_4}{R_3\left(1+\omega^2 R_4^2 C_4^2\right)} \times \dfrac{C_2}{\omega^2 R_4^2 C_2 C_4}}$$

---

[1] Compare the expressions for R_1 and C_1 with those obtained for the Wien's bridge. Once again, when checked, the expressions are found to be dimensionally correct.

which simplifies to

$$\tan \delta = \omega\, R_4\, C_4 \quad \text{and} \quad \delta = \tan^{-1}(\omega\, R_4\, C_4)$$

The expression for C_1 can be written in terms of "δ".

Thus,
$$C_1 = \frac{R_4}{R_3} C_2 \cos^2\delta$$

since
$$\cos^2\delta = \frac{1}{1 + \tan^2\delta}$$

If the loss angle of the capacitor is very small, $\delta \to 0$ and $\cos^2\delta \to 1$, and C_1 can simply be given by

$$C_1 = \frac{R_4}{R_3} C_2$$

which is the same as for the "low-voltage" bridge, with a series representation of the loss component.

The phasor diagram of the bridge at balance is shown in Fig.11.25[1].

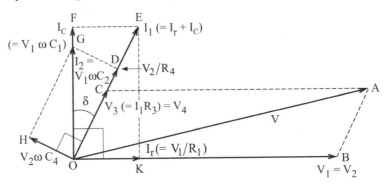

Fig.11.25 : Phasor diagram of (HV) Schering bridge at balance

---

[1]The diagram can be explained as follows:

OA = V = voltage applied to the bridge, resultant of V_1 and V_3 (or V_2 and V_4)

OB = V_1 = voltage drop across arm II, equal in magnitude and phase to PD across arm I ($=V_2$)

OC = V_3 = PD across arm III, equal in magnitude and phase to PD across arm IV ($=V_4$)

OE = I_1 = current through arms I and III

OG = I_2 = current through arms II and IV

OF and OK = components of I_1 through C_1 and R_1

OD and OH = components of I_2, split through R_4 and C_4

GENERAL CONSIDERATIONS ABOUT BRIDGES

To a greater or lesser extent, all the bridges dealt with in the preceding sections are 'idealisations'. In practice, however, the behaviour of the bridge circuits may depend, in some cases seriously, on various factors and considerations, indirectly related to balancing of the bridge. If not accounted for, the end result will be either great difficulty in balancing the bridge itself or obtaining false or inaccurate value of the un-known quantity.

Sources of Error

These constitute the most important aspects related to any of the bridges and may usually go un-seen or un-accounted. The various possible sources of error, encountered in a particular bridge or common to most, are

(i)   stray-conductance effects due to, say, imperfect insulation;

(ii)   mutual inductance effects due to magnetic coupling between various components and current-carrying parts of the bridge;

(iii)   stray capacitance effects, caused by electrostatic fields between conductors at different potentials, esp. in bridges employing high-voltage supply;

(iv)   errors due to 'loose' contacts and haphazardly laid out or criss-crossing wires carrying "heavy" currents.

Errors introduced by some of the above effects may be quite appreciable in, for example, high-frequency and high-voltage bridges and great precautions may be necessary to eliminate or minimise them; or else account them in the final derivations/calculations.

In general, good results can be achieved by resorting to the followings:

Organisation and Layout of the Bridge

The usual bridge-circuit diagrams pertaining to various bridges have been shown above to represent the potential distribution and current flow in the circuit; they do not indicate the optimal or best *physical* lay-out. Each component, for example a coil or capacitor under test, the decade resistance boxes (representing variable resistances), the detector and source, should be arranged so that these are connected to the appropriate 'corners' or 'nodes', comprising the bridge, preferably by their own individual leads. If additional wires or leads are used, these must be *as short in length as required*, of sufficient cross-section, in twisted-pair form and connected so as not to result in large closed loops.

A typical layout of the key components of a 4-arm bridge comprising main requirements may look like that shown in Fig.11.26, with external leads or wires running closely parallel, or better still as twisted pair to minimise any "pick-up", esp. at high frequency[1].

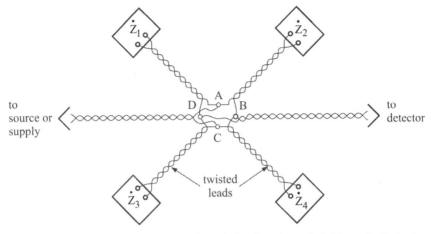

A, B, C, D correspond to conventional 'nodes' or junctions of a bridge as in the text

Fig.11.26 : Typical layout of a 4-arm AC bridge

Screening

The undesirable effects of stray capacitance(s) in a bridge may be reduced to a great extent by suitable, effective screening of the components and leads[2]. The stray capacitances may not only exist between the various bridge components, depending on the difference of potential levels, but even between the components and the operator which may vary with the movement of the latter. The worst aspect of these capacitances is their un-predictability, necessitating their elimination rather than being accounted for in the analysis. Depending on the type of a bridge, and operating frequency, it may be necessary to screen individual components or the entire bridge circuit as a whole. It is usually sufficient to ground the screen at one point only to eliminate any ground loop currents.

Components

Many components or quantities, variously referred to in the text as "standard", "loss-free" or "non-inductive" etc., and which appear in the expressions for the un-knowns – inductance, capacitance and associated resistance – have been assumed to be without any error of their own. In practice, this may not be so unless very "high-quality" components are used. If not, the accuracy with which the un-known quantities are obtained may be greatly affected by the accuracy of the "known" components or devices. In

---

[1]This is discussed in detail later in Appendix IX: "Instrumentation"

[2]See, for example, the high-voltage Schering bridge.

this context, as mentioned, the sensitivity of the detector and stability of the supply also play an important role.

The Final Balance

Apart from the quality and sensitivity of the detector, the final balance may be affected by the environment around the experimental set-up and owing to subjective errors, the latter may particularly apply when a telephone or head-phone is used as a detector when the hearing ability of the operator may form a source of error. In many cases, therefore, repeatability of the test may form a sound basis for acceptance (or otherwise) of the final results.

Numerical Solutions

In numerical problems, bridges comprising four or more branches, are described in terms of the various impedances, joined to form the bridge. In dealing with the solution of such problems, the following tips may prove to be helpful:

(a) form the bridge circuit or network around the ABCD (and E) nodes as described, by actually drawing the circuit diagram and marking the locations of the detector and the source;

(b) with the knowledge of component values at balance as given out in the problem write down the "equations at balance";

(c) simplify the equations by elimination of branch currents, for example, to obtain the final expressions, usually containing real and imaginary terms;

(d) in nearly all cases, the un-known quantities, L,C and R, will be obtained from equating read and imaginary parts of the expression;

(e) check whether the expressions for the un-known quantities are dimensionally balanced;

(f) a phasor diagram at balance showing the various currents and PDs in proper phasor relationship, preferably drawn to scale, may then conclude the solution.

WORKED EXAMPLES

1. In an AC bridge measurement, the four arms, at balance, comprise:

    (i)   AB: an un-known inductance L_1, having an internal resistance, R_1;

    (ii) BC: a non-inductive resistance of 1000 Ω;

    (iii) CD: a capacitance of 0.5 μF, shunted by a resistance of 1000 Ω;

    (iv) DA: a resistance of 1000 Ω.

    Derive the values of R_1 and L_1.

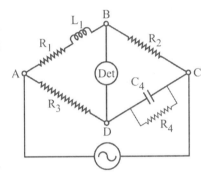

The bridge is drawn as described and can be identified as Maxwell's inductance-capacitance bridge

Given

$$R_2 = R_3 = 1000 \ \Omega$$

$$R_4 = 1000 \ \Omega$$

$$C_4 = 0.5 \ \mu F$$

At balance

$$\frac{R_1 + j \omega L_1}{R_2} = \frac{R_3}{\dfrac{R_4}{\left(1 + j \omega R_4 C_4\right)}}$$

or $$\frac{R_1 + j \omega L_1}{R_2} = \frac{R_3 \left(1 + j \omega R_4 C_4\right)}{R_4}$$

Since $R_2 = R_4 \ (= 1000 \ \Omega$ each$)$, equating real and imaginary parts

$$R_1 = R_3 = 1000 \ \Omega$$

$$L_1 = R_3 R_4 C_4$$

$$= 1000 \times 1000 \times 0.5 \times 10^{-6}$$

$$= 0.5 \ H$$

Dimensionally

$$[L_1] = [R_3 R_4 C_4]$$

$$= L^2 T^{-2} \mu^2 L^{-1} T^2 \mu^{-1} = L \mu = [L]$$

2. The four arms of an AC bridge are arranged as follows:

AB : an imperfect capacitor C_1, with an equivalent series resistance, R_1

BC : a non-inductive resistance, R_3

CD : a non-inductive resistance, R_4

DA : another imperfect capacitor, C_2, with an equivalent series resistor, r_2 in series with a variable resistance, R_2

A supply at 450 Hz is connected across AC and a telephone detector across BD.

The bridge is balanced with $R_2 = 4.8\ \Omega$, $R_3 = 2000\ \Omega$, $R_4 = 2850\ \Omega$, $C_2 = 0.5\ \mu F$ and $r_2 = 0.4\ \Omega$.

Calculate the values of R_1 and C_1 and the loss angle of the capacitor.

The bridge is drawn as shown and is similar to De-Sauty's bridge.

@ 450 Hz

At balance,

$$\frac{R_1 + \dfrac{1}{j\omega C_1}}{R_3} = \frac{(r_2 + R_2) + \dfrac{1}{j\omega C_2}}{R_4}$$

or $\dfrac{1 + j\omega R_1 C_1}{j\omega C_1 \times R_3} = \dfrac{1 + j\omega (r_2 + R_2) C_2}{j\omega C_2 \times R_4}$

or $R_4 C_2 + j\omega R_1 R_4 C_1 C_2 = R_3 C_1 + j\omega R_3 (r_2 + R_2) C_1 C_2$

Equating real and imaginary parts

$$R_4 C_2 = R_3 C_1$$

and $R_1 R_4 C_1 C_2 = R_3 (r_2 + R_2) C_1 C_2$

giving $C_1 = \dfrac{R_4}{R_3} \times C_2 = \dfrac{2850}{2000} \times 0.5$

$= 0.712\ \mu F$

and $R_1 = \dfrac{R_3}{R_4} \times (r_2 + R_2) = \dfrac{2000}{2850}(0.4 + 4.8)$

$= 3.65\ \Omega$

The phasor diagram of the bridge at balance is as shown in the adjoining figure.

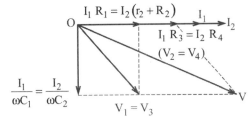

It is seen that the expressions of R_1 and C_1 are dimensionally correct.

$$\frac{I_1}{\omega C_1} = \frac{I_2}{\omega C_2}$$

Loss factor of the capacitor

$$\tan \delta = \omega R_1 C_1$$

$$= 2\pi \times 450 \times 3.65 \times 0.712 \times 10^{-6}$$

$$= 0.007344$$

and loss angle, $\delta = \tan^{-1}(0.007344) = 0.42°$

Comment

At power frequency, that is, 50 Hz, the loss angle will be

$$\delta = \tan^{-1}\left[\frac{0.007344}{450} \times 50\right] = 0.00082$$

and the "loss" in the capacitor proportionally much less.

---

3. In the Anderson bridge for measurement of inductance, the arm AB consists of an un-known impedance $(R + j\omega L)$ representing an ordinary coil; BC a known, variable resistance, R_3; CD and DA, fixed standard resistances, R_2 and R_4 each of 600 Ω; DE a known, variable resistance, r and CE a standard capacitance, C of value 1 μF. A suitable detector is connected across B and E. The bridge is balanced with r = 400 Ω and R_3 = 800 Ω.

Calculate the parameters of the coil, R and L.

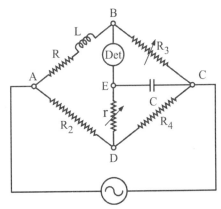

At balance,

$$R_2 = R_4 = 600 \ \Omega$$

$$R_3 = 800 \ \Omega$$

$$r = 400 \ \Omega$$

$$C = 1 \ \mu F$$

The bridge composition is shown in the adjoining figure.

Now, as derived in the text, R is given by

$$R = \frac{R_2 R_3}{R_4} = \frac{600 \times 800}{600}$$

or $R = 800 \, \Omega$

Also $L = \dfrac{CR_3}{R_4}\left[r(R_2 + R_4) + R_2 R_4\right]$

$$= \frac{1 \times 10^{-6} \times 800}{600}\left[400 \times (600 + 600) + 600 \times 600\right]$$

$$= \left(\frac{8}{6}\right) \times 10^{-6}\left[480,000 + 360,000\right]$$

$$= 1.12 \, \text{H}$$

4.   A capacitor bushing forms branch AB of an LV Schering bridge whilst DA consists of a standard capacitor of 500 pF. The arm BC, R_3, is a non-inductive resistance of 300 Ω. At balance, the arm CD comprises a resistance, R_4, of 72.6 Ω, in parallel with a capacitance, C_4, of 0.148 μF. A supply of power frequency is connected across AC and a vibration galvanometer across BD. Calculate the capacitance and resistance representing the bushing.

The bridge is drawn in the adjoining figure.

At balance

$C_2 = 500$ pF

$R_3 = 300 \, \Omega$, $R_4 = 72.6 \, \Omega$

$C_4 = 0.148 \, \mu$F

The equations at balance are

$$\frac{R_1 + \dfrac{1}{j\omega C}}{R_3} = \frac{\dfrac{1}{j\omega C_2}}{\left(\dfrac{R_4}{1 + j\omega R_4 C_4}\right)}$$

or $\dfrac{1 + j\omega R C}{j\omega R_3 C} = \dfrac{1 + j\omega R_4 C_4}{j\omega R_4 C_2}$

or $R_4 C_2 + j \omega R R_4 C C_2 = R_3 C + j \omega R_3 R_4 C C_4$

Equating real and imaginary parts

$$R_4 C_2 = R_3 C, \text{ giving } C = \left(\frac{R_4}{R_3}\right) \times C_2$$

or $\qquad C = \left(\frac{72.6}{300}\right) \times 500 = 121 \text{ pF}$

and $\qquad R R_4 C C_2 = R_3 R_4 C C_4$

giving $\qquad R = \dfrac{C_4}{C_2} \times R_3 \quad$ or $\quad \dfrac{0.148 \times 10^{-6}}{500 \times 10^{-12}} \times 300 = 88.8 \text{ k}\Omega$

5. A balanced Schering bridge, ABCD, has AB as a standard capacitor of 100 pF with the loss factor of 2×10^{-4}; BC a 1300 Ω resistor in parallel with a 24 pF capacitor; CD a 1000 Ω resistor and DA the capacitor under test with its loss component represented by a series resistance.

Determine the capacitance and loss angle of the test capacitor. The supply frequency is at 1000 Hz.

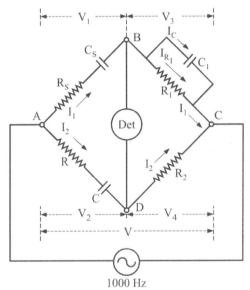

The connection of the bridge is as shown.

At balance, it is given that

$C_S = 100 \text{ pf}$ and $\omega R_S C_S = 2 \times 10^{-4}$ $\qquad [\omega = 2\pi \times 1000]$

$$R_1 = 1300 \ \Omega, \ C_1 = 24 \text{ pF}$$

$$R_2 = 1000 \ \Omega$$

R and C represent the test capacitor.

The equations at balance are

$$\frac{R_S - \dfrac{j}{\omega C_S}}{R - \dfrac{j}{\omega C}} = \frac{\dfrac{R_1}{(1 + j\,\omega\,R_1\,C_1)}}{R_2}$$

or $\left(R - \dfrac{j}{\omega C}\right)\dfrac{R_1}{(1 + j\omega R_1 C_1)} = \left(R_S - \dfrac{j}{\omega C_S}\right)R_2$

or $\qquad \left(R - \dfrac{j}{\omega C}\right) = \left(R_S - \dfrac{j}{\omega C_S}\right)(1 + j\omega R_1 C_1)\dfrac{R_2}{R_1}$

$$= \left(R_S + j\omega R_1 R_S C_1 - \frac{j}{\omega C_S} + \frac{R_1 C_1}{C_S}\right)\frac{R_2}{R_1} \quad \ldots(A)$$

Equating the imaginary terms

$$-\frac{j}{\omega C} = \left(-\frac{j}{\omega C_S} + j\omega R_1 R_S C_1\right)\frac{R_2}{R_1}$$

or $\qquad \dfrac{1}{C} = \left(\dfrac{1}{C_S} - \omega^2 R_1 R_S C_1\right)\dfrac{R_2}{R_1}$

giving $\qquad C = \left(\dfrac{C_S}{1 - \omega^2 R_1 R_S C_1 C_S}\right)\dfrac{R_1^1}{R_2}$

Now $\omega\,R_S\,C_S = \tan \delta_S$, the loss factor of the standard capacitor,

$$= 2 \times 10^{-4}$$

$\therefore \qquad C = \left(\dfrac{C_S}{1 - \omega\,R_1\,C_1 \tan \delta_S}\right)\dfrac{R_1}{R_2}$

---

[1]The expression for C is dimensionally correct, reducing to

$$[C] = \left[\frac{1}{\omega^2 R_1 R_S C_1}\right],$$

and by substituting dimensions of ω, R, C etc.

$$= \frac{1}{T^{-2}L^{-2}T^{-2}\mu^2 L^{-1}T^2\mu^{-1}}$$

$$= L^{-1}\,T^2\,\mu^{-1} = [C],$$

– dimensions of a capacitor.

Substituting the values of $\omega\, R_1\, C_1$ etc. in the denominator

$$1 - \omega\, R_1\, C_1 \tan \delta_S = 1 - 2\pi \times 1000 \times 1300 \times 24 \times 10^{-12} \times 2 \times 10^{-4}$$

$$= 1 - 3.92 \times 10^{-8}$$

making the second term negligible *in the present case.*

Thus, C is simply given by

$$C = C_S \times \frac{R_1}{R_2}$$

$$= 100 \times 10^{-12} \times \left(\frac{1300}{1000}\right)$$

$$= 130 \text{ pF}$$

Equating the real terms [of expression (A)]

$$R = \left(R_S + \frac{R_1 C_1}{C_S}\right) \frac{R_2}{R_1}$$

and the loss factor of the test capacitor

$$\tan \delta = \omega\, C\, R = \omega\, C_S \frac{R_1}{R_2}\left(R_S + \frac{R_1 C_1}{C_S}\right)\frac{R_2}{R_1}$$

using the approximate expression for C as above.

or
$$\tan \delta = \omega\, C_S \left(R_S + \frac{R_1 C_1}{C_S}\right)$$

$$= \omega\, R_S\, C_S + \omega\, R_1\, C_1$$

$$= 2 \times 10^{-4} + 2\pi \times 1000 \times 1300 \times 24 \times 10^{-12}$$

$$= 2 \times 10^{-4} + 1.96 \times 10^{-4}$$

$$= 3.96 \times 10^{-4}$$

Phasor diagram

With the various branch currents and potential drops shown in the circuit, the phasor diagram at balance can be drawn as shown:

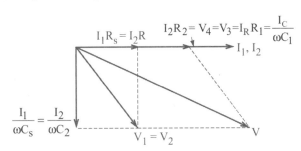

6.   In an AC bridge, the four branches are arranged as follow:

AB :   a choke coil wound on a closed iron core having an inductance of L and resistance r

BC :   a non-inductive resistor R

CD :   a mica capacitor C in series with a non-inductive resistor S

DA :   a non-inductive resistor Q

When the bridge is fed from a source of 500 Hz, balance is achieved with

$$R = 750 \ \Omega, \ Q = 2410 \ \Omega \text{ and } S = 64.5 \ \Omega$$

The capacitor C has a capacitance of $0.35 \ \mu F$ with an equivalent series resistance of $0.4 \ \Omega$ to represent the loss.

Calculate the inductance and resistance of the choke coil under the above conditions.

The bridge is connected as shown.

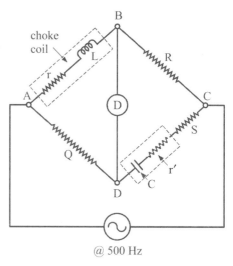

At balance

$$\frac{r + j \omega L}{R} = \frac{Q}{(S + r') + \dfrac{1}{j \omega C}}$$

where r' is the loss component of capacitor C, such that

$S + r' = 64.9 \ \Omega$.

Substituting the values of R, S, Q etc. at balance and rearranging

$$(r + j \omega L) \times \left(64.9 - \frac{j}{1099 \times 10^{-6}} \right) = 2410 \times 750$$

or $64.9 \ r - \dfrac{j \ r}{1099 \times 10^{-6}} + j \ 64.9 \ \omega L + \dfrac{\omega L}{1099 \times 10^{-6}} = 2410 \times 750 \ldots(A)$

Equating real and imaginary parts

$$64.9 \ r + \frac{\omega L}{1099 \times 10^{-6}} = 2410 \times 750$$

and
$$\frac{r \times 10^6}{1099} = 64.9\omega\, L = 64.9 \times 2\pi \times 500\, L$$

From the second equation

$$r = 223.96\, L$$

Substituting in the first equation (of balance), (A), and solving for L gives

$$2.8645\, L = 1.8075$$

or
$$L = 0.63\ H$$

whence
$$r = 223.96 \times 0.63$$
$$= 141\ \Omega$$

7. In a Schering bridge, a sample of insulation forms the branch AB. The bridge is balanced using a supply at 50 Hz. The other arms of the bridge are found to be as follows:

branch BC   : a loss-free capacitor of 109 pF

branch CD   : a non-inductive resistance of 309 Ω in parallel with a loss-free capacitor of 0.5 μF

branch DA   : a non-inductive resistance of 100 Ω

Determine the capacitance, equivalent series resistance and power factor of the capacitor.

The configuration of the bridge is as shown

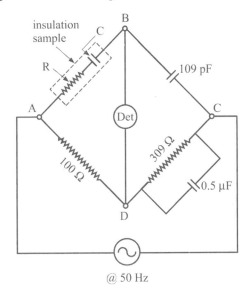

@ 50 Hz

At balance

$$\frac{\left(R + \dfrac{1}{j\,\omega\,C}\right)}{\dfrac{1}{j\,\omega \times 109 \times 10^{-12}}} = \frac{100\left(1 + j\,\omega \times 309 \times 0.5 \times 10^{-6}\right)}{309}$$

or $\quad 309 \times \left(R + \dfrac{1}{j\omega C}\right) = \dfrac{1}{j\omega \times 109 \times 10^{-12}} \times \left[100 \times \left(1 + j\omega \times 309 \times 0.5 \times 10^{-6}\right)\right]$

$$309R + \frac{309}{j\omega\,C} = \frac{100 + j\omega \times 100 \times 309 \times 0.5 \times 10^{-6}}{j\omega \times 109 \times 10^{-12}}$$

or $\quad 309R - \dfrac{j\,309}{\omega\,C} = -j\,29.22 \times 10^{8} + 29.22 \times \omega \times 309 \times 0.5 \times 10^{-6} \times 10^{8}$

Equating real and imaginary parts

$$309\,R = 29.22 \times 314 \times 309 \times 50$$

whence $\qquad\qquad R = 0.4587\ \text{M}\Omega$

Also,

$$\frac{309}{\omega\,C} = 29.22 \times 10^{8}$$

from which $\qquad\qquad C = \dfrac{309 \times 10^{-8}}{314 \times 29.22}$

$$= 337\ \text{pF}$$

The loss factor $(\tan\delta)$ for the series equivalent of the capacitor (sample) is

$$\tan\delta = \omega\,C\,R = 314 \times 337 \times 10^{-12} \times 0.4587 \times 10^{6}$$

$$= 0.0485$$

EXERCISES

1.   The four arms of an AC bridge are as follows:

arm AB : R_1; arm BC : R_2 and L_2 in series; arm DA : R_4; arm CD : an "un-known" coil. The bridge is balanced with R_1 = 200 Ω, R_2 = 100 Ω, L_2 = 0.1 H and R_4 = 400 Ω.

Calculate the values of the coil parameters.

[200 Ω; 0.2 H

2.   A Maxwell's bridge comprises:

AB and BC as non-inductive resistances of 100 Ω each; DA a standard variable inductor in series with a resistance of 32.7 Ω and CD a standard variable resistor R in series with a coil of un-known impedance. If the balance is obtained with the inductor having an inductance of L = 47.8 mH and R = 1.36 Ω, determine the resistance and inductance of the coil.

[31.34 Ω; 47.8 mH

3.   An Owen's bridge is used to measure the properties of a sample of sheet steel at 2 kHz. At balance, arm AB consists of the test specimen having a resistance R_1 in series with an inductance L_1; arm BC is R_3 = 100 Ω; arm CD a pure capacitor C_4 = 0.1 μF and arm DA is a resistance R_2 = 834 Ω in series with a capacitance C_2 = 0.124 μF. Calculate the values of R_1 and L_1.

[80.7 Ω; 8.34 mH

4.   The four arms of an AC bridge, ABCD, supplied with a sinusoidal voltage, comprise

AB   :   200 Ω resistor in parallel with a 1 μF condenser;

BC   :   a 400 Ω resistance;

CD   :   a 1000 Ω resistance; and

DA   :   a resistance R in series with a 2 μF condenser.

Determine (i) the value of R and (ii) the supply frequency at which the bridge is balanced.

[400 Ω; 398 Hz

5.   A typical 5-nodes AC bridge is connected as follows:

branch AB and BC : a resistance of 1500 Ω each;

branch CD   : an un-known coil of inductance L in series with a resistance R;

branch DA   : a 'pure' resistance of 327 Ω;

branch AE   : another resistance of 529 Ω;

branch EB   : a 'pure' capacitance of 20 μF.

A vibration galvanometer is connected across nodes E and C and an AC power supply at 50 Hz is connected across B and D.

Identify the bridge network and determine the values of the coil constants.

[A variation of Anderson's bridge; R = 327 Ω; L = 29 H

6.   An AC bridge consists of

AB : a choke coil, r and L, wound on a closed iron core

BC : a non-inductive resistor, R

CD : a mica condenser C, in series with a non-inductive resistor, S

DA : a non-inductive resistor, Q.

With the bridge fed from a source at 500 Hz, the balance is obtained with R = 750 Ω, Q = 2410 Ω and S = 64.5 Ω.

The condenser C has a capacitance of 0.35 μF and an equivalent series resistance of 0.4 Ω.

Calculate the inductance and resistance of the choke at this frequency.

[0.63 H; 140.5 Ω

7.   A sample of Bakelite was tested by Schering bridge at 25 kV, 50 Hz supply across A and C terminals of the bridge. Balance was obtained with a standard capacitor of 106 pF across AD; a capacitor of 0.4 μF in parallel with a non-inductive resistance of 318 Ω across CD and another non-inductive resistance of 120 Ω across BC.

Determine the capacitance and equivalent resistance of the sample.

[280.9 pF; 0.453 MΩ

QUIZ QUESTIONS

1. In an AC bridge, it is enough to balance the magnitude of PDs across the appropriate branch.

    □ true   □ false

2. A dimensional balance provides a good check on the expression for un-known quantity in an AC bridge.

    □ true   □ false

3. The AC supply in an AC bridge is always at power frequency.

    □ true   □ false

4. A bridge generally used for measuring capacitance is

    □ De-Sauty's bridge   □ Owen's bridge

    □ Wien's bridge   □ any of these

5. An AC bridge that may be used for measuring frequency in the bridge is

    _____

6. An AC bridge uses a supply of 2 kHz. A suitable detector for the bridge would be

    □ a D'Arsonval galvanometer   □ a vibration galvanometer

    □ a head-phone   □ none of these

7. At radio-frequency supply, a suitable detector in an AC bridge is

    _____

8. In high-voltage AC bridges, screening of components is essential to

    _____

9. The loss component of an imperfect capacitor is accounted for by

    _____

10. The Schering bridge is ideally suited to measure

    □ self-inductance of a coil

    □ mutual inductance of a coil

    □ supply frequency of the bridge

    □ dielectric loss and p f of an insulator

XII : Magnetic Measurements

XII

MAGNETIC MEASUREMENTS

RECALL

The most important physical quantity in electrical science and engineering is the magnetic flux density, B; for example, its significance can be realised by its role in the 'production' of EMF by

(i) the law of induction in a *time-varying* magnetic field where the EMF e is given by

$$e = (-)\frac{dB}{dt} \left(or - \frac{d\emptyset}{dt} \right)$$

[Ø being the flux corresponding to B in a given cross-sectional area]

– a law first enunciated by Michael Faraday as early as 1831 and later expressed by James C. Maxwell as one of the four classic field equations, and fundamental to the operations of *all* transformers,

or

(ii) the "flux-cutting" rule where

$$e = B\mathnormal{l}v \qquad \text{[as discussed later]}$$

forming the basis of working of all electric generators.

Indeed, the working principle of most electric *motors* is couched in terms of the stressed magnetic field in the *airgap* of the machine, a concept that is analytically relatable to what is called Maxwell's stress. Therefore, an accurate knowledge of magnetic flux density and the corresponding magnetising field at a point in a magnetic circuit, or rather the distribution in a region (for example a part of a machine) is of utmost importance for the designers and analysts of electric machinery. Most magnetic measurements in practice are aimed to achieve just that. It is to be noted that the primary quantity of interest, and hence measurement, is flux density, B, and NOT the magnetising field, H; the latter being simply the cause to produce B in the magnetic or non-magnetic region(s) of interest.

MAGNETIC MEASUREMENTS WITH DC AND AC

Magnetic tests or measurements can be divided into two broad categories:

A. Those performed using direct current. These are generally aimed to obtain B/H characteristics of magnetic materials; only the B-H curve in the case of soft magnetic materials and B-H curve as well as B-H loop of permanent-magnet materials. In this case, the current and hence the magnetic (or magnetising) field is steady or non-time-varying.

B. Those performed using alternating current of a given frequency[1] to determine experimentally, in most cases, the "*iron*" *losses* in the sample which may be in the form of strips or laminations. The total loss may be further separated into hysteresis and eddy current loss. The essence of tests in this category is that the magnetic field is time-varying.

    Both categories of test necessitate a number of devices and instruments, many of which may be common to both, but some may be necessary either for DC or AC tests.

Magnetic and Related Quantities

DC testing

As stated, the basic magnetic quantity to be measured is the flux density B which may be related to the corresponding magnetising field H by the expression.

$$B = \mu_0 \mu_r H$$

where $\mu_0 \left(= 4\pi \times 10^{-7} \right)$ is the 'absolute' permeability and μ_r the relative permeability for the given material. The relationship between B and H will be a straight line if μ_r were constant. However, the B-H curve (or variation) of a magnetic material in practice is inherently highly non-linear and thus the above relation is true only for one value of B and H; else, there would be 'infinite' values of B for the corresponding H and μ_r, the latter varying as

$$\mu_1, \mu_2 \ \cdots \ \mu_\infty .^{2}$$

Magnetic Flux

From the knowledge of flux density, the flux in a magnetic circuit or part of a machine can be obtained by multiplying the flux density with the area of cross section of the part, or

$$\emptyset = B \times A$$

---

[1]Usually the power frequency, but may also be audio or radio frequency.

[2]In fact, this is what makes most analyses highly complicated where magnetic circuits are involved.

This expression would hold only if the flux density distribution throughout the area is *uniform* and perpendicular to the surface of cross section[1]. If not so, the flux through the section well be given by

$$\emptyset = B_1 \times A_1 + B_2 \times A_2 + \ldots + B_n \times A_n \ldots + B_\infty \times A_\infty$$

where it may be assumed that B_1 is 'nearly constant' flux density in, and normal to, a very small cross section A_1, B_2 in A_2 and so on.

In a few measuring schemes, it may be that the flux, $\emptyset$, rather the flux density is measured. Then the flux density may be deduced from the measured flux based on similar considerations.

One special aspect of "DC" measurements is that the tests are usually performed on solid materials or samples. Also, the methods used generally border on classical. In many of such methods the magnetising *field H* is first computed in terms of the (DC) excitation current in a coil of N turns which is then related to corresponding flux density, B, when deriving, for example, the B-H characteristic of the materials, dealt with later in this chapter.

AC testing

An important aspect of magnetic measurements using alternating supply is that all the three quantities, viz., B, H and $\emptyset$, may invariably be *time-varying*: sinusoidally if the 'excitation' (generally meaning the current in the circuit) is sinusoidal, in the 'linear' range of B v/s H variation. A time-varying field would result in induction of EMF and hence of current and loss (the "eddy-current" loss) in the magnetic material or the sample used for the testing. The actual loss will depend on electrical and magnetic properties of the material; for example, electrical resistivity, the permeability as well as the frequency of the supply. Whilst the method(s) of testing the sample with AC will yield the two losses (that is, eddy-current and hysteresis) measured by a wattmeter, say in terms of loss per cubic metre or kg, that is, the loss density, the final effect of the losses in terms of heating and temperature rise of the material will not be known which may need further 'treatment'. In terms of cause-effect relationship, therefore, the various quantities of interest can be identified as shown in Fig.12.1.

In actual machines where it is desired to estimate the temperatures during operation at different parts, measurements may have to be made of the "basic" quantity, B or H, the induced currents (distribution) and the temperature rise(s) with time. The appropriate mathematical relations then form the basis for these measurements[2].

---

[1]As shown later, this has an important bearing on one type of magnetic (flux density) measurement.

[2]See Appendix VI for these relations.

alternating H produced by AC supply

↓

alternating B = $\mu_0 \mu_r H$

↓

alternating $\emptyset = BA$

↓

induced eddy EMF, e $= -\dfrac{d\emptyset}{dt}$

↓

induced eddy currents, $\mathrm{i} = \dfrac{e}{r}\left[r = \rho\left(\dfrac{l}{a}\right)\right]$

↓

eddy-current loss, $\mathrm{p} = \mathrm{i}^2 \mathrm{r}$

↓

heat equivalent of eddy current loss, Q

↓

temperature rise due to Q, θ
localised or overall

Fig.12.1: Alternating magnetic field leading to temperature rise[1]

ACCESSORIES, MEANS AND DEVICES USED FOR MAGENTIC MEASUREMENTS

These constitute, in general, one or more of essential requirements for DC or AC testing or both.

Magnetometer

A magnetometer represents a historic device and simplest to measure earth's magnetic field, often the horizontal component of the field. In laboratories, these may also be applied to measure flux density of (permanent) bar magnets, the advantage being that they measure the actual value of flux density in the specimen as distinct from some methods which measure a *change* in flux density.

The apparatus essentially consists of a suspended (or pivoted) magnetic needle in a glass case with a very small control torque on the needle. The needle may be aligned to a given magnetic field, say H_x, in a horizontal plane. When acted upon by another field, H_y, in the same plane, but at right

---

[1]Most magnetic materials used in electrical machines and equipment (e.g. a transformer) use a variety of silicon steel (in laminated form), having low hysteresis loss and relatively more eddy-current loss. The temperature rise is thus largely a consequence of induced eddy currents.

angles to the first field, the needle is deflected by an angle, θ, from its initial position. The two fields are then related to each other by the expression

$$H_y = H_x \tan \theta$$

Thus, knowing H_x and the angle θ, H_y can be determined.

Solenoid

A solenoid which forms an important means for calibration of a number of practical magnetic measuring devices as discussed later, essentially comprises a large number of turns of a thin, insulated wire, wound on a round non-metallic, non-magnetic 'tube' as the former. A typical, long solenoid – in which the length of the solenoid is much more (usually more than four times) compared to its 'mean' diameter – is shown in Fig.12.2. The turns are arranged (very) close to each other, each carrying a direct (or alternating) current in clockwise or anticlockwise direction when viewed from an end. The important aspect of a solenoid is that, when "sufficiently" long, its field is fairly uniform in the mid-section as indicated in Fig.12.2(b).

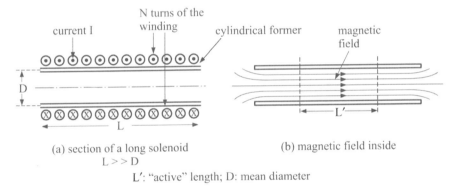

(a) section of a long solenoid
L >> D

(b) magnetic field inside

L': "active" length; D: mean diameter

Fig.12.2 : A long solenoid and its magnetic field

If the solenoid is wound with N turns and is excited by a current I ampere, the magnetic field in the *middle* of the solenoid is given by

$$H = \frac{N\,I}{L} \text{ A/m}$$

where L is its length in metre.

Toroid and Ring Specimen

Toroid

A toroid can be identified as a long solenoid bent in a circular manner so that its two ends close on to each other. Thus, in a toroid the field is directed *along its internal 'length' or axis,* all along its axial length, the value of the field being given by the same expression as for the (long) solenoid where N is the number of turns arranged in close proximity with each other, I the current through the winding and L now represents the *'mean'*, circumferential length of the device[1].

Ring specimen

In the toroid, the winding may be supported on a non-metallic former as in the case of a solenoid, but, more often, is wound around a   ring (of rectangular or circular cross section) of a magnetic material of which magnetic properties are to be tested. In such cases, the device is known as a "ring specimen" as shown in Fig.12.3 and forms an essential accessory in all DC as well as a few AC tests.

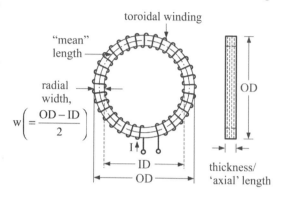

Fig.12.3 : A ring specimen and toroidal winding

The sample in a ring form, making the "core" of the ring specimen, is preferred against a bar or rod as the flux inside the section is *continuous*, free from any self-demagnetising (end) effect inherent to bar specimens. For DC testing, the ring is usually of solid, rectangular cross-section, with the radial width (w) being much smaller compared to the outside diameter (or OD) to ensure 'uniform' magnetisation of the sample and justify the deduction of flux density from the measured value of flux by dividing the latter by the ring's cross-sectional area. A ratio of OD to w of about 15 is considered adequate, with the width, w, being about 2 cm to 3 cm.

Specimen for AC Tests

When used for tests with alternating current, the specimen is formed by stacking annular punchings, *cut randomly* from the (silicon steel) sheets, to achieve the desired thickness or axial width (indicated by dotted lines in the right-side sketch in Fig.12.3). This is necessary to ensure, as much as possible, a uniform magnetic composition of the specimen along the circular length.

---

[1]Clearly, the mean *magnetic* length for the flux path inside the toroid will be different from its mean geometric length [given by $\pi \times \left(\dfrac{D_i + D_o}{2} \right)$: D_i being inside and D_o outside diameter of the ring], depending on saturation of the core, and skin effect in the case of AC excitation.

Preparation

Since some kind of "machining" is unavoidable when preparing the sample in either form, any burrs and deformaties should be carefully removed and samples suitably annealed to achieve good, uniform magnetic properties throughout its volume.

Winding

The core is wound 'uniformly' with large number of turns, placed side-by-side in close proximity, using enamel-insulated copper wire, usually of 20 to 24 SWG. This is known as the "excitation winding" and carries the excitation current from the supply. In most cases, the specimen carries another – or a secondary – winding of fewer number of turns (identified as a "search coil"). This is wound first, followed by the main or excitation winding. The number of turns, N, the mean length, L (in m) and the "safe" excitation current, I, of the excitation winding should be so worked out as to give a value of H $\left(= N\,I/L \right)$ of the order of 2×10^4 A/m, found satisfactory in practice to drive most specimens into saturation.

A disadvantage of making of such specimen is the tedious process of winding large number of turns with hand requiring threading of the wire around the core for each turn. However, toroidal winding machines are now available for the purpose; but it may still be advantageous to hand-wind the toroid to obtain a 'good' specimen.

Search Coils[1]

These constitute an important accessory in a number of magnetic measurement processes. In their simplest form, search coils are essentially one- or multi-turn windings of a variety of design, used mainly for measuring flux density and magentising field in numerous electrical devices and machines. Typical applications include calibration of a ballistic galvanometer (described later) and AC testing of laminated samples in a magnetic "square" for which they form an integral requirement.

Depending on the application, a search coil may comprise

(a) ideally a single-turn "loop" of as small size or area as possible; for example, a 5 mm × 5 mm square coil, wound using thinnest possible wire (up to 49 SWG), positioned with the plane of the coil being at

---

[1]As the name implies, these are special-purpose coils to variously "search" for the presence and measurement of flux in electric/electromagnetic devices, machines and equipment. The "basics" and theory of search coils is discussed in Appendix VII.

right angles to the direction of flux to be measured so as to ensure maximum flux linkages. Then the EMF induced in the coil is simply given by

$$e = -\frac{d\emptyset}{dt}\ [N=1]$$

where $\emptyset$ is the flux linking the coil, the EMF thus being a statically induced voltage;

(b) a winding of more than one turn, from about five to several hundred, to form the secondary winding in some of the measurements; for example, in the ring specimen as discussed earlier, linking the flux produced by a magnetising winding carrying the excitation current.

Several important factors may have to be considered when employing a search coil in a given application/measurement to ensure utmost accuracy and 'correct' interpretation of results. An important aspect of the use of search coil is that it would provide the value(s) of

• H when linking the flux in air, or non-magnetic region, and

• B, directly, when enclosing a magnetic path/part in the device

in terms of the induced EMF, e, in either case, from the expression

$$H \text{ or } B = \int e\,dt$$

Hall Generators, Hall Probes and Gauss Meters

Hall generators can be termed as small devices of miniscule design of historic importance that now form important accessories of many simple and elegant instruments, increasingly used for 'quick' and yet fairly accurate measurement of flux density or rather the magnetising field.

The Hall generator[1] has its genesis in the Hall effect, named after its discoverer E. H. Hall who observed (circa 1879) through extensive experimentation that when a (direct) current flows in a conductor, its surface being perpendicular to a magnetic field, a potential difference would appear across the opposite edges, or surfaces, which are at right angles to both the current and the magnetic field; the phenomenon thus being three dimensional. An elementary depiction of the effect is shown in Fig.12.4.

---

[1]The name implies that the device 'generates' a small voltage proportional to the magnetic field being measured when 'excited' by a current.

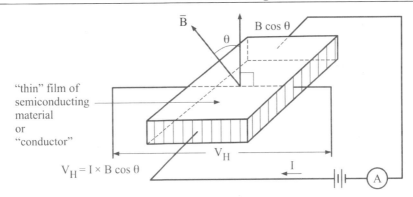

Fig.12.4 : Principle of a Hall generator

A number of metals/conductors such as carbon, copper, iron, bismuth, silicon, tin etc exhibit the property of Hall-voltage generator, but the EMF may be very small even for 'large' excitation current and strong magnetic field. However, materials like germanium and gallium-arsenide (GaAs), now commonly used, can produce reasonable output for currents of a few milliampere and flux densities as low as 1 to $10\,\mu T$.

Hall probe(s)

The actual Hall generator may be very small in size, usually about 3 mm × 3 mm as the active area and about 0.5 mm thick, and is thus quite fragile. It may, however, be constructed to take the form of a long probe, by encapsulating it suitably, and the two pairs of leads fixed to its two mutually perpendicular surfaces – one to 'feed' the excitation current and the other to pick up the induced EMF – and brought out for external connections. The third surface of the generator is meant to be exposed to the *transverse* magnetic field, desired to be measured, as shown in Fig.12.4. A probe commonly designed to measure only the transverse field is known as a "transverse" probe. However, by suitable design, it is possible to have 2-axis, 3-axis or even "axial" probes to measure, respectively, a 2-dimensional, 3-dimensional or axial flux in a magnetic region or machine.

Gauss meter[1]

A gauss meter is an instrument for 'direct' measurement of flux density. The instrument has the provision for connecting a given (Hall) probe of required length (from about 2 m to 10 m). It is provided with a stabilised, well-designed *constant-current* source to 'excite' the generator. The output from the probe (or rather the generator) is fed to an analogue meter, calibrated to read the flux density values on a scale. In some instruments, a A/D converter may be incorporated to read the output in digital form, and for digital processing of

---

[1]See Appendix VIII for full details about Hall generators and gauss meters.

data later. A typical gauss meter may have the appearance as shown on Fig.12.5.

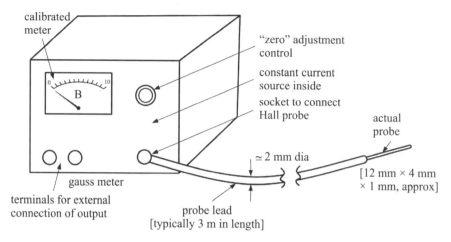

Fg.12.5 : A typical gauss meter with a probe

> ### *Comment*
>
> Because of its 3-dimensional property and operation, a Hall generator may be used as a "multiplier"; for example, to "measure" power in a circuit. In this case, the excitation current may be a small fraction of the load current whilst the flux may be proportional to the voltage in the circuit. The output would then represent the power consumed and may be so calibrated. Such devices, for measurement of small power, free from various usual sources of error, are now commercially available.

Magneto-resistors

These are special-type resistors which are sensitive to change in magnetic field to which they are exposed, other parameters remaining unaltered. Thus, these can be used to measure the magnetic field directly, or those effects related to the measurement of other quantities by means of a magnetic field variation. The usual material for fabricating a magneto-resistor is permalloy, an alloy of 20% iron and 80% nickel.

The principle of operation of a magneto-resistor is similar to that of the Hall generator, except that no excitation is necessary; the magneto-resistor forming part of a circuit. When exposed to a changing magnetic field, its resistance would change (ideally in direct proportion) and may be incorporated; for example, in a Wheatstone bridge to "measure" the field indirectly, or by calibration.

Ballistic Galvanometer

In terms of its importance, an instrument almost invariably used in many magnetic measurements is known as the Ballistic galvanometer[1].

The galvanometer essentially measures the *quantity of electricity* that passes, or discharges, through a coil that forms a part of its moving system. This quantity, in magnetic measurement, is owing to a *suddenly* induced EMF in a circuit, or a search coil, connected to the galvanometer terminals, and in which the flux linking the coil undergoes a sudden change, usually following a "human" action. The quantity of electricity is thus proportional to the EMF and hence the change of flux.

A ballistic galvanometer is similar in construction to the D'Arsonval type, except that it does not show a steady deflection when in use, but gives a "throw" which is proportional to the electricity 'instantaneously' passing through the coil. A schematic of the galvanometer is shown in Fig.12.6.

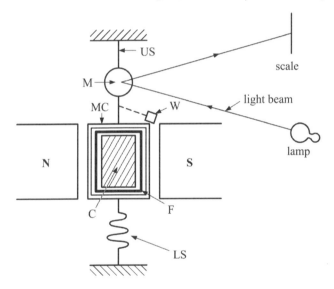

US : upper suspension [flat ribbon of phosphor-bronze]
N, S : permanent magnet [having radial poles]
MC : moving coil (N turns)
C : soft-iron core
M : (concave) mirror
F : non-metallic former
LS : lower suspension
W : additional weight (if required)

Fig.12.6 : Schematic of a ballistic galvanometer

---

[1]The term "ballistic" associated with the name, perhaps, derives from the fact that the first throw of the moving system of this galvanometer, which is crucial for measurement, resembles the shape of trajectory of a ballistic missile, as shown later.

Constructional and other features

In addition to the constructional details shown in Fig.12.6, a ballistic galvanometer is characterised by the following features.

- It has practically no control torque and the moving system is free to *oscillate*.

- Its moving system has a high moment of inertia, sometimes enhanced by adding extra weight, and hence a large time constant which may be 15 to 20 s; the system is thus under-damped, that is, has negligible damping.

- The deflection in the instrument, or strictly the first throw from the rest position, corresponds to the charge that has passed through the moving coil and forms the most important action of the instrument; hence this must be carefully monitored and recorded[1].

- The suspension strip should be carefully chosen and mounted to avoid any "set", that is, a tendency to remain in a position, rather than returning to "normal".

- All the parts, such as terminals, coil and connections within, should be of copper to eliminate errors due to thermo-electric EMFs at the junctions.

- The moving coil should be free from any magnetic material to avoid induced currents and consequent damping produced by them.

The galvanometer is usually provided with a lamp and scale arrangement, at the appropriate distance, and the scale may be graduated to read directly the value(s) of the flux being measured.

Theory: equation of motion

The theory of operations of the galvanometer, leading to equation of motion and expression for deflection(s), is based on the various assumptions as stated above, viz.,

- the quantity of electricity being measured MUST have been discharged through the moving coil in a *very short time*;

- moment of inertia of the moving *system* is sufficiently large so that its movement from the zero position can be considered negligible during the above "very short time";

- following that, the passage of electricity through the instrument provides the moving system energy which is dissipated *gradually* in

---

[1]The discharge of electricity through the coil is assumed to have taken place *before* the commencement of the throw. For this reason, the time elapsed in the induction of EMF and the time taken for the discharge must be absolutely minimum; theoretically zero.

friction and electromagnetic damping; the same being very small to allow convenient measurement of successive deflections[1].

Let

a : moment of inertia of the moving system

b : damping constant ⎤ [note that both these are extremely small,
c : control constant ⎦ but present in practice]

θ : deflection of the moving system of the galvanometer in radians

t : time in second

G : deflection/displacement constant of the galvanometer

[= NBA or NB(ld), where N is the number of turns of the moving coil, B the radial flux density in the gap and A the 'vertical' area of the coil]

Now the deflection torque of the coil when a current i is flowing through it will be

$$T_D = Gi \ Nm$$

Under the above assumptions, during the actual motion, the *deflection torque is zero* and the equation of motion can be written

$$a\frac{d^2\theta}{dt^2} + b\frac{d\theta}{dt} + c\,\theta = 0$$

The 'standard' solution of this equation can be expressed as

$$\theta = A\ \epsilon^{K't} + B\ \epsilon^{K''t}$$

where A and B are constants to be determined and

$$K' = \frac{-b + \sqrt{b^2 - 4ac}}{2a}, \quad K'' = \frac{-b - \sqrt{b^2 - 4ac}}{2a}$$

Since damping, and hence b, is very small

$b^2 << 4ac$, making both K′ and K″ imaginary that can be written

$$K' = -k_1 + j\,k_2 \quad \text{and} \quad K'' = -k_1 - j\,k_2$$

where $\quad k_1 = \dfrac{b}{2a} \quad$ and $\quad k_2 = \dfrac{\left(\sqrt{4ac - b^2}\right)}{2a}$

$\therefore \qquad \theta = A\ \epsilon^{(-k_1 + j k_2)t} + B\ \epsilon^{(-k_1 - j k_2)t}$

$$= \epsilon^{-k_1 t}\left[A\ \epsilon^{j k_2 t} + B\ \epsilon^{-j k_2 t}\right]$$

Using the trigonometric identities

$$\epsilon^{jpx} = \cos px + j\sin px \quad \text{and} \quad \epsilon^{-jpx} = \cos px - j\sin px$$

---

[1] Although, almost invariably, observing/recording deflections after the first one, or the first throw, would hardly be necessary.

$$\theta = \epsilon^{-k_1 t} [A(\cos k_2 t + j \sin k_2 t) + B (\cos k_2 t - j \sin k_2 t)]$$

or $$\theta = \epsilon^{-k_1 t} [(A+B) \cos k_2 t + j (A - B) \sin k_2 t]$$

$$= \epsilon^{-k_1 t} [P \cos k_2 t + Q \sin k_2 t]$$

by writing $A + B = P$ and $j (A - B) = Q$

Then θ can be written

$$\theta = \epsilon^{-k_1 t} [F \sin (k_2 t + \alpha)]$$

where F is another constant, equal to $\sqrt{(P^2 + Q^2)}$ and may be evaluated from the knowledge of the initial conditions of the motion and α is an arbitrary 'constant'.

The above equation of θ is analogous to the 'standard' equation of motion in circular configuration if the factor $k_2 \left[= \left(\dfrac{\sqrt{4ac - b^2}}{2a} \right) \right]$ can be written/identified as an angular frequency, ω, of the moving system, with the 'frequency' of vibration (or oscillation) being

$$f = \frac{\omega}{2\pi} \text{ or } \frac{k_2}{2\pi}$$

Substituting for k_1 and k_2, the expression for θ can finally be written as

$$\theta = \epsilon^{-\left(\frac{b}{2a}\right)t} \times F \times \sin \left(\frac{\sqrt{4ac - b^2}}{2a} t + \alpha \right).$$

The term $\epsilon^{-\left(\frac{b}{2a}\right)t}$ shows that the deflections of the moving system are damped with respect to time, the successive deflections being of decreasing amplitudes as the time from the instant of occurrence of the first throw increases. If the damping is too small or negligible ($b \to 0$), the term $\left(b^2/2a \right)$ can be neglected altogether in comparison with $\left(4ac/2a \right)$ in the sine term and θ expressed by

$$\theta = \epsilon^{-\left(\frac{b}{2a}\right)t} F \sin \left[\left(\sqrt{\frac{c}{a}} \right) t + \alpha \right]$$

Initial Conditions

When $t = 0$, $\theta = 0$, giving $\alpha = 0$[1]

---

[1] This is obvious since the first throw of the galvanometer is assumed to commence at $t = 0$.

Also, if i is the current (in ampere) in the coil at any instant during the discharge of electricity, the deflection torque will be G i and must equal the deflection term in the equation of motion, that is,

$$a \frac{d^2\theta}{dt^2} = G i$$

Integrating with respect to time

$$\int_0^\tau G i \, dt = \int_0^\tau a \frac{d^2\theta}{dt^2} \, dt$$

where τ denotes the time in seconds of discharge of electricity through the coil.

or
$$G \int_0^\tau i \, dt = a \left. \frac{d\theta}{dt} \right|_{t=\tau}$$

Now $\int_0^\tau i \, dt$ = the *quantity* of electricity discharged in time τ

$$= Q \text{ coulomb (say)}$$

Then

$$G Q = a \left. \frac{d\theta}{dt} \right|_{t=\tau}$$

$$\left. \frac{d\theta}{dt} \right|_{t=\tau}^{1} = \frac{G}{a} Q, \ \tau \simeq 0$$

Now from the expression for θ derived earlier

$$\frac{d\theta}{dt} = -\frac{b}{2a} \in^{-\left(\frac{b}{2a}\right)t} \times F \times \sin\left[\left(\sqrt{\frac{c}{a}}\right)t + \alpha\right]$$

$$+ \in^{-\left(\frac{b}{2a}\right)t} \times F\left(\sqrt{\frac{c}{a}}\right)\cos\left[\left(\sqrt{\frac{c}{a}}\right)t + \alpha\right]$$

From this, at t = 0

---

[1]Here, $\left. \frac{d\theta}{dt} \right|_{t=\tau}$ denotes the 'velocity' of the moving system at the end of time τ second, or almost at the beginning of the first deflection since τ is very small compared to the time period of the system.

$$\frac{d\theta}{dt} = -\frac{b}{2a}F\sin\alpha + F\left(\sqrt{\frac{c}{a}}\right)\cos\alpha$$

and since α is also equal to 0,

$$\left.\frac{d\theta}{dt}\right|_{t=0} = F\left(\sqrt{\frac{c}{a}}\right) = \frac{G}{a}Q$$

giving

$$F = \frac{G}{a}\sqrt{\frac{a}{c}}\,Q$$

$$= G\,\frac{1}{\sqrt{ac}}Q$$

And, hence, finally

$$\theta = \frac{G}{a}\times Q \times\ \in^{-\left(\frac{b}{2a}\right)t}\sqrt{\frac{a}{c}}\sin\sqrt{\frac{c}{a}}\,t$$

The deflection of a ballistic galvanometer at any time is thus proportional to Q and the motion is oscillatory.

If $\left(\sqrt{\frac{c}{a}}\right)$ in the sine term can be written as an angular frequency ω (to

express $\sin\left(\sqrt{\frac{c}{a}}\right)t \equiv \sin\omega t$), the frequency of oscillations of the moving

system, f, will be

$$f = \frac{\omega}{2\pi} = \frac{1}{2\pi}\sqrt{\frac{c}{a}}$$

and the time period of the motion will be given by

$$T = \frac{1}{f} = 2\pi\sqrt{\frac{a}{c}}\ s$$

which will be a constant for a given galvanometer.

A typical graph of the oscillatory motion of a ballistic galvanometer is depicted in Fig.12.7, the successive diminishing maxima corresponding to the various times (or angles of deflection).

With $T = 2\pi\sqrt{\frac{a}{c}}$, the oscillations of decreasing amplitudes occur at angles given by

$$\theta_1 = \frac{T}{4},\ \theta_2 = \frac{3T}{4},\ \theta_3 = \frac{5T}{4}$$

and so on.

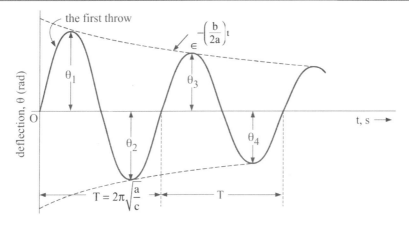

Fig.12.7 : Damped oscillations of a ballistic galvanometer

Substituting these instances in the expression for θ^1,

$$\theta_1 = \frac{G}{a} \times Q \times \in^{-\frac{\pi}{4}\left(\frac{b}{\sqrt{ac}}\right)} \sqrt{\frac{a}{c}}$$

$$\theta_2 = \frac{G}{a} \times Q \times \in^{-\frac{3\pi}{4}\left(\frac{b}{\sqrt{ac}}\right)} \sqrt{\frac{a}{c}}$$

$$\theta_3 = \frac{G}{a} \times Q \times \in^{-\frac{5\pi}{4}\left(\frac{b}{\sqrt{ac}}\right)} \sqrt{\frac{a}{c}}$$

etc.

Thus, $$\theta = K\,Q\,\in^{-\left(\frac{b}{2a}\right)t} \sin\,\omega t$$

where $K = \dfrac{G}{a}\sqrt{\dfrac{a}{c}}$ can be identified as the "galvanometer constant" obtained from the known design values (N, B, A, a and c) of the instrument. Then, by observing deflection θ at given intervals, Q can be deduced, from which e and hence B can be determined.

Calibration of a ballistic galvanometer

In terms of a "direct" application of a ballistic galvanometer to obtain the charge and hence flux and flux density related to the charge, and considering

---

[1]If there were no damping, that is, b = 0, all the deflections would be of constant amplitude, given by $\theta' = \dfrac{G}{a} \times Q \times \sqrt{\dfrac{a}{c}}$.

that it is really the first throw of the moving system corresponding to the angle θ that is a measure of the charge, the equation of the galvanometer operation can be expressed as

$$Q = K\,\theta$$

where K is truly the "constant of the galvanometer."

Thus, if the constant K be known, the charge can be derived by knowing/measuring the deflection θ corresponding to the first throw, being better achieved on the scale (of the lamp-and-scale arrangement) in terms of the movement of the spot of light from its rest position. The experimental determination of the constant K is called the calibration of the ballistic galvanometer and can be carried out in many ways.

- Charged Capacitor Method

In this method, a capacitor (e.g. an electrolytic type) of known capacitance is charged by a battery of known voltage and connected across the galvanometer terminals in series with a current limiting resistor and a key as shown in Fig.12.8.

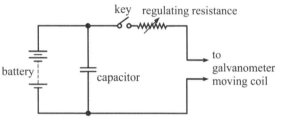

Fig.12.8 : Calibration of a ballistic galvanometer using a charged capacitor

When the key is pressed, the capacitor is discharged through the galvanometer moving coil resulting in a throw and movement of the light spot on the scale. Since the value of charge in this case is known, the value of K can be determined. For example, if $C = 2\ \mu F$ and battery voltage is 50 V, the charge in the capacitor at the instant of closing the key is $Q = CV$ or $100\ \mu C$. If the 'deflection' caused on the scale were 10 cm, the constant K will simply be

$$K = 10\ \mu C/cm \text{ or } 10^{-5} \text{ coulomb/cm.}$$

- Calibration Using an Air-core Solenoid

This method is more accurate, and rather elaborate, being commonly employed for calibration. The main requirement is a long[1], air-core solenoid, wound uniformly on a non-metallic round former, the number of turns and conductor size being such that a magnetising field of the order of 10 kA/m can be produced at the 'centre' (or mid-length) of the solenoid when carrying maximum designed current, but without any appreciable heating.

---

[1]Usually about 1m long, with the "mean" diameter being about 10 cm.

Then, for a long solenoid, it is known that the magnetising field at the centre, being uniform across the axial length, will be

$$H = \frac{N I}{L} \text{ A/m}$$

where   N = no. of turns on the solenoid

I = current in ampere through the winding

and   L = axial length of the solenoid in m.

In the "mid-part" of the solenoid is wound another coil of several (hundred) turns of thin wire, called the secondary winding or the "search" coil, the axial length of the coil being much smaller, say one-tenth of the main winding length. The secondary winding is usually wound beneath the main winding.

The secondary or the search coil is connected to the galvanometer moving coil in series with a current limiting resistor, R_1, whilst the main or excitation winding is connected to a battery of ample capacity via a reversing switch, an ammeter to measure the excitation current and a current-adjusting resistor, R_2, as shown in Fig.12.9.

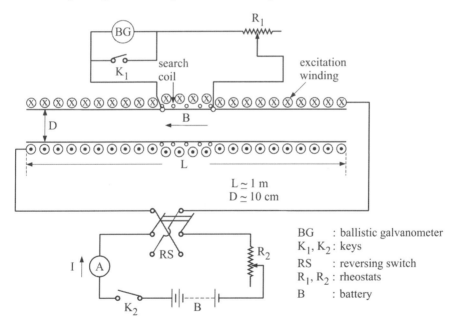

Fig.12.9 : Calibration of a ballistic galvanometer using a long solenoid

Principle of calibration

An EMF is induced in the secondary winding when the excitation current is reversed in the main circuit, producing a throw in the galvanometer. The number of flux linkages in the secondary winding, responsible for the throw can be related to the magnetising field at the centre produced by the (reversing) current and the number of turns and dimensions of the secondary winding, as follows:

The magnetising field in the mid section or the 'centre' of the solenoid

$$H = \frac{N\,I}{L}\ A/m$$

resulting in a (uniform) flux density inside the solenoid given by

$$B = \mu_0 H\ T$$

If D is the inside diameter of the solenoid, also very nearly the diameter of the secondary winding, the flux linking the latter will be

$$\varnothing = \left(\frac{\pi}{4}\right) D^2 \mu_0 H\ Wb$$

resulting in induction of an EMF in the winding when this flux is changed from $+\varnothing$ to $-\varnothing$, or $2\varnothing$ in terms of magnitude, when the current is reversed in the main winding. If the time of reversal, being very small, is Δt s, the induced EMF will be given by

$$e = \frac{2\varnothing}{\Delta t} N_1,$$

N_1 being the number of turns in the secondary winding, producing a current $i = e/R$ in the moving-coil circuit of the galvanometer, where R is the *total* resistance of this circuit (including R_1).

The charge 'passed on' to the galvanometer will be

$$Q = G\,i\,\Delta t$$

$$= \frac{2G}{R} \times \varnothing \times N_1 \qquad \text{by substituting for i}$$

This produces a deflection, θ, whereby

$$\theta = K'Q$$

Knowing both θ (as measured) and Q (as calculated above), K' can be determined, usually in terns of $N_1 \varnothing$, or weber-turns per unit of deflection.

Note that an accurate knowledge of the galvanometer circuit resistance, that of the moving coil and R_1 in series, is essential, to be used in the above calculations. Usually, R_1 is kept as small as possible, mainly for protection of the moving coil so that the "charging" current can be adequately high.

In the above analysis, θ is related to Q and hence the flux linkages by

$$\theta = K' \frac{2G}{R} \times \varnothing \times N_1, \quad \text{giving} \quad K' = \frac{R\,\theta}{2\,G\,\varnothing\,N_1}$$

Now define a constant K'' given by

$$K'' = \frac{1}{K'} \times \frac{R}{2\,G\,N_1}$$

Then for any measured deflection on the scale, θ, the flux as measured by the galvanometer can be obtained as

$$\emptyset = K'' \, \theta \text{ Wb}$$

Comment

A quick check on the method can be made by exciting the main winding by an alternating current of sinusoidal waveform and connecting the search coil to vertical input of a sensitive CRO. The peak-to-peak value of the induced EMF measured on the CRO, preferably by switching off the time base such that the EMF output appears as a vertical trace, should match the calculated value.

Referring to Fig.12.9, the DC supply is replaced by an appropriate AC supply with the ammeter to be capable of reading AC current. The reversing switch is no longer required. Then, if the current is I(rms) A, its maximum value will be given by

$$I_{max} = \sqrt{2}\,I$$

and the magnetising field, flux density and flux linking the search coil will be, respectively,

$$H_{max} = \frac{NI}{L} \times \sqrt{2}$$

$$B_{max} = \mu_0 H = \sqrt{2}\,\mu_0 \frac{NI}{L}$$

and

$$\emptyset_{max} = \left(\frac{\pi}{4}\right) D^2 \sqrt{2}\,\mu_0 \frac{NI}{L}$$

the time variation of the flux being

$$\emptyset = \emptyset_{max} \sin \omega t$$

$$= \left(\frac{\pi}{4}\right) D^2 \sqrt{2}\,\mu_0 \frac{NI}{L} \sin \omega t$$

This flue linking with the secondary winding will produce an EMF

$$e = N_1 \frac{d\emptyset}{dt}$$

$$= N_1 \left(\frac{\pi}{4}\right) D^2 \sqrt{2}\,\mu_0 \frac{NI}{L}\, \omega \cos \omega t$$

with its maximum value being

$$E_{max} = \sqrt{2}\left(\frac{\pi}{4}\right) D^2 \mu_0 \omega\, I \frac{N\,N_1}{L} \text{ V}$$

and the vertical trace on the CRO should correspond to $2 \times E_{max}$, peak-to-peak, according to the vertical scale.

An Alternative

As an alternative to winding a search coil under the main winding as in Fig.12.9, a separate coil of N_1 turns, wound axially may be placed *inside* the solenoid at mid-length so as to ensure linking of *the* uniform field. This affords the simplicity in that the search coil can be wound independently of the main winding and may be re-used, if necessary; however, its positioning exactly at mid-length may require some efforts. The terminals of the coil may then be connected to the ballistic galvanometer and the calibration carried out as before. It is necessary to know accurately the diameter or area of cross-section of the coil. It may be desirable to align the coil such that it is co-axial with the solenoid; for this, the coil any be just sliding type. The advantage is that the wire used for winding can be as fine as possible and even without a former, so that the mean diameter of the coil is the true diameter used in calculations.

- Calibration Using a Permanent Magnet

Permanent magnets in 'standard' design and of "massive" construction (for 'magnetic' stability) are available in various forms, two of which are shown in Fig.12.10. It is essential that flux density in the gap of the magnet is accurately known; for example 1 tesla, to be uniform and being directed appropriately across the gap.

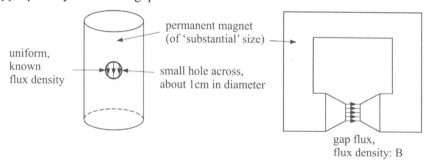

Fig.12.10 : Permanent magnets of known uniform flux density

A 'small' search coil of known number of turns and area of cross-section is inserted inside the gap/hole of the permanent magnet to occupy the central position, its terminals being connected to the ballistic galvanometer to be calibrated. No separate excitation is required. The search coil is rapidly pulled out, thereby *delinking* the original flux, inducing an EMF in the coil and causing a throw in the galvanometer. The electricity discharged in the moving-coil circuit can then be used to obtain deflection constant of the galvanometer as before[1].

---

[1]Clearly, the accuracy of results would depend on the alignment of the search coil plane vis-a -vis the direction of magnetic field inside; the positioning of the coil must be such as to make the plane of the coil to be at right angles with the field.

The advantage of the method is that it does away with any long solenoid and its excitation from a DC source. However, a disadvantage is a change in the value of flux density in the gap/hole with time (the ageing effect) which may go unnoticed.

- Calibration Using Hibbert Magnetic Standard

This device, due to Hibbert, provides a quick and convenient means of calibration of a ballistic galvanometer.

The standard consists of a round magnet of a suitable diameter, arranged vertically (the polarity at the top being north or south), fitted inside a circular soft-iron yoke with a narrow annular gap between them.

The device incorporates a brass tube, carrying a (search) coil of given number of turns *wound at one end*. The tube is designed to just slide vertically through the gap. The ends of the search coil are brought out on top of the tube for external connections. The gap flux density and length of the gap must be known for calibration. The "cross-sectional" schematic of the standard is shown in Fig.12.11[1].

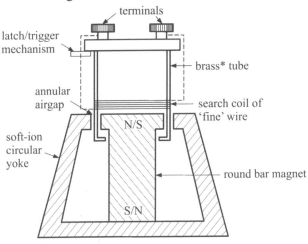

*could also be aluminium

Fig.12.11 : The schematic of the Hibbert magnetic standard

Operation
- The brass tube is raised to its specified top position and held there with the help of a trigger mechanism such that the coil is outside the airgap as shown in the figure.
- The terminals of the coil are connected as usual to the ballistic galvanometer being calibrated.

---

[1]The device is somewhat similar in design and operation to that just described, using permanent magnets with known airgap flux density, but more practical and accurate.

- The trigger mechanism is then operated so that the tube falls down under gravity, sliding through the airgap.
- In the process, the coil cuts through the magnetic field of the magnet, inducing an EMF in the coil and producing a charge which is passed over to the galvanometer.
- The throw of the galvanometer is observed and, knowing the design parameters of the device, the galvanometer constant can be determined as follows:

Let N be the number of turns of the coil on the brass tube

B the average, *radial* flux density in the airgap in tesla

A the 'area' of the gap in sq. metre

Then the flux linking the coil is

$$\emptyset = B\,A \ \text{Wb}$$

and the flux linkages producing the EMF and the charge are $N\emptyset$.

If the deflection of the galvanometer is θ, then

$$\theta = K N \emptyset$$

Knowing θ and $N\emptyset$ (as above), K can be determined.

The time of flux cutting in the gap, which is constant every time the tube drops, does not figure in the operation, or the result. For the desired accuracy, the exact knowledge and constancy of B is essential; that is, the flux density must not vary with time.

Flux Meter

An important variation of a ballistic galvanometer, commonly used for flux measurements, is called the flux meter, akin to the various indicating instruments. Although not as accurate and sensitive as a 'properly' calibrated ballistic galvanometer, it possesses the following advantages:

(a) The instrument is compact in construction, portable and designed with a scale calibrated to read directly flux-turns, or weber-turns; alternatively, the flux meter may be designed to operate in conjunction with a lamp-and-scale arrangement for enhanced accuracy.

(b) The operation and indication of a flux meter is independent of the time for the flux-change (although it should be limited to about 30 s) – an important advantage with highly inductive circuits where the flux-change may be relatively slow.

Constructional details

A typical flux meter is designed to have a very small controlling torque and relatively heavy damping, usually of the electromagnetic type. The instrument comprises a coil of small cross-section, similar to that in a D'Arsonval galvanometer, suspended from a spring support by means of a single thread of silk which also act as support for the 'current' lead and to provide only a very small control torque, there being no hair/control springs. The coil is wound without a former so as to have negligible inertia. The coil hangs vertically with its parallel sides inside narrow airgap of a permanent magnet system and is free to rotate as shown in Fig.12.12.

The instrument is usually fitted with a pointer attached to the moving system and a scale graduated in terms of flux-turns (or weber-turns); alternatively, the moving system may be fitted with a small mirror to read the deflection on a metre scale using reflected light from a lamp.

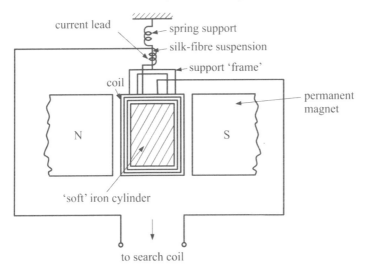

Fig.12.12 : Constructional features of a flux meter

Operation

When the terminals of the flux meter are connected to a search coil, of a given design and number of turns, and the flux linking the search coil is changed – for example, by reversing the magnetising field or quickly removing the coil from a known magnetic field – the instrument coil is deflected through an angle θ which is a measure of the change of flux turns. The instrument coil moves during the whole period of this flux change, but comes to rest as the flux change ceases owing to high damping, achieved by keeping the resistance of the search coil circuit as low as possible (say, a few ohms). Thus, since the deflection is caused by actual discharge of electricity from the search coil, the time of discharge is not critical as in the case of a ballistic galvanometer. Having measured flux-turns, that is $\emptyset N$, on the instrument and knowing N, the number of turns in the search coil, $\emptyset$ can be derived.

Theory of deflection

In the flux meter, the control torque is negligible and can be assumed to be zero for the analysis, and the *air damping* is also negligibly small. The electromagnetic damping is accounted for and included in the equation for current that flows during the entire deflection, through the moving-coil circuit comprising the search coil and any series resistance. Hence, the equation of motion of the moving system can be written

$$J\frac{d^2\theta}{dt^2} = G\,i$$

where J is the moment of inertia of the moving system, G the galvanometer *design* constant $(= BAN)$ and i the coil current at any instant.

Let

R and L	:	the total resistance and inductance, respectively, of the circuit comprising the instrument and the search coil
Ø	:	the flux linking the search coil
N	:	the number of turns on the search coil

EMFs in the Coil Circuit

These can be described as

(a)   EMF due to change of flux linking the search coil

$$e_1 = N\frac{d\text{Ø}}{dt}$$

(b)   motional EMF owing to the movement of the *instrument* coil in its field

$$e_2 = K\frac{d\theta}{dt} \quad (\text{K a constant}), \text{corresponding to a deflection}, \theta$$

(c)   EMF due to presence of inductance in the circuit

$$e_3 = L\frac{di}{dt}$$

The last two EMFs, acting as 'back' EMFs, will actually oppose the search coil EMF and also represent the electromagnetic damping.

The coil-circuit current at any instant will thus be

$$i = \frac{N\left(\dfrac{d\text{Ø}}{dt}\right) - K\left(\dfrac{d\theta}{dt}\right) - L\left(\dfrac{di}{dt}\right)}{R},$$

assuming the effect of circuit reactance to be negligibly small.

Substituting in the equation of motion

$$J\,R\frac{d^2\theta}{dt^2} = G\left[N\frac{d\varnothing}{dt} - K\frac{d\theta}{dt} - L\frac{di}{dt}\right]$$

or

$$N\frac{d\varnothing}{dt} = \frac{J\,R}{G}\frac{d^2\theta}{dt^2} + K\frac{d\theta}{dt} + L\frac{di}{dt}$$

$$= \frac{J\,R}{G}\frac{d\omega}{dt} + K\frac{d\theta}{dt} + L\frac{di}{dt}$$

where $\omega\left(=\dfrac{d\theta}{dt}\right)$ is the *angular* velocity of the instrument at any instant.

Now if T is the time for the complete flux change, the above equation can be written

$$\int_0^T N\frac{d\varnothing}{dt}\,dt = \int_0^T \frac{J\,R}{G}\frac{d\omega}{dt}\,dt + \int_0^T K\frac{d\theta}{dt}\,dt + \int_0^T L\frac{di}{dt}\,dt$$

or

$$N\left(\varnothing_2 - \varnothing_1\right) = \frac{J\,R}{G}\left(\omega_2 - \omega_1\right) + K\left(\theta_2 - \theta_1\right) + L\left(i_2 - i_1\right)$$

where the suffixes 1 and 2 indicate, respectively, the values of the various quantities at the beginning and end of the flux change.

Since the angular velocity and the current are zero at both the beginning and end of the change (coil being at rest at the two positions and current about to flow and off), that is, $\omega_2 = \omega_1 = 0$ and $i_2 = i_1 = 0$, the above expression simplifies to

$$N\left(\varnothing_2 - \varnothing_1\right) = K\left(\theta_2 - \theta_1\right)$$

that is, the (change of) deflection $(\theta_2 - \theta_1)$ is proportional to the change of flux-turns $\left(\varnothing_2 - \varnothing_1\right)$ associated with the search coil. The instrument, therefore, has a uniform scale to read flux-turns. Also, the motional EMF in the instrument coil will depend on the magnetic field, B, in the gap, its number of turns and (width of) its coil sides. The constant K is thus the same as G, the so-called "galvanometer constant".

Therefore, finally,

$$N\left(\Delta\varnothing\right) = G\,\theta, \quad \theta \text{ being } \Delta\theta \text{ or } (\theta_2 - \theta_1);$$
$$\theta_1 = 0, \text{ deflection starting from rest,}$$

or

$$\Delta\varnothing = \frac{G}{N}\theta^1$$

---

[1]In the flux meter, the quantity (G/N) is built into it whilst the instrument is calibrated using a "standard" search coil of given leads length, provided invariably with the instrument.

Magnetic Potentiometer

This is a simple magnetic device that can be used in an easy manner to measure *magnetic* potential difference across any two parts in a magnetic circuit. As shown in Fig.12.13, a typical such device consists of a "long", thin strip of flexible non-magnetic, non-metallic material (for example, even PVC) wound uniformly and uni-directionally with a large number of turns using thin wire, from end to end. The flexibility allows it to be positioned, bending suitably if required, in 'firm' contact at any two points, A and B, distant L metre apart, on the surface of a part of magnetic circuit as shown, it being assumed that the width, w, of the strip is less than the width of the magnetised surface.

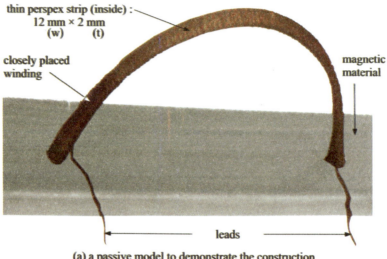

(a) a passive model to demonstrate the construction

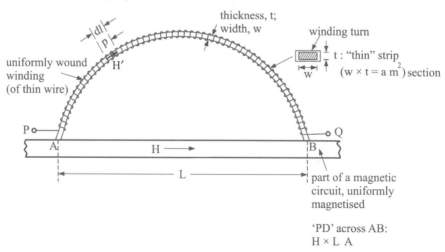

'PD' across AB:
H × L A

(b) schematic to explain operation

Fig.12.13 : Details of a magnetic potentiometer

The ends P and Q of the potentiometer winding are connected to a ballistic galvanometer in the usual manner, whilst the magnetising field, H, in the magnetic circuit, or the specimen, can be reversed by reversing the magnetising current. Reversal of the magnetisation will cause a deflection of the galvanometer which can be related to the magnetising field H in the specimen.

Theory of operation

Referring to the figure, let H' be the magnetising field in the strip at a 'point' p, directed along the 'length' of the strip as shown. The corresponding value of the flux density will be $B = \mu_0 H'$, and if the area of cross section of the strip is a m^2 and the number of turns per meter is n, the flux linkages with the elementary length dl will be

$$d\psi = \mu_0 H' \times a \times n \times dl \quad \text{Wb-turns}$$

Integrating over the entire length of the potentiometer, the total flux linkages will be

$$\psi = \mu_0 \times a \times n \times \int H' \, dl$$

But $\int H' dl$ is the magnetic potential across points A, B of the strip,

or
$$\int H' dl = \frac{\psi}{\mu_0 \, a \, n} \quad \text{A/m}$$

Since, on reversal of magnetisation in the magnetic circuit, ψ can be measured by the ballistic galvanometer, $\int H' dl$ can be determined. Now the magnetic potential difference in the *specimen* is given by HL and since the points A and B are common with, or parallel to, the specimen and the potentiometer, $\int H' dl$ must be proportional to HL. The potentiometer may be calibrated by simply forming it into a *closed loop*, linked with a coil of N turns carrying a current I, Then

$$\oint H' dl = NI$$

and the galvanometer deflection corresponding to $2 \times NI$ can be obtained by reversing the current I. From this, $\oint H' dl$, and hence $\int H' dl$ can be deduced when the length of the strip is known, providing the magnetic potential across AB in the magnetic circuit.

Permeameters

These are specially designed and constructed magnetic devices, frequently used for DC magnetic testing, mainly to obtain the B-H characteristics of specimens in *rod* or *bar* forms when it is difficult to obtain these in ring form, or to wind them to prepare a ring specimen of the usual type. The specimen/sample may also consist of a bundle of tightly-packed strips of sheet magnetic material, care being taken that the strips are cut from different directions in the sheet so that they represent a nearly isotropic sample. The rods and bars cannot normally be tested by simple methods as it is difficult to ensure their uniform magnetisation owing to the demagnetising effect at the 'poles' (formed at their 'ends') even if the specimen length is considerable.

A permeameter affords to provide near uniform magnetisation of a good proportion of the specimen length by 'completing' the magnetic circuit. For this, the rod specimen is clamped in a relatively massive yoke of laminated ferro-magnetic material, provided with an excitation or magnetising winding and an arrangement for reversal of the magnetising current. A variety of permeameters are commercially available. Of these, a simple one due to Fahy is shown in Fig.12.14[1].

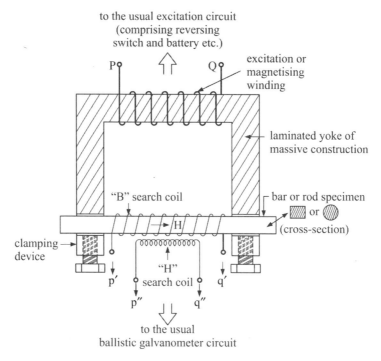

Fig.12.14 : Fahy-Simplex permeameter

---

[1]Later, a device or special permeameterr to measure B-H characteristics of a *permanent*-magnet material in strip form, or an *annulus*, using a calibrated Hall probe is also described.

As shown, the permeameter consists of a massive laminated yoke, carrying a magetising winding. The other end of the yoke is designed to receive a bar specimen of appropriate cross-section that can be clamped to the yoke to form as close and continuous magnetic circuit as possible. Along the length of the specimen, provision is made of a pair of coils, known as "H" and "B" coils to measure the magnetising field H and flux density B, respectively, *inside* the specimen based on the principle of operation described below.

The "H" and "B" search coils

"H" Search Coil

This comprises a coil of fairly large number of turns, wound on a non-magnetic, non-metallic former of small cross-section and is placed *externally* in close contact with the specimen as shown [see Fig.12.14]. It measures the magnetising field *inside* the specimen, H, on the basis of the "law of continuity" of *tangential* component of magnetic field across an air-iron boundary. Since the specimen is assumed to be 'uniformly' magnestised and the "H" coil is in close proximity with the specimen surface, the magnetising field linking the search coil is the same as that inside the specimen.

Then, if the coil has N_1 turns and a cross-section of a_1 m^2, its flux linkages corresponding to a given magnetising current will be

$$\psi_1 = \mu_0 H a_1 N_1 \text{ Wb-turns}$$

which can be measured with the help of a ballistic galvanometer, and H deduced therefrom.

"B" Search Coil

This is the coil which is wound/provided in the permeameter so as to *enclose* the specimen as shown in the figure. Thus, the coil links with the flux *inside* the specimen, given by

$$\emptyset = \mu_0 \mu_r a_2 H \text{ Wb}$$

where a_2 is the area of cross-section of the coil (assumed to be same as that of the specimen) and μ_r represents the relative permeability of the specimen material corresponding to the given magnetising current[1].

The corresponding flux-linkages of this coil will be

$$\psi_2 = \mu_0 \mu_r a_2 H n_2 \text{ Wb-turns}$$

where ψ_2 is the number of turns in the coil. Again, ψ_2 can be measured on the ballistic galvanometer from which $\emptyset$ and hence B can be determined, given by $B = \emptyset/a_2$.

---

[1]That is, a point represented on the B-H curve of the specimen.

The measurement process

The test is performed by setting the magnetising current at the chosen value and connecting the "H" search coil to the calibrated ballistic galvanometer. On reversing the current, the flux linkages and hence the value of H can be determined from the expression of flux linkages as discussed. Next, *without altering the main winding current*, the galvanometer is connected to the "B" search coil and flux linkages measured as before by reversing the current. This would provide the value of B inside the specimen corresponding to H derived above.

The test is then repeated by adjusting different values of the magnetising current to obtain corresponding values of H and B. Although a convenient means for testing specimens in solid rod or bar forms, a source of error in the method can be the assumption of the magnetising field inside the specimen being the same as that in the "H" coil and a correction may have to be applied for better accuracy.

Magnetic Squares

These magnetic devices, shaped as a "square" have been ingeniously developed as a substitute for ring specimens to get over the 'problems' associated with the latter, viz., obtaining the sample in the form of a ring and tedious process of winding a large number of turns required for testing.

Magnetic squares, a variety of which are available, have been developed mainly to test magnetic materials in *strips*, or *laminations,* form assembled to yield a magnetic square. The devices have been particularly used for testing of (magnetic) sheet materials[1] with alternating current, one important requirement being to obtain experimentally the iron-loss density of the material. Although mainly employed for AC testing, there is no reason why DC tests cannot be performed in terms of the design, construction and principle of operation.

The essential requirement of a magnetic square is the provision of FOUR fixed magnetising windings and search coils wound uniformly on a hollow non-magnetic former (for example, made of wooden batons of given width and thickness), making a side each of the square.

The strips are slipped into the hollow, inner part of the winding sets and clamped when tightly packed. The ends of the strips are joined together to form a nearly continuous magnetic 'corner' with minimum "airgap effect". In one type, the plane of each strip is in the plane of the square with simple interleaved joints (as in a transformer core construction) as shown in Fig.12.15(a). In the other type, more commonly used arrangement, the plane of each strip is perpendicular to the plane of the square. The corner joints are

---

[1]Extensively employed in the manufacture of nearly all electrical equipment like generators, transformers, motors etc.

made using a number of right-angled corner pieces, bent into form, as shown in Fig.12.15(b).

Lloyd-Fisher square

A frequently used magnetic square, employing latter form of corner construction is the Lloyd-Fisher square, the schematic details of which are shown in Fig.12.16.

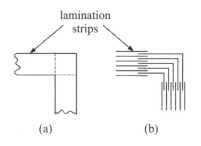

Fig.12.15 : Corners in magnetic squares

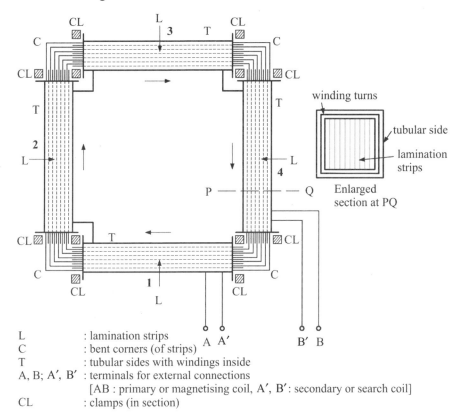

L	: lamination strips
C	: bent corners (of strips)
T	: tubular sides with windings inside
A, B; A′, B′	: terminals for external connections
	[AB : primary or magnetising coil, A′, B′: secondary or search coil]
CL	: clamps (in section)

Fig.12.16 : Lloyd-fisher magnetic square

The square consists of four identical coils − 1, 2, 3 and 4 − of inner, square-shaped 'tubular,' hollow section to form the four sides of the square. Each coil is wound with a given number of primary, or magnetising, winding and a secondary winding, or search coil of large number of turns beneath the primary, contained in the tubular sides and protected by a covering. All the windings are connected in series and final terminals are brought out, at A, B and A′, B′ for external connections.

The insides of the coils are packed with strips of the sample to be tested, each strip being nearly 25 cm long and about 5 or 6 cm in width. The strips are cut half in the direction of rolling of the sheet material during manufacture and half perpendicular to this direction so as to make the overall assembly as homogeneous or isotropic as possible. Each coil (side) contains the *same* number of laminations.

The corners of the square are made of bent pieces of the same material and the strips as in Fig.12.15(b) with corner pieces clamped tightly at four ends as shown (a very small number of strips are shown for the sake of clarity). It is a common practice to weigh one set of strips before assembly to determine its area of cross-section from the knowledge of its length and density[1].

Hysteresigraph

As implied by the name, this is a special apparatus designed to perform magnetic measurements on permanent magnet materials, or rather their samples, in *solid* form to obtain the B-H characteristic and, more importantly, the hysteresis or B-H loops of the material by simple control of the magnetising field in the forward and reverse directions. With easy manipulation, even the recoil loops, both major and minor, can be obtained whilst the apparatus is in operation. The tests are performed with direct current. The samples or specimens of the material are usually of round cross-section, about 15 to 20 mm in diameter, or square having sides about 15 mm each. The 'length' can be typically 12 to 15 mm.

A typical hysteresigraph

It essentially consists of

(a)   A "massive" electromagnet, excited by direct current passing through a winding of large number of turns that can carry sufficiently high current; the poles of the electromagnet are of special conical shape, about 8 to 10 cm in diameter at the lower end. A built-in device allows them to move up and down in a vertical direction to vary the gap between them to receive the sample[2].

(b)   A control mechanism to vary the excitation current from zero to a very large value so as to produce a magnetising field of up to 300 to 500 kA/m and drive the sample material into complete saturation, or demagnetise it if required, and to reverse the direction of current to apply magnetising field in the opposite direction.

---

[1]Application of the Lloyd-Fisher square for AC testing is described later in the chapter.

[2]Mostly alloys of nickel, iron and cobalt such as ALNICO or ferrites; and now more modern alloys containing rare earth metals.

(c)   Devices or means to measure magnetising field and corresponding flux density in the sample. The former is usually measured by means of a calibrated Hall probe, positioned appropriately in the vicinity of the sample, whilst the latter is invariably measured with the help of specially designed "B coils", embedded in one of the poles (usually the lower pole), or by winding a search coil around the sample.

(d)   An X-Y plotter or recorder to which are fed the outputs from the Hall probe and "B-coil" to the X and Y terminal pairs, respectively, to plot the B-H curves and loops as the excitation current is varied. The schematic diagram of a hysteresigraph is given in Fig.12.17 whilst the details of an actual device and results obtained from a sample are described later.

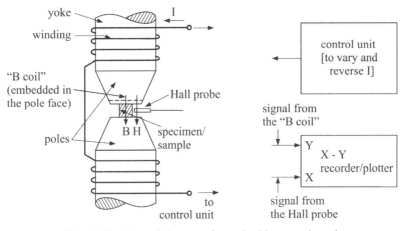

Fig.12.17 : Essential parts of a typical hysteresigraph

Comment

In addition, there are a few more accessories and apparatuses such as

- AC bridges
- AC potentiometer
- Cathode ray oscilloscope

which are employed for a variety of magnetic measurements. The design and operational aspects of these have already been discussed variously; their actual applications as appropriate are discussed later in the chapter.

MAGNETIC MEASUREMENTS WITH DIRECT CURRENT

Using Ring Specimens

For magnetic measurement using direct current to obtain B-H characteristics, a specimen consisting of a ring of the material under test with a toroidal magnetising winding and a B- or search-coil beneath it yields results of the best accuracy. It is assumed that both the windings, the former in more than one layer if required, are *uniformly* wound with the adjacent turns to be closely placed. With this ensured, and provided that the annular width is not greater than about 1/8th of the mean diameter[1], both the magnetising field (H) and flux density (B) in the specimen will be sufficiently uniform across the cross-section for the 'mean' value of B in the section to be taken as corresponding to the value of H at the mean radius along the circumference. The magnetising field H is expressed in terms of the number of turns of the magnetising winding, the current through it and the length along the mean circumference as discussed earlier.

Core of the Specimen

The core of a ring specimen may take one of the following forms:

Laminated

This is the usual form for isotropic sheet materials, each lamination having been punched from the sheet. The ring should be assembled with the rolling directions of the individual lamination arranged one above the other radially at random so that the test may yield information of the 'mean' properties of the material.

Clock-spring

A long strip of material may be wound spirally to form a ring, similar to a tightly wound clock spring, except that the previously specified ratio of radial thickness to mean diameter must be adhered to. This form is particularly suited for tests on anisotropic materials; for example, where the magnetic properties in one particular direction, say in the direction of rolling are required.

Solid

In this form, the specimen may be cast, forged or machined (from a cylinder of the given material of large diameter; for example, mild steel), forming the desired ring. The edges may be carefully smothered, to be free from burrs, and the ring annealed prior to winding to de-stress it mechanically and magnetically.

The three forms are illustrated in Fig.12.18.

---

[1]In most cases the diameter may vary from about 15 cm to 20 cm whilst the width is about 2 to 2.5 cm; the thickness (or the 'axial length') being 1.5 cm to 2 cm.

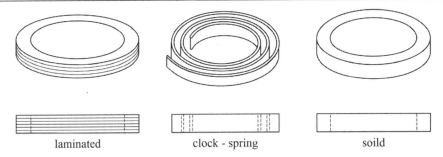

| laminated | clock - spring | soild |

Fig.12.18 : Various forms of ring-specimen cores

Preparation of the Specimen

Before winding, the dimensions of the ring must be determined. When the ring consists of laminated material, it may be helpful to determine the *effective* cross-section from the weight of the ring and using the appropriate value of density or specific gravity (usually 7.8 gm/cc for steel); this is necessary to eliminate inaccuracy on account of the "air space" trapped between individual laminations.

A layer of thin insulation tape is then wound on the ring and a search coil comprising 'sufficient' number of turns of thin wire, well insulated, is wound over the tape, the number of search-coil turns depending on the sensitivity of the ballistic galvanometer, and these must, of course, be noted. The search coil is covered with another layer of (protective) tape over which the magnetising winding is uniformly wound using (double) enamelled or cotton-covered copper wire of appropriate gauge. This enables the "B" or search 'coil' to be close to the magnetic sample and very nearly of the same cross-section.

Determination of magnetisation or B-H curve using ballistic galvanometer

The knowledge of this characteristic, that is, the B-H curve forms an important, basic requirement to help the designers of electric machines and equipment. However, the relationship between magnetising force H and resulting flux density B in a ferro-magnetic material can be highly non-linear in general when the 'operating' point is beyond the initial linear part or region; though the determination of this part being desirable for reasons of economy of the design. This necessiates a good number of points on the curve to be obtained experimentally.

Two methods are commonly employed:

A. Method of Reversals

This is the most accurate and frequently used method, the circuit diagram of which is shown in Fig.12.19.

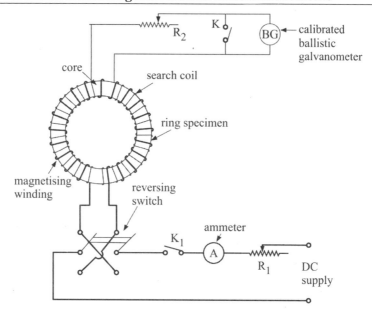

Fig.12.19 : Circuit for B-H measurement using the method of reversals

The essence of this method, and the key behind its high accuracy, is that the points on the B-H curve are obtained as the locus of tips of the hysteresis loops for various increasing values of the magnetising flux densities in the specimen, as shown in Fig.12.20. This process ensures stable magnetic state and repeatability of the sets of H and B points on the curve. For a "soft" magnetic material, the loops may not be as 'fat' as shown in the figure.

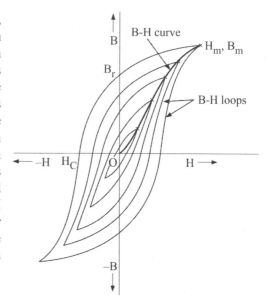

Fig.12.20 : B-H curve as the locus of tips of hysteresis loops

Demagnetisation

Before commencing the actual test for the B-H curve, it is essential to demagnetise the sample to eliminate any residual magnetism that might be present from the previous tests or for other reasons; for example, owing to an accidental exposure to a strong, stray magnetic field of a conductor carrying large current.

For demagnetisation, and referring to Fig.12.19, whilst keeping the galvanometer key K closed, the magnetising winding is given such a value of current so that the magnetising force H acting on the specimen is much in excess to the maximum value to be used in the test. The magnetising current is then *gradually* reduced whilst at the same time throwing the reversing switch forward and backward a few times for each setting in order to let the specimen pass through as many cycles (or loops) of magnetisation as possible during this process. As the current finally reaches 'zero' value, or the minimum possible, the specimen would (deem to) have been demagnetised.

Process of measurement

To begin the test after demagnetisation, the magnetising current is set at the lowest value to provide a small magnetising force, say 100 A/m. With the galvanometer key K closed, the reversing switch is thrown forward and backward some twenty or more times to bring the specimen into what is called "reproducible cyclic magnetic state", or the 'unique' H,B state. The key K is next opened and the flux in the specimen corresponding to the value of H set above is measured in terms of the throw of the galvanometer (assumed to have been calibrated previously) when the reversing switch is quickly reversed. Whilst the value of H is deduced from the expression $H = (NI)/L$ (where N is the number of turns on the magnetising winding, I the current setting as above and L the mean circumferential length of the specimen), the value of corresponding B is obtained by dividing the measured flux by the already known area of cross-section.

The process is repeated for various successively increasing values of magnetising current in small steps to a sufficiently high value so as to drive the specimen well into saturation[1]. The B-H curve may then be plotted from the measured values of B corresponding to the various computed values of H.

B. B-H Curve by "Step-by-step" Method

In this, rather simplified method, the part of the circuit comprising the ring specimen and ballistic galvanometer remains the same. However, the reversing switch is replaced by a special potential divider as shown in Fig.12.21. An advantage of the method, although it may not be as accurate as the method of reversals, is that it does away with the reversal of magnetising current or field; this is particularly helpful when a heavy current is used to effect very high value of magnetising force, the reversal of which by means

---

[1]It is important, of course, that for each new value of H, the specimen is brought to the reproducible cyclic magnetic state by moving the reversing switch forward and backward some twenty times or more as described previously, before making the measurement of flux.

of reversing switch may pose problem of sparking etc. at the contacts. The various values of H are still obtained from the magnetising current and the specimen design constants as before, but the values of B are measured as *increments* from the previous to the next value. Assuming that the sample is completely demagnetised, the various steps of measurement are

- Keep switch S_1 open.

  Then $I_{exctn} = 0$ and H = 0.

- Move selector switch S_2 to step 1 of the potential divider. This corresponds to the setting of a small[1] magnetising force H_1.

- Now close the switch S_1 swiftly, applying the magnetising force H suddenly to the magnetising winding and thus causing a throw in the ballistic galvanometer. From this, derive the increment of B, ΔB_1.

- Move S_2 quickly to step 2, thereby increasing H to $H_2 = H_1 + \Delta H$. The corresponding throw in the BG will provide B_2,

  equal to $B_1 + \Delta B$.

- Move S_2 in quick succession to steps 3, 4, . . . , to effect sudden changes in H, increasing it in magnitude to H_3, H_4, . . . , H_m, and corresponding flux densities to B_3, B_4, . . . , B_m.

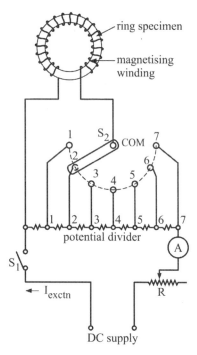

Fig.12.21 : Circuit for step-by-step method

- These values of H and B are then plotted to give the desired B-H curve of the sample as depicted in Fig.12.22 by the curve 0, 1, 2, 3, 4, 5, 6, 7 (H_m, B_m), showing the increaments of H (ΔH_1, . . . , ΔH_6) and that of B (ΔB_1, . . . , ΔB_6).

Determination of hysteresis or B-H loop

A. Step-by-step Method

The determination of hysteresis loop of the sample by this method is carried out using the same test set-up and in a similar manner as for the B-H curve.

---

[1]There is no rigid basis of designing or deciding the various steps (or taps) on the potential divider. The steps could be chosen to effect 'equal' increments of H, or in an arbitrary manner, to effect approximately equal increments of B.

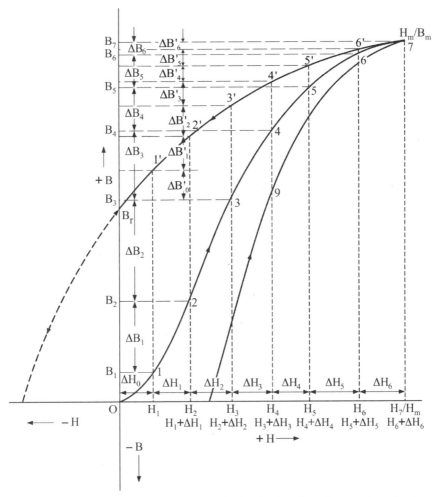

Fig.12.22 : B-H characteristic and loop obtained using "step-by-step" method

The steps to be followed are

- First obtain the B-H curve as described above.
  [If necessary, the test could be repeated if the results in the first instance are not satisfactory][1].
- After attaining the H_m/B_m point, start reducing H by moving S_2 *backward* from its last position (for example mark "7" in Fig.12.21) till H = 0.
- Negative values of H can then be obtained by reversing the switch in the main circuit (added for the purpose, for this test) and increasing H in steps by moving switch S_2 from position 1 to 7 as in the case of the B-H curve.

---

[1] It is, of course, necessary to demagnetise the sample as described previously before repeating the test for the B-H curve.

- Once the point $-H_m$, $-B_m$ is reached, the right hand part of the loop can be constructed (by symmetry) as reversed mirror image of the left-hand part.

Observe that whilst obtaining the points for the B-H *loop*, the *decrements* of B, corresponding to *same* decrements of H, that is, $(-)\Delta H_6,\dots, (-)\Delta H_1$, $\Delta H_0,\dots$, will NOT be same as $\Delta B_6,\dots, \Delta B_1$ (used for the B-H curve) owing to hysteresis property of the sample/material, but $\Delta B_6',\dots, \Delta B_0'$ as shown to yield the portion of the loop marked by 7 (H_m, B_m), 6′, 5′ etc., up to B_r. Similar reasoning applies to the portion of the loop in the II and III quadrants, corresponding to negative values of H.

B.   Method of Reversals

This method is considered superior to the step-by-step method as it yields more accurate results and can even provide recoil loops for the sample. The test circuit and the B-H loop obtained from it are shown in Fig.12.23.

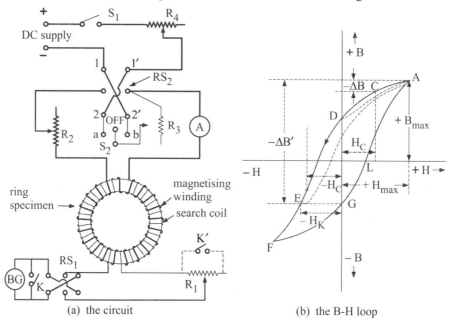

(a) the circuit   (b) the B-H loop

Fig.12.23 : Plotting B-H loop using the method of reversals

The circuit

R_1	:	resistance to control current in the galvanometer circuit, to be shunted by key K' for maximum sensitivity
R_2, R_4	:	resistances to control excitation current in the magnetising winding
R_3	:	a variable resistance to shunt the magnetising winding as required

RS_1 : reversing switch in the galvanometer circuit

RS_2 : reversing switch in the main circuit

S_2 : a special 3-way switch, designed to move to positions a and b and OFF as shown

The procedure

The ring specimen, constructed in the usual manner, is first demagnetised thoroughly as described.

The test consists of the following steps:

- Set the switch S_2 to OFF position and switch RS_2 in $1-1'$ position.
- Adjust resistors R_2 and R_4 such that the current in the magnetising winding corresponds to maximum value of magnetising force, H_m, (to be) encountered during the test[1]. 'Open' switch RS_2.
- Adjust galvanometer circuit resistance R_1 such that a convenient deflection is obtained when I_m is reversed.
- The resistor R_3 is adjusted such that as it shunts the magnetising winding, a 'suitable' *reduction* of current in the magnetising winding is obtained when R_3 is introduced in the circuit; this being one of many settings during the test as seen later.
- Open galvanometer key K and quickly throw switch RS_2 to position $1-1'$ (from its 'OFF' position). This results in the current I_m to flow in the magnetising winding and the galvanometer throw provides the value of flux density B_m, corresponding to point A (H_m, B_m) on the hysteresis loop.
- Next throw switch S_2 quickly over to contact "b" from OFF position, thereby shunting the magnetising winding (corresponding to a given setting of R_3) and reducing the magnetising force to H_C (say). [See Fig.12.23(b)]. The corresponding reduction in flux density, $-\Delta B$, is obtained from the galvanometer throw. This provides point C on the B-H loop.
- The key K is now closed (to 'discharge' the ballistic galvanometer) and switch RS_2 reversed on to contacts $2-2'$.
- Switch S_2 is moved back to OFF position and RS_2 thrown back to contacts $1-1'$. This step passes the specimen through the cycle of mangnetisation and back to point A on the loop[2].

---

[1]Once again, derived from the expression $H_m = \dfrac{(NI_m)}{L}$ with the usual notation.

[2]The reversing of RS_2 to contacts $2-2'$, and then to $1-1'$, may be avoided by simply moving to $1-1'$ whilst S_2 is in OFF position. This 'moves' point C back to A through a minor recoil loop as shown (dotted) in Fig.12.23(b).

- Repeat the above steps by successively adjusting R_3 (actually increasing to effect corresponding reduction(s) of H) to obtain various points along section AD of the loop.

To obtain section DEF

- Close key K, move S_2 to the OFF position and throw RS_2 on to contacts $1-1'$. This brings the operation again to paint A of the loop, traversing the recoil loop from D to A.

- Move S_2 to contact 'a'. This would introduce (a given value of) R_3 again in the circuit, shunting the magnetising winding and reducing excitation current from I_m, *but in the reverse direction*, effecting a reduction of magnetising force from H_m to $-H_K$ (say), depending on the value of R_3.

- Open key K and reverse RS_2 to contacts $2-2'$, observing the corresponding throw in the galvanometer. This operation corresponds to measuring a change in flux density, $-\Delta B'$, for H_m reducing to $-H_K$ and providing point E on the loop.

- To bring the magnetisation of the specimen back to point A, close key K open S_2 (to OFF position) and throw RS_2 quickly on to contacts $1-1'$; again traversing a recoil loop from E to A.

- By repeating the above steps for different values of R_3, so as to effect changes in the magentising force in reverse direction from H_m, other points on section DEF of the loop can be obtained.

- The section FGLA of the loop can then be obtained by drawing in reverse the section ADEF, since the two values are identical.

Hysteresis loss

By measuring the area of the hysteresis loop so obtained – for example, by means of a planimeter – and expressing this area in terms of appropriate units of B and H, the hysteresis loss for the material may be obtained since hysteresis loss per cycle per cubic metre, in joule

$$= \text{area of the B-H loop}$$

when B is in tesla and H in A/m.

MAGNETIC MEASUREMENTS USING ALTERNATING CURRENT

Almost invariably, magnetic measurements with alternating supply or current[1] are performed to access the so-called "iron loss" in the material

---

[1] An alternating current at a given frequency, if allowed to pass through a magnetising circuit of test set-up will result into a time-varying magnetising field and corresponding magnetic flux.

under test. When a ferrous material is subjected to an alternating magnetic field a loss of power occurs due to

(a) hysteresis effect, and

(b) eddy currents induced in the iron which is electrically conducting.

Although the hysteresis loss per cycle may be determined from the hysteresis loop obtained by DC test as stated above, this may be somewhat different under the actual alternating magnetisation conditions to which the material may be exposed in practice, esp. at high flux densities. In "soft" magnetic materials, the hysteresis loss is not significant; and the eddy-current loss can be much reduced by using material in laminated form – a universal practice being followed for the manufacture of transformers and electric machines. Nevertheless, a knowledge of both the above losses is of prime importance for the designers of electrical apparatuses and hence all manufacturers of sheet steel invariably follow various schemes of "iron" loss measurement.

Iron Losses : Hysteresis and Eddy-current

Both hysteresis and eddy current losses depend on a number of parameters, such as the maximum flux density, frequency of the supply and its waveform and, in the case of eddy-current loss, on the thickness of laminations and whether these are grain-oriented or otherwise. Typically, the hysteresis loss due to alternating currents (and corresponding magnetic field) is given by the expression, commonly called the Steinmetz's law.

$$W_h = K \ f \ B_m^k$$

where W_h is the (hysteresis) loss per m^3 of the material, f the frequency of the supply, B_m the maximum value of the flux density, k an index (usually varying from 1.6 to 2.0) and K a constant related to metallurgical and physical properties of the material, known as Steinmetz's coefficient.

Likewise, a generally followed expression for the eddy current loss is

$$W_e = K' \ k_f^2 \ f^2 \ t^2 \ B_m^2$$

where W_e represents the eddy-current loss per m^3, K' a constant dependent on the material under test, k_f the "form factor' of the alternating supply (equal to 1.11 for 'pure' sinusoidal supply[1]), f the supply frequency, t the thickness of the sheet material and B_m the maximum flux density value to which the material is exposed.

---

[1]Clearly, if the waveform is distorted, that is, non-sinusoidal containing appreciable harmonics, the form-factor may be higher and eddy current loss correspondingly much higher being proportional to square of the form-factor.

Separation of losses

In any scheme of measurement, if the *total* iron loss including both the above losses is obtained experimentally, a separation of losses in its components may be warranted in almost all cases.

There are two ways in which it can be achieved.

A.   Separation by Variation of Frequency

If in the above expressions of the two losses, the maximum flux density and form-factor (with laminations of a given thickness, t, of course) are maintained constant, the two losses can be expressed as

$$W_h = P\,f \qquad \text{and} \qquad W_e = Q\,f^2$$

with P and Q being 'new' constants given by

$$P = K\,B_m^k \quad \text{and} \qquad Q = K'\,k_f^2\,t^2\,B_m^2$$

and the total loss can be written

$$W_i = P\,f + Q\,f^2$$

or

$$\frac{W_i}{f} = P + Q\,f$$

This shows that if the *total* iron loss in the material is measured at *two* different, known frequencies, the ratio of total loss and frequency will follow a straight line against the frequency values as shown in Fig.12.24. The intercept of the line on the ordinate (for f = 0) directly yields the value of the "constant" P; whereas the slope of the line would provide the value of Q, as shown. Knowing P and Q, the two losses can be obtained at any frequency, for given values of maximum flux density and form factor.

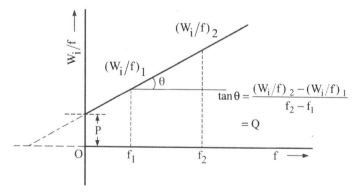

$$\tan\theta = \frac{(W_i/f)_2 - (W_i/f)_1}{f_2 - f_1}$$

$$= Q$$

Fig.12.24 : Separation of losses by variation of frequency

B. Separation by Variation of Form-factor

In this case, the frequency of the supply and B_m are kept constant whilst the form-factor is varied. If the total loss is now measured for various values of form-factor, the same can be expressed as

$$W_i = A + B \; k_f^2$$

where A and B are 'new' constants for this case, given by

$$A = K \, f \, B_m^k, \qquad B = K' \, f^2 \, t^2 \, B_m^2$$

If W_i is now plotted along the ordinate against k_f^2 along the abscissa, the constants A and B may be obtained; the constant A being simply the intercept on the ordinate and B the slope of the straight line as shown in Fig.12.25. Since form-factor of the supply is the ratio of the rms to average value, it is essential to measure both these values suitably during the test whilst the waveform of the supply is varied.

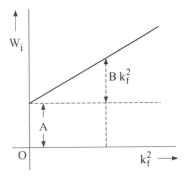

Fig.12.25 : Separation of losses by variation of form factor

Methods of Measurement of Iron Loss

Tests using ring specimens

A ring specimen prepared using laminated punchings as discussed earlier yields highest possible accuracy of measurement and must be used whenever feasible. The circuit for carrying out tests for loss measurement using such a specimen is shown in Fig.12.26. The core in ring form carries a magnetising winding M and *two* secondaries or search coils (or "B" coils) B_1 and B_2, all being uniformly wound, that is, with the adjacent turns closely placed and the winding encompassing the entire length of the core. The magnetising winding is fed from a well-designed, single-phase alternator, driven by a variable speed DC motor, to provide the magnetising current of sinusoidal waveform and *at variable frequency*.

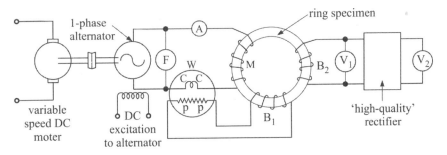

Fig.12.26 : Iron loss measurement on a ring specimen at variable frequency

The loss is measured using a precision wattmeter, with its current coil in the supply circuit and in series with the magnetising winding, but the voltage coil connected across one of the search coils instead of across the supply. This eliminates the copper loss of the magnetising winding from being measured by the wattmeter. The wattmeter reading is thus a measure of iron loss in the specimen when the number of turns on M and B_1 are equal. However, the reading must be corrected for the power loss in the voltage circuit of the wattmeter and the loss in the two voltmeters if these cannot be disregarded[1].

Two voltmeters are included in the circuit as shown. The one across search coil B_2 is a precision type to read rms voltage induced in the coil. The other, V_2, is connected at the output terminals of a high-quality rectifier and reads the 'true' DC voltage of the AC voltage across B_2. The ratio of the two voltmeter readings is monitored during the test to ensure constancy of the form-factor.

Then, following the process already discussed, the test is performed by measuring the iron loss at different supply frequencies obtained by varying speed of the alternator, whilst B_m is maintained constant by adjusting the magnetising current, by suitably varying the excitation to the alternator. The value of B_m at any step is derived as follows:

If the area of cross section of the specimen is a (m^2), the flux corresponding to B_m will be B_m a. The expression for induced EMF in a coil of N turns and at a frequency f Hz is

$$E_{rms} = 4k_f \emptyset_m f N \ V$$
$$= 4.44 \ \emptyset_m f N \ V$$

when k_f = 1.11 for sinusoidal supply.

Thus, by measuring E_{rms} on voltmeter V_1, the flux $\emptyset_m$, and hence B_m, can be deduced using the above expression for every step of measurement from the knowledge of f and N, the number of turns on the search coil B_2. The frequency is measured using a frequency meter connected across the supply, or 'derived' by measuring speed of rotation of the alternator.

If the test is to be performed at varying form-factors instead of varying frequency, the supply is obtained from an alternator at constant frequency. The waveform of the supply is changed with the help of a variable inductor and resistor connected in the supply circuit as shown in Fig.12.27. The two are varied/adjusted in such a manner that the reading of V_2 remains constant at all steps whilst the reading of V_1 varies as widely as possible. Constancy of V_2 implies constancy of B_m since f is also constant as shown below:

---

[1]This may be waranted since the iron loss in the sample itself may be small and may be even comparable to "I^2R" loss in the instruments.

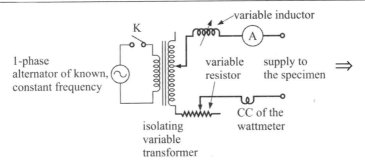

Fig.12.27 : Supply to ring specimen for test at varying form factor

For number of turns in B_2 to be N and the flux linking it changing from $(+B_m a)$ to $(-B_m a)$ in half-a-cycle, the *mean* value of the induced EMF for a half cycle is given by

$$E_{mean} = \frac{2(B_m a)N}{(1/2)\,f}\ V$$

and is read by the voltmeter V_2.

Iron loss measurement using Lloyd-Fischer square

For frequent tests on a variety of magnetic sheet materials, it is more convenient to use the Lloyd-Fischer magnetic square described earlier. [See Fig. 12.16]. The sides/coils of the square are identically packed tightly with enough number of laminations (to minimise 'air space' trapped between the laminations as much as possible) with interleaving corners as required, the laminations having been carefully weighed to obtain the area of cross section.

The connections of various windings for tests at varying frequency, assuming an essentially sinusoidal supply, are given in Fig.12.28. The primary, or magnetising winding is connected either directly, or through a transformer having a variable secondary to an alternator having an output

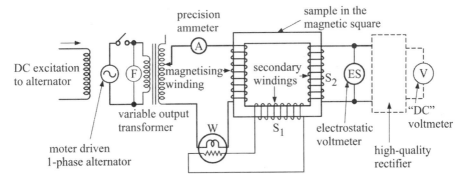

Fig.12.28 : Iron loss measurement using the Lloyd-Fischer magnetic square

nearly exactly sinusoidal and driven by a variable-speed (DC) motor to effect variation of supply frequency. The magnetising current in the square, and hence the flux density, is varied by adjusting the excitation of the alternator rather than by variation of a series resistance thereby avoiding any alteration of the waveform.

The connections of the two secondaries[1] to include a wattmeter and a (rms) voltmeter are made similar to that in the case of the test using a ring specimen. The test is also carried out as before by changing frequency and adjusting B_m to be constant at every step. The constants P and Q are obtained and the total loss split into its components using the expressions

$$W_h = P\,f \qquad \text{and} \qquad W_e = Q\,f^2$$

as before.

If, instead of variable frequency, the test is to be carried out at varying waveform, a rectifier and "DC" (moving coil) voltmeter are connected as shown dotted in the figure to deduce the form-factor from the readings of ES and V, and the supply to the magnetising winding is obtained by modifying the supply circuit as shown in Fig.12.27.

Correction for the area of cross section of the sample

Whilst deducing the value of B_m from the rms value of the induced EMF as read on the (electrostatic) voltmeter ES, it was assumed that actual area of cross-section of the sample was the same as that of the coil, the value of B_m being given by

$$B_m = \frac{E}{4.44\,f\,N_2\,A}$$

where E is the voltmeter reading, f the frequency, N_2 the number of turns on the secondary winding, S_2, and A the area of cross section of the sample/coil.

In practice, it may be necessary to correct the above expression, esp. at high values of B_m approaching saturation of the sample. This is due to the fact that the coil S_2 would enclose some "air flux" as well as the flux in the sample, since the inner cross-sectional area of the coil must necessarily be somewhat greater than that of the sample itself to accommodate the laminations, having been previously obtained by weighing the sample.

---

[1]The three windings are shown discretely in the figure on separate sides of the square for clarity of connection(s). Actually, these are distributed on all four coils/sides of the square and connected in series as described earlier. [See Fig. 12.16]. Usually, the number of turns may also be equal and known.

Let A_C =   the cross-sectional area of the coil, m^2

A_S =   the actual-cross sectional area of the sample, m^2

H_m =   the maximum magnetising force in the air space within the coil; this may equal that derived in terms of the magnetising current (or from the B-H curve of the sample)

B_m =   the actual maximum flux density in the sample

B'_m =   'apparent' maximum flux density, used in the above expression and related to the measured rms voltage

Then the total flux within the coil is

$$\emptyset = B_m\,A_S + \mu_0\,H_m\,(A_C - A_S) = B'_m A_S$$

Thus,

$$E = 4.44\,f\,N_2\,[B_m\,A_S + \mu_0\,H_m\,(A_C - A_S)]$$
$$= 4.44\,f\,N_2\,A_S\,B'_m$$

Knowing B'_m, B_m can be deduced.

Core loss measurement using an AC bridge

An AC bridge configuration may be adapted to the measurement of iron loss and effective relative permeability[1] of magnetic samples. Such methods are very useful when the samples are small in size; for example, in a ring form comprising laminations and wound uniformly with a winding. The tests are carried out at fairly low values of flux density, avoiding saturation. A modification of 'standard' Maxwell bridge for the purpose is shown in Fig.12.29 to demonstrate the method: several variations may be possible.

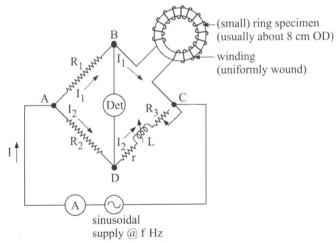

Fig.12.29 : AC bridge network for measurement of iron loss

---

[1]Effective permeability and resistance relate to the effect of alternating current in magnetising the sample and resulting in iron loss.

In the bridge, R_1, R_2, R_3 are non-inductive resistors, R_3 being variable. A variable inductor, L, having a small internal resistance r is connected in series with R_3. The sample in the form of a ring specimen is inserted in the fourth branch, its winding having an inductance L_S and an effective series resistance R_S. The effective resistance contains an iron loss component, the actual resistance of the sample being R_W (actually the DC resistance of the winding, measured suitably). Det is the detector; a vibration galvanometer or a telephone depending on the supply frequency. The bridge is fed from an alternator having a 'pure' sinusoidal waveform, having frequency f. The bridge is balanced by adjusting R_3 and L.

Theory of Operation

When the bridge is balanced, nodes B and D are at the same potential.

$\therefore$ PD across R_1 = PD across R_2

or $\dot{I}_1 R_1 = \dot{I}_2 R_2$

        [this means that $\dot{I}_1$ and $\dot{I}_2$ are in phase]

    Also, the PD across branch BC (the ring specimen) = PD across CD

or $\dot{I}_1 (R_S + j\omega L_S) = \dot{I}_2 [(r + R_3) + j\omega L]$

From the first equation,

$$\frac{\dot{I}_1}{\dot{I}_2} = \frac{R_2}{R_1}$$

And in the second equation, equating real and imaginary parts

$$\dot{I}_1 R_S = \dot{I}_2 (r + R_3) \quad \text{and} \quad \dot{I}_1 L_S = \dot{I}_2 L$$

Solving for R_S and L_S gives

$$R_S = (r + R_3) \times \frac{R_1}{R_2} \quad \text{and} \quad L_S = \frac{R_1}{R_2} \times L$$

As seen, the expression for L_S is independent of ω. Also, R_1 and R_2 appear in the form of "ratio" arms and thus can be used to enhance the accuracy of obtaining R_S and L_S in much the same way as in the usual Wheatstone bridge.

With the current I_1 through the winding on the sample, the *iron* loss in the latter is given by

$$W_i = I_1^2 R_S - I_1^2 R_W = I_1^2 (R_S - R_W)$$

where $I_1^2 R_W$ represents the copper loss in the winding, R_W having been measured separately.

Also, the input current is $I = I_1 + I_2$, measured on the ammeter A.

or
$$I = I_1 + \frac{R_1}{R_2}I_1 = I_1 \times \frac{R_1 + R_2}{R_2}$$

(substituting for I_2 from above)

That is,
$$I_1 = \frac{R_2}{R_1 + R_2} \times I$$

Hence, the iron loss in the sample is given by
$$W_i = I^2 \times \left[\frac{R_2}{(R_1 + R_2)}\right]^2 \times (R_S - R_W)$$

with the R_S as obtained above and I and R_W measured already.

If the number of turns on the specimen winding be N, its mean circumferential length l (in m) and a the area of cross-section of the core (in m^2), its inductance can be expressed as
$$L_S = \frac{N^2 a}{l}\mu_0\mu_r \quad \text{henry}$$

where μ_r represents the "effective" relative permeability of the sample and is thus given by
$$\mu_r = \frac{l}{N^2 a \, \mu_0} \times L_S$$

Measurement of iron loss using AC potentiometer

A co-ordinate type AC potentiometer, the operation of which has already been described in Chapter IX, provides a convenient method of measuring iron loss in a magnetic sample at low flux densities; for example, on the CT cores. The connections of the test set-up are shown in Fig.12.30.

The sample which is in the ring form carries two windings, wound uniformly; the 'primary' having N_1 turns whilst the 'secondary' N_2 turns. The supply to the primary is derived from a regulating transformer which itself is fed from a 1-phase alternator designed to generate EMF of sinusoidal waveform, at a known frequency, f. The alternator also supplies the "in-phase" and "quadrature" slide wires of the potentiometer through step down transformers and phase splitting circuit as usual.

The current in the primary winding will also be sinusoidal at low values of flux density and hence the core flux of the specimen. Supply to the slide wires from the same alternator ensures the same frequency and waveform of the slide-wire currents. A standard resistance of known, low value is connected in the primary winding for the purpose of measuring the excitation current in the winding, in terms of magnitude and phase of the voltage drop across the resistance. A two-way switch, S, is incorporated in the circuit which allows either the primary or secondary winding to be connected to the slide wire circuits.

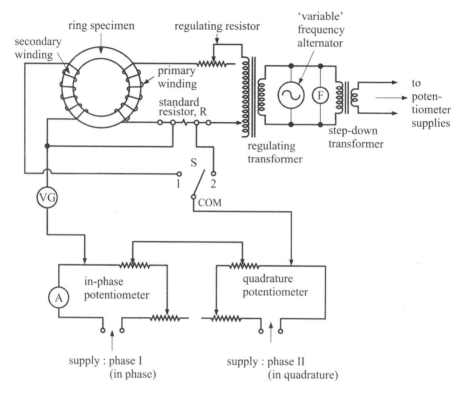

Fig.12.30 : Iron loss measurement using an AC potentiometer

Procedure

Let the maximum value of flux density in the sample, assumed to be uniformly distributed, be B_m (T). If the area of cross section of the core is A (m^2), the flux linking the secondary will be B_m A (Wb) and the induced EMF given by

$$E = 4.44 \, N_2 \, \varnothing_m \, f \quad V$$

at the supply frequency of f (Hz) and assuming a form-factor of 1.11 for the sinusoidal waveform.

- The EMF is measured by moving the switch S to contact 1 so that primary winding is in circuit, setting the quadrature potentiometer to zero and adjusting (only) the in-phase potentiometer slide-wire to obtain null in the galvanometer[1].

---

[1]It is, of course, assumed that the potentiometer has already been standardised in the usual manner and made direct reading.

The reading of the potentiometer then provides the value of E, related to B_m as above.

- The switch S is then moved over to contact 2 and both the in-phase *and* quadrature potentiometers are adjusted to obtain null in the galvanometer.

  The reading of the in-phase potentiometer in this case gives the value of I_e R where I_e is the loss component of the primary winding current (in phase with E) and R being the value of the standard resistance.

  The reading of the quadrature potentiometer likewise provides the value of I_m R where I_m is the magnetising component of the primary current (in phase quadrature with E).

- The iron loss in the sample is then proportional to E I_e. It may be separated into its components, hysteresis and eddy current losses, by performing the test at two different frequencies as was described for the test using a magnetic square.

'Measurement' of B-H loop using a CRO

A quick, visual 'measure' of the B-H loop of a magnetic material can be obtained on a CRO of good sensitively. The sample may be arranged in the form of a ring specimen, uniformly wound with a primary or magnetising winding and a secondary winding or search coil[1]. The connection scheme is shown in Fig.12.31. The magnetising winding is fed from an adjustable AC supply, preferably of known frequency and waveform. A standard non-inductive resistance, R, is connected in series to obtain a potential drop proportional to the excitation current.

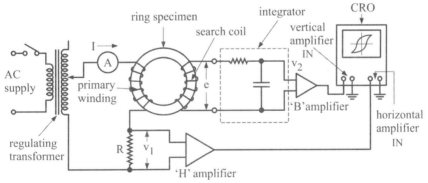

Fig.12.31 : Connections for obtaining B-H loop on a CRO

---

[1]If it is not convenient to prepare a ring specimen, or if only a *qualitative* check for B-H loop is required, the specimen, esp. in laminated form, may be used in a magnetic square.

Operation

The time base of the CRO is disabled and the output v_1 across the resistor R is connected to the horizontal amplifier terminals of the CRO. If necessary, the signal across R may be amplified using a preamplifier before connecting to the CRO, if the latter is not very sensitive[1].

The output from the search coil is an induced EMF given by

$$e = N_2 \frac{d\varnothing}{dt} \text{ V}$$

where $\varnothing$ is the flux in the sample and N_2 the number of turns of the coil. When integrated, the output v_2 is equal to $\int e\, dt$, or proportional to $\varnothing$, and is connected to vertical amplifier terminals of the CRO as shown.

On being excited suitably, the resulting trace on the CRO is the desired B-H loop of the sample, the size of which can be adjusted by adjusting the gains of the amplifiers, and may be photographed appropriately.

MEASUREMENTS ON PERMANENT MAGNETS OR MATERIALS

The tests on permanent magnets, or materials, are generally characterised by the requirement of obtaining their hysteresis loops, including major and minor recoil loops, in addition to the B-H curve. The testing methods may vary at times depending on the size and shape of the magnet and, at times, according to the purpose for which the magnet is to be used[2].

The "special" requirements of testing methods here involve

(a) Measurement of magnetising force *inside* the sample/magnet that cannot be measured in the manner(s) employed for soft magnetic materials, owing to the self demagnetising force of the magnet, and the necessity of providing extremely large magnetising force so as to drive the material well into saturation[3]. Typically, these materials also posses very high degree of coercivity and retentivity.

---

[1]If R is a 4-terminal, standard resistance having a value of a few milliohm, the output v_1 would be in mV, without exceeding the magnetising current too much to avoid overheating of the sample.

[2]For example, a small ALNICO magnet used as "brake magnet" in 1-phase or 3-phase energy meter.

[3]Nearly all permanent magnet materials are, in general, 'mechanically' extremely hard, produced by casting or sintering processes, with no machining of the magnets normally possible, and require enormous magnetising force of the order of several hundred thousand A/m.

(b) Measurement of internal flux density in the sample corresponding to the applied field; this may usually be achieved by arranging a "B" coil enclosing the sample cross section, or employing new techniques as discussed latter.

(c) Special devices or means for controlling and reversing (enormous) magnetising field, incorporated in the apparatus, esp. if it is desired to obtain recoil loops of the material, major and/or minor, in addition to the main hysteresis loop.

In what follows, a few methods used in practice are described considering special requirements and applications.

Tests Using a Hysteresigraph

In continuation with the previous discussion, the schematic diagram of an actual test set-up using a hysteresigraph is shown in Fig.12.32[1].

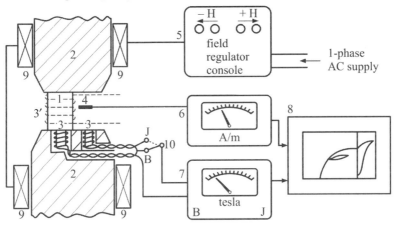

Description

1	:	Sample under test
2	:	Poles of the electromagnet
3	:	Pole (B) coils
3′	:	"B" coil around the sample
4	:	Hall probe for measurement of magnetising force
5	:	Regulator for controlling magnetising force in "+ve" and "–ve" direction
6	:	Indicator for magnetising force, A/m
7	:	Meter to read flux density 'inside' the sample
8	:	"X-Y" plotter, or recorder, to plot B-H curve and loop(s)
9	:	Magnetising winding
10	:	Switch to adjust pole coil connection

Fig.12.32 : Schematic of a hysteresigraph with various detail

[1]Developed and manufactured by Dr. Steingroever, GMBH, (of) Magnet-Physik, Germany.

Process of measurement

The test piece, or the sample, is positioned between the parallel poles of the (massive) electromagnet, the yoke part of which carries the excitation winding designed to produce extremely large magnetising force. The upper pole of the magnet is movable in a vertical direction and can be clamped in a particular position to suit the vertical height of the sample, the upper and lower surfaces of which are thus in firm contact with the poles.

The magnetising force in the sample is sensed and measured with the help of a flat, (transverse) calibrated Hall probe, located as close to the side of the sample as possible, the output being monitored on a meter, usually a gauss meter.

The flux density in the sample corresponding to a given magnetising field can be measured in two different ways:

(a)   by making use of a closely fitting search coil of a suitable number of turns, N, around the sample, (3′ in the figure), the flux density B being given by B = Ø/A where A is the area of cross section of the sample and Ø the flux produced in it and linking the coil, as the magnetising field is varied and measured on a calibrated flux meter or otherwise;

(b)   by means of "pole coils" embedded in the lower, removable pole piece of a cross-sectional area, or rather the diameter, smaller than that of the sample, shown as 3, 3 in the figure.

Basis of flux measurement by pole coils

The provision of pole coils in the apparatus does away with the necessity of winding a search coil around the sample every time a new sample is to be tested, making the apparatus more versatile and quick in operation. The only necessary requirement is that the cross sectional area, or the diameter of the flat surface of the sample in firm contact with the pole, should be bigger than that of the pole coil so as to cover it completely during the test.

Assuming that the pole pieces are made of very high permeability material[1] so that they do not get saturated even at the highest value of magnetising field used for measurement, the pole surface would be a (magnetic) equipotential and hence the normal component of *flux density* to be continuous across the common surface as shown in Fig.12.33(a). The *tangential* component of magnetising field is likewise continuous along the air-iron boundary by the side of the sample and hence there is no error in it being sensed and measured by a Hall probe.

---

[1]For example, Permendur.

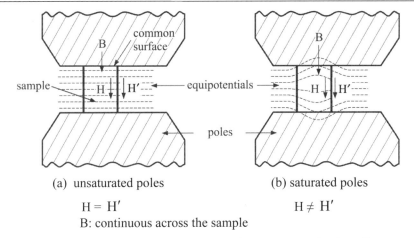

(a) unsaturated poles   (b) saturated poles

$H = H'$ $H \neq H'$

B: continuous across the sample

Fig.12.33 : Measurement of flux density by means of pole coils

The *lower* pole piece is provided with two identical coils (having same number of turns and cross-sectional area), connected to a flux meter via a special switch, 10, so as to allow either one or both coils connected in circuit for the purpose described later.

The outputs of B and H measuring instruments are fed to an X-Y recorder, or graph plotter, to obtain or plot the B-H curve and/or the un-interrupted B-H and recoil loops.

Operation

The sample of appropriate size[1] and having been demagnetized completely is placed between the poles of the electromagnet, the lower surface fully covering one of the pole coils (usually the left coil). The field regulator is adjusted to slowly increase the magnetising field in "+ve" direction, plotting the B-H *curve* of the sample. At the "limiting" value of H, given by $+B_{m_2}$, with corresponding flux density $+B_m$, the current is reversed and decreased slowly to reach the residual flux density position $(H = 0)$, and then to coercivity $(B = 0$ value$)$. The magnetising field is continued to 'increase' in negative direction till the point $-H_m, -B_m$ is reached. Following this, the current direction is again reversed and increased (in the positive direction) to reach the $+H_m, +B_m$ point although this part of the loop is not generally required to be plotted[2].

---

[1]Usually round in shape, about 2 to 2.5 cm in diameter and 1 cm in 'height'; the flat surfaces being machined very carefully and polished so as to make firm, good contact with the pole surfaces.

[2]The part of the B-H loop that is really important for the designers is the "demagnetisation" curve, in the II quadrant of the graph. A series of these curves are plotted corresponding to various values of H_m, B_m and B_r, the residual flux density.

If the process is followed correctly, smoothly and *without a pause*, the end point at the close will terminate at the start (H_m, B_m) point.

Plotting a Recoil Loop

If at any stage of plotting of the B/H loop, a major or minor recoil loop is desired to be plotted, the direction of magnetising current, and hence the magnetising field, is reversed at the given point on the B-H loop and increased in the "positive" direction. The process of plotting the B-H loop can then be continued *without any break* in the variation of the current.

Test Results

Two sample pieces, one of ALNICO and the other Samarian cobalt, measuring about 2.5 cm in diameter and 1 cm in height were tested using the Magnet-Physik hysteresigraph. The results are shown on Fig.12.34.

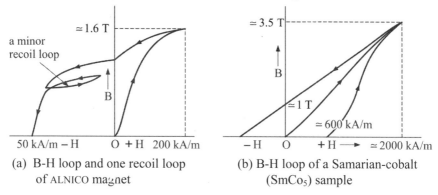

(a) B-H loop and one recoil loop   (b) B-H loop of a Samarian-cobalt
 of ALNICO magnet   (SmCo$_5$) sample

Fig.12.34 : B-H curves and (upper half) loops of ALNICO and Samarian-cobalt
magnets

Measurement of J-H curve of a permanent magnet

For the development of magnetic materials, it is often important to measure the *intrinsic* magnetisation, J, with respect to magnetising field, H. Since J is related to the flux density in the material by

$$J = B - \mu_0 H,$$

the measurement of J is accomplished on the same hysteresigraph by using *both* the pole coils; one measuring the flux density B whilst the other measuring the field strength, or rather the "airgap" flux density, $\mu_0 H$. For this, the two coils 3,3 [see Fig.12.32] are connected in *series opposition* with the help of switch 10, with only the left-side coil covering the test sample. If both the coils are identical in design (which is usually the case), the output to the flux meter, and hence the plotter along the Y-axis, will be proportional to J. The plotting of the J-H curve and loop can then be carried out as in the case of the B-H curve. However, for direct reading or measurement, a fresh calibration of the apparatus may be necessary.

Measurements on samples in powder form

Many magnetic materials that are sintered to be produced in permanent magnets in various forms, for example ferrites, may have to be tested in "powder" form to ascertain 'basic' magnetic properties of the material. For this, the material available as powder is tightly packed in a non-magnetic annular ring of about 3 cm OD and 1 cm height, made of stainless steel or brass, the underneath of which is sealed by a very thin foil to retain the powder in the flat-surface form. The sample so prepared is positioned over a pole coil as shown in Fig.12.35 and tests performed as before. The powder may be mixed with a thermo-plastic resin as a bonding agent in order to prevent movement of the powder particles during measurement.

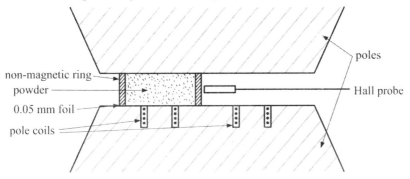

Fig.12.35 : B-H measurement on a sample in powder form

Calibration of the hysteresigraph set-up

When used with flexible, varying settings of controls on the H and B instruments (including the field regulator) and the scales of the X-Y recorder adjusted for measuring a variety of samples, a quick overall calibration of the hysteresigraph may be necessary for various given settings.

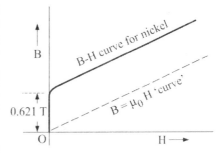

Fig.12.36 : "Saturation" characteristic of pure nickel

A method to achieve quite reliable calibration is to use a sample of usual dimensions made of pure nickel which has a residual flux density of 0.621 tesla at 'room' temperature, after magnetised to well-saturated state. This property of nickel is used by testing the sample in the usual manner and obtain its 'B-H loop' on the X-Y recorder/plotter as shown in Fig.12.36. The B-H loops of other materials can then be related to the B-H curve of nickel so obtained, taking into consideration the settings of scales on the recorder. It is to be noted that the B-H *curve* for nickel rises sharply with only slight increase of excitation, reaches 'saturation' and then follows almost the "air" magnetisation curve which is traced back without deviation to yield the unvarying residual flux density.

A Permagraph for Measurement of B/H Characteristics of Annular Magnets

When the sample is in the form of an annulus or ring, for example made of thin strips of permanent magnet material[1], wound round like a clock spring, the making of a "ring specimen" in the conventional manner to apply usual methods of measurement[2] may not be practical. The difficulty may be on account of the requirement of having to wind extremely large number of turns of the magnetising winding – a tedious process by itself – to produce enormous magnetising field necessary to drive the material into saturation. Apart from the cumbersome process, the winding consisting of large number of turns, arranged in several layers and carrying high to very high excitation current may result in excessive heating of the sample during the test and consequent severe temperature rise may prove detrimental to the very magnetic properties of the material.

In such cases, a permagraph[3] of a unique design that can be used conveniently for samples in annular form is shown in Fig.12.37.

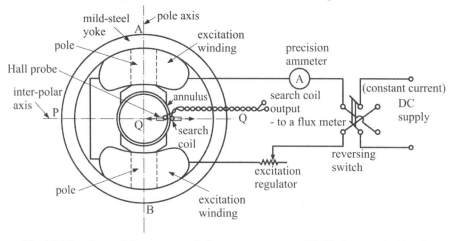

Fig.12.37 :   A special permagraph for measurement of B-H characteristics of a sample in annulus form

A two-pole DC electromagnet is used to provide magnetisation of the sample, the flux being uniformly distributed in the parts of the sample located at the interpole diameter, assuming the axial length of the poles

---

[1]Such as Vicalloy – an alloy of vanadium, iron and cobalt, Remalloy or Comol, Cunico and Silmanal.

[2]For example the method of reversals.

[3]Developed by the author during the course of experimental research.

being 'sufficiently' more than the width/'axial' length of the annulus. Only a single-turn search coil (for best accuracy), wound round this part of the sample, may be required to obtain the flux linked by the coil and hence the flux density in the sample by dividing the flux by the measured area of cross-section of the annulus. The flux can be measured on a flux meter in the usual manner.

The magnetising field in the specimen at a given excitation is obtained by measuring the magnetic field *externally* to the sample (on both sides) with the help of a calibrated (transverse type) Hall probe. For this, the plane of the Hall probe is positioned at right angles to the pole axis, the sensing area of the probe being at mid-height of the sample width/height. With the magnetising current held constant at a given value, the Hall probe is traversed away horizontally from as close the sample surface as possible along the inter-polar axis, PQ, both outside and inside the sample. The corresponding flux density in the sample is measured on the flux meter in the usual manner using the search coil.

The Hall probe outputs along the axis PQ at various excitations are plotted as shown on Fig.12.38. The value of magnetising force *within* the specimen is obtained by extrapolating the measured H variations in air close

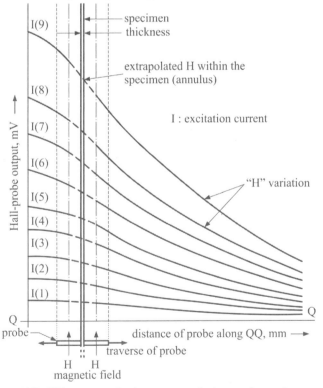

I(1), I(2), . . . : excitation currents in increasing order

Fig.12.38 : "H" variations measured using a Hall probe

to the internal and external surface[1]. Although a single reading of the Hall probe adjacent to the annulus side might suffice, the process as described would yield more accurate results, esp. if the annulus thickness is appreciable, and at higher excitation(s).

The B-H curve can then be plotted using the values of B and H so obtained. The technique may be extended to also obtain B-H loop of the sample by using, say, the "step-by-step" method. The characteristics for a Vicalloy sample in annulus form as measured using the above permagraph are shown in Fig.12.39 which also shows the hysteresis loops for the alloy.

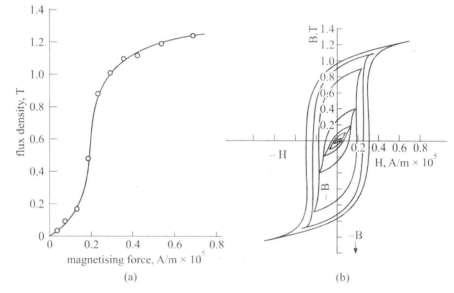

(a)   (b)

Fig.12.39 : B-H curve and hysteresis loops for Vicalloy

MISCELLANEOUS MEASUREMENTS

Magnetic Field Distributions

Magnetic field patterns, or field plots in general in regions of interest, in visual form are aimed to provide an spatial indication of flux lines and are of immense importance and help to machine designers and analysts. Various *experimental* means are available to obtain such plots.

Magnetic field patterns using iron filings

This is the well-known, "classic" method generally used to study the field patterns of permanent magnets in bar form and consists of sprinkling, as

---

[1]As in other methods, for example using a hysteresigraph, the basis of measurement of H is the continuity of its tangential components on the "air-iron" surface or boundary of the sample.

uniformly as possible, iron filings in the region of interest[1], on a horizontal, nonmagnetic support at a suitable height or in a plane. Under the action of excitation, or otherwise in the case of a permanent magnet, as the support is gently tapped, the iron filings move into the shape of flux distribution in the region. When dealing with an excited machine, care must be taken to prevent rushing of iron filings into the airgap or other magnetised parts of the machine.

To record the patterns for future reference, the easiest means is to photograph the patterns using a camera. However, this may not be easy and straightforward and can be time consuming. As an alternative, a photographic paper, for example DUOSTAT, cut to required shape may be used as the non-magnetic support referred to above. The iron filing patterns can then be obtained directly on this paper, the whole process being carried out in a suitable dark room and the paper processed photographically by exposing it briefly to light and developing and 'fixing' it in the usual manner[2].

A flux pattern, obtained using iron filings as above, showing the field distribution in the inter-polar region of an unsaturated, 2-pole salient field system (alone) is shown in Fig.12.40

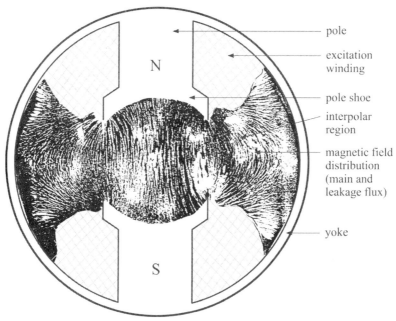

Fig.12.40 : Magnetic field distribution in the inter-polar region of a 2-pole field system using iron filings

---

[1]For example, typically, the inter-polar region of a salient pole machine showing the useful as well as leakage flux.

[2]If a "digital" camera is available, the pattern can be photographed and stored digitally on a PC and a print-out obtained later as required.

which depicts typical shape of flux lines in such a machine, indicating considerable leakage in the absence of a rotor in between the poles, and also appreciable flux reaching the overhang.

Magnetic flow lines using Teledeltos plots

Teledeltos or conducting-paper plots present another useful means of showing visual, distinct flux distribution usually in the non-magnetic regions of electric machines. These papers are coated with a special, electrically-conducting material of a few micron thickness, thus possessing only surface resistivity (or conductivity). When "energised" appropriately with a potential difference along two boundaries, currents flow in the conducting surface across the boundaries, the equipotentials of desired, regular difference being at right angles to the current paths. Thus, when the region of interest is modeled appropriately, by cutting the paper to proper shape and size, maintaining a suitable scale, and a DC potential applied, the equipotential lines, measured with the help of an electronic voltmeter, provide a visual image of the flux (or flow) lines in the region.

The method was applied to the study of flux pattern in the inter-polar region of the same machine as in Fig.12.40. Symmetry was assumed and plot was obtained for one-quarter of the field system only. The iron surfaces were treated as flow (or flux) line boundaries and the excitation winding was replaced by a current "kernel," so that equipotential lines would correspond to lines of H (or B) in the region. The position of the kernel was determined by the extrapolation of the iron filing plots shown in Fig.12.40 into the winding overhang. A potential of $\emptyset = 1$ ($\equiv 10$ V) was maintained at the kernel and zero at the pole axis. The results are shown in Fig.12.41. Note the concentration of flux at the pole tips, typical of a salient-pole field system with pole shoes, and an appreciable proportion of pole flux reaching the yoke, as depicted in the iron-filing plots, too.

Another example of the use of Teledeltos-paper plot is the pattern of induced currents across transverse or 'cross' slots[1] in the pole surface region of the rotor of a 2-pole alternator. The region is modeled by considering one-quarter of the pole surface and a to-scale representation of the cross slots which are maintained at the potential of $\emptyset = 0$, the other 'edge' common to the adjacent axial slot being maintained at $\emptyset = 1$ (or, again, 10 V). The equipotential lines then represent the contours of surface current as shown in Fig.12.42[2].

---

[1]These slots, typically shaped as crescent (and hence at times referred to as "moon" slots), are invariably machined into the pole regions for 'balancing' of the rotor, essentially to equilise the rotor inertia on the pole and inter-pole axes, the two being different on account of axial slots in the inter-polar region, cut to receive the excitation winding.

[2]Such currents are typically induced in pole regions, as well as other metallic parts, of the rotor during the un-balanced operation of the alternator as discussed later.

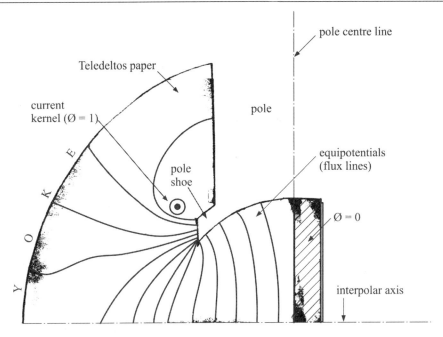

Fig.12.41 : Teledeltos paper plot of flux distribution in the inter-polar region
of the 2-pole field system

Typically, instead of flowing directly across, axially on the rotor surface,
the currents are squeezed around the cross-slot ends, resulting in much
higher current density, heating and temperature rise at the transverse slot
ends.

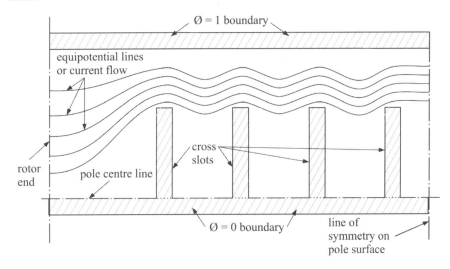

Fig.12.42 : A qualitative view of the Teledeltos plot of surface currents in the rotor
of a 2-pole alternator

Radial flux-density distribution

Often a quantitative pattern of *radial* flux density along the pole surface or in the airgap of a machine is required for design or analysis. In most cases, this can easily be achieved by using 'graded'[1] search coils, fixed circumferentially along the pole surface. An arrangement of such search coils along one half of the pole surface of the field system of Fig.12.40 is shown in Fig.12.43, which also shows a plot of flux density measured from the search coils in the usual manner[2].

As indicated by the plots of Figs.12.40 and 12.41, the flux density value peaks at the pole tips and is relatively low in the mid part of the pole.

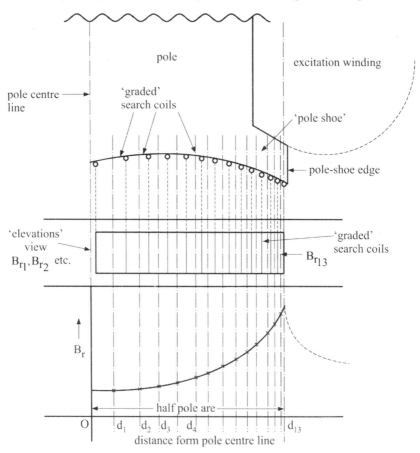

Fig.12.43 : Radial flux density distribution along pole surface

---

[1]Narrower towards the pole tips; relatively wider closer to pole centre line.

[2]A calibrated Hall probe could also be used for the purpose; however, search coils provide a better spatial reference.

Measurement of Induced Currents Caused by Time-varying Magnetic Field

A consequence of time-varying magnetic flux/field is to induce (eddy) EMFs, and currents, in all metallic parts, ferrous or non-ferrous, of an electric machine or equipment. A classic example is that of currents induced in the rotor of a 3-phase induction motor during "single-phasing". A more serious case pertains to induction of high-magnitude currents in various parts of the rotor of an alternator during its steady-state or transient unbalanced operation. Then, induced currents of nearly double-the-supply frequency (or the "negative-sequence") confine to the rotor surface, leading to extreme heating and temperature rise at key locations, over and above the designed operating temperature, leading to permanent, irreparable damage of the rotor.

In such cases, the induction of current(s) in metallic media is governed by the expression

$$\dot{J} = \omega \, \sigma \, \dot{A} \quad A/m^2$$

where $\dot{J}$ = induced current density (in complex form) in A/m^2

ω = angular frequency, $2\pi f$ rad/s

σ = electrical conductivity in S/m

and $\dot{A}$ = the (alternating) magnetic vector potential (in complex form) in Wb/m

The depth of penetration of currents *into* the media is given by

$$\delta = \frac{1}{\sqrt{\omega \sigma \mu}} \quad m$$

where $\mu = \mu_0 \, \mu_r$ is the permeability of the medium. Clearly, when the iron parts are not saturated and μ_r is high in value, δ can be only a fraction of a mm (i.e. almost the surface) even when f is approximately 100 Hz; hence intense heating of the surface(s).

J – probes

The measurement of such surface currents can be successfully carried out by means of specially designed sensors called "J-probes", fixed to a 'small' part of the region of interest, keeping in view the direction of current flow. Two examples of such probes to measure induced currents along one and in two perpendicular directions are shown in Fig.12.44.

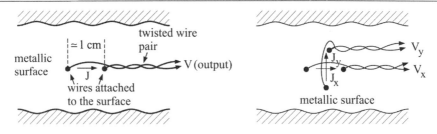

Fig.12.44 : J–probes for measurement of induced current density

The potential drop V (or V_x and V_y in the two directions) can be measured appropriately (for example on a calibrated potentiometer) and interpreted in terms of current density from the knowledge of electrical conductivity and permeability of the material.

The probes, using a thin insulated wire (for example, 24 SWG enamelled wire) can be attached to the metallic surface using a suitable adhesive, ensuring good *electric* contact. As an alternative, a charged capacitor may be employed to attach the bare wire end to the surface by discharging the capacitor through a tiny airgap as the wire end is quickly brought 'close' to the surface and a spark jumps across. The joint can then be made 'firm' by covering the area with an adhesive. Since the potential drop across two ends of the probe may be only a few millivolt, it is essential to twist the wire pair to avoid any pick up or stray-field effect en-route, as shown in the figure.

Measurement of Residual Flux of a Permanent Magnet

The magnetism or residual flux in a permanent magnet resides due to the magnetic polarisation of dipoles within the magnet. Under the influence of an externally applied field, the dipole move, tending to 'align' or orient in the direction of the applied field when the latter is increased beyond the saturation value. On removal of the field, however, the dipoles exhibit inherent hysteresis, leaving the magnet (partially) 'magnetised, this being known as "residual" magnetism[1]. The degree of residual flux or magnetism depends on a number of factors such as the composition of the material, the process of manufacture and magnetisation, and forms an important consideration for the design and application of a permanent magnet in a given device.

The concept of residual magnetism, in a rectangular permanent magnet (being mostly used in a variety of applications) is best explained by reference to Fig.12.45.

---

[1]In contrast, in a "soft" magnetic material, the dipoles 'fall back' to the random state on removal of the magnetising field, leaving little residual magnetism.

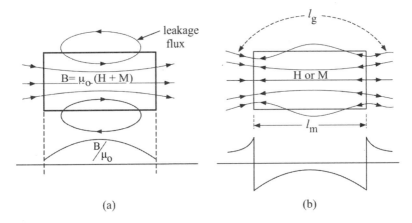

Fig.12.45 : B and H variation in a rectangular permanent magnet

The magnetic flux distribution will be of the nature of Fig.12.45(a) where the lines of B are continuous. However, the variation of H along the magnet axis will be as depicted in Fig.12.45(b) given by the field equation

$$\oint \bar{H} \cdot d\bar{l} = 0$$

where the length, l_m, within the magnet is as shown and l_g for the path in air can be chosen arbitrarily. Fig.12.45(b) thus points to an important feature of a permanent magnet: *lines of B and H coincide outside the magnet, but are oppositely directed inside.* The H, or rather M (the intensity of magnetisation), within the magnet is the seat of residual magnetism.

Clearly, the residual flux (and 'performance') of a permanent magnet can be much improved by 'containing' the air-path length by providing a soft iron yoke round the magnet, leaving only a small airgap as required by the application; for example, the brake magnet used in AC energy meters[1]. This would 'control' the demagnetisation of the magnet, move the working point up along the demagnetisation curve and thus mitigate the leakage of flux.

Residual flux of an ALNICO magnet

The principle of measurement in this case is based on the use of a search coil in which an EMF is induced due to change of flux linkages. The output of the search coil is measured in the usual manner by means of a calibrated ballistic galvanometer or a flux meter. The change of flux linkages can easily be effected by a quick movement of the search coil from around the magnet to the airgap and then away from it to a 'sufficiently' long distance as illustrated in Fig.12.46.

---

[1]See Chapter VIII: Measurement of Energy.

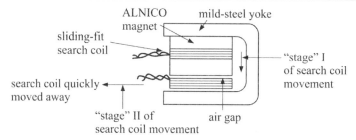

Fig.12.46 : Principle of measurement of residual flux of a rectangular ALNICO
            magnet

The search coil, pre-wound and consolidated, must be slightly loose fit around the magnet cross-section for its easy and quick movement from the magnet to the airgap down below. Since the cross-section of the magnet is 'sufficiently' large, the error due to loose-fitting search coil in obtaining the 'average' flux density from the measured flux may not be serious.

WORKED EXAMPLES

1. A conventional flux meter comprises a deflection coil of 30 turns and an area of cross section of 9.5 cm^2. It moves in the airgap of a magnet having a radial flux density of 10 mT. The meter is used with a search coil of 20 turns and of cross-sectional area of 3 cm^2. Calculate the deflection of the flux meter when the search coil is reversed in a magnetic field of 40 mT.

In its simplest form, the flux meter deflection is related to the change of flux by the expression

$$G \, \Delta\theta = N \, \Delta\varnothing$$

where G is the flux meter deflection constant and N represents the number of turns of the search coil.

Here $\qquad G = NBA$

$$= 30 \times 10 \times 10^{-3} \times 9.5 \times 10^{-4}$$

$$= 2850 \times 10^{-7}$$

Also,

$$\Delta\varnothing = 2 \times [3 \times 10^{-4} \times 40 \times 10^{-3}]$$

$$= 240 \times 10^{-7}$$

$$\therefore \quad 2850 \times 10^{-7} \times \Delta\theta = 20 \times 240 \times 10^{-7}$$

[since no. of search coil turns = 20]

or $\qquad \Delta\theta = \dfrac{4800}{2850}$ rad

$$= \dfrac{4800}{2850} \times \dfrac{180}{\pi} \text{ or } 96.5^\circ$$

2. A ring specimen having a mean diameter of 0.1 m and a cross section of 33.5 mm^2 is wound with a primary winding of 320 turns and a secondary winding of 220 turns. On reversing a current of 10 A in the primary winding, a BG gives a throw of 272 divisions. When used with a Hibbert magnetic standard having a coil of 10 turns and airgap flux of 0.25 mWb, the BG results in a throw of 102 divisions. Find the flux density in the specimen.

The flux linkages cut in the Hibbert standard

$$\Psi = N \varnothing$$

$$= 10 \times 0.25 \times 10^{-3}$$

$$= 2.5 \times 10^{-3} \text{ Wb-turns}$$

This causes a deflection of 102 divisions in the BG

$$\therefore \quad \text{BG constant} = \frac{2.5 \times 10^{-3}}{102} = 24.5 \times 10^{-6} \text{ Wb-turns/div}$$

When used with the search coil on the specimen, the BG gives a deflection of 272 div.

$\therefore$ *change* in flux linkages of the search coil

$$\Delta\psi = 272 \times 24.5 \times 10^{-6}$$

$$= 6664 \times 10^{-6} \text{ Wb-turns}$$

$\therefore$ flux linkages of the coil,

$$\psi = \frac{\Delta\psi}{2}$$

$$= 3332 \times 10^{-6} \text{ Wb-turns}$$

and the flux linking the search coil

$$= \frac{\psi}{N}$$

$$= \frac{3332 \times 10^{-6}}{220}$$

$$= 15.1 \times 10^{-6} \text{ Wb}$$

$\therefore$ the flux density in the specimen, $B = \dfrac{15.1 \times 10^{-6}}{33.5 \times 10^{-6}}$

$$= 0.45 \text{ T}$$

3. An iron ring specimen of 400 mm^2 cross-section and a mean length of 1 m is wound with a magnetising winding of 100 turns. A secondary winding of 150 turns, wound on the ring, is connected to a ballistic galvanometer having a constant of 2 μC/div. The total resistance of the secondary circuit is 500 Ω. On reversing a current of 10 A in the 'primary' winding, the BG shows a deflection of 150 divisions. Calculate the mean flux density in the specimen and the relative permeability corresponding to same.

MMF of the magnetising winding

$$N_1 I_1 = 100 \times 10 \qquad \text{or} \qquad 1000 \text{ A}$$

$\therefore$ mean magnetising field $H = \dfrac{N_1 I_1}{L} = \dfrac{1000}{1} \qquad \text{or} \qquad 1000 \text{ A/m}$

Charge through the galvanometer for a deflection of 150 div,

$$Q = 2 \times 150 \qquad \text{or} \qquad 300 \text{ μC}$$

Assume the flux through the ring corresponding to the above magnetising field be Ø Wb. Then flux linkages of the search coil will be

$$N\,Ø \text{ or } 150{\times}Ø \text{ Wb-turns.}$$

When the current is reversed, change of flux linkages

$$= 2{\times}150\,Ø \quad \text{or} \quad 300\,Ø$$

$$= \Delta\psi$$

∴ EMF induced in the search coil,

$$e = \frac{\Delta\psi}{\Delta t}$$

where Δt is the time of reversal.

or $\qquad\qquad e = 300\dfrac{Ø}{\Delta t} \text{ V}$

and current through the coil

$$i = \frac{e}{R} = 300\frac{Ø}{\Delta t\ R} \text{ A}$$

∴ the charge through the coil

$$Q = i\,\Delta t = 300\frac{Ø}{500} \text{ C}$$

Equating the two expressions for the charge

$$\frac{300\,Ø}{500} = 300{\times}10^{-6}$$

or $\qquad\qquad Ø = 500{\times}10^{-6} \text{ Wb} \quad \text{or} \quad 500 \text{ μWb}$

∴ the flux density in the specimen

$$B = \frac{Ø}{A}$$

$$= \frac{500{\times}10^{-6}}{400{\times}10^{-6}} \quad \text{or} \quad 1.25 \text{ T}$$

and relative permeability, $\mu_r = \dfrac{B}{H}{\times}\dfrac{1}{\mu_0}$

$$= \frac{1.25}{1000}{\times}\frac{1}{4\pi{\times}10^{-7}}$$

$$= 995$$

4.   A ballistic galvanometer with a constant of 0.1 micro-coulomb/div and total circuit resistance of 5000 Ω is connected to a search coil of

2 turns, wound round the field coil of a DC machine. It is then connected to a 3-turn search coil placed on the armature such that it embraces the total flux/pole entering the armature. When the field current is suddenly switched off, the BG readings for the two cases are 113 and 136, respectively. Calculate the flux/pole and leakage coefficient of the machine.

Let Ø be the flux linking any search coil of N turns.

Then linkages of the coil = N Ø Wb-turns. If these linkages are brought to zero in Δt s by switching off the magnetising current, the change in linkages will be

$$\Delta\psi = N\,Ø$$

and EMF induced in the circuit linking with Ø

$$e = \frac{\Delta\psi}{\Delta t} = \frac{N\,Ø}{\Delta t}$$

and the current in the circuit

$$i = \frac{e}{R} = \frac{N\,Ø}{\Delta t\; R}$$

where R is the total resistance of the BG and search coil circuit.

Then the charge in the circuit

$$Q = i\,\Delta t \quad \text{or} \quad \frac{N\,Ø}{R}$$

$$= K\,\theta \text{ (say)}$$

where K is the BG constant and θ the deflection corresponding to Q.

∴ $$K\,\theta = \frac{N\,Ø}{R} \quad \text{or} \quad Ø = \frac{R\,K\,\theta}{N}$$

I : Armature flux, $$Ø_A = \frac{5000 \times 0.1 \times 10^{-6} \times 136}{3}$$

$$= 0.0226 \text{ Wb}$$

II : flux/pole $$= \frac{5000 \times 0.1 \times 10^{-6} \times 113}{2}$$

$$= 0.0282 \text{ Wb}$$

and leakage coefficient of the machine

$$= \frac{\text{total flux/pole}}{\text{useful flux (flux through the armature)}}$$

$$= \frac{0.0282}{0.0226}$$

$$= 1.25$$

5. A specimen of 'iron' stampings weighing 10 kg and having a cross-sectional area of 1680 mm^2 is tested using a magnetic square which has both secondary windings, S_1 and S_2, of 515 turns, each having a resistance of 40 Ω. A sinusoidal supply of 50 Hz is given to the primary winding with an excitation of 0.35 A and an electrostatic voltmeter across a secondary indicates 250 V. The resistance of the voltage coil of the wattmeter is 80 Ω. If the wattmeter connected to the primary side reads 80 W, calculate the flux density in the specimen and the iron loss/kg.

From the induced EMF formula in an alternating field

$$E = 4.44 \, f \, B_m \, A \, N_2 \, V,$$

for the sinusoidal waveform [$K_f = 1.11$].

∴
$$B_m = \frac{250}{4.44 \times 50 \times 1680 \times 10^{-6} \times 515}$$

$$= 1.3 \, T$$

Total 'measured' loss, $P = 80$ W

Taking into account the resistance of the wattmeter pressure coil etc., iron loss in the stampings, P_i, is given by

$$P_i = P \left[1 + \frac{r_c}{r_p} \right] - \frac{E^2}{r_p + r_c}$$

where

P = total loss

r_c = resistance of the coil/winding

r_p = resistance of wattmeter pressure coil

E = voltage in coil S_1 or S_2

  = voltmeter reading

∴ substituting

$$P_i = 80 \times \left(1 + \frac{40}{80,000} \right) - \frac{250^2}{40 + 80,000}$$

$$= 80.04 - 0.78 \quad \text{or} \quad 79.26 \, W$$

∴ iron loss/kg $= \dfrac{79.26}{10}$ or 7.926 W/kg

6. In a loss test on a ring specimen, performed at constant flux density and form-factor, the following results were obtained:

Frequency, f (Hz)	30	40	50	60	70
loss, W (watt)	2.7	3.8	5.0	6.3	7.7

Readings of mean and rms values of induced voltage in a 140-turn search coil were 10.0 and 12.0 V, respectively, at 50 Hz. The ring has the mean circumference of 40 cm, weight 0.9 kg, and the specific gravity of the iron is 7.5. Estimate (a) the value of maximum flux density, (b) the total iron loss/kg at 50 Hz, the flux variation being sinusoidal with a peak value of flux density the same as in the test.

(a)   Volume of the specimen

$$= \frac{0.9 \times 1000}{7.5} = 120 \text{ cm}^3$$

∴   net iron area $= \dfrac{120}{40}$ or 3 cm^2

Now, from EMF equation,

$$E_{mean} = 4 \, f \, B_m \, a \, N \quad \text{V, with the usual notation}$$

∴ $B_m = \dfrac{10.0}{4 \times 50 \times \left(3 \times 10^{-4}\right) \times 140}$,

using mean value of EMF

$$= 1.19 \text{ T}$$

(b)   By division, values of W/f for the frequencies given in the above table are 0.09, 0.095, 0.1, 0.105, and 0.11, respectively. Plotted against f (as explained in the text), the intercept on the W/f axis is 0.075. Therefore, hysteresis loss at 50 Hz is 0.075 × 50 or 3.75 W, and thus the eddy current loss is 5.0 − 3.75 = 1.25 W.

Now, test form factor

$$= \frac{12.0}{10.0} \quad \text{or} \quad 1.2$$

For sinusoidal variation(s), the form factor is 1.11.

Eddy current loss at form factor of 1.11

$$= \left(\frac{1.11}{1.2}\right)^2 \times 1.25 \text{ or } 1.07 \text{ W}$$

$$\left[\because \text{ eddy-current loss } \propto k_f^2\right]$$

$$\therefore \quad \text{total loss} = 3.75 + 1.07 = 4.82 \text{ W}$$

$$\text{and loss/kg} \quad = \frac{4.82}{0.9} = 5.35 \text{ W}$$

7. In the magnetic circuit shown in the figure, the mutual inductance M is 9 mH. The galvanometer is connected in series with the "B-coil" on the specimen, the coil having 20 turns. A reversal of current of 3 A in the primary of the mutual inductance produces a deflection of 60 divisions. The cross-section of the specimen is 5 cm^2. Calculate the flux density in the specimen if reversal of this flux density (by reversing the magnetising current) results in a galvanometer deflection of 36 divisions.

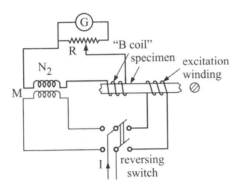

If I is the current in the primary of a mutual inductance M and the flux linking the secondary of N_2 turns is Ø, then for any change in I with time, the EMF induced in the secondary will be given by

$$e_2 = M \frac{dI}{dt}$$

$$= N_2 \frac{dØ}{dt}$$

giving

$$N_2 \, \Delta Ø = M \, \Delta I$$

where ΔØ and ΔI represent the changes in the flux and current, respectively.

If the galvanometer constant is K flux-linkages/div, then

$$N_2 \, \Delta Ø = K \, \theta_1 = M \, \Delta I$$

$$= 0.009 \times (2 \times 3) \text{ [multiplied by 2 to show reversal]}$$

where θ_1 is the deflection, giving

$$K = \frac{0.009 \times 6}{60} = 9 \times 10^{-4} \ \text{Wb-turns/div}$$

When the deflection is 36 division, the *change* of flux linkages will be

$$\Delta\psi = 36 \times 9 \times 10^{-4} \ \text{Wb-turns}$$

and the flux linking the coil of 20 turns will be

$$\varnothing = \frac{36 \times 9 \times 10^{-4}}{2} \times \frac{1}{20}$$

$$= 8.1 \times 10^{-4} \ \text{Wb}$$

and the corresponding flux density

$$B = \frac{8.1 \times 10^{-4}}{5 \times 10^{-4}}$$

$$= 1.62 \ \text{T}$$

EXERCISES

1. The constant of a form of magnetic potentiometer is obtained with the help of a coil of 3000 turns and, when a current of 0.6 A is reversed in the excitation winding, is found to be 157 divisions. It is then used to measure the magnetic potential difference between two ends when it gives a throw of 304 divisions. Determine the potential difference under measurement.

[697 A

2. A solenoid of length 1 m is wound with 1000 turns, together with a search coil of 100 turns of a cross-sectional area of 5 cm^2 placed at the mid-length. When a current of 5 A is reversed in the excitation winding, a deflection of 10 divisions is observed in the galvanometer connected to the search coil. Determine the galvanometer constant.

[0.628 × 10^{-4} Wb-turns/div

3. An iron-ring specimen of mean diameter of 30 cm and cross section of 6 cm^2 is uniformly wound with an excitation winding of 1000 turns. It also carries a secondary winding of 500 turns, connected to a ballistic galvanometer through an adjustable resistance, the total BG circuit resistance being 500 Ω. When a current of 2 A is reversed in the excitation winding, the galvanometer gives a deflection of 20 divisions. If the galvanometer constant is 100 μC/div, calculate the mean flux density in, and relative permeability of, the specimen material.

[1.667 T; 625

4. A test for measurement of hysteresis and eddy-current loss on an iron sample provided the following sets of results:

   A. applied voltage: 230 V; supply frequency: 50 Hz; total iron loss: 175 W

   B. applied voltage: 100 V; supply frequency: 40 Hz; total iron loss: 50 W

   Calculate for each case the hysteresis and eddy-current loss.

[A: 142.6 W, 32.4 W; B: 43.8 W, 6.2W

5. In the iron test on a sample, the total measured loss is 300 W at the given AC supply and at 50 Hz frequency. It is noted that the eddy-current component is 5 times of the hysteresis loss at this frequency. Determine the frequency at which the total iron loss will be 600 W, the flux density in the sample having been maintained as before.

[72.5 Hz

6. A single-phase transformer is tested for the iron loss. The total loss at frequencies of 50 Hz and 40 Hz while the maximum flux density in the core is maintained constant is 200 W and 150 W, respectively. Determine the hysteresis and eddy-current loss at the 'standard' supply frequency.

   If the constant for hysteresis loss is 0.075, calculate the maximum flux density in the core.

   [137.5 W, 62.5 W; 1.627 T

QUIZ QUESTIONS

1. The word "ballistic" in a ballistic galvanometer relates to the science of ballistics.
   □ true   □ false

2. The ballistic galvanometer used in magnetic measurements is
   □ a vibration galvanometer   □ a D'Arsonval type meter
   □ an electrostatic volt meter   □ none of these

3. A ballistic galvanometer essentially measures
   □ current in the circuit
   □ potential difference in the circuit
   □ quantity of electricity
   □ vibrations

4. A typical ballistic galvanometer has a period of about
   □ 10 ms   □ 1 or 2 seconds
   □ 10 to 15 seconds   □ one hour

5. The theory of deflection of a ballistic galvanometer is governed by
   □ a linear equation of motion
   □ a first-order differential equation
   □ a second-order differential equation
   □ none of these

6. The usual ballistic galvanometer can be calibrated by means of
   □ a capacitor   □ a long solenoid
   □ Hibbert magnetic standard   □ any of these

7. A flux meter is essentially
   □ a CRO   □ a voltmeter
   □ an ammeter   □ a ballistic galvanometer

8. The control torque in a typical flux meter is
   □ small   □ high
   □ very high   □ practically zero

9. The reason for using ring-type specimens for magnetic measurements is
   □ they are free from end effects
   □ they give relatively more accurate value of B or H
   □ they afford economy of measurement
   □ all of these

10. The B/H curve of a typical magnetic material signifies
    - □ its composition
    - □ magnetic properties including saturation
    - □ electrical properties
    - □ none of these

11. In the flux measurement method by ballistic galvanometer, the flux density in the specimen is derived from
    - □ an ammeter in the circuit
    - □ PD across a resistor in the circuit
    - □ a search coil wound round the specimen
    - □ EMF of the supply to the circuit

12. The B/H loop of a typical magnetic material provides information about
    - □ magnetic properties of the material
    - □ power loss per cycle of magnetisation
    - □ energy content of the material per unit volume
    - □ all of these

13. The two methods generally employed for B/H loop measurement are
    a._____ b. _____

14. The coordinates of a typical hysteresis loop are
    - □ V and I   □ B and I
    - □ H and I   □ –H to +H and –B to +B

15. The magnetic tests on a magnetic material in strips or laminates form are best performed by using
    - □ a ring specimen   □ a transformer
    - □ a magnetic square   □ none of these

16. The search coil used in AC magnetic measurements actually measures
    - □ flux density, B
    - □ field strength, H
    - □ hysteresis loss
    - □ EMF induced corresponding to B in the specimen

17. A typical magnetic square method is also useful for separation of

    _____

18. A magnetic potentiometer is an easy device to measure

    _____

19. The loss component of no-load current in a specimen for measuring iron-loss using a co-ordinate potentiometer is read by
    ☐   in-phase potentiometer
    ☐   quadrature potentiometer
    ☐   either of the two
    ☐   none of these

20. A Gauss meter can be used to measure
    ☐   current in a circuit   ☐   PD in a circuit
    ☐   magnetic flux density   ☐   none of these

21. The essential component of a typical Gauss meter is a
    _____.

22. An apparatus generally used for testing permanent magnet materials is called a hysteresigraph
    ☐ true   ☐ false

23. The measurement of magnetic field, H, in many tests is based on the magnetic property of _____

24. The special probes used to measure induced surface currents are known as _____

XIII : Measurement of Non-electrical Quantities

XIII

MEASUREMENT OF NON-ELECTRICAL QUANTITIES

RECALL

Measurement of non-electrical quantities by electrical means forms an important requirement in various facets of application nowadays.

As a typical example of measurements in this category, consider the case of measuring "speed", say, of a moving car. To begin with, speed may imply

(a) the speed of the car engine in revolutions per minute (or per second);

(b) the angular speed of the wheels of the car if moving round a curve (or in a circle) in rad/sec;

(c) the lateral speed along a (straight) track or path in km/hr or m/s.

The three may, of course, be related to each other.

Thus, in the above example, imagine that the car moves round a circular track of radius 3.0 m with the wheels rotating at 2 revolutions per second (being suitably connected to the engine by means of gears, propeller shaft and differential).

What is the

(i) angular speed, ω;

(ii) period, T;

(iii) speed (or velocity), v, of the car?

To work these out

(i) for 1 revolution, angle turned, $\theta = 2\pi$ rad. When the wheels make 2 revolution per second, the angular speed, $\omega = 2 \times 2\pi$ or 4π rad/s

(ii) period T = time for 1 rev $= \dfrac{2\pi}{\omega} = \dfrac{2\pi}{4\pi} = 0.5$ s

$$\left[\text{else, } T = \frac{1}{\text{rps}} = \frac{1}{2} \text{ or } 0.5 \text{ s} \right]$$

(iii)   (linear or lateral) speed of the car along the track,

$$v = r \, \omega$$

$$= 3.0 \times 4\pi \quad \text{or} \quad 38 \text{ m/s [nearly 137 km/hr]}$$

Thus, "speed" in a given application, may mean one of the above and may require a given process to measure it *electrically*.

The other applications may pertain to measurement of many mechanical quantities, like force, pressure, stress, torque etc; in thermodynamics and industry, like temperature; in hydraulics, like flow and level of liquids; in physics, like illumination and sound, and so on by employing appropriate electrical means or devices.

Yet another field of measurement in which electrical (and/or electronic) devices are being increasingly employed is medicine or medical diagnostics. This includes measuring blood pressure of a patient, recording his/her ECG, and monitoring many more vital functions of the body.

GENERAL REQUIREMENTS

In general, the measurement of (a variety of) non-electrical quantities by electrical means involves two distinct processes:

A.   The conversion of the mechanical or other non-electrical quantity to be measured to an (equivalent) electrical signal; the device for effecting this is known as a *pick-up*, or a *transducer*.

B.   The conversion of the electrical signal so obtained to the information presented on a dial, or a similar device for displaying and/or recording the measured value, usually involving appropriate calibration.

Transducers

Depending on the requirement or application, transducers may be of a great many possible variations. They may have a single output; alternatively, the output may be of differential type, that is, the quantity to be measured 'cancels' the balanced condition of two transducer elements (acting in 'opposition'). Further, they may produce unidirectional (or "DC") or alternating EMF at a given frequency, or may *operate* in circuits in which the currents are unidirectional or alternating. Some transducers may be suitable for slow changes in the variable, others for rapid changes or vibrational conditions. The operating principles and selection criteria of a variety of transducers are discussed below: they are grouped under headings according to the type of measurement to be made.

Operational aspects of transducers

In most cases, the variation(s) of a non-electrical quantity may manifest in terms of a corresponding, proportional or otherwise, variation of an electrical circuit element such as a resistance, inductance (self or mutual) or capacitance in the electrical measurement set-up. For example, a small linear or rotational movement of a body may be made to move the variable contact of a linear or rotary potentiometer which may be one of the four arms of a Wheatstone bridge and the out of balance EMF or PD across the appropriate terminals may be interpreted (and calibrated) to indicate the movement of the body. Alternatively, a magnetic circuit may be altered in terms of the variation of the quantity to be measured and the magnetic change may be 'translated' into an electric output to represent the variation. The following examples illustrate application of various types of transducer for the measurement of some of the common non-electrical quantities.

MEASUREMENT OF SPEED

The speed to be measured may be rotational or linear.

Rotational Speed Measurement

Tachometers

The simplest means of measuring rotational speed, say of a shaft of a motor, is to relate the rpm of the shaft to the EMF induced in a small DC or AC generator coupled to the shaft. If the excitation is maintained constant, for example using a permanent magnet, the output of the generator will usually be proportional to the rotational speed and this can be calibrated to read the speed directly on a suitable voltmeter connected across the generator terminals. The device is usually called a "tacho-generator", or "tachometer", and is commercially available in various ranges of rpm.

In the most compact form, the generator and the (calibrated) voltmeter are arranged in a single case and the shaft of the generator, fitted with a conical rubber grommet, can be connected manually to a matching recess at the shaft end, the measurement being made in hand-held position, by 'firmly' pressing the grommet into the shaft recess. A tachometer may be designed to have an accuracy of 1 percent or better for a speed range of a few hundred to a few thousand rpm.

An Electronic Tachometer

For 'moderate' rotational speed measurement, a tachometer in electronic form may be devised and used, providing an accuracy as high as 0.05% or better.

The working principle of such a tachometer is illustrated in Fig.13.1.

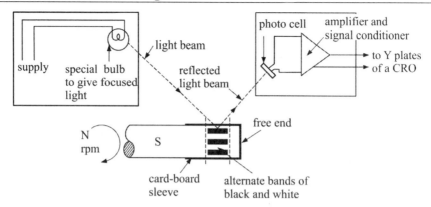

Fig.13.1: Principle of operation of an electronic tachometer

In the figure, S is the rotating shaft of which the rotational speed (or rpm) are to be measured. At its free end, it is firmly fitted with a card-board sleeve, painted with alternate bands of black and white as shown. A focused beam of light from a lamp is made to fall on the bands, getting reflected from the *white* bands, thus generating a regular trail of pulses of light which is made to fall on a photo cell. The electrical output of the cell, also in the form of pulses, is fed to the Y or vertical plates of a CRO after appropriate amplification (if the signal strength is low) and conditioning to produce a nearly sinusoidal output. To the X or horizontal plates of the CRO is connected a variable-frequency, high-accuracy oscillator, after deactivating the time base of the oscilloscope. The two signals, applied simultaneously, would produce a Lissajous figure on the CRO screen (for example a stationary ellipse) which in terms of the oscillator frequency can be related to the speed of the shaft. Any change in the speed of the shaft could be detected easily as it would result is a precession of the Lissajous trace[1] on the CRO.

A Digital System

A simple digital method of rotational speed measurement esp. at low rpm, for example < 100 rpm, consists of a stationary reed switch, or relay, activated by a small, but 'reasonably' powerful, rectangular permanent magnet suitably attached to the rotating shaft, or fixed on a disc rigidly mounted on it. The schematic of the method is shown in Fig.13.2. As the rotating magnet sweeps across the reed switch, a rectangular pulse is generated by virtue of make and break of current in an appropriately designed circuit as shown, which may be connected to, and measured on a CRO in the form of a train of pulses. Alternatively, the pulses may be recorded on a digital pulse counter. The 'first' pulse may also be used to

---

[1]See also Chapter IV.

activate a timer so that the number of pulses counted in a given time will provide the shaft speed in rpm[1]

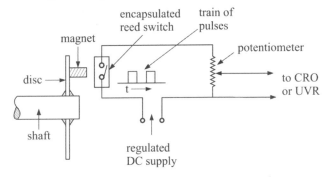

Fig.13.2 : Digital system of speed measurement

Stroboscopic methods

A stroboscopic device presents a quick 'indirect' method of measuring rotational speed, at times with a high accuracy of measurement. The device essentially comprises a disc, mounted on the rotating shaft and painted with simple geometrical patterns as depicted in Fig.13.3.

Fig.13.3 : Examples of stroboscopic discs

Principle of Operation

The disc is illuminated by a series of rapid flashes, obtained from a "flasher" unit[2], each of the flashes being of very short duration (or the interval for which the light is falling on the disc) compared with the time interval between successive flashes. If the speed of rotation of the shaft, and hence the disc, is such that each point of the pattern on the disc moves 'forward' a distance of one "point-pitch", p, during the time interval between successive illuminating flashes, the pattern will appear to be stationary. If the two time intervals differ from each other, the pattern will appear to move forward or backward depending on the relative difference of speeds, and slow or fast, as per the 'degree' of the difference.

---

[1]The accuracy of measurement in the method thus depends on the accuracy with which both the time interval and number of pulses can be measured.

[2]Such units are commercially available, with the range of 'frequency' of flashes adjustable in steps, or being continuously variable.

Clearly, the same effect would be obtained if the speed of rotation is exactly two, or three, times, or any other multiple, of this speed.

Theory

When the pattern of the disc appears to be stationary, the speed of rotation of the shaft is given by

$$N = \frac{P}{p} \text{ rpm}$$

where P = no. of flashes from the flasher unit per minute

and   p = no. of "points" on the pattern

To eliminate the confusion of multiplicity of the speed, that is twice, thrice etc., the speed of the machine may first be estimated approximately by using a tachometer or otherwise; for example, the rpm of a 4-pole induction motor can never exceed its "synchronous" speed of 1500 rpm for a supply frequency of 50 Hz and a given slip[1].

Measurement of Linear Speed

A common example of measurement of linear speed is to measure the speed of a moving vehicle, say a car, and the device commonly employed for the purpose is called a speedometer, essentially fitted in every vehicle to indicate its speed in km/hr.

The first stage of working of a typical speedometer is the pick up of the rotational speed of the engine, wheels or the propeller shaft which is transmitted to a magnetic rotor assembly of the speedometer with the help of a drive cable, typically a flexible mandrel twisted with steel wires to make it sturdy and at the same time flexible enough to slightly bend corners where necessary, the cable being housed in a strong PVC sleeve (known as outer cable) to provide mechanical protection.

The magnetic rotor assembly itself comprises an aluminum cover, called the "cup" owing to its shape, and an internal magnetic rotor. The drive cable connects to this rotor. When the former rotates, corresponding to a given rotational speed of the vehicle, the motion is transmitted to the magnetic rotor. The schematic of such a device is shown in Fig.13.4.

---

[1]Interestingly, a qualitative case of stroboscopic phenomenon may be observed when light from a fluorescent tube (pulsating at 50 Hz, but appearing stationary to naked eye due to persistence of vision) falling on a fan blades results in the blades to appear stationary or slowly moving forward or backward, depending on actual speed of the blades at a given instant.

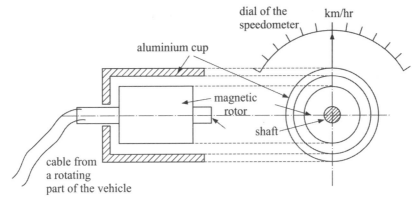

Fig.13.4 : Schematic of a car speedometer

Operation

The rotating magnet produces an opposite magnetic field in the aluminum cup according to Lenz's law, resulting in the induction of eddy currents on the cup surface. These interact with the magnetic field to produce a deflecting torque in the cup, similar to that in an indicating instrument. A control spring is suitably incorporated to produce a control torque. A steady deflection of the pointer (or needle), attached to the cup, is obtained when the two torques are equal. The pointer moves along the dial of the speedometer, calibrated to directly indicate the vehicle speed in terms of kilometre per hour. The faster the speed of the vehicle, the faster will be the rotation of the magnet, larger the developed torque and higher the deflection of the cup or the pointer.

MEASUREMNET OF PRESSURE

Pressure, or the force per unit area, is frequently measured by allowing it to produce mechanically a small displacement and measuring the same appropriately. The displacement may result in a corresponding change in, say, resistive property of a transducer that may be utilised to cause an electrical unbalance in a Wheatstone bridge, or produce a proportionate electrical output or signal that can be measured suitably and calibrated to read the pressure directly on the scale of an indicating instrument.

Transducers

The commonly used transducers for the purpose are:
1. strain gauges
2. piezo-electric (quartz) crystals
3. electromagnetic means or devices
4. inductive or capacitive sensors

Typically, the pressure may arise from 'solid' bodies, fluid or gaseous matter, necessitating the use of an appropriate senor or transducer.

Strain Gauges

Historically developed and used during the World War II, strain gauges constitute the most frequently used "resistance-type" transducers for the measurement of strain and stress caused by the variation of pressure in any manner. During measurement, these usually form one or two arms of the simple Wheatstone bridge.

Construction

'Standard' gauges are constructed from extremely thin Eureka or Nichrome wire. Gauges may also be made from stainless steel wire which has higher sensitivity factor and lower hysteresis. Additionally, gauges can be made to order from iridium-platinum wire (10% iridium) and these have even higher sensitivity factor than the steel gauges, but are more expensive[1].

Examples of a few commercially available "foil"-type strain gauges are shown in Fig.13.5. A typical gauge has an active length of about 6 mm and a nominal resistance of 120 Ω ($\pm$ 0.2 Ω) and has the property of being usually self-temperature compensated.

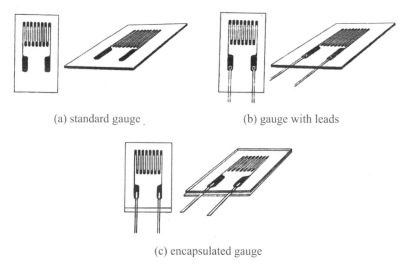

(a) standard gauge   (b) gauge with leads

(c) encapsulated gauge

Fig.13.5: Various forms of commonly used strain gauge

Operation

The wire-resistance strain gauge as above depends for its operation on the change in its resistance which takes place when a metal wire is extended/ stretched or compressed, this change bearing a definite relationship to the change in length of the wire.

---

[1]The cost factor becomes somewhat significant when large number of strain gauges are used in some applications.

The gauge is bonded to the test surface and partakes of its strain resulting from a stress or pressure[1]. Measurement (of the strain) is effected by the change of electrical resistance, ΔR, as a result of change in dimensions, and specific resistance of the wire when the test specimen is strained. Under ideal circumstances, the relationship between the *change* of resistance and strain will thus seem to be a straight line. The ratio of proportionate change of resistance to the mechanical strain is defined as the "sensitivity factor" and reflects the quality of a strain gauge and strived to be as high as possible for good results.

Bonding Process

It is most important that the strain gauge is attached or bonded very *firmly and uniformly* to the surface under test. Lack of rigidity will falsify the test results. For 'good' bonding the surface under test should be 'slightly' rough and must be thoroughly degreased to form a clean base for proper adhesion. For self adhesive gauges, the gauge should be dipped quickly in acetone, shaken to remove the surplus liquid and 'quickly' affixed firmly and evenly to the test surface. The use of a suitable roller, or rubbing lengthwise with the thumb, with a piece of (tissue) paper between the thumb and the gauge, will facilitate firm and 'uniform' attachment. Even though special adhesives are available, non-(self) adhesive strain gauges can be fixed on the test surface by means of a *very thin* film of Araldite and following the same procedure as above, full twenty-fours being allowed to ensure proper adhesion. In most applications, this would result in reliable, satisfactory fixing of the gauge(s).

Temperature Characteristic

Primarily comprising an electrical resistance of a given dimension and specific resistance, strain gauges are prone to temperature variations of the ambient or otherwise and it is essential to compensate for the same to ensure reliable results. In most cases, this is achieved by the use of a "dummy" gauge, of identical type and mounted in exactly the same way as the test gauge and subject to the same temperature variation, *but free from strain*. The dummy gauge usually forms the adjacent arm of the Wheatstone bridge, used for the measurement[2].

---

[1]Or due to any other physical quantity such as torque developed in a shaft as discussed later.

[2]To eliminate errors caused by self-heating effects, the operating current through the gauge when in the bridge circuit is limited to be as small as possible; usually it is kept to a maximum of about 35 mA, a normal value being 5 mA, preferably even lesser.

Load Cells

An important accessory for measurement of pressure (and, of course, force) in a vast majority of cases, esp. dealing with large pressures, employing foil-type strain gauges is called a "load cell". This is a transducer incorporating strain gauges and a suitable diaphragm that converts the pressure into a proportionate electrical signal in a circuit, usually a Wheatstone bridge. The essentials of a typical load cell are depicted in Fig.13.6.

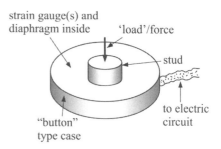

Fig.13.6 : A load cell using strain gauges

Construction and Operation

The cell comprises a flat cylindrical case, about 25 to 40 mm in dia., enclosing the strain-gauge assembly, attached to a flat diaphragm. The latter is fitted with a stud projecting upward to which is applied the pressure/'load' to be measured. The leads from the strain gauge assembly are brought out through a sleeve for external connections.

The conversion of pressure into corresponding electric signal is achieved by the physical deformation of strain gauges in the usual form. The gauges are bonded onto the circular diaphragm that deforms when weight or pressure is applied to it via the stud. In most cases, four identical strain gauges are used to obtain maximum sensitivity and temperature compensation. Two of the gauges are usually in tension and two in compression, and are wired with a suitable compensation adjustment. With this arrangement, an accuracy of 0.03% to 0.25% can be achieved.

An Application of Load Cells

Electric weigh bridge

A much useful and interesting example of application of load cells is the "electric weigh bridge" to measure the weight of loaded trucks or vehicles[1].

The weigh bridge incorporates a number of load cells, mounted appropriately underneath the platform which is free to move downward by a very small distance under the weight/pressure of the loaded truck. The pressure acting on the load cell is then 'transmitted' from the combined action of the load-cells/strain gauges to the measuring electrical circuit. The schematic location of two such load cells under the steel platform is shown in Fig.13.7.

---

[1] These 'facilities' can be seen variously along the sides of roads or highways where the truck can drive over to a huge steel platform and position itself suitably for being weighed.

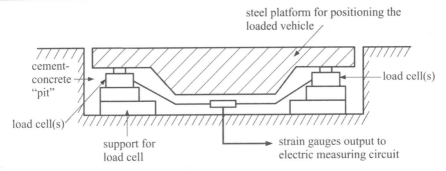

Fig.13.7 : Schematic of an electric weigh bridge

Piezo-electric Transducers

Piezo-electric materials, for example piezo-oxides, are polycrystalline dielectric materials, made from smaller crystallites, and can perform either of two functions:

(a)   under the action of mechanical 'energy', they result into a source of electrical energy;

  or

(b)   where electrical energy is available, the action of the device will be to produce a source of mechanical energy.

Hence, when acted upon by the pressure to be measured, a piezo-electric device constructed in a suitable form would produce an electrical signal that can be conditioned and measured (in analogue or digital form) to provide the value of the un-known pressure. An example of a typical piezo-electric pressure sensor assembly in common use is shown in Fig.13.8(a).

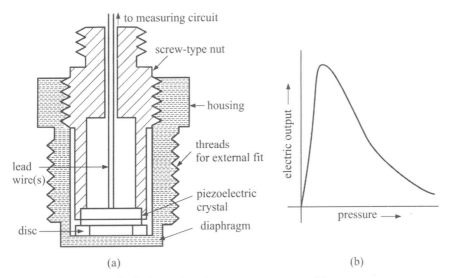

Fig.13.8 : A typical piezo-electric pressure sensor and its response curve

This type of sensor has a very rapid response/characteristic as indicated in Fig.13.8(b). However, a serious disadvantage is that the sensor would response to only *dynamic* (or fast changing) pressures, with practically no electric output for constant, static pressures[1].

Electromagnetic or Electromechanical Sensors

A common type of pressure transducer that is based on electromagnetic principle of operation is the "Linear Variable Differential Transformer", abbreviated LVDT. In effect, it is an electromechanical device that produces an electrical output which is *linearly* proportional to the displacement of a movable core in the device[2].

Essentially, it consists of a primary coil with two secondary coils placed on either side of a hollow cylindrical former. A cylindrical, rod-shaped soft (iron) magnetic core inside the coil assembly provides a path for the magnetic flux linking the coils. When the primary coil is energised by an alternating current, appropriate voltages are induced in the two secondary coils. These coils are connected in series with the start of each winding being connected together. See Fig.13.9.

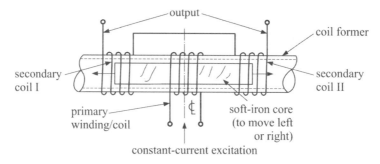

Fig.13.9 : Schematic of an LVDT transducer

As seen, the secondary coils are connected *in "phase" opposition* to produce a net zero signal output when the induced voltages in the two coils are equal (and opposite). This condition occurs when the core is centrally disposed between the two secondaries; that is, symmetrical about the LVDT mid-section or about the primary coil.

A movement of the core leads to an increase in magnetic coupling to the coil in the direction of movement and a (corresponding) reduction in magnetic coupling to the other coil producing a net output signal from the connected secondaries. Movement in the opposite direction produces an

---

[1]This 'property' makes the sensor ideally suited to applications where pressure changes cyclically with time as in audio devices, discussed later.

[2]For this reason, the device is also used in a variety of applications as a "displacement transducer" to measure a small, linear displacement or movement variously.

identical signal output, but of opposite sign. Thus, if the iron core is not saturated, the electric output signal will be linearly related to the displacement in a certain range as indicated in Fig.13.10. It is important to avoid saturation of the core by using a low, within-limits, constant current excitation for the primary coil and also use a core of high-permeability material.

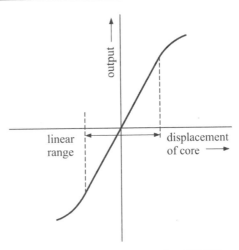

Fig.13.10 : Response curve of an LVDT transducer

The distinct advantage of using an LVDT transducer is that the moving core does not make contact with other electrical components (or circuits) of the assembly, as in the case of other types. This enables an LVDT transducer to offer high reliability and long life.

To form a pressure transducer, the core displacement of the LVDT is produced by the movement of a metallic pressure-responsive diaphragm or any other mechanism. Two (general) types are discussed here[1].

A. Bellows

These elements called "bellows" are cylindrical in shape and contain many circular folds of leather or similar material of decreasing 'diameters'. They deform in the axial direction (indicating compression or expansion) with changes in pressure. The pressure that needs to be measured is applied to one side of the bellows while the atmospheric pressure acts on the opposite side. [See Fig.13.11]. 'Absolute' pressure can be measured by evacuating the interior space of the bellows and then measuring the pressure on the opposite side.

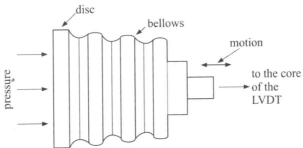

Fig.13.11 : Schematic of a bellows element

---

[1]Other, numerous possibilities exist depending on the ingenuity of the user.

B.   Bourdon tube

The principle behind a Bourdon tube is that an increase in pressure on the inside of the tube in comparison to the outside pressure causes the oval or flat shaped cross-section of the tube to try to achieve a 'deformed' circular shape as depicted in Fig.13.12. The phenomenon causes the tube to straighten itself out.

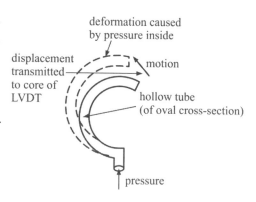

Fig.13.12 : A "C-type" Bourdon tube

This deformation can then be transmitted to the core of an LVDT in the form of a displacement and measure the corresponding output signal that can be related to the applied pressure.

Inductive Sensor(s)

These operate based on the principle of electromagnetic induction.

An example of an inductive pressure sensor incorporating a diaphragm is shown in Fig.13.13[1]. For this type of pressure sensor, 'chamber' 1 acts as the reference chamber with a reference pressure P_1, for example the atmospheric pressure. The coil L_1 located in this chamber surrounding the core is excited with a 'reference' current. When the pressure in the other chamber changes due to the pressure being measured, the

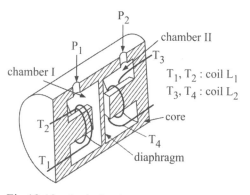

Fig.13.13 : An inductive type pressure sensor

diaphragm moves laterally, thereby inducing an EMF in the other coil, L_2 by virtue of *its* movement about the core. This voltage output may be used to give a measure of the pressure.

Capacitive Sensor(s)

A capacitive pressure sensor, in general, consists of a parallel plate capacitor, coupled with a diaphragm that is usually metallic and exposed to

---

[1]Likewise, an LVDT may also come in the category of an inductive sensor for the movement of iron core results in the change of relative inductances.

the pressure being measured on one side and a reference pressure on the other side. A typical sensor of this type is shown in Fig.13.14, depicting the essential parts.

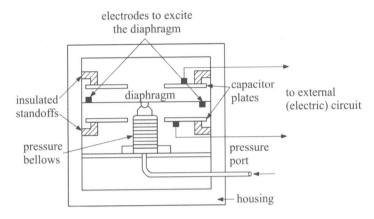

Fig.13.14 :   Schematic of a capacitive pressure sensor

Two electrodes are attached to the diaphragm and are excited by a high frequency oscillator to provide a steady-state 'reference' for the capacitor. Any movement of the diaphragm, caused by the pressure through the bellows, is reflected as a change of capacitance that is detected by an appropriate external circuit which results in an output voltage proportionate to the change of pressure.

MEASUREMENT OF TORQUE

An essential requirement to be met frequently for most rotating machines (or machinery), electrical or non-electrical, is the measurement of torque developed at the shaft when delivering a load. The developed torque may be *static* or *dynamic*. The former pertains to the one at constant speed, with no acceleration; for example, that of a synchronous motor, run at synchronous speed, or the torque transmitted to the wheel axles of a car in a given gear position and constant rpm. A common example of dynamic torque is that produced by a car engine at the crank shaft as the cylinders fire in succession. The most common means to measure torque has been, and still is, to use strain gauges, appropriately bonded to the rotating shaft, although other methods might also have been developed in the recent past.

Torque Measurement Using Strain Gauges

In one scheme, the principle of measurement of torque by strain gauge(s) is based on the *angular twist* developed in the shaft under the action of torsional stress, and corresponding strain sensed by the gauges. The gauges used for the purpose are either directly mounted on the shaft or on a torque 'tube', firmly fitted on to the shaft for the purpose of measurement, and experiencing the same torque[1]. The output from the strain gauges is fed to a (Wheatstone) bridge circuit as usual and the out-of-balance electric signal calibrated to read the torque directly on a suitable indicating instrument.

For best results, two identical strain gauges of 'standard' foil-type are used, bonded with their axes at 90° with respect to each other as shown in Fig.13.15; that is, each at 45° with respect to the axis of rotation of the shaft.

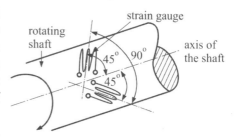

To ensure proper alignment (at 90° to each other), specially designed and manufactured strain gauges, *in a pair*, are now available, called "rosettes", as shown in Fig.13.16[2]. This is bonded to the shaft surface with *its* axis aligned with the shaft axis, and has the advantage that only two terminals are used for connections.

Fig.13.15 : Mounting of strain gauges on rotating shaft

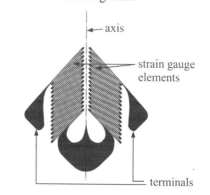

For more 'effective' measurements, *two* rosettes may be mounted on diametrically opposite surfaces of the shaft to provide two outputs. Such an arrangement will result in

Fig.13.16 : A (45°) rosette strain gauge

(a)   maximum output for a given torque, and hence maximum sensitivity in the bridge circuit;

(b)   self compensation against temperature increase or variation.

Transmission of Output

Since the shaft would be rotating, it is imperative that the 'output' from the strain-gauge(s) assembly is taken to the external stationary circuit by means of slip rings mounted on the shaft or the torque tube, and the use of stationary brushes pressing on the rings. Although straightforward in application, a great disadvantage of this arrangement is the "noise" (electrical as well as mechanical) produced at the slip rings and brush contacts which can become worse with the wear of the brushes.

---

[1]Other novel 'techniques' may also be employed; for example, one described later in the chapter, using a "torque arm".

[2]The term rosette in general describes a gauge with two or more 'standard' elements, mounted on a common base or backing and used for special measurements.

To overcome this problem, the output from the gauges is sometimes transmitted using

(a) a rotating transformer

(b) FM telemetering

as developed variously.

Torque Measurement Using Strain Gauges and a Torque Arm

A convenient means of measurement of torque, esp. in small or medium-size machines, is to use a "torque-arm" fitted to the shaft such that the *rotor is held stationary* against a rigid support, that is, restrained from any rotation, whilst the torque developed by the machine strains the strain gauges mounted on the torque arm. The arrangement is particularly suited to experimental tests on a machine in the development stage where the developed torque is a function of stator excitation, producing 'static' torque, leading to the plotting of typical "torque-excitation" curve of, say, an electric motor[1]. The schematic end-view of the general arrangement is shown in Fig.13.17.

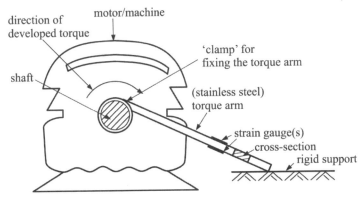

Fig.13.17: Schematic of measuring developed torque using a torque arm

The torque arm may simply be a stainless steel bar of rectangular cross-section and appropriate length, on the opposite flat sides of which are bonded two standard foil-type strain gauges. The use of stainless steel for torque arm is most suitable as it exhibits negligible mechanical hysteresis. Under the action of the developed torque in the assumed direction, the upper gauge is acted under compression whilst the lower is under tension, producing reasonably high signal/output and inherent temperature

---

[1]The scheme was quite successfully applied to derive torque-excitation characteristic of a hysteresis coupling in which the excitation was produced by a mechanically-rotated salient-pole field system, the torque being developed in the stationary rotor fitted with an annulus of hysteresis material and the rotor restrained from rotation by the use of a torque arm.

compensation. The 'output' from the gauges is fed to a bridge circuit as usual and the later calibrated to read the torque directly for a given excitation[1].

As seen, the scheme has the advantage of the torque arm being 'stationary' with no moving parts, requiring no slip rings or brushes. It, therefore, ensures better accuracy and repeatability. Also, if made detachable, the same torque arm can be used for measuring torque in a variety of machines where the requirement is to derive a developed torque v/s input characteristic[2].

MEASUREMENT OF TEMPERATURE

Measurement of temperature in a multitude of machinery, devices and equipment forms an important requirement related to their design, development and performance.

For the purpose of measurement, the variation of temperature(s) can be broadly classified in two categories:

  A.   steady-state, or
  B.   transient

Steady-State Temperature Rise

Steady temperatures are reached when the *final* rate of heat dissipation in the equipment matches the heat input or rate of heat generation on account of any physical changes. For example, in an electric machine it is the various losses that give rise to heating and temperature rise with time. After the elapse of "sufficient" time, the machine acquires a final, steady temperature under 'constant' working conditions as a result of action of cooling and ventilation[3].

Transient Temperature Rise

Under transient condition, in simple terms, the temperature of the machine may continue to rise with time; for example, the initial time interval of the

---

[1]Standard "torque meters" are commercially available which can be fed from the strain gauges mounted as above, and capable of indicating the developed torque in Nm on the dial.

[2]If required, to increase the signal level or sensitively of the circuit, the flat sides of the arm may be fitted with *four* strain gauges, two on either surface, and the leads taken to the bridge network appropriately.

[3]In practice, some of the heat generated is also 'absorbed' in the equipment in its various parts depending on the specific heat and other factors so that, for steady temperature rise, the heat balance equation will take the form

Rate of heat generation in the machine	=	Rate of heat absorption variously	+	Rate of heat dissipation from cooling surface(s) of the machine

steady-state (temperature) condition as depicted in Fig.13.18. In this context, the final, steady temperature (rise) may be reckoned as the 'end' stage of the phenomenon of heating.

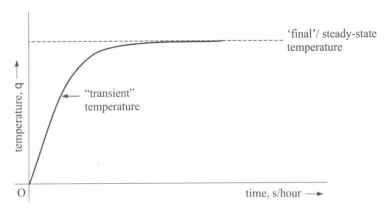

Fig.13.18 : Typical temperature v/s time curve of heating in a machinery

Strictly speaking, the transient temperature (rise) condition may be considered as one in which the *rate of rise of temperature*, that is dθ/dt, is 'extremely' high so that the temperature reaches a high value in a very short time interval, usually milliseconds. A simple example of this case is rise of temperature in an electric machine, over and above its steady temperature, in the event of a severe fault occurring in the machine, or the supply system to which it is connected, when high to very high fault current starts to flow. The temperature at various 'points' in such cases depends on various factors and may be best described by the "transient heat conduction equation", or the Poisson's equation

$$\frac{\partial}{\partial t}\left(\kappa \frac{\partial T}{\partial t}\right) + Q = C\frac{\partial T}{\partial t}$$

in which

κ = thermal conductivity

T = temperature

Q = generated heat

and   C = specific heat

pertaining to the various media/regions.

Clearly, in this case it is the *gradient* or rate of rise of the temperature that can play the dominant role in the operation and performance of the machine[1].

---

[1]An important case is the *transient* unbalanced operation of a turbo-alternator in which transient temperatures may rise to several hundred °C in a few *milliseconds* at key parts.

For the measurement of transient temperatures, the selection of appropriate transducers and process of measurement may generally depend on the kind of temperature (rise) to be measured.

Transducers for Temperature Measurement

The most commonly employed transducers for the measurement of temperature, usually *both* the steady-state and transient, from a few tens of degree to several hundred degrees, can be generally divided into the following three types:

A. resistance or resistance-dependent types

B. thermocouple (effect) types

C. infra-red (radiation) based devices

Resistance-type Transducers[1]

The basic principle of operation of this type is the effect of temperature on the resistance of the device making up the transducers, the resistance of which may vary with the temperature to which they are exposed, and which is required to be measured.

The resistance based transducers may be further classified as

(a) metallic, comprising a given metal that is most sensitive to temperature variations and possesses some other qualities;

(b) non-metallic, like thermistors which have the same or similar properties as the metallic ones in terms of resistance v/s temperature relationship.

Law of resistance variation

The law of *increase*[2] of resistance of a material with temperature, forming the basis of its selection for temperature measurement, has been found to be

$$R_t = R_0 \left(1 + \alpha t - \beta t^2\right)$$

where

R_t = the resistance at temperature $t°C$

R_0 = the resistance at $0°C$

and α and β are constants which depend on the physical properties of the given material.

---

[1] Also usually classified as Resistance Temperature Detectors, or RTDs.

[2] Unlike in most other applications, it is desirable in the case of temperature measurement that the increase in resistance with temperature is as much as possible; although an RTD would, in general, respond to any *change* in temperature.

In many cases, for simplicity, a single constant k, called the "fundamental constant", may be introduced which can be used to express the resistance at $t^{\circ}C$ by the simple expression

$$R_t = R_0 (1 + k\,t)$$

General requirements

It is imperative that the resistance-based sensors or transducers used for temperature measurement meet certain functional requirements for their reliable and satisfactory operation. Some of these are

- the relationship of temperature v/s resistance, or rather the rate of increase of resistance ($\Delta R/R$), of the sensor should be as much *linear* as possible[1]
- there should be no "set" effect; that is, the temperature reading should return to initial value when the temperature (effect) is removed
- the response of the detector should be as fast as possible
- the "ageing" effect should be negligible; that is, the "base" value of the resistance should be same even after years of use
- the physical condition, and hence the resistance properties, should not be affected by the environment of the equipment where temperature is being measured; for example, inside a furnace/oven by way of melting/oxidation or corrosion
- the device should be easy to calibrate and use, in general

Metals for temperature sensors

The most commonly employed metals for temperature measurement are platinum and nickel, although copper, too, has been in use for low temperatures, up to $100^{\circ}C$.

Platinum Resistance Detector(s)

Typically, a platinum-resistance detector comprises a *bare* platinum wire, about 0.2 mm in diameter, wound on a mica frame and enclosed within a porcelain or fused silica tube[2], the latter may further be encapsulated within a steel tube for (mechanical) protection. The leads from the platinum sensing element are taken to a (4-arm) bridge, or measuring circuit, which is often located at a considerable distance from the site of temperature measurement and necessitate the use of 'long' leads. Clearly, resistance changes with temperature in the leads may introduce appreciable error of measurement in most cases unless suitable compensation means are employed.

---

[1]However, this becomes less significant if a "point-to-point" calibration of the indicating device/instrument is carried out.

[2]All these materials are typically suited to withstand high temperatures ($>1000^{\circ}C$) for which the sensors may primarily be used.

An arrangement used in practice employing platinum-resistance sensors is illustrated in Fig.13.19, incorporating a Wheatstone bridge, A B C D, and compensating leads.

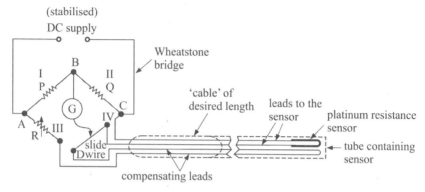

Fig.13.19 :   A Wheatstone bridge network for temperature measurement using platinum resistance sensor

In the circuit, P and Q are "ratio" arms of the bridge as usual whilst in the third arm is a variable resistor, R, in series with the compensating leads. The actual (platinum) sensor coil with its leads is contained in the fourth arm, with a slide wire in between these last two arms as shown.

If properly compensated, the Celsius temperature θ_p measured by the platinum sensor can be given by

$$\theta_p = \frac{R_\theta - R_0}{R_{100} - R_0} \times 100°C$$

where R_θ, R_0 and R_{100} are the resistance values of the platinum RTD at the un-known temperature θ, the ice-point and the 'steam' point, respectively. With the resistance values R_0 and R_{100} as constants, R_θ is linearly related to θ_p by $R_\theta = k_1 \theta_p + k_2$; k_1, k_2 being derived constants. Following the knowledge of all the resistances, or rather the constants k_1 and k_2, the bridge-circuit can be calibrated to read θ_p, in accordance with the measurement of R_θ.

Nickel Temperature Sensors

These are similar to foil-type strain gauges and are particularly suited for *surface* temperature measurements in the range of –200°C to 300°C, particularly the cryogenic temperatures. They are produced by a photo-etching process in exactly the same way as a typical strain gauge, having a resin backing with a thickness of about 0.02 to 0.04 mm. Two typical forms of nickel temperature sensors are shown in Fig.13.20. Note that whilst the strain gauges are designed to give outputs dependent on the strain and (ideally) independent of temperature variation, nickel sensors are highly sensitive to thermal effects as expected.

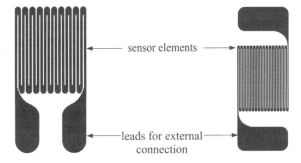

Fig.13.20 : Typical examples of nickel temperature sensor

Nominal resistance of a typical nickel sensor would be 50 Ω at 24°C. Since the temperature coefficient of nickel is high, relatively large resistance changes can occur; for example, the resistance value would double to 100 Ω when a temperature of 190°C is reached. A typical variation of change of resistance as a percentage of the 'base' resistance, that is $(\Delta R/R) \times 100$, with temperature is shown in Fig.13.21. An advantage of these sensors is the extremely low thermal inertia owing to the fine wire size, leading to high sensitively and fast response in transient temperature measurement.

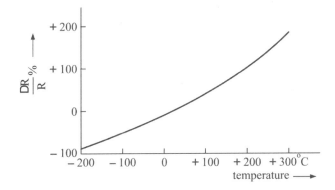

Fig.13.21: Variation in resistance with temperature of a nickel sensor

Thermocouple Transducers[1]

The basis of working of a thermocouple, widely used in steady-state and transient temperature measurements, is the thermoelectric effect produced in junctions of two *dissimilar* metals when maintained at different temperatures. The mechanism that converts heat into electricity was discovered by Seeback in 1822 and hence the effect is also known as Seeback effect. In his experiments, Seeback connected a plate of bismuth

---

[1]The word "thermocouple" for the device derives from the fact that its principle of operation is based on the use of a pair, or a 'couple', of two dissimilar metals with their junctions exposed to thermal effects.

between copper wires, leading to a galvanometer as shown in Fig.13.22. He found that if one of the bismuth-copper junction was heated whilst the other was kept cool, a current flowed through the galvanometer, indicating production of EMF across the junctions.

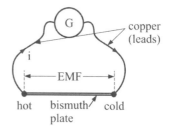

Fig.13.22 : Seeback's experiment

The (thermoelectric) effect was observed to manifest in any pair of *dissimilar* metals forming two junctions, one maintained at higher temperature relative to the other. The metals exhibiting the effect may be copper, iron, constantan, bismuth, antimony, zinc, lead, platinum and so on. For example, the arrangement using copper and iron wires (a commonly used combination) to form a thermocouple will look like a device shown in Fig.13.23, to demonstrate the thermo-electric effect for measuring temperature of boiling water[1].

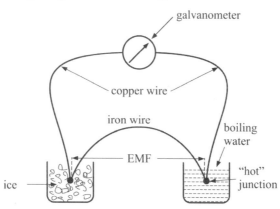

Fig.13.23 : A copper-iron thermocouple

The law of thermoelectric EMF

A generally used expression, obtained experimentally, for the EMF produced by way of variation of the temperature *difference* between hot and cold junctions of a thermocouple may be written as

$$E_{th} = \alpha \, (T - T_0) + \beta \left(T^2 - T_0^2 \right) \text{ V}$$

where T and T_0 denote the absolute temperatures ($^{\circ}$C + 273) of the hot and cold (usually at 0°C) junctions, respectively, and α and β are constants which depend upon the metals forming the thermocouple. The law holds for most

---

[1]This arrangement is fundamental to measurement of temperatures using thermocouples in all applications, and continues to be so.

thermocouples. After selection of a pair of metals, which may be based on several considerations, the thermo EMF may be pre-estimated for a given range of temperatures, $(T - T_0)$, using the above formula.

Thermocouples for steady-state and transient temperature measurements

As seen from the above law, the EMF produced for any given temperature difference between the hot and cold junctions for a chosen pair of metals, is constant at a given hot junction temperature, regardless of the size (or diameter) of the two wires as well as of the area of contact at the junction. For steady-state temperatures, such as in furnaces or ovens where measurement of only the final, average temperature is of interest, a thermocouple may be made of thick wires, usually up to 2.0 mm in dia. (that is, #12 or 13 SWG), to provide a sturdy construction. What is important is a good 'joint' (or the junction) of the two wires and thermal contact with the body or surface of which the temperature is to be measured. The metals used for the wires may be iron and constantan for temperatures up to 900°C. To protect the wires, each wire may be inserted in a porcelain sheath (or tube) up to the junction, separated by mica washers as shown in Fig.13.24.

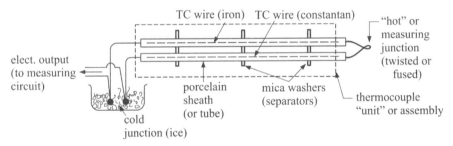

Fig.13.24 : Construction of thermocouples for industrial applications

The junction itself may be formed by either firmly twisting or fusing (for better 'contact') the two wires together.

Transient Temperature Measurement

When it is required to measure transient temperatures in the region of interest, with the rate of rise of temperature being hundreds of degrees per second; for example, surface temperatures in some applications, it is imperative to use as thin a wire-pair as possible so that the hot junction responds to the temperature rise 'spontaneously'.

Time constant and response time of a thermocouple

For transient measurements, time constant is the most important aspect of dynamic performance of a thermocouple (or any temperature transducer) and is defined as "the time required for the thermocouple output to reach 63 percent of its final, steady value, following a 'step' input of heat"[1]. However, its specification such as "time constant : 100 ms" is meaningless unless accompanied by a statement of the two limits of the step change in temperature and the type and flow rate of the medium or measured fluid; for example, "from still air at $25° ± 5°C$ to distilled water at $80° ± 2°C$, moving at 1 m/s".

Response time is "the time required to reach a definite percentage of the final, steady value (for example 90, 95, or 98 percent)" and should be stated accordingly. The term is specified under the same test conditions as applicable to time constant.

Another time of interest is "the time in which the output of the thermo couple changes from a small to a large, specified percentage of the final value" and is called the rise time; for example, time from 10% to 90 % of the final value. As an illustration, the output/response of a copper-constantan thermocouple, constructed using 36 SWG wires, under different test conditions is shown in Fig.13.25. Clearly, the response would be much faster if finer wire were used; however, the construction would be rather tedious and fragile.

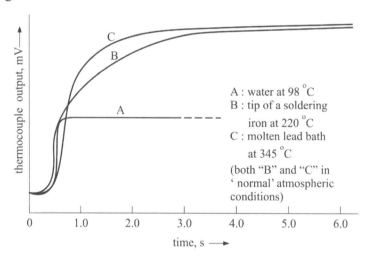

Fig.13.25 : Transient response curves of a 36 SWG copper-constantan thermocouple, recorded on a UVR

The various (recorded) response times to 100% are:

A. still air to boiling water at 98°C   : 0.7 s

B. still air to tip of a soldering iron at 220°C   : 3.3 s

C. still air to molten lead bath at 345°C   : 4.4 s

---

[1]This follows from an *exponential* rise of temperature as sudden heat is applied.

Thermocouple output

The output from a thermocouple may vary from a few microvolts to millivolts, depending on the difference of temperatures in the region and, of course, on the type of metal pair. This can be measured

(i)   directly with the help of a galvanometer
(ii)   on a potentiometer
(iii) on a CRO
(iv)   using a bridge circuit
(v)   by recording on a UVR, with some (pre)amplification of the signal, if required.

The measuring device which may in many cases be an indicating meter, can be calibrated to read the temperature directly. The calibration may be carried out by reference to boiling or melting points of various substances; for example, for low temperatures, boiling point of pure, distilled water which would correspond to 100°C at NTP.

Infra-red Radiation Based Devices

Pyrometers

Pyrometers are specially designed and manufactured instruments, or pieces of apparatus, which are generally used for direct measurement of high to very high temperatures, particularly under circumstances where it may not be possible to employ ordinary thermometers, that is, RTDs, thermocouples etc., in the usual manner.

Whilst for temperatures up to about 1200°C, the (internally located) sensors for the pyrometers may still be of resistance-type or thermocouples, for temperatures above 1200°C, for example those encountered with arc furnaces, it is more convenient to employ some method of measurement in which the measuring apparatus is not subjected to the usual, full heating effect of the source of which the temperature is to be measured. Pyrometers for such measurements depend for their action on the heat *radiated* from the source, for example inside of the arc furnace, and are called "radiation" pyrometers.

Two commonly-used types are considered here.

Fixed-focus Type Pyrometers

The schematic of a "standard" fixed-focus type radiation pyrometer is shown in Fig.13.26. It essentially consists of a long metallic tube AB containing a small concave mirror M at the far end of the tube as shown, positioned to focus the heat rays which pass through the tube, from the source XY via the narrow aperture PQ at the 'near' end of the tube. The image formed at the point T by the heat rays, radiated from the source, is a "heat image" and it is always in focus, requiring little adjustment over a wide range of distances,

D, between the pyrometer and the source of heat, this property being brought about by making the tube long and the aperture small; then D being not so critical. A sensitive thermocouple is located at T, the focal point of the mirror, to 'capture' the heat, originally emanating from the heat source.

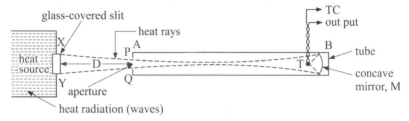

Fig. 13.26 : Fixed-focus type of radiation pyrometer

The EMF produced in the thermocouple is measured in the usual manner as previously described and the output calibrated to read the temperature directly. For improved functioning, the 'surroundings' at T is blackened so as to approach as closely as possible the theoretical "black body", to absorb almost all of the radiation falling upon it.

Following Stefan-Boltzmann law of radiation[1], the temperature rise sensed by the thermocouple is proportional to the fourth power of the absolute temperature of the hot body whose temperature is to be measured. For this reason, the scale of the indicating instrument used with radiation pyrometers is cramped at the lower end and sparse at the upper end; for specific applications, the pyrometer is selected to read beyond the cramped part of the scale.

Infra-red Pyrometers (or Thermometers)

Instead of a light beam emitted from a hot body or surface (or a furnace enclosure), the alternative that a highly-heated body produces infra-red radiation (sometimes called "blackbody" radiation), in the form of electromagnetic waves of appropriate wave-length (less than 8000 AU) forms the basis of infra-red pyrometers, now very commonly in use. At the same time, a distinct colour of the heated/hot object in the visual spectrum is indicative of the *order* of temperature reached as given in Table 13.1, pertaining to a particular infrared radiation[2].

---

[1]The law can be stated as

$$W = K\left(T_2^4 - T_1^4\right),$$

where W is the energy in J/cm^2/s, K a constant and T_1 and T_2 are absolute temperatures of the 'cold' and 'hot' body, respectively, applicable to the thermocouple in this case.

[2]Again, generally governed by the Stefan-Boltzmann simplified expression $J = \varepsilon \, \sigma \, T^4$ where J denotes thermal radiation, ε the emissivity and σ a constant.

Table 13.1: Body temperature and acquired colour

Temperature, °C (approx.)	Colour
480	faint red glow
580	dark red
730	bright red
930	bright orange
1100	pale-yellowish orange
1300	yellowish white
>1400	white

Each colour (or shade) corresponds to a given wavelength (or frequency of radiation) in the infra-red spectrum of electromagnetic radiation.

Principle of operation

The basic design of an infra-red pyrometer (or thermometer) consists of a lens to focus the infra-red (thermal) energy on a detector, for example a suitable thermocouple, similar to that in a radiation pyrometer, which converts the energy to an electrical signal that can be related to, and displayed as, the temperature to be measured. This configuration facilitates temperature measurement from a distance *without any contact with the object* of which the temperature is to be measured, thereby being most advantageous under circumstances where thermocouples or other probe-type sensors cannot be used, or do not produce accurate data for various reasons.

Some special circumstances of the scheme are where the object of which the temperature is to be measured is moving, or surrounded by an electromagnetic field (as in induction heating); or where the object is contained in vacuum or other controlled atmosphere[1].

It is possible to achieve an accuracy up to ±2°C using infrared thermometry.

The process of operation of a common infra-red measuring system is shown schematically in Fig.13.27.

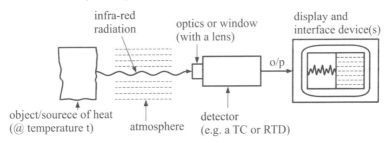

infra-red radiation   optics or window (with a lens)   display and interface device(s)

o/p

object/source of heat (@ temperature t)   atmosphere   detector (e.g. a TC or RTD)

Fig.13.27 : An infra-red measuring system

[1]An important, modern application of infra-red thermography is in the measurement of (human) body temperature using what are known as (and commercially available) clinical thermometers.

Some aspects of infra-red thermometry are

(a) the object (or the target) must be *optically* visible to the thermometer; high levels of dust or smoke may flout the measurements;

(b) concrete obstacles, such as a closed metallic reaction vessel, allow for only *surface* temperature measurement: the inside temperature cannot be measured;

(c) the optics of the sensor or detector must be protected from dust and condensing liquids for accurate measurements.

For the convenience of "quick" measurements, hand-held devices are now available that can be 'aimed' at the heated region/spot and the temperature indicated directly as a digital display, esp. in the case of clinical thermometers.

MEASUREMENT OF FLOW OF LIQUIDS

Measurement of flow across vessels or through the conduits can be interpreted in two ways:

(a) measuring *quantum* of liquid or volume flowing per unit time, for example m^3/s; or

(b) measuring the *velocity* of the flow which can be related to the volume if area of cross section of the conduit is known.

The measurements are best carried out by measuring a derived non-electrical quantity dependent on the flow rate. For example,

(a) the *cooling effect* of the liquid as it passes a heated element that is measured as a temperature change proportional to the flow-rate;

(b) the *differential* pressure developed across a calibrated orifice, indicating the flow rate and which is measured using a pressure transducer connected in its differential mode;

(c) the *rotational* speed of a small turbine-type rotor that may be taken as an indication of the flow of liquid passing over it[1].

In all the above schemes, the final stage is a suitable transducer or sensor that would convert the non-electrical quantity into a proportionate *electrical* signal or output, to be measured/monitored and related to the flow of the liquid[2].

---

[1]The rotational speed can be related to velocity of flow and, by multiplying by the area of cross-section, yield the volumetric flow rate.

[2]It is important that the type of liquid dealt with for the flow measurement is an *incompressible* fluid; that is, changes of pressure cause practically no change in the fluid density at various stages of flow. Such a liquid would also follow the common 'laws'/expressions of motion.

Bernoulli's Principle

About 1740, Bernoulli brought out a relation between pressure and velocity at different parts of a moving incompressible fluid that forms the basis of various flow measuring schemes. If the viscosity of the fluid is negligible, for example that of water, so that the frictional forces are negligibly small, then Bernoulli's principle can be stated as "the work done by the pressure difference per unit volume of a fluid flowing along a pipe steadily is equal to the gain in kinetic energy per unit volume PLUS the gain in potential energy per unit volume". That is,

$$P_1 - P_2 = \frac{1}{2}\rho \left(v_2^2 - v_1^2 \right) + \rho\, g\!\left(h_2 - h_1 \right)$$

↑	↑	↑
pressure difference	KE (per unit volume)	PE (per unit volume)

where

P$_1$ and P$_2$ are the pressures of the liquid at the beginning and end of the pipe;

v_1 and v_2 are the respective velocities;

ρ the density of the liquid; and

h$_1$ and h$_2$ are the respective heights measured from a fixed level at the beginning and end of the pipe.

[For a horizontal pipe, h$_2$ = h$_1$ and PE term will be zero].

Venturi meter

This device is frequently used to measure the volume of a liquid per second through a conduit. The operation of the meter is based on the Bernoulli's principle and is illustrated in Fig.13.28.

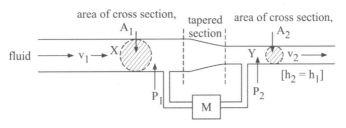

Fig.13.28 : Principle of operation of a venturi meter

Construction and Principle of Operation

A venturi meter is of tubular construction having two parts of different diameters, joined by a tapered section as shown. The wider section X has a cross-sectional area of $A_1(m^2)$ whilst the narrow section has the area $A_2(m^2)$. The meter is held *horizontally* so that both the sections are at the same height above the ground. Following the property of an incompressible fluid, since the velocity v_2 at Y is greater than the velocity v_1 at X, the pressure P_2 of the fluid at Y is less than the pressure P_1 at X. A manometer M is attached to the meter as shown to measure the difference in levels, H, of a liquid of density ρ', say mercury.

Suppose Q is the volume per second of the liquid flowing at X or Y. Then $Q = A_1 v_1 = A_2 v_2$. Also, from the Bernoulli principle

$$P_1 - P_2 = \frac{1}{2}\rho\left(v_2^2 - v_1^2\right) = H\,\rho'g, \text{ for the horizontal meter,}$$

that is, the pressure differential measured by the manometer.

But $\qquad v_2 = \dfrac{Q}{A_2} \text{ and } v_1 = \dfrac{Q}{A_1}.$

Therefore, by substituting for v_1 and v_2

$$H\rho'g = \frac{1}{2}\rho\left(\frac{Q^2}{A_2^2} - \frac{Q^2}{A_1^2}\right) = \frac{1}{2}\rho\,Q^2\left(\frac{A_1^2 - A_2^2}{A_1^2 A_2^2}\right)$$

or $\qquad Q = \sqrt{\dfrac{2H\rho'g\,A_1^2 A_2^2}{\rho\left(A_1^2 - A_2^2\right)}}$

from which Q can be determined from the known constants of the venturi meter and the liquid, and measured value of H. Since $Q \propto \sqrt{H}$, an experiment can be carried out to calibrate the difference in levels H of the manometer in terms of flow volume per second, using the above expression.

Example

Water flows steadily along a horizontal pipe at a volume rate of 8×10^{-3} m³/s. If the area of cross section of the pipe is 40 cm², calculate

 (i) the flow velocity of the water;

 (ii) the total pressure in the pipe if the static pressure in the horizontal pipe is 3.0×10^4 Pa, assuming that water is incompressible, non-viscous and its density is 1000 kg/m³.

 (i) Velocity of water $= \dfrac{\text{volume per second}}{\text{area of cross section}} = \dfrac{8 \times 10^{-3}}{40 \times 10^{-4}}$

$\qquad\qquad\qquad\qquad = 2$ m/s

(ii)   Total pressure $= \text{static pressure} + \dfrac{1}{2} \rho v^2$

$$= 3.0 \times 10^4 + \dfrac{1}{2} \, 1000 \times 2^2$$

$$= 3.0 \times 10^4 + 0.2 \times 10^4$$

$$= 3.2 \times 10^4 \text{ Pa}$$

Measurement of Flow Rate in Practice

Although innumerable techniques are available for measurement of flow in practice, two examples are described as illustrations.

Flow meters

Flow meters are instruments that are specially designed to measure flow of liquids and indicate it in practical units in terms of litres or cubic metre.

Residential "Water Meter"

This meter is inserted in the residential pipe line for measuring consumption of water over a period of time (usually one month) by the user, as depicted in Fig.13.29.

Fig.13.29 : A typical residential water meter

Most meters in the water distribution system in residential townships, or even in commercial complexes, are designed to measure only the volume of 'cold', potable water. However, there are specialty water meters manufactured for other, specific uses; for example, 'hot' water meters are designed with special materials that can withstand high temperatures.

The water meters most commonly used to measure 'moderate' flow rates, typically for residential and small commercial consumers, are the "displacement type", also referred to as Positive Displacement or "PD" meters. Two common methods of producing this positive displacement are by using (i) an "oscillating piston", and (ii) a "nutating disc" (based on revolving mechanism). Either method relies for its operation on the water to physically displace the moving measuring element in direct relation to the amount of water that passes through the meter. The piston or disc used in the meter for producing displacement moves a magnet that drives a register[1].

The meters are normally provided with a built-in strainer to filter out any small pebbles or debris that may be accompanying the water and which may cause damage to the mechanism. As expected, such meters are inherently uni-directional. The meters normally have bronze, brass or plastic bodies with internal measuring chambers made from moulded plastics or stainless steel to avoid corrosion, and maintain accuracy of measurement for sufficiently long period of usage.

Electromagnetic Meters

These are special-category flow meters for better accuracy of measurement and provide an *electrical* output proportional to flow *rate* that can be related to fluid flow. Apart from mechanical flow meters, these are the most common type, offering various advantages and are also known as "mag meters".

Technically, these are velocity-type meters, except that they use electromagnetic properties to determine the water flow velocity, rather than mechanical means such as jet or a (small) turbine.

The meters work on the principle of Faraday's law of electromagnetic induction that states that a voltage will be induced when a conductor moves through a magnetic field which itself may be AC or DC. In this respect the liquid of which the flow is to be measured serves as a conductor; the magnetic field being created by energised coils outside the flow tube as shown in Fig. 13.30[2]. The flowing fluid through the magnetic field produces an electric potential difference across two electrodes mounted on the pipe a small distance apart. Thus, for a constant field, the PD across the electrodes is proportional to the flow rate of the liquid. Clearly, the fluid passing through the meter must be *electrically* conducting (water being one such fluid) having as high electrical conductivity as possible to obtain high electrical output.

---

[1] The registration referred to here is mechanical in nature using a gearing mechanism to indicate the water consumption by means of dials or in 'windows', accurate enough for most applications. For more accurate results, the motion of the magnet may be related to an electrical output and digital indication.

[2] As an alternative, 'good' quality, high-'energy' permanent magnets, suitably shaped may be used for a sturdy construction.

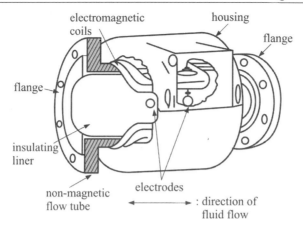

Fig.13.30 : Details of an electromagnetic flow meter

Electromagnetic flow meters have the advantage of being equally capable of handling water that may be raw (untreated/unfiltered) as well as other corrosive liquids and slurries since the liquid does not come in contact with the measuring mechanism. There are no moving or oscillating parts that would require to be activated by the fluid.

The meters may also be used to measure the flow of conducting gases since viscosity of the fluid does not come into picture. Another important advantage is that the meters can measure forward as well as reverse flow with equal accuracy; that is, the meters are bi-directional and can be fitted into the pipe network in either position.

Since stray electrical potential due to, for example, surrounding stray, strong magnetic field(s) may affect the accuracy of readings, most magnetic meters are installed with either grounding rings or a grounding electrode to 'divert' stray PD away from the electrodes inside the flow tube, used to measure the flow. The part of the tube carrying the fluid inside the meter has to be non-magnetic and electrically insulating as shown.

An added advantage of these meters is that the output being an electric PD can be monitored or measured at a distance, if required. Their one disadvantage is that they cannot be used with electrically non-conducting liquids.

MEASUREMENT OF LIQUID LEVELS

The measurements related to liquid *levels* may generally belong to two requirements.
- (a) to *continually* monitor the level of liquid in, say, a reservoir or tank as the liquid pours into it from a pipe or other means;
- (b) to *control* the level of the liquid in the tank when it reaches a given height, for example to stop the input flow to avoid overflow, or to

recommence the flow of liquid into the tank when the level falls below a given height/depth[1].

In either case, special sensors or transducers are required to perform the desired function and produce an output corresponding to the desired condition to lead to appropriate action, mostly an automatic control through an electric/electronic action.

Resistance/capacitance Level Controller

Some of the conventional means used, esp. in the first requirement, may be based on the "resistance" or "capacitance" principle that may vary according to the liquid level.

Resistance type controller

This device works by employing the liquid surface as the 'sliding' contact moving over a set of fixed contacts of a resistance divider. Clearly, this is applicable for conducting liquids only such as water. The principle of operation of such a set-up is illustrated in Fig.13.31 and is meant to operate for *discrete* water levels. The arrangement shown in the figure pertains to up to five 'fixed' levels of the fluid; for nearly continuous level monitoring, a multitude of electrodes will have to be inserted. Whilst simplest in construction, the scheme can become cumbersome when the number of electrodes is too large and their relative positional adjustment difficult to carry out. Also, for corrosive liquids the life of electrodes may be greatly reduced and the calibration may be adversely affected if the conductivity of the liquid changes with time or temperature. In the circuit, "R_s" may be different as R_1, R_2, ... etc., for varying heights or levels in the tank.

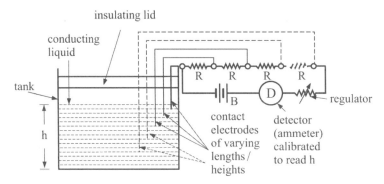

Fig.13.31 : Liquid level measurement by stepped-resistance method

---

[1] A common application being in a residential overhead water tank where it is desirable to automatically switch off the pump motor when the water reaches a high mark, and restart the motor when the water level goes down, to nearly empty tank depth.

Capacitance type level controller

The capacitance-type controllers operate on the basis of effecting *change* of capacitance in the control circuit as the liquid level rises or falls. The two practical devices can be based on variation of capacitance between

(a) two fixed, vertical electrodes as the space between them is filled with more, or less, liquid *which must be electrically insulating*;

(b) the surface of the liquid and a fixed electrode above it, the scheme being applicable for only conducting liquids, as illustrated in Fig.13.32.

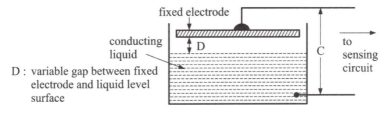

Fig.13.32 : Capacitance-based level detector

In this case, the 'effective' capacitance between the fixed electrode and upper, conducting surface of the liquid depends on the gap between the two and the 'common' area, with the 'trapped' air being the dielectric. It is implied that, for an appreciable capacitance, the shape and area of the electrode should match the liquid surface as far as possible. Also, for good results, the surface of the liquid must be 'clean', 'still' and not undulating.

Level Detector Using Ultrasonic Sensor(s)

A modern detector to monitor continually-varying level of a liquid – conducting or non-conducting – comprises an ultrasonic sensor. These sensors incorporate an analogue signal processor, a microprocessor and an output driver circuit. "Transmit" pulses and a gate signal from the microprocessor route through the analogue signal processor to the sensor which sends an ultrasonic beam to the

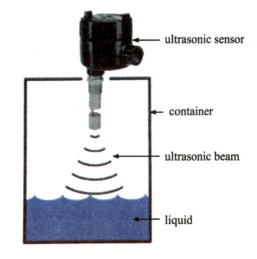

Fig.13.33 : Non-contact ultrasonic level detector

liquid surface as illustrated in Fig.13.33. The sensor detects the *echo* from the surface and routes it back to the microprocessor for a digital interpretation of the distance between the sensors and the liquid surface level. Through constant updating of received signal, the microprocessor calculates averaged values to provide a measure of the liquid level and its digital indication.

On-off Type Water Level Controller

These are automatic operating detectors-cum-switches to control the "low" and "high' levels of a particular liquid.

Float-"switch" Controller

One of the most common types of scheme comprises a simple float, generally employed for closing and opening of pressure-activated water tap in a tank, controlling a switch to make or break an electrical contact in an external "control" circuit. A simple arrangement and all details of such a scheme for controlling a pumping motor is shown in Fig.13.34[1].

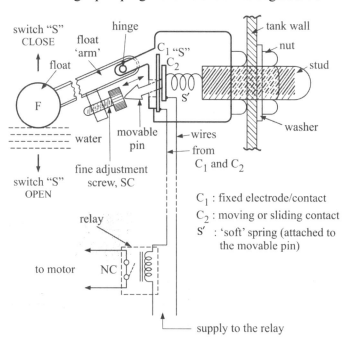

Fig. 13.34 : A simple float-type ON/OFF switch

[Clearly, by fitting the assembly "up side down", the ON/OFF action can be reversed and "S" can be wired appropriately to operate the motor directly, that is without the relay, to switch it off when the water level has risen to the desired height]

---

[1]Developed by the author to switch on and off the electric supply to the 1-phase motor used to pump water from an underground sump to an overhead tank.

Construction and Operation

The float F is hinged about a pin attached to a box-type body fitted to the tank wall, and is free to move up and down with the water level. When the latter is lower than the desired height the switch "S" is open or "OFF" as the contacts C_1 and C_2 are separated against the pressure of the spring S, by the action of the float 'pushing' the pin back.

The switch is connected to a relay, the "normally closed" (NC) contacts of which control the pumping motor. When the desired level is reached (for example when the tank is "full") the contacts C_1 and C_2 close by the spring action, the relay is energised, opening (NC) contacts and disconnecting the motor. The operation of the switch with respect to the water level can be adjusted by turning the screw SC.

Float Switch for Low AND High Liquid Level Control

A compact, encapsulated (or sealed) float switch offering NO and NC operation for "low" and "high" liquid level control incorporates a float activating a reed relay or switch (type SPST). The switch can be used at the top or bottom of the tank, remaining submerged for the "bottom" action. A permanent magnet fitted to the float moves up and down according to the liquid level triggering the reed relay.

The operation for the two situations for a typical switch of the above type is shown schematically in Fig.13.35 (a) and (b). The reed relay switch can be set to normally open (NO) or normally closed (NC) action simply by rotating

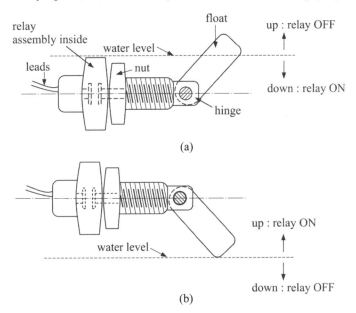

Fig.13.35: Compact, submersible float switch assembly

the sensor through 180 degrees (or one half turn) making it easy to operate with systems which turn on/off when the liquid reaches a certain level (float rising), or turn on/off when the liquid falls below a certain level (float falling).

Thus, the same relay unit can be used for either operation. The actual arrangement of the reed contacts and permanent magnet, of annular make and capable of sliding about the central limb, is shown in Fig.13.36. The reed switch contacts close or open depending on the relative position of the magnet, which itself depends on the liquid level gauged by the float resting on the liquid surface. The reed switch of the above type may be rated for currents up to 1A and voltages up to 100 V DC or 250 V AC[1].

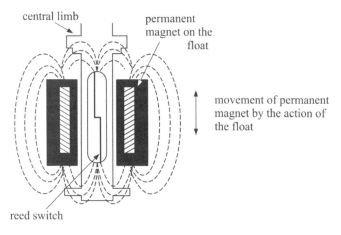

Fig.13.36 : Hermetically sealed reed switch assembly

The float-type assembly may be fitted to the tank by drilling an appropriate hole through the side of the tank at the optimum height, pushing the float switch stud through and tightening a nut, the whole fixture then being sealed to make the fitting leak proof.

Thermal Level Switches

A novel type of liquid level sensor or transducer is the thermal level switch that senses either the difference between the temperatures of *vapour* space (above the liquid in a closed tank etc.) and the liquid, or more commonly, the increase in thermal conductivity as a probe becomes submerged in the liquid in question.

One of the simplest thermal level switch design consists of a temperature sensor heated with a constant amount of heat input (for example by external heating). As long as the probe is in the vapour space, the probe remains at a

---

[1]The switch in turn may control the actual contactor relay, rated to suit the switching or operating current of a motor or other device.

high temperature, because low-conductivity vapours do not conduct much heat away from the probe. When the probe is submerged, the liquid absorbs more heat and the probe temperature drops, actuating the switch.

Another type of thermal sensor uses two resistance temperature detectors (RTDs), both mounted at the same elevation inside the tank. One probe is heated and the other provides an unheated reference. The outputs from the two probes are fed into a Wheatstone bridge circuitry as shown in Fig.13.37. Whilst the sensor is in the vapour phase, the heated probe will be warmer than the reference probe and the bridge (circuit) will be unbalanced. When both probes are submerged in the liquids their temperatures will approach that of the liquid. Their outputs will nearly be equal and the bridge balanced.

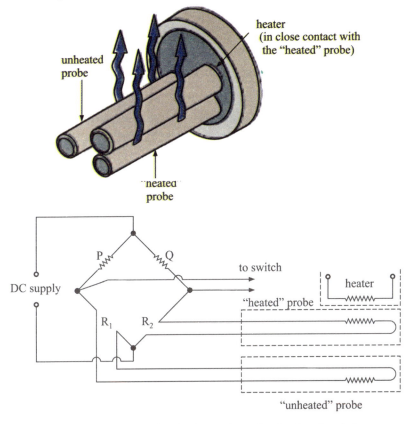

Fig.13.37 : Schematic of a thermal-conduction level switch

Since all liquid materials have their characteristic heat transfer coefficients, thermal level switches can be calibrated to detect the presence or absence of any fluid. Thus, these switches can be used for several purposes, such as interfaces and to detect the levels of slurry and sludge etc.

An important advantage of the scheme is that the switches have no mechanical or moving parts and hence may be rated for pressures up to

3000 psig and liquid temperatures as low as –75°C. When detecting level(s) of water, the response time is typically 0.5s and accuracy of water level within 2 mm[1].

Vibrating Switches

Vibrating level switches detect the dampening that occurs when a vibrating probe, usually in the form of reed, is submerged in the liquid. The reed switch consists of a paddle, a driver coil and a pick up. The driver coil induces vibrations in the paddle, of the order of 120 Hz, that is damped out as soon as the paddle gets covered by the liquid and activates the pick up. The switch can detect both rising and falling levels of the given fluid, only its actuation depth (the fluid depth over the paddle) increases slightly as the density of the process fluid decreases.

A Simple Overflow Alarm System

An alarm system to indicate the overflow of water in the overhead tank, a common occurrence in residential buildings, to alert the residents to manually switch off the pump motor is illustrated in Fig.13.38. The system makes use of conducting property of water, to act as an electrical link in a pre-wired circuit as shown.

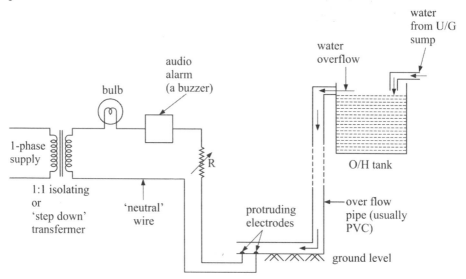

Fig.13.38 : Water overflow alarm system

The system comprises two electrodes, fitted into the horizontal part of the PVC *overflow pipe* (usually about 5 to 8 cm in dia.), protruding about a cm

---

[1]Clearly, the response time would depend on the viscosity of the liquid, being relatively high for high viscosity liquids.

inside the pipe as shown. As the water from the overhead tank begins to flow across the electrodes, shorting them electrically, the circuit on the secondary side of the transformer is completed causing the alarm to sound. A bulb of suitable rating may also be included in the circuit providing a visual indication of the overflow. A wire-wound variable resistor, R, may be inserted in the circuit as shown to limit the current to the desired value. It is important to choose the electrodes material to be free from corrosion; in practice, these may simply be stainless steel screws with nuts, to be replaced in due course, if necessary.

The system is particularly suited when the overhead tank is inaccessible for some reasons and may not be able to be fitted with other devices.

MEASUREMENT OF VIBRATIONS

Vibrations may occur variously and frequently in static or dynamic systems or equipment and may have to be detected and measured as required. The vibrations may be mechanically induced; for example, in a turbogenerator when its rotor is not dynamically balanced, or induced electromagnetically such as in a transformer. In most cases, vibrations manifest as sound (or noise) at a certain level, and if not detected may result in serious (even irreparable) damage whilst being a nuisance to the environment.

Piezoelectric Transducers or Pickups

Almost all modern vibrations-detecting transducers are piezoelectric type. When certain materials are subjected to electric 'stresses', related mechanical stresses will occur in the material. Conversely, mechanical stresses applied to the material will usually manifest in electrical stresses. This phenomenon is termed the "piezoelectric effect"[1] and forms the basis of measuring vibrations in the range of a few Hz to kHz. The effect or piezoelectricity occurs in a number of natural crystalline materials, but a few (fabricated) ceramic materials manifest a very high degree of the effect and are commonly used in practice.

Two common applications

Sonic Detector

A sonic sensor or detector is generally used in remote control of devices that are triggered by a signal vibrating, or producing an output, at a given frequency. The detector consists of a disc of piezoelectric material mounted centrally on an aluminum diaphragm (or base) as shown in Fig.13.39.

---

[1]The effect was predicted and shown experimentally by Pierre and Jacques Curie in 1880, but was though to be of very little practical importance then.

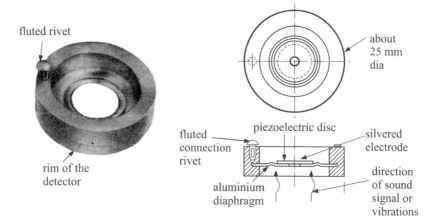

(a)   (b)

Fig.13.39 : A piezoelectric sonic detector

One lead of the detector is connected to the centre of the top face of the piezoelectric disc, the other to the rivet fluted to the rim of the assembly. The disc, tuned to the frequency of vibrations, utilises the piezoelectric effect to transform the vibrations into an electrical output, being maximum at the frequency of interest and decreasing sharply at other (lower or higher) frequencies. The electrical output can then be made to "trigger" an appropriate circuit into

- starting or stopping an electric motor;
- activate a safety device; or
- operate any other control device at a remote location.

The detector, in general, has the advantage of being unaffected by moisture, large temperature changes, or by any adjacent magnetic field; the form of construction also ensures mechanical robustness in use.

Gramophone Pick-up

Piezoelectric pick-ups have been in universal use in record players or changers, the basic principle of operation of such a pick up being the same: to convert vibrations caused by the movement of the pick-up needle inside the grooves of the record into corresponding electrical signals[1]. The process of picking up the vibrations and final reproduction of sound from the record is depicted in Fig.13.40.

---

[1]However, the present-day, modern recordings of music, usually in digital form, and their reproduction, based largely on digital technique(s), may obviate the need of a pickup based on mechanical vibrations and, indeed, the devices like record players.

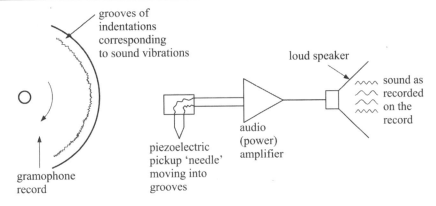

Fig.13.40 : A piezoelectric gramophone pick up

Hand-held vibration meters

Further applications of a piezoelectric probe in the form of a strip to pick up transverse vibrations is in the hand-held vibration meters which essentially comprise the probe at one end of a long lead (similar to that of a gauss meter), the output being capable of measurement or monitoring on either a portable digital (display) meter or for remote sensing as indicated in Fig.13.41, or even on-line monitoring. The meters are extensively used to detect and monitor vibrations from bearings and rotating shafts.

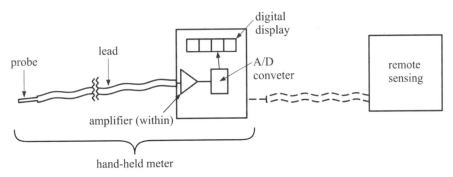

Fig.13.41 : Schematic of a hand-held vibration meter

MEASUREMENT OF SOUND

Sound, a form of energy, manifests itself as a travelling wave which is an oscillation of pressure transmitted through a solid, liquid or gaseous medium composed of frequencies of different range(s) and at a sufficient level that can be perceived as a sensation in a given organ.

Sonic and Ultrasonic

When the frequency range and level is such that it can be heard by a 'normal' human ear, the sound is known to be audible or sonic; if beyond that range, it is generally reckoned to be ultrasonic, or supersonic which may not be audible to humans, but to other species or animals. The discussion in this section is limited to the former type.

Perception of sound

For humans, the sensation due to sound waves is simulated in the ears and the range of frequency of such waves lies between about 12 Hz to 20 kHz, at least theoretically[1]. Just for comparison, dogs' ears are sensitive to vibrations higher than 20 kHz.

Comment

As a signal perceived by one of the major senses – the others being sight, smell or touch – sound is used by the various species for detecting danger, navigation, predation and communication. To this end, earth's atmosphere, water, and virtually any physical phenomenon such as fire, rain, wind, surf or an earthquake produces (and is characterised by) its unique sound. Furthermore, humans have developed culture and technology, such as music, telephone and radio, that allows them to generate, record, transmit and broadcast sound in a variety of ways.

Physics of Sound

From the point of sensing and measurement of sound, the three properties in addition to pressure and resulting vibrations, that characterise sound are

  (i)   pitch, related to the *frequency* of the sound wave;

  (ii) loudness or *amplitude* that shows how 'strong' are the vibrations[2]

  (iii) the *timbre* or quality which in a practical sense relates to sound's quality or, conversely, if there is a noise associated with the sound, and may sometimes help in identifying the source of the sound.

Speed of sound

A factor that may usually be associated with measurement of sound is its speed which distinctly depends on the medium through which the sound waves pass or travel, and its physical properties such as the ambient temperature. For example

---

[1]The actual range may depend on a particular person's ears' sensitivity, usually lower in limits, and may also be age-related. Note that the profile of external ear of a person is such as to receive as much sound as possible and channelise it to the middle ear for 'processing'.

[2]A term commonly used in practice, related to sound recording and reproduction is called "volume" or "volume level".

(a) the speed of sound through air at $20°C$ is approximately 343 m/s (or 1230 km/hr);

(b) in fresh water, also at $20°C$ the speed of sound is approximately 1482 m/s (or 5,335 km/h);

(c) in steel, the speed is about 5,960 m/s (or 21,460 km/hr), notwithstanding any attenuation that generally accompanies a sound signal as it travels.

Sound pressure level

Many transducers used for detection of sound are based on *pressure* that leads to vibrations and finally to an electrical signal corresponding to the sound to be measured. Clearly, in terms of the range of audible sound frequencies, there are corresponding pressure levels *which are not linearly related*. The sound pressure in practice is often measured as a level on a logarithmic scale. Thus, the "sound pressure level" (SPL) or L_P is defined as

$$L_P = 10 \log_{10}\left(\frac{P^2}{P_{ref}^2}\right)$$

$$= 20 \log_{10}\left(\frac{P}{P_{ref}}\right)$$

where P is the *actual* sound pressure and P_{ref} is a reference sound pressure.

Commonly used reference sound pressures, defined in the standard ANSI S1.1-1994, are 20 micro Pa (or $20×10^{-6}$ N/m²) in air and 1 μPa in water.

The reference in air related to human hearing corresponds to the lowest sound audible or discernable to a 'normal' human ear. On the higher side, a pressure of about 0.632 Pa (or N/m²) is normally the maximum sustained level that a human ear can tolerate or withstand before extreme pain or damage to the ear and is known as the "upper threshold" of hearing.

When defined as above, the unit of L_P is called the "decibel" or dB[1]. Thus, the upper threshold expressed in dB would be 90 dB; although, 'momentarily', the human ear is capable of tolerating a sound of even 120 dB in extreme circumstance.

---

[1]From the actual unit "Bel", named after the inventor of telephone, Alexander Grahm Bell.

Schemes of Sound Measurement

Incorporating the various aspects as discussed, the measurement of sound is not straightforward and may depend on which aspect is of interest.

Frequency response of a sound system

In some applications, for example an audio amplifier, the quantity of interest may be the frequency response of the amplifier; that is how does the gain of the amplifier varies with the frequency, covering its entire range so that the "sound" in terms of other characteristics is reproduced faithfully by the loudspeaker at the output stage. It is desirable that this response should be a near flat curve as shown in Fig.13.42; however, it usually deviates from the "ideal" in the very low- and high-frequency domains owing to inherent limitations of the electronic components and circuitry.

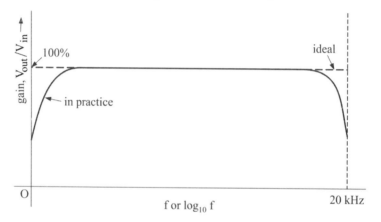

Fig.13.42 : Frequency response of an audio amplifier

The measurement scheme in this case simply consists of measuring the output in mV or V for a given input signal in the same unit at a given, known frequency, usually obtained from a signal generator complying to the frequency-range of interest or a little beyond that.

Measuring "loudness"

A much common requirement is to measure the loudness of sound encompassing the entire audio frequency spectrum. In this case the input quantity is 'sound', emanating from a source at a given level, to be measured suitably as output on a device in terms of the level expressed in dB. The measurement does not reveal the frequency content and also may not be able to distinguish between the sound of interest and any noise associated with it unless special means are adopted to do so.

Sound Measurement Using Microphones

The transducer most suited for measurement of sound, and in common use, is the microphone, a variety of which are now available.

A microphone is essentially a pressure transducer that converts acoustic or sound energy in the form of pressure waves into proportionate electrical energy, the output to be used for measurement and interpreted in terms of sound level. In a microphone, the sensing element is invariably a diaphragm, responding to the incoming sound pressure. A desirable feature of the diaphragm is that its action should be linear over a wide range of amplitudes and frequencies.

Types of Microphone

Carbon (button) microphone

The basic construction of this microphone is shown in Fig.13.43. A PVC or non-metallic/non-magnetic cup-type enclosure is 'filled' with carbon 'powder' (or any other electrically conducting particles), with a diaphragm fitted in the front as shown. The acoustic pressure due to sound waves causes the diaphragm to vibrate to-and-fro, with the amplitude of vibrations being proportional to loudness of the sound resulting in change of physical consistency of the powder material and hence its electrical resistance which forms a part of an external resistive circuit. The change of microphone resistance due to sound volume results in change of current in the external circuit which can be a measure of the sound level or pressure.

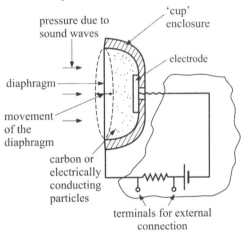

Fig.13.43 : A carbon microphone

Such microphones are commonly used in telephone receivers or hand-sets. The frequency response may be 'DC' to 5 kHz. It is clear that even when the external voltage source is DC, the current variation, monitored across a fixed resistance, for example, would be oscillatory or alternating owing to the nature of sound waves.

Capacitive or condenser microphone

In this type, the diaphragm in the form of a thin metallic disc, is used as one plate of a parallel plate capacitor, the construction of the microphone being similar to that shown in Fig.13.43. Another metal disc fitted in the vicinity to the diaphragm forms the second plate of the capacitor. A constant charge is maintained across the capacitor so formed from an external, stabilised supply of 200 to 500 V (DC). When an acoustic (or sound) wave strikes the diaphragm, it oscillates causing the capacitance between the plates to vary in accordance with the sound level, and generating an AC output in the external circuit.

Such microphones may be expensive, but have the advantage of high sensitivity (expressed as mV output per unit change of acoustic pressure) and wide frequency response, up to 50 kHz. The high voltage supply in the external circuit may, however, pose a problem in some cases.

In order to achieve faithful reproduction, to convert low- to ultra-low-frequency sound variations, a high-frequency voltage in the form of a carrier, may be applied across the plates. The output signal will then be the modulated carrier.

Moving-coil or dynamic microphone

Moving-coil microphones work on the principle of electromagnetic induction and are in common use in public address (PA) systems. The construction of a typical dynamic microphone is shown in Fig.13.44. It consists of a conical membrane of a special shape, one (hollow) end being provided with a coil of large number of turns placed closed to each other and in one layer. The coil is designed to move freely, acted upon by the field of a permanent magnet of annular construction as shown.

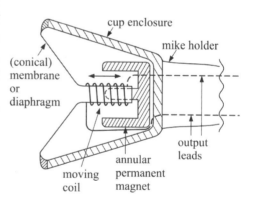

Fig.13.44 : A moving-coil microphone

The membrane acts as a diaphragm, capable of moving back and forth over the central limb of the magnet, under the pressure generated by the incoming sound, resulting in lateral movement of the coil attached to it. The (AC) voltage thus developed in the coil is proportional to, and a measure of, the acoustic pressure on the membrane, to be monitored externally. Unless the coil is wound with very large number of turns and the magnet is extremely powerful, the sensitively of the microphone may be low and the output may have to be amplified suitably before measurement. The frequency response is typically up to 20 kHz, thus covering the entire audio range.

Piezoelectric microphone

These microphones are especially suited for very high frequency range, extending up to 100 kHz and thus useful for ultrasonic applications. The basic construction of a piezoelectric microphone is shown in Fig.13.45.

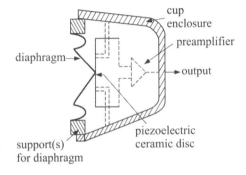

The diaphragm is attached to the piezoelectric ceramic or quartz crystal which is free to vibrate under the action of the oscillating

Fig.13.45 : A piezoelectric microphone

diaphragm, which itself responds to the acoustic pressure. Since the (mV) output of the microphone may be low, many times a built-in preamplifier is provided inside the cup to enhance the output. This arrangement may also improve the signal-to-noise ratio and reduce output impedance. The accuracy is good, the response being linear over a wide range of amplitudes, thus making these microphones widely useful in sound measurement systems.

Sound Output Meters

These are specially-designed and calibrated instruments, analogue or digital type, to read the sound output (level) in dB to indicate the all-encompassing loudness or intensity of sound, taking into account the amplitude and frequency spectrum of the impinging sound.

There are also the portable, hand-held digital output meters to monitor spatial sound levels in any desirable locations and from different directions, sometimes to make a comparative study of sounds emanating from rotating machines, engines, automobiles or the ambient atmosphere at large. Clearly, their operation depends on the use of a suitable built-in transducer meant to first convert the sound level into a proportionate electrical output.

Testing of Loudspeakers

A common application of measurement of sound is the testing of loudspeakers, used in almost all audio systems, with respect to their audio output corresponding to the entire audio-frequency range.

In this case, the loudspeaker is fed from a signal generator, capable of providing a note of a given frequency (e.g. 1 kHz) and fixed amplitude as shown in Fig.13.46. The reproduced sound from the loudspeaker is then measured on an output meter or analysed in other suitable form.

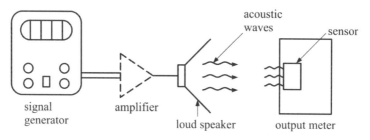

Fig.13.46 : Schematic for testing a loudspeaker

Clearly, it is advantageous if such tests are performed in an en-echoic room or chamber.

MEASUREMENT OF LIGHT

In general terms, light is defined as that part of the energy radiated from a 'body', to be identified as "a source of light", which produces the sensation of light as perceived by the human eye, to allow the eye to 'see'. Thus, light is a *form of energy*. In scientific term, light is a kind of electromagnetic radiation in the frequency domain of 3.75×10^{14} Hz to 7.5×10^{14} Hz, with the corresponding wave-lengths of 8000 Angstrom (units), or AU[1], to 4000 Angstrom $\left(1 \text{ AU} = 10^{-10} \text{ m}\right)$, having a velocity of (very nearly) 3×10^{8} m/s, identical to any electromagnetic wave. These limits correspond to the well-known "visible spectrum" of electromagnetic waves with reference to lower and higher spectra as shown in Fig.13.47.

Unlike sound, light can travel only through air or vacuum and a few 'materials' such as fibre-optics or some semiconductors.

---

[1]Universally adopted unit of wavelength in EM spectrum.

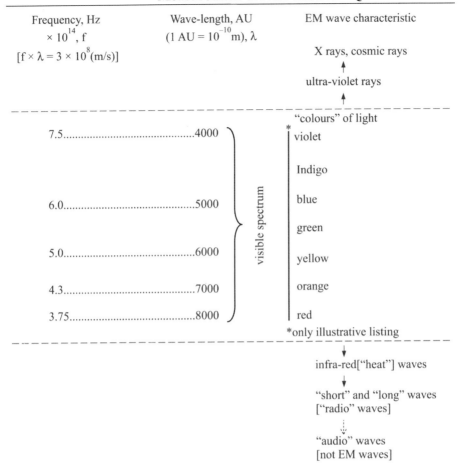

Fig.13.47 : Electromagnetic radiations and light

Concepts Related to Light

The followings are some of the most widely applicable and important concepts related to light which might also influence the means of its measurement.

- ## Luminous flux

  This is a measure of 'total' *quantity* of light emitted from a light source, similar to magnetic flux from, say, a permanent magnet, and is expressed in **lumen** which is also the unit of luminous flux. For example, an ordinary 60 W filament (or incandescent) lamp will have an output of about 700 lumen (abbreviated as lm). It may be noted, however, that light emitted from a given source, the 60 W lamp in the present case, is a *3-dimensional* phenomenon and has its implications when considering methods of measurement.

- ## Illumination

    This follows from the "cause and effect" principle applied to a source of light, in that the light from the source manifests as "illumination". Thus, the illumination on a *surface* is the amount of light or luminous flux falling *normally* on that surface and is, therefore, conveniently defined as "lumens per unit area"; in MKS units as lm/m^2. Another name for this unit is "lux" $\left[1 \text{ lux} = 1 \text{ } lm/m^2 \right]$.

- ## Luminous intensity

    This term is used to express *directional* property of propagation or influence of light, the proper definition being "luminous flux per unit *solid* angle."

Solid Angle

A solid angle, frequently associated with the phenomenon of light – qualitatively as well as quantitatively – differs from a plane angle as depicted in Fig.13.48.

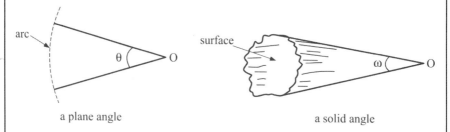

a plane angle   a solid angle

Fig.13.48 : A plane and solid angle

A plane angle subtended at the point O is represented by the 'space' between the two converging *lines*, bound by an arc at the other end, and represented by θ, α or ϕ etc. The unit of a plane angle is *radian* (convertible to degrees). In contrast, a solid angle is subtended by an 'infinite' number of lines meetings at point O, bound by a *surface* and enclosing a finite volume. The unit of a solid angle is **steradian**, denoted by ω, and can be related to a plane angle, and interpreted as follows:

The unit of a plane angle, radian, is defined as the angle subtended at the centre of a circle by an arc whose length is equal to the radius; thus

$$\text{radian} = \frac{\text{arc}}{\text{radius}}$$

The unit steradian is defined as the solid angle subtended at the centre of a sphere by an area on the surface of the sphere which is numerically equal to the (radius)^2; thus

$$\text{steradian} = \frac{\text{area}}{(\text{radius})^2}$$

Since the area of the surface of a sphere of radius r is $4\pi r^2$, it follows that the total solid angle subtended by a *point* in all directions is 4π steradians.

The luminous intensity of a source of light is specified by comparing it to that of a "standard" lamp. Earlier, the standard used to be a candle, made up in accordance with a rigid specification; for example, similar to a standard cell for EMF. Such a source of light when viewed in a horizontal direction was said to have a luminous intensity of one *candle power* which was used to define the unit of luminous intensity, too, abbreviated as (cp). The term "candle power" is still used, with the luminous intensity expressed in terms of cp, but the standard is now maintained in terms of carbon-filament vacuum lamps kept at the national laboratories of the USA, the UK and France[1].

A more common unit of luminous intensity, now universally used (in MKS system), is "candela" or cd, such that one candela is equal to one candle power.[2]

- **Brightness (or luminance)**

  In general terms, an area of a surface would appear brighter if the quantity of light falling on it from above increases progressively. In terms of luminous intensity of the source, and taking into account the directional property, when the source of light covers an appreciable area; for example, a cinema screen so that an observer sees only the *reflected* light, the term "brightness" (also called luminance) is employed and can be defined as "luminous intensity per unit area". Accordingly, luminance is expressed in candela per square metre (cd/m^2). Since it involves luminous intensity, it, too, is a directional facet of light.

- **Mean spherical candle power**

  Considering that phenomenon of light is truly three-dimensional, the luminous intensity from a source of light, expressed in terms of

---

[1]For a qualitative understanding, an ordinary 60 W lamp, when viewed from the floor will have a luminous intensity of about 70 cp; a search light viewed from a distance along the beam may have a luminous intensity of as much as a million cp, being intensely directional.

[2]In practice, it may be stated that a 'common' candle emits light that may roughly be assumed to have a luminous intensity of one candela. 'Officially', the candela is defined "as the luminous intensity of a source that emits *monochromatic* radiation of frequency 540×10^{12} Hz", a radiation intensity of 1/683 W per steradian.

candle power, may vary appreciably in different directions, or the various planes. A term that is widely used, and associated with a light source, to account for this variation is called "Mean Spherical Candle Power" (abbreviated MSCP). It is defined as the mean or average of luminous intensities (in cp) in all directions and in all planes from the source of light,

or $\qquad$ MSCP $= \dfrac{\text{total flux in lumens } (\text{from a source})}{4\pi}$

where 4π represents the solid angle formed around the source of light (assuming it to be a "point" source)[1].

Laws of Illumination

There are two laws of illumination which form the basis of many measurements and calculations related to light.

Inverse square law

According to this law, the illumination of a surface in perpendicular direction is inversely proportional to the square of distance between the source and the surface, under the assumption that the size of the source is much smaller compared to the distance so that the former can be regarded as a "point" source. Referring to Fig.13.49, consider such a source of light, S, having an intensity of I lumen/steradian, or a lumen output F.

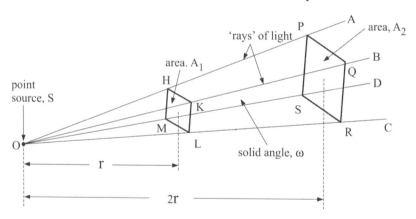

Fig.13.49 : Demonstration of the inverse square law

---

[1]Alternatively, if the intensities I_1, I_2, ... I_n, were known in n different directions encompassing the entire three-dimensional space around the (point) source of light, with n being as large an integer as possible, theoretically approaching infinity, an average can be found to give MSCP as

$$\text{MSCP} = \dfrac{I_1 + I_2 + \cdots + I_n}{n} \ \text{cp}$$

The intensity pertains to the solid angle, ω, enclosed by the four lines or 'rays' of light OA, OB, OC and OD.

The whole of the light from S within this solid angle will be falling on the surface HKLM, having an area A_1, *placed at right angles to the axis of the beam*, the distance of A_1 from O being r metre. Another surface of area, A_2 (PQRS), is assumed at a *perpendicular* distance 2r, or twice the distance of A_1. Also, the two surfaces are enclosed in the same solid angle, ω, as shown.

The intensity of light is also the same for the two surfaces or areas. In terms of I, the total flux falling on *either* surface will be

$$F = I \times \omega \quad lm$$

Also, the solid angle, $\omega = \dfrac{area}{(distance)^2} = \dfrac{A_1}{r^2} \text{ or } \dfrac{A_2}{(2r)^2}$

Therefore $\qquad F = I\dfrac{A_1}{r^2} \text{ or } I\dfrac{A_2}{(2r)^2}$

Now, the illumination on any surface is given by

$$illumination = \frac{flux \text{ in } lm}{area \text{ in } m^2}$$

$\therefore$ illumination on surface HKLM

$$E_1 = \frac{IA_1}{r^2}\frac{1}{A_1} = \frac{I}{r^2}$$

and illumination on surface PQRS

$$E_2 = \frac{IA_2}{(2r)^2}\frac{1}{A_2} = \frac{I}{(2r)^2}$$

Hence

$$E_1 : E_2 :: \frac{I}{r^2} : \frac{I}{(2r)^2}$$

This shows that illumination on a surface is inversely proportional to the square of the *normal* distance of the surface from the (point) source of light, that is

$$illumination \propto \frac{1}{d^2_{(normal)}}$$

Also, the intensity of light remains the same over the two surfaces.

Lambert's cosine law

In the above case, the surface upon which light was falling was assumed to be at right angles to the axis of the beam. However, in practice this condition may not always be met and the surface may be inclined at an angle to the direction of the beam or luminous flux as shown in Fig.13.50. The illumination on the inclined surface is then

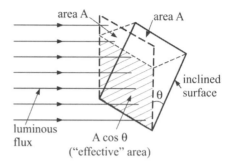

Fig.13.50: Demonstration of Lambert's law

governed by another law known as "Lambert's cosine law". According to this, the illumination on such a surface is given by "the illumination if the surface were normal, multiplied by cosine of the angle of inclination" as shown. That is, the illumination is proportional to cos θ.

If the luminous flux from the source is F lm, the illumination on a normal surface of area A m^2 will be

$$E = \frac{F}{A}$$

When the surface is inclined to the original, normal surface by an angle θ, the new illumination will be

$$E' = E \times \cos \theta,$$

the area remaining the same in both cases.

The cos^3 θ Law

This law applies to illumination on a given surface due to a "point" source at some height, and at a point off the vertical line defining the source location as depicted in Fig.13.51.

Let the point of interest on the surface be P, at a slant distance d from the source of light, the latter located at a height h metre from the surface. If the source has a luminous intensity I (cp), the illumination at P according to the inverse square law will be

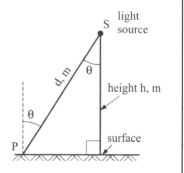

Fig.13.51: Illumination at point P

$$E_P = \frac{I}{d^2}$$

and according to second law,

$$E_P = \frac{I}{d^2}\cos\theta$$

Since

$$d = \frac{h}{\cos\theta}$$

it follows that

$$E_P = \frac{I}{h^2}\cos^3\theta$$

This expression is particularly useful in carrying out "illumination calculations".

Measurement Methods

In general, the process of measurement of light in its various aspects discussed above is known as "photometry" and involves two common principles:

  (a) comparison methods, using a "standard" source of light and a device called "photometer" to effect the comparison;

  (b) direct methods, usually in the form of conversion of light from a source into an equivalent electrical output that can be suitably measured and related to the source of light.

The Measurement of candle-power

In this method of measurement, the intensity of light of a given lamp, in cp, is obtained by comparing it with the intensity of a standard lamp which is assumed to be known. For this purpose, an apparatus comprising a "photometer-bench" and a "photometer-head" is employed, in conjunction with a standard lamp and the lamp of which the intensity is to be measured. The principle of operation of the method is based on the inverse square law of illumination. The schematic of the method is shown in Fig.13.52.

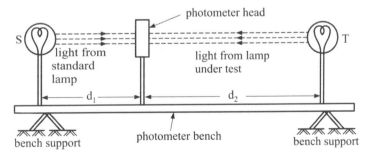

Fig.13.52 : Schematic of a photometer measurement

In the figure, let S represent the standard lamp of known cp and T the lamp under test of comparable candle power. Both lamps are arranged such that their light reaches opposite faces of a photometer head horizontally as shown. The position of the photometer head is adjusted by moving or sliding it laterally between the lamps till the illumination from the two lamps is equal as indicated by the photometer head according to its very design and construction. Then, from the inverse square law

$$\frac{\text{candle power of T}}{\text{candle power of S}} = \frac{d_2^2}{d_1^2}$$

where d_1 and d_2 are the distances of the photometer head from S and T, respectively, when *optical* balance is obtained.

Photometer Bench

A common photometer bench consists of two parallel rods, placed horizontally some 15 cm apart and about three to four metre in length, with provision for two stands or saddles at the ends to hold the two lamps at the same heights and a suitable carriage, positioned between the two lamps, to hold the photometer head that can be moved to left or right and clamped at the position of balance. One of the rods carries a brass strip which bears a scale graduated in millimeters to indicate the distances of the photometer head from the respective ends, related to the locations of the lamps.

Both the lamps and the active surfaces of the photometer head must be at the same height, in a horizontal line of sight, such that the light falls on the photometer at right angles. Also, it is important that no light, other than that from the lamps under comparison, shall reach the screen of the photometer head. For this reason, the measurement is carried out in a special room, usually darkened with the walls painted 'dead' black. Clearly, the bench should be placed on a rigid support or table of 'convenient' height to carry out the experiment(s).

Photometer Head

Most photometer heads are designed and constructed such that the illumination on their opposite surfaces, one illuminated by the standard lamp and the other by the lamp under test, may be compared distinctly and easily by the observer without movement of the eye.

Bunsen head

This is the most commonly used device and essentially consists of a piece of thin opaque paper which has a translucent "spot" (round or square) at its 'centre', this being achieved by treating the area of the spot with oil, grease or wax as shown in Fig.13.53[1]. When the screen is held between the two lamps being compared, with its plane perpendicular to the "line of sight", the opaque part of the screen on ether side will be illuminated from one lamp only whilst the translucent spot will be illuminated from *both* the lamps at the same time. If the two illuminations are not identical, the spot would appear dark to the eye as indicated in Fig.13.53(i). However, when the screen

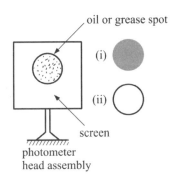

Fig.13.53: Bunsen photometer head

is identically illuminated, showing that the two intensities are same – the condition of optical balance – the spot will be seen 'clear' or just perceptible as depicted in Fig.13.53(ii). The candle-powers of the two lamps will then be related as per the expression given above.

Measurement of Illumination

A common requirement in homes, industry, commercial establishment or even outdoors such as streets and roads, is to measure illumination on surfaces – working or general – provided by the various sources of light, sometimes under controlled conditions. Such measurements usually employ special sensors, calibrated to read the illumination directly in lux, built into what are called "photometers" or "light meters". These are, in general, portable instruments to facilitate the measurement where required[2].

Principle of photometers

Most photometers detect the light using some kind of sensor that responds to the *level* of illumination and then can be a part of a given electric circuitry. In some cases, to *analyse* the 'light', the photometer may measure the light after it has passed through a filter or through a monochromer for determination of defined wavelengths or for analysis of the spectral distribution of light.

Detectors or Sensors

The various sensors used in photometers are

   (a)   photoresistors

   (b)   photodiodes

---

[1]Owing to its design, the device is often called the "grease-spot photometer."

[2]The photometers referred to here are different from the Bunsen photometer (head) just described and would generally depend on their operation based on a variety of "sensors".

(c)   photomultipliers

(d)   photo cells

(also sometimes called photo voltaic cells or PVCs and may serve double purpose)

Photoresistor

A photoresistor or light dependent resistor (also known as cadmium sulfide – CdS – cell) is a special resistor of which the resistance *decreases* with increasing incident light or illumination. When used to control conduction of currents in a part of the circuit, it can also be referred to as a photoconductor[1].

A typical photoresistor and its symbol in an electric circuit is shown in Fig.13.54.

Fig.13.54 : A photoresistor and its symbol

A particular application of a photoresistor in the form of a simple light detector is in the automatic control of street lights. The photoresistors, activated by the quantum of light controls the current flowing through a heater which opens the main power contacts (comprising a bimetallic mechanism) at day break. At night, the bimetallic device cools, closing the power contacts and energising the street lights. The heater or bimetallic mechanism may also be designed to provide a built-in time delay. Being sensitive to frequency of radiation, lead sulfide or indium antimonide photoresistors may be used for the mid-infrared spectral region.

In general, when incorporated in a photometer, a photoresistor may be used in a circuit similar to the Wheatstone bridge such that the out-of-balance current resulting from the change of resistance can be calibrated in terms of the illumination falling on the photoresistor.

Photodiode

A photodiode, whose function is similar to that of a photoresistor, is a PN-junction semiconductor. When light strikes the diode, photons of sufficient energy excite electrons in outer orbits thereby creating a mobile

---

[1]A photoresistor is made of a high-resistance semiconductor such that if an EM radiation falling on it is of high enough frequency (for example, 'normal' light), photons absorbed by the resistor give bound electrons enough energy to jump into conduction mode, resulting in conduction of electricity and thereby lowering of resistance.

electron and a positively charged hole. The holes so released move towards the cathode and electrons toward the anode, and a 'photocurrent' is produced.

Photodiodes can be used in applications similar to photoresistors, and generally have a better, more linear response compared to the latter.

Some of the common applications include

- camera light or exposure meters
- remote controls for VCRs and TVs
- compact disc player
- street light control
- CT scan

Being more sensitive and having faster response than photoresistors, these are often used for accurate measurement of light when incorporated in photometers.

Photomultipliers (or PMTs)

These are essentially a special type of vacuum tubes (of the past) and more specifically photo tubes, and are extremely sensitive detectors of 'light' in its entire electromagnetic spectrum, from ultraviolet, visible to near-infrared. Significantly, these detectors multiply the current produced by incident light by as much as 100 million times – hence the name – enabling, for example individual photons to be detected even if the incident flux of light is extremely faint. The combination of high gain, owing to inherent multiplication of signal, low noise, high-frequency response and large 'area' of collection has resulted in photomultipliers being extremely useful in applications that include

- nuclear and particle physics
- medical diagnostics, including blood tests
- astronomy
- motion picture film scanning

Structure and operating principle

The conventional photomultipliers consist of a (hard) glass tube or envelope with a high vacuum inside. The tube houses a photocathode, several "dynodes" – special-type electrodes held at increasingly positive charge – and an anode as shown in Fig.13.55.

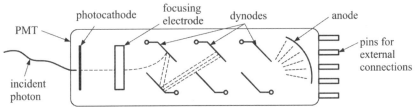

Fig.13.55 : Schematic of a photomultiplier tube

Incident photons strike the photocathode which produces electrons as a consequence of the photo electric effect. These electrons are directed by the focusing electrode towards the dynodes which act as electrons multipliers in stages as shown. This is achieved by the process of secondary emission; the successive dynodes being held at higher positive potential than the previous ones. As the electrons move from one dynode to the next, they are accelerated by the electric field and arrive at the approaching dynode with much greater energy.

The geometry of the dynode chain is such that a cascading effect occurs with an ever-increasing number of electrons being produced at each stage. Finally, the multitude of electrons in high energy reach the anode where the accumulation of charge results in a sharp pulse indicating the arrival of a photon at the photocathode[1].

Usage considerations

Photomultipliers typically operate at 1000 to 2000 V, to accelerate electrons through the chain of dynodes. The negative terminal of the supply is connected to the cathode whilst the highest positive terminal to the anode. In between the two limits, voltages to the various dynodes, in a graded manner, are distributed by a resistive potential divider. The divider design which influences frequency response or rise time (of interest in transient applications) can be selected to suit varying requirements. In some cases, provision may be made to vary the anode voltage to control the 'gain' of the device. Clearly, the overall 'amplification' of the incident light level would depend on the number of dynodes used and the operating voltage across the cathode and anode.

Silicon or solid-state photomultipliers

These are the latest, modern developments which replace the earlier photo-multiplier tubes. By integrating low-power CMOS electronics into silicon photomultiplier chip, a digital silicon photomultiplier is now available in which each photon detection is converted directly into an ultra high-speed *digital* pulse that can be directly counted by on-chip counter circuitry, thus resulting in extremely fast response – an important factor in applications such as medical imaging, scanners and high-energy nuclear particle detectors.

Photo-emissive and Photo Voltaic Cells

It was the French physicist Edmund Bequerel who, in 1839, discovered the photo-electric effect, the first module built by Bell Laboratories in 1905. The

---

[1]There are two common photo multiplier orientations: the "head-on" or "end-on" design as shown in Fig.13.55 where the light enters the flat, circular top of the tube, and the "side-on" where light enters at a particular spot on the side of the tube. The choice would depend on a particular application.

photo-electric devices, commonly known as "photo-cells", find extensive use in photometry – as an integral part of the photometer – in industrial control and counting operations.

Photo-electric cells of earliest design were based on *electrons* to be emitted when light would fall on the active part of the cell and were confined to earlier, limited applications. The modern types which are now in common use produce an *EMF* consequent to the light falling on the active surface and can therefore provide a direct measure of the incident light or provide a current in the external circuitry.

A typical photo-electric or photovoltaic cell, fitted behind a glass cover or window, consists of a copper disc, oxidised on one face (that is Cu_2O/Cu 'pair') as shown in Fig.13.56. Over the exposed surface of the oxide, a film of gold (Au) is deposited by evaporation in a vacuum, the film being so thin that (incident) light can pass through it. A thin layer of selenium is interposed between the conducting (gold) film and the oxidised copper surface, the former acting as a 'cathode'.

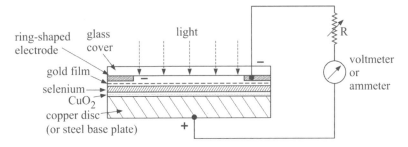

Fig.13.56 : Schematic of a typical photovoltaic cell

Light falling on the cell, passing through the outer glass cover and the gold film, causes a movement of the electrons from the selenium (film) towards the gold coating/film; the selenium acting as an anode, together with the ring-shaped electrode for making external connection as shown.

The EMF generated in the cell can be collected as a voltage output or the current in a circuit. The EMF, or the current, is in general proportional to the light flux, but the relationship may not be linear[1].

When calibrated with the help of a 'standard' source of light, photocells provide the most effective, 'practical' and accurate means of measuring illumination, either in controlled laboratory environment or as portable devices for "on-the-spot" measurements in the form of illuminometers.

---

[1]For "large-scal" production of *electricity* from solar radiation, it being the source of light, multitudes of PVCs are connected in various series-parallel combination to provide "modules" and "arrays".

Measurement of Luminance (or Brightness[1])

Luminance is defined as luminous intensity per unit area, the unit being candela/m^2. This points to the implicit inclusion of

    (a)   luminous flux (lm), and

    (b)   a solid angle (ω)

           which define luminous intensity, and

    (c)   the area of interest of the surface

to comprehend the term "luminance".

Also, the "area of interest" above is actually the *projected* area on the surface in a given direction, usually orthogonal to the rays of light. It is the light falling on this projected area, and reflected from it, that is the key to the perception of brightness to the eye and essence of measurement of luminance by suitable means.

To understand the term projected area, refer to Fig.13.57 which shows, in the single line of vision, a source of light, an "aperture" (that is, a small pin-hole in a piece of cardboard in this case), a convex lens of 'convenient' focal length and a screen, the plane of which is perpendicular to the line of sight.

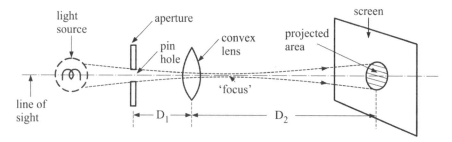

Fig.13.57 : Schematic to show the concept of projected area

As shown, the light passing through the aperture and focused by the lens, falls on the screen in the form of a circular spot as the "image" of the pin hole. This spot would then be known as the projected area, corresponding to the luminous flux received from the lamp, producing the bright image on the screen as seen or perceived by the eyes. Clearly, the "size" of the projected area would depend on the distances D_1 and D_2 (related to the focal length of the lens) and the pin-hole size, or the aperture.

When it is required to *measure* the luminance or brightness of the projected area, rather than simply perceived by the eyes, an arrangement as

---

[1]Brightness, earlier used to describe luminance – the correct term now in use – is a rather qualitative phenomenon as perceived by the eyes, by the action of rod and cone cells, and is not a measurable quantity. In comparison, luminance, an effect of light similar to brightness, can be measured, as described here.

shown in Fig.13.58 can be contemplated which forms the basis of working of a luminance photometer. As seen, this has many features common to those of Fig.13.57.

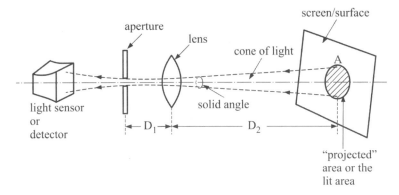

Fig.13.58 : Basis of working of a luminance photometer

Here, the arrangement of Fig.13.58, although still incorporating the aperture and the lens, is "reversed", in that the projected area A now represents a 'source' of light or the lit area from a source of which the luminance is to be measured[1]. The light from A, now 'traversing' back through the lens and the aperture reaches a light sensor or detector as shown, with the "intensity" of light sensed by the detector being a function of the light emanating from A, or how 'bright' is this area. Once again, this involves the distances D_1 and D_2, related to the focal length of the lens. The cone of light, say over the distance D_2, forms a solid angle and with the amount of light (or the light flux) on A from the source, incorporates the essence of the term luminance.

When related to measurement of luminance of a large surface or screen, or part thereof, the above concept can be extrapolated greatly where each sub-area like A in Fig.13.58, approaching a point, subtends a cone or solid angle, encompassing the lens periphery and thus representing intensity of light for each point that would reach the detector. The latter would integrate (or average out) the light intensities from all the points of the projected area and, converting the total input from the area of interest into input per unit area and calibrated accordingly, would provide a measure of luminance of the (projected) area of interest. Again, the distance D_2 and lens specification (size and focal length) are involved. A large surface or screen may not always be illuminated uniformly and hence its luminance may vary from one part to the other, and may be measured accordingly, using for example, a hand-held photometer, or a luminance meter.

---

[1]For example, this may simply represent a part of the cinema screen, illuminated by the light from the projection unit.

It is significant that the human eye is remarkably capable of perceiving luminance over a great range of phenomena, expressed in terms of candela/m^2. This is achieved since the sensitivity of the eyes *decreases* with luminance; this shows that brightness, perceived by a complex function of cone and rod cells of the eyes, is NOT the same as luminance pointed out earlier. A comparative measure of luminance from the various sources and phenomena of light to bring out above premise is provided in Table 13.2. However, the figures indicated in the table are only illustrative and must be viewed with caution.

Table 13.2: Luminance of different sources/phenomena

Source/ phenomena	Luminance, cd/m^2
SUN	900,000,000
TUHGSTEN FILAMENT AT 2700K	3,000,000
	300,000
UPPER LIMIT OF VISUAL TOLERANCE	30,000
FRESH SNOW ON CLEAR DAY	
FLUORESCENT LAMP	3,000
SURFACE OF MOON	
SKY, HEAVILY OVERCAST DAY	300
WHITE PAPER IN GOOD READING LIGHT	
	30
NEON I AMP	
	3
1/4 HOUR AFTER SUNSET, CLEAR	
	0.3
SNOW IN FULL MOON	
WHITE PAPER IN MOONLIGHT	0.03
FAIRLY BRIGHT MOONLIGHT	
	0.003
MOONLESS CLEAR NIGHT SKY	
SNOW IN STARLIGHT	0.0003
GRASS IN STARLIGHT	0.00003
THRESHOLD OF VISION	0.000003

Hand-held or portable luminance meters

These are specially designed instruments for quick and direct measurement of luminance of 'small' areas or applications requiring 'directional' measurements; for example, traffic lights, airport lighting lamps, LEDs lit screens, picture tubes etc.

A typical such meter is shown in Fig.13.59. This is a single lens reflex (SLR) optical system capable of precise targeting of the exact area, however small, whose luminance is to be measured at a given distance. The handle, similar to a portable movie camera, makes handling of the instrument secure and stable for accurate, 'spot' measurement. With the help of a add-on close-up lens, small areas – even up to about 0.4 mm in diameter – can be measured with good accuracy. The meter is capable of measuring the luminance as low as 0.001 cd/m^2, nearly very dark surfaces which may be of special interest.

Fig.13.59 : A portable luminance meter

Measurement of Mean Spherical Candle Power (MSCP)

Since the phenomenon of light emanating from a source, for example a naked tungsten-filament lamp hung freely in space without a reflector, is truly three-dimensional, an important requirement is to measure the intensity of light of the source in *all* directions which is identified as its "mean spherical candle power", numerically given by

$$\text{MSCP} = \frac{\text{total or integrated value of luminous flux from the source}}{4\pi}$$

as explained earlier.

The key to the measurement of MSCP of a source, for example a lamp, is in the *experimental summation* of light from the source accounting for all directions in space. Theoretically, this would imply measuring the intensities of light of the lamp – the candle powers – in an "infinite" directions, sum them up and then divide by the number of such measurements.

Measurement of MSCP by means of an "integrating" sphere

This is the commonly employed method for the measurement of MSCP. The principle of measurement essentially consists of measuring total flux of light radiated by the lamp under test in the apparatus and MSCP obtained by dividing the sum by 4π. The candle-powers of the lamp in all directions is thus accounted for. The property of light to produce proportional brightness on a given surface, or a 'window' and its evaluation forms the basis of the measurement.

The integrating sphere

This is a hollow sphere, about a meter or more in diameter, being much larger than the size of the lamp under test and having a very smooth inner surface painted with high-quality reflecting white paint, such that when the lamp is lit, hung at the centre of the sphere, the light is so diffused (by successive reflections) that a uniform illumination is obtained over the entire (inner) surface of the sphere and the space within. On one side, the sphere contains a 'small' window of translucent glass (similar to "ground" glass) which is illuminated by reflection of the light from the inner surface.

The source of light, or the lamp, is positioned at the centre of the sphere whilst a small screen is interposed between the lamp and the window, in the line of sight, to prevent the light from the lamp reaching the window directly. Clearly, this arrangement of using a very much large sphere ensures that the light reaching the window is the *sum total* of all the light produced by the test lamp, being successively reflected from the inner surface, accounting for all directions in the enclosed space. The arrangement of the test is shown schematically in Fig.13.60.

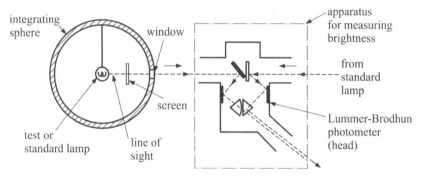

Fig.13.60 : Integrating sphere arrangement for measurement of MSCP

The measurement entails the following two steps:

1. The lamp under test is first placed, or held suitably, at the center of the sphere and the brightness of the window as produced by the light from the lamp is measured by suitable means.

2. The lamp is then replaced by a standard lamp whose MSCP is already known and the brightness of the window is again measured.

Since the MSCPs of the two lamps are proportional to the corresponding brightness(es) as measured from the window, the MSCP of the test lamp can be obtained in terms of that of the standard lamp.

Some form of an illuminometer may be used for measurement of the window brightness and compared with the brightness of a surface whose illumination can be varied and is known. A very reliable means of comparing the two brightnesses is the Lummer-Brodhun photometer head, an improved version of the Bunsen grease-spot photometer head, which can effectively provide the relation between the MSCPs of the test and standard or sub-standard lamp.

LUMMER-BRODHUN PHOTOMETER HEAD

The essential construction and schematic of this device – of "equality-of-brightness" type – is shown in Fig.13.61.

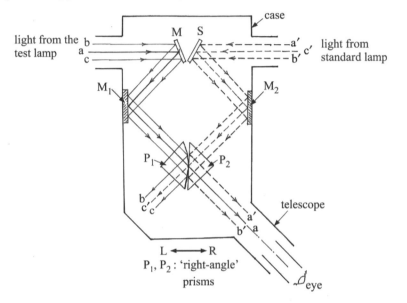

Fig.13.61: Operational features of Lummer-Brodhun photometer head

The photometer head comprises mirrors M, M_1 and M_2 and the reflecting screen S, located in the instrument case as shown. P_1 and P_2 are two right-angled glass prisms, the principal surface of one of these, P_1, is spherical, but has a small flat portion at its centre to make good optical contact with the other prism, P_2, of usual 'standard' design.

[See Fig.13.62(a) for an enlarged view of the compound prism so formed].

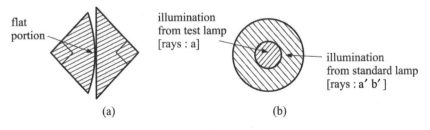

Fig.13.62 : Compound prism and image through the telescope

Light entering from the test lamp (rays a,b,c) get reflected first from mirror M and then M_1, enters the prism P_1. From this prism, only the light represented by ray(s) "a" which strikes the flat part of the compound prism is allowed to pass through the telescope, the rest of the light, represented by b and c, being reflected back. On the other hand, light (beam) from the standard lamp, a',b',c', after getting reflected from S and M_2 falls on the full-flat surface of prism P_2. Here, only the outer part of the light, represented by a', b', gets reflected from P_2 and passes through the telescope as shown.

The effect is that the telescope, as viewed by the eye, shows the central portion (of the prisms in contact) in the form of a small circle illuminated by the test lamp and a surrounding, circular area that is illuminated from, and owing to, the standard lamp as shown in Fig.13.62(b). During the test, the photometer head is moved to the left or right (L $\leftrightarrow$ R in Fig.13.61) until the dividing line between the two images of illumination disappears or merged into one single circle of illumination. This would show that the intensity of light from the two lamps, in terms of their respective lumen output and relative distances from the photometer head are equal.

In the test for MSCPs, employing the Lummer-Brodhun photometer head to assess the brightness of the test and standard lamps, the first balance of comparison produces a distance D_1 of the test lamp to the photometer head. The second distance, D_2, is obtained when the test lamp is replaced by the standard lamp of known MSCP.

Then,

$$(MSCP)_{test} \propto D_1^2$$

$$(MSCP)_{standard} \propto D_2^2$$

or
$$\frac{(MSCP)_{test}}{D_1^2} = \frac{(MSCP)_{standard}}{D_2^2}$$

from which MSCP of the lamp under test can be determined in terms of measured distances D_1 and D_2 and MSCP of the standard lamp. An accuracy of within 1% can be obtained by the above method and using the Lummer-Brodhun photometer head, even allowing for the level of sensitivity of the eye used as a kind of detector.

MEASUREMENTS IN MEDICINE BY ELECTRICAL MEANS

Basic Principles

There are innumerable parameters and human-body functions which are now monitored and recorded with the help of instruments and apparatuses dependent on the use of electricity, and electronics, incorporating electric circuits and electronic devices, in various forms. Likewise, countless diagnostic procedures make use of similar, electricity-related techniques. Of the many, some **key** monitoring and recording devices are discussed in the following sections.

Monitoring and Measurement of Blood Pressure (BP)

Accurate monitoring and measurement of blood pressure of a patient – or as a routine for other 'healthy' persons – is extremely important not only in cases of hypertension (BP above 'normal') or hypotension (BP below 'normal'), but even as an indicator of smooth functioning of the heart and other vital organs of the body such as the kidney, and to preclude any damage to them in course of time. Any significant persisting deviation of BP from the normal would call for a proper course of treatment.

Common procedure

The commonest and non-invasive method in use to measure the blood pressure is known as the "Ausculatory" method in which the patient/person is made to sit on a chair[1] with his arm resting on the table in slightly bent position, roughly at the same vertical height as the heart. An inflatable cuff is wrapped smoothly and snugly around the upper arm of the person through which brachial artery passes. It is essential that the size of the cuff is just right for the patient: too small a cuff can result in too high a pressure whereas too large would result in too low pressure. Usually, it is the right arm that is used for the purpose of measuring BP.

The conventional, non-electric instrument which has been in use ever since – and still in common use and preferred owing to its simplicity and good accuracy – is known as sphygmomanometer[2] (or sometimes just known as manometer) which consists of the inflatable cuff (as above), a pressure measuring unit (usually a mercury manometer or the aneroid, dial-type gauge) and an inflation bulb made of special rubber with a screw-operated valve.

The cuff is inflated to about 180 mm (pressure) of mercury, at which the brachial artery is completely occluded, by pumping the bulb and the value closed. A stethoscope is placed under the inflated cuff in close contact with the arm surface near the elbow. Listening with the stethoscope the 'examiner', a doctor or a trained paramedic, slowly releases the pressure in

---

[1]At times, in lying conditions; for example, when admitted to a hospital in serious state of injury or otherwise.

[2]The word *sphygmos* derives from Greek (meaning pulse) while manometer simply means a "pressure meter". The device was invented by Samuel Siegfried Karl Ritter von Basch in 1881, but it was not until circa 1901 that it was popularised.

the cuff by gently opening the valve. As the pressure in the cuff falls, a "whooshing" or pounding sound is heard in the ear plugs showing the start of first flow of blood in the artery. The pressure in the manometer at the instant this sound is heard is recorded as the upper 'limit', or the "systolic" blood pressure. The cuff pressure is further released until the sound can no longer be heard; the pressure at this instance is recorded as the lower 'limit' or the "diastolic" blood pressure. In simple terms, the two stages correspond, respectively, to the heart pumping blood and 'resting' over each cycle.

Clearly, measuring the blood pressure by the above method is a very delicate, and *even subjective*, process and only a well-trained and much-experienced examiner can do justice to the measurement, and arrive 'correctly' at the two blood pressure levels[1]. This is necessary to avoid any psychological 'set-back' to the patient, or the person being examined.

Electronic devices for measurement of BP

With a view to mitigate the subjective error of measurement based on hearing of sound(s) on the part of the examiner, electronic, semiautomatic devices are now in vogue which indicate the two pressure levels on a meter, usually a digital one. The device(s) also have the advantage in that they may not require the help of an experienced doctor and the patient himself/herself can use them, especially when frequent monitoring of BP is required as instructed by the physician and can be used even in a 'noisy' environment.

One type of electronic BP instrument which also requires an inflatable cuff (usually a wrist type), is based on the oscillometric technique and uses the oscillometric detection to calculate or *derive* systolic and diastolic blood pressures. Essentially, it records the oscillation of the contraction and expansion of the heart, whereby the maximum point of oscillation corresponds to the *mean* of the arterial pressure. The systolic and diastolic pressures thus refer to the opposite ends of this oscillation. The instrument then derives the BP 'reaching' by an algorithm and displays the same digitally on the meter. In this sense, these devices do not *actually* measure the blood pressure and thus provide rather approximate values, which must

---

[1]The 'normal' levels of BP, esp. in adults in medium age group are: the systolic pressure may vary between 100 to 130 mm of Hg; the corresponding diastolic range is 70 mm to 90 mm of Hg – the average normal figures being 110 or 120 (systolic)/ 80 (diastolic). Any persisting deviation above the systolic limit indicates a condition called hypertension; a deviation below the diastolic limit is indicative of "below normal" or hypotension condition.

Depending on a number of factors and circumstance, such as physical state of body or mental tension etc., the BPs of even a 'healthy' person is seldom 'steady' and may show noticeable variation(s) during the 'day'. As such, the physicians usually recommend monitoring of BP several times during a given intervals, (that may be 24 hour) and obtain an 'average' figure.

be viewed with a degree of caution, even as the instrument may be user friendly and 'accurate' in itself.

Unlike the mercury manometer-type BP device which provide absolute measurement and are far more accurate in the hands of an *experienced* doctor, the electronic devices may require frequent calibration (with the help of the manometer type or otherwise) to make their readings meaningful or dependable. These are therefore, largely 'indicative' of the trend of variation of BP over a period, say 24 hours.

Direct Blood Pressure Meter

This is a novel variation of the common electronic blood-pressure instruments in that it converts the blood pressure directly into a corresponding electric signal. The transducer or detector for the purpose consists of two highly-sensitive *silicon* strain gauges in direct contact with a flat membrane about 0.1 mm thick and having a 'working' diameter of about 10 mm. The membrane is arranged to sense the pressure as it materialises at the two levels and activates the strain gauges connected electrically in differential mode. The strain gauges form two branches of a 4-arm Wheatstone bridge in which the out of balance PD is measured and related to the blood pressure.

Such meters are now commercially available with analogue or digital display and compensating adjustments/controls to take care of temperature variation etc.

Recording Electro-cardiogram (ECG or EKG)

Monitoring and recording the activity of human heart constitutes a very important aspect of functioning of human body and it is now possible to electrically monitor it.

The graphical representation of the working of heart at its various locations inside the body is called the "electrocardiogram" (in short ECG or EKG, in German) and the apparatus used for the purpose is called the Electrocardiograph. It is known that the functioning of heart has direct bearing on the blood pressure – or vice versa – and thus it is usual to also monitor and record the blood pressure of the patient/person whilst the ECG is being recorded.

Basis of heart's electrical activity

In its most basic form, human heart is a muscle, comprising countless cells, the primary function of the heart being to pump blood towards various parts of the body, the pumping action itself being derived from the expansion and contraction of the heart, or rather the heart muscle cells. The electrocardiogram is based on the *electrical* activity of these cells.

All cells are known to maintain electrical potential gradient across their cell membrane, the potential being accomplished by a kind of ionic pump on the cell surface; this PD being of the order of a few mV, with the inside of the cell being negative with respect to its surroundings. These potentials, depending on their location of generation cause 'weak' currents to flow in the body resulting in a small potential difference across a pair of points on body surface, that is, skin near the heart.

Several distinct points in the vicinity of the heart have been indentified, the PDs across such successive points, detected suitably, provide a graphical reflection of the normal, or otherwise, function(ing) of the heart. A typical example of this aspect is shown in Fig.13.63. The resistance R represents the 'body' resistance across the location of externally connected electrodes between which a potential difference appears due to the current flow as a result of electrical activity of the heart. Since the PD, or the signal is essentially weak, it has to be amplified suitably before it can be monitored or recorded. Usually, (at least) one leg of the body should be grounded (by appropriate external connection) to obtain the signal in conjunction with the instrumentation.

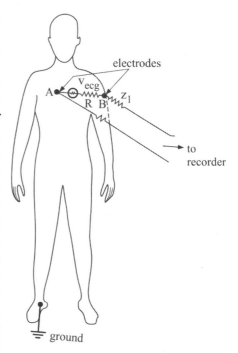

Fig.13.63 : Electrical activity of heart

In terms of location of the various leads to sense the condition of heart, the electrical activity of the heart can be best approximated by a dipole (a vector drawn across two oppositely charged electrical particles) with time varying amplitude and orientation as it obtains in the heart. For this simple model, the cardiac activity of the heart can be represented by a vector $\overline{M}$ as shown in Fig.13.64. If two electrical leads are connected to human body at two different locations such as A and B in

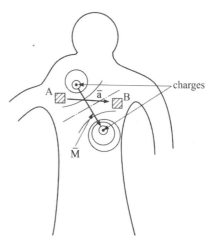

Fig.13.64 : Principle of production of ECG voltage, v_{ecg}

Fig.13.63, another vector in space, $\overline{a}$, can be drawn from one electrode to the other. Then, electrical voltage developed between these electrodes is given by

$$v = \overline{M} \cdot \overline{a}$$

and represents the voltage v_{ecg} to be measured. [See also Fig. 13.63].

The "standard" ECG

The heart being located centrally below the collar bone in the human body, the leads for recording ECG are located as shown in Fig.13.65. Of the ten, six electrodes are positioned in the chest, covering the heart region whilst four in the arms (near the wrists) and legs (towards the heels) as shown. The time-varying electrical signals from the various electrodes are fed to a suitable amplifier and later to a band-pass filter (to eliminate noise and improve signal-to-noise ratio) before passing on to an analogue (paper) recorder.

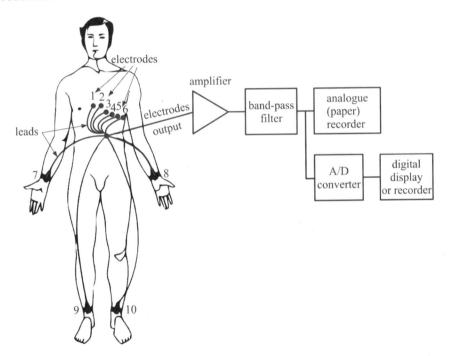

Fig.13.65 : Location of electrodes/leads on human body and schematic of recording ECG in analogue or digital form

For LED display and/or digital recording, the amplified signals may be routed through an A/D converter. The amplifier used is invariably a special instrumentation amplifier, having a programmable gain from around 100 to 2500 and a very high CMRR (common mode rejection ratio), of the order of 130 dB (for gains of 500 to 1000) which is essential in view of small (about

5 mV) signals, usually accompanied by a large AC common-mode component as well as a large variable DC component.

It is important to maintain good electrical contact between every electrode and the skin. For this, a special gelatin solution is first applied at the points of contact before the re-usable electrodes are snapped on the body. As a standard practice, a test of the quality of contacts, on the calibrated paper graph may be desirable before commencing the test to ensure that the recording is faithful and reliable for later examination by the cardiologist. The paper used for recording is divided in small $(1 \text{ mm} \times 1 \text{ mm})$ and large $(5 \text{ mm} \times 5 \text{ mm})$ squares for easy and direct correlation with the recorded ECG. Some features of a typical ECG recording paper are shown in Fig.13.66 which also depicts the 'reference' signal as a means of calibration. Translated in terms of paper movement, the latter moves at a speed of 25 mm/s in a typical electrocardiograph.

Clearly, the entire process of attaching the various electrodes and later recording of the ECG needs special care and only trained personnel are engaged to carry out the test.

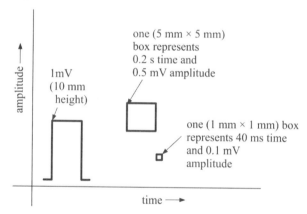

Fig.13.66 : Some features of ECG recording paper

A typical ('normal' or "within normal limits") ECG recording, pertaining, for example to v_{ecg} in Fig. 13.63, is shown in Fig.13.67 in which the various distinct moments of time, standardised as P, Q, R, S and T, represent appropriate functioning of the heart in terms of the electrical signals produced at the cells level[1]. These points also relate to "systole" and "diastole" intervals of the heart activity, also reflected in monitoring of the BP.

---

[1]The discussion in general pertains to what is called the "Rest ECG", the patient in lying position. More intricate instrumentation and procedure may be employed for the "Stress ECG"; for example, the one being recorded as Tread-Mill Test or TMT.

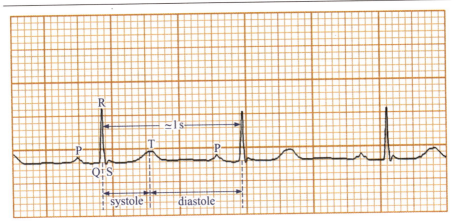

Fig.13.67 : An actual ECG showing systolic and diastolic intervals of the heart activity

Acoustical monitoring

The instrument for this type of monitoring is a variation of the standard electrocardiograph that may be used with chronic heart patients whose heart condition is to be nearly constantly monitored for any sudden stress(es) developed in the heart which may lead to near fatal conditions. The device essentially consists of a detector fitted to the patient, almost all the time, near the heart to detect its functioning and reflect it acoustically in the form of "peeps" at regular intervals; even activate an audible alarm in an emergency.

Recording CT Scan

CT Scan, also popularly known as CAT scan, is the term used to express **Computer Axial Tomography** and is especially prescribed by doctors in cases of internal injuries/ailments of a body which are difficult to be diagnosed otherwise. Its main advantage is that it is a non-invasive medical test.

Invented first in its practical form by Hounsfield in early 1970s, CT scanning combines a special equipment incorporating sophisticated X-rays with add-on computers to produce *multiple* images of the inner areas of the body, across the sections. These cross sectional images of the area having been produced by the 'base' equipment are processed in the computer to produce appropriate 'images' for examination by the physician on the screen or in printed form, later.

The basic operative

In simple terms, the process of scanning consists of passing X-rays through the patient's body or part from one end and obtaining information by means of a detector on the other side as depicted in Fig.13.68 for a CT scan of brain. The X-ray source and the detector are inter-connected suitably, the assembly being capable of rotation *around the patient* during the process of scanning, the latter itself being carried out for a number of "slices" of the part of the body being scanned. Dedicated digital computer then processes the vast data that are so obtained and 'integrate' it to produce a cross-sectional image, called a "tomogram", commonly known as a "CT scan", that is displayed on the computer screen while also being photographed or digitally stored for later retrieval and use as required.

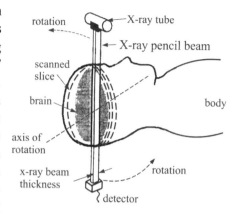

Fig.13.68 : Basics of CAT scanning

Comment

X-rays which have been in extensive, wide-spread use for decades for detection of bone injuries and fractures are one form of electromagnetic radiation or waves, having a wave length of about 1 AU and frequency of about 3×10^{18} Hz. The main reason why X-rays are employed for CT scan, or even routine diagnosis of fractures etc., is because various substances and tissues in the body differ in their ability to absorb X-rays when exposed to same. Some substances are more permeable to X-rays whilst others are not, resulting in different tissues being seen differently when the X-rays' exposed films are developed. For example, dense tissues such as the bones appear white on the film whereas the soft tissues like brain or kidney would appear gray. In contrast, cavities filled with air such as the lungs would appear black, and so on.

Parts of the Body That can be Scanned

Practically any part of the body, or even the entire body, can be scanned by the CT scan equipment to get the desired information. However, the parts often scanned, for examination or to confirm the diagnosis are mainly

- Brain   : for tumors and strokes, and aneurysms
- Head   : for paranasal sinuses, temporal bones
- Neck   : for cervical spondelysis
- Abdomen/Pelvis : for internal injuries or bleeding
- All joints   : for exact fracture(s) location and extent of damage, ligament damage
- Spine   : for spinal-cord injury/damage
- All organs   : heart, kidney, liver, lungs, spleen etc.

In all these cases, the images produced by a CT scan are more detailed and informative than those from ordinary X-rays, even from various angles.

The process of scanning

A commonly used CT scanning machine is illustrated in Fig.13.69 which essentially comprises a circular "gantry" housing the X-ray sources (and associated equipment), detectors and data acquisition system (DAS), and a special table capable of sliding horizontally, to receive the patient and move him/her into and out of the gantry.

Preparation for CT scan

During a CT scan, the patient in loose and comfortable clothing lies *still* on his/her back on the table, with no jewelry or metal parts on the body. The table slides into the CT gantry which is like a small tunnel, with the circular gap being sufficient to receive the patient. It is important that during the scan the patient makes no body movement and in some cases even hold the breath if so required. For this, the doctor may sometimes give a mild sedative to restless or anxious patients to prevent inadvertent movement[1]. Communication with the patient can be maintained throughout the procedure which is totally painless and quiet.

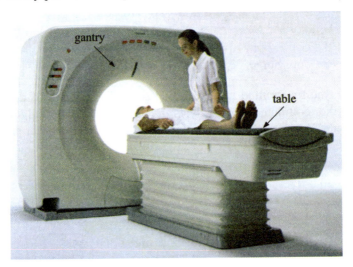

Fig.13.69 : A modern CT scan machine

The procedure

The technologist begins by positioning the patient by sliding the motorised table in and out depending on the body part required to be scanned as

---

[1]This is warranted to obtain clear, unfazed images of the body part with good contrast.

prescribed by the doctor. At the same time, the generator is energised to produce the X-rays by the scanner inside the gantry. The scanner rotates slowly around the body directing the X-rays through a small slice (or cross section) of the part being scanned[1]. A set of detectors rotates in synchronism on the (opposite) far side of the patient (See Fig.13.68). The X-ray source in the scanner produces a narrow fan-shaped beam, with width ranging from 1 to 20 mm. In axial CT scan, commonly used in many diagnostic centres, the table is stationary during one rotation of the scanner, after which it is moved along inside for the next 'slice' of the body part[2].

In modern, 'helical' CT scan, which is commonly used for full-body scans and is much faster, the table moves continuously at a slow pace whilst the X-ray source and detectors rotate in synchronism, producing a spiral or helical scan, operating side-by-side so that a large number of slices (up to 64) can be imaged simultaneously reducing the overall scanning time considerably. The schematic of this arrangement is illustrated in Fig.13.70.

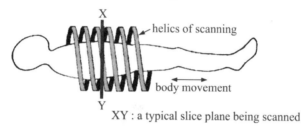

Fig.13.70 : Schematic of helical body scanning

Finally, the transmitted radiation, picked up by a series of detectors are fed into the computer for analysis by a mathematical algorithm and reconstructed as a tomographic image. A 'block' diagram of the entire process is shown in Fig.13.71.

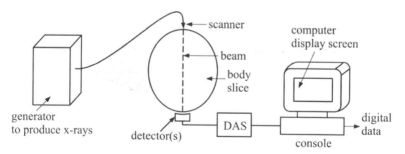

Fig.13.71: Main facets of CT scan arrangement

---

[1]This process is what is called tomography.

[2]The part of the body being scanned is thus "divided" into various (axial) slices, similar to a sliced bread, each slice being scanned (or X-rayed) in succession.

A typical scan of a cross-section of brain is shown in Fig.13.72. The scan shows a darkened area, in contrast with the surroundings, in the brain which indicates that the blood supply to the area in question is blocked. This confirms the initial diagnosis of a stroke in the brain, arrived at, for example, by a routine X-ray.

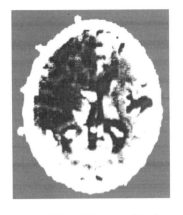

Fig.13.72 : CT scan of brain

Side effects

Although the wide-spread use of CT scans represents perhaps the single, most significant advance in diagnostic radiology, the process is not without its side effects. Prolonged exposure to X-ray radiation, esp. in comparison to routine, ordinary X-raying, may enhance the chance of cancer for some patients. The effective radiation dose, in fact, from a single scan or procedure can be about the same as what an average person may receive from general background radiation in three to five years. Children and pregnant women may be particularly susceptible to radiation and consequent damage. In general, a CT scan should be resorted to only when essential and strictly under prescription of experienced physicians.

Another drawback is the cost of a scan involving great many X-rays which can be considerably high compared to ordinary X-ray(s).

Magnetic Resonance Imaging: MRI

Magnetic Resonance Imaging (MRI), or Nuclear Magnetic Resonance Imaging (NMRI), constitutes one extraordinary modern imaging techniques – an invention being reckoned as most revolutionary after the invention of X-rays more than 100 years ago – is now commonly employed in radiology to visualise comprehensive internal structure of human body and functioning of nearly all body parts[1].

In comparison to CT scanning, MRI procedure *does not use any ionizing radiation*, that is, X-rays, and yet provides much greater contrast between the different soft tissues of the body, or its various parts, making it especially

---

[1]Originally invented in late 1930s by Dr. Isider Rabi – a Nobel laureate in 1944. Nearly three deades later, the first MRI was demonstrated by Lauterbur in 1973, initially known as Nuclear Magnetic Resonance, with the acronym NMR. However, the procedure's name was later changed to Magnetic Resonance Imaging, probably due to the 'negative' connotation attached to the word "nuclear", associated with harmful radiation etc.

suited for oncological imaging for detection of cancer, 'working' at cell levels. See, for example, a sagittal MR image of the knee shown in Fig.13.73 which may be difficult to obtain in a CAT scan with such clarity.

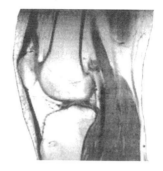

Fig.13.73 : MR image of a knee

Basis of operation

The machine or equipment used for MRI is similar to that used for CT scan, viz., a motorised table for the patient to lie on (on his back) and hollow tunnel-type gantry into which the table can be slid in or out. However, the gantry instead of housing the X-rays source, fed from a generator, and detector at the other end, is provided with a (powerful) magnet, an RF (radio-frequency) transmitter and associated equipment.

The RF field, under interaction with the field from the magnet, is used to systematically alter the alignment of hydrogen atoms in human body to produce a rotating magnetic field that is detected by a scanner. The signals thus obtained are then processed in the computer to provide an image of the body, or of a given part as desired. Thus, the first requirement to perform an MRI is to provide an as strong a magnetic field as possible, technically or otherwise. The magnets may be

(i)   permanent magnets, using rare-earth alloys;

(ii)   electromagnets, with large number of turns and high current, the conductor used being hollow or tubular to allow for flow of a coolant to control the temperature;

(iii) super-conducting magnets, now most commonly used, still employing coils to provide a horizontal magnetic field, but using liquid helium at $4°$ Kelvin, to result in negligible winding resistance and consequent heat loss and temperature rise.

Whilst the last alternative is now universally used and provides magnetic field of flux density as high as 3 to 10 T, the initial cost of equipment may be very high, in addition to the complexity of operation and maintenance.

Molecular structure of human body and "electric" basis of MRI

Human body is largely composed of water molecules (H_2O), comprising up to 70% of body weight, each molecule containing two hydrogen nuclei or protons. When a patient is positioned under the influence of the field of the magnet, the magnetic moments of the protons, being the seat of electric *charge*, align with the direction of the field. In the next stage, a pulse of a radio-frequency electromagnetic field of a given frequency, and *time-varying*, is turned on for a brief moment. This results in altering of the protons alignment relative to the applied field. When the RF field pulse is turned off, the protons would fall back to original alignment. The application of the RF field is, in effect, to make the protons to precess in phase with the field, causing a resonance action, the frequency of resonance depending on the strength of the magnetic field and moment(s) of the given proton(s).

The changes in the proton alignment, while under the influence of the magnetic field, produce an electric signal according to Faraday's law of electromagnetic induction which is detected by the scanner. Apart from water content of the body, since hydrogen is also present in fat and various other tissues (and cells) in the body, the varying molecular structures and the amount of hydrogen in various tissues affect how the (hydrogen) protons would behave in the given magnetic fields – both from the super-conducting magnetic and the RF field. This results in the contrast of images of different parts as obtained from the computer processing. Diseased tissues, such as those of tumors, are detected in MRI because of the basic property that protons in different tissues returns to their equilibrium state, or re-alignment, at different rates, producing corresponding electric signals. By changing the various parameters on the scanner, including the processing algorithm in the computer, this effect is used to create the desired contrast between different types of body tissues, and thus help in the diagnosis by the physicians. A simple block diagram depicting various interactive aspects of a typical MRI equipment is shown in Fig.13.74.

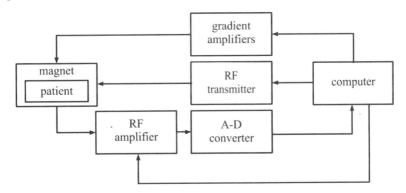

Fig.13.74 : Schematic of an MRI system

The computer directs all the actions of the MRI system from acquisition to processing of data, including controlling gradient amplifiers and the RF transmitter, and when to turn it on and off as per the proper pulse sequence. The RF (receiver) amplifier is also controlled by the computer and relays the signals received by the RF coil from the patient to the A/D converter that digitises the signals to be processed by the computer to reconstruct the body image.

Owing to its capability of sensing any part or organ of body with great finesse, MRI is particularly useful for studying neurological conditions, for disorders of the muscles and joints, for evaluating tumors and for showing abnormalities in the heart and blood vessels. However, patients with some metallic implants or cardiac pacemakers may not be allowed to go in for an MRI due to possibility of harmful effects caused by strong magnetic fields inherent to the MRI equipment.

For various advantages and superiority, an MRI is always preferred over a CT scan. However, the procedure is much more expensive compared to the latter[1].

Recording Electroencephalogram: EEG

An electroencephalogram – acronym EEG – is another important diagnostic process, directed mainly towards recording activity of the brain. Essentially, it is a graphical recording, or an image on a VDU (visual display unit), of *electrical* activity of human brain produced by the firing of neurons[2].

The phenomenon of existence of electric currents, and hence the potential at various points of the brain, was first noted by an English physician Richard Caton around 1875 while experimenting on the brains of rabbits and monkeys. From 1920s and 1930s, the process was gradually being developed for humans and emerged as one of the standard diagnostic technique from 1980s. In modern clinical context, an EEG refers to the recording of the brain's electrical activity with time, typically over a period of 20 to 40 minutes.

Brain wave types

The electrical signals or outputs produced in a typical brain are oscillatory, or time-varying, in nature and relate to state of "working" of the brain. Four

---

[1]A commercial MRI equipment is now available from SIEMENS that operates at only 1.5 T and claimes to be quite 'cost-effective', fastest and the gantry being designed to preclude clauster-phobia (fear of closed spaces), common to some patients.

[2]The entire process is very complex in recording and more so in interpretation in the absence of a unique, "standard" EEG or brain-wave pattern. The process being quantitatively subjective, only basic aspects are discussed here, mainly from electrical/electronic view point.

major frequency ranges or types have been found to cover the brain function in general. These are, for a 'healthy' individual:

a. alpha waves   : range – 8 to 13 Hz
   - pertain to relaxed mind, with closed eyes

b. beta waves   : range – 13 to 30 Hz
   - correspond to awake condition, eyes open

c. theta waves   : range – 4 to 8 Hz
   - represent drowsy state, or light sleep

d. delta waves   : range – 0.5 to 4 Hz
   - correspond to deep sleep

It is observed that the frequency of brain signals progressively decreases as the condition of mind changes from full alertness (wide awake and, perhaps, busy with some activity) to complete relaxedness (deep sleep), and would be so expected.

To an extent, the above frequency range has a bearing on the instrumentation used for the recording.

Process of electrical potential production and amplitude

It is the neurons or nerve cells in the brain that are electrically active and are primarily responsible for carrying out the brain's functions. Neurons produce discrete electrical signals, called "action potentials", that 'travel' within the brain and cause the release of the chemical called "neurotransmitter" that is the source of electrical (brain) currents, resulting in a net potential difference across two given points of the brain. The electric potentials produced by the action of single neurons are at nano level, not being able to be picked up easily. EEG activity in practice, therefore, invariably reflects the *summation of the synchronous* activity of millions of neurons that have similar spatial orientation, *being radial to the scalp*. The peak-to-peak amplitude of signals may typically vary from 0.5 to 100 µV and must be amplified by nearly 1000 or more before these could be recorded into an EEG[1].

Recording procedure

Human EEG is recorded with the help of a number of special electrodes, typically 0.4 to 1.0 cm in diameter and held in place at various

---

[1]The apparatus or recorder used in practice is called an electroencephalograph and the process is known as electroencephalography.

specific/standard locations on the scalp using special, electrically conducting paste, as depicted in Fig.13.75. In the absence of a good contact, the "electrode impedance", greater than about 5000 Ω, can result in lower transmitted signal and high noise. In standard clinical practice of recording, 19 electrodes are incorporated, placed uniformly over the scalp, as per the "international 10-20 system", as illustrated in Fig.13.76. The labeling of the electrodes shown in the figure follows a standard nomenclature as follows:

F : frontal

C : central

T : temporal (sides)

P : posterior

O : occipital (back side)

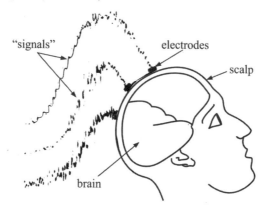

Fig.13.75 : Electrodes on the scalp for an EEG

Typically, the letters are numbered as odd for the left side and even for the right side.

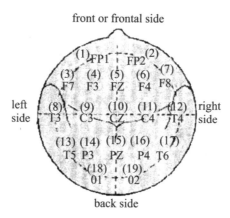

Fig.13.76 : Plan projection of scalp viewed from the top, showing location of various electrodes

Comment

The recommended location of electrodes has its significance in that different brain areas may be related to different functions of the brain, and each scalp electrode is located near a particular brain activity centre. For example

- F7 is located near centre for rational activities
- FZ near intentional and motivational centre
- F8 close to sources of emotional impulses
- C3, C4 and CZ locations deal with sensory and motor functions
- P3, P4 and PZ contribute to activities of perception and differentiation
- T3 and T4 represent emotional processors
- T5 and T6 refer to memory functions
- O1 and O2 point to primary visual functions

and so on.

However, these activities v/s the location of various electrodes must be viewed with caution due to the limitation(s) on account of non-homogeneous properties of the skull which can themselves be highly subjective in nature.

Reference nodes

For successful, effective instrumentation and recording, 'absolute' potential as available at each of the electrodes is not enough; it is essential to provide one or two "reference" electrodes and these are usually located on ear lobes. In addition, a "ground" electrode may often be placed on the nose. The potentials for recording are then collected between each of the electrodes and the fixed reference node(s).

However, the distinction between the "recording" and "reference" electrodes is rather artificial since both types of electrode involve PDs between body tissues, allowing flow of current through tissues and EEG apparatus.

Electrode caps

For patients who may be averse to the idea of fixing electrodes on their scalps, and may develop complexes that might affect the EEG, ready-made, special caps are now available which have the requisite number of electrodes already fitted to the cap, with the various leads brought out for external connections. All that the patient has to do is to wear the cap *firmly*, with the electrodes making good electrical contacts with the desired locations of the scalp. The latter may have to be 'prepared' carefully to ensure good contacts and keep the electrode impedance as low as possible. Two examples of such EEG caps are shown in Fig.13.77.

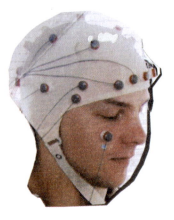 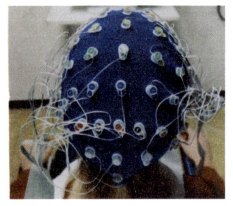

Fig.13.77 : Examples of caps with electrodes for recording of EEG

Instrumentation for EEG

The essential requirements of instrumentation for EEG would comprise

- pre-amplifiers and amplifiers
- filters: analogue and/or digital
- A/D converters
- digital storage and display units
- paper recorders, for detailed examination by a neurologist

The amplifiers are usually of differential input type, with gains varying from 1000 to 100,000, having high input impedance (>100 MΩ) and very high CMMR (100 dB or better). Each electrode is connected to one of the inputs of the differential amplifier; a common system reference electrode being connected to the other input terminal.

The incoming multi-channel data are fed to digital computers for analyses using algorithms that estimate potentials on the brain surface, taking into account the distortions caused by intervening tissue and the physical separation of electrodes themselves from the brain inside. Before proceeding with the actual measurements, the EEG system should be tested as a whole. The inter-channel calibration with known input signal parameters should be checked and should not display significant discrepancies, with particular reference to the noise caused by the analogue amplifier circuitry and A/D converters.

The recording itself may be best performed in specially, electrically shielded rooms to minimise the impact of external electric field(s) and interference, particularly the power-frequency line noise.

Applications

The EEG provides a convenient and effective window on the mind, revealing synaptic action of the brain that can be correlated to the EEG recording. Most EEG signals originate in the brain's outer layer, called the cerebral cortex and believed to be largely responsible for human thoughts, emotions and behaviour on *individual* basis.

In neurology, one of the main diagnostic applications of EEG is in cases of epilepsy where an epileptic activity is reflected as clear abnormalities when compared to a "standard" EEG of a 'normal' person. A typical EEG of an epileptic patient, collected through some of the electrodes is shown in Fig.13.78, the inputs to the amplifier being in differential modes. [See also Fig.13.75].

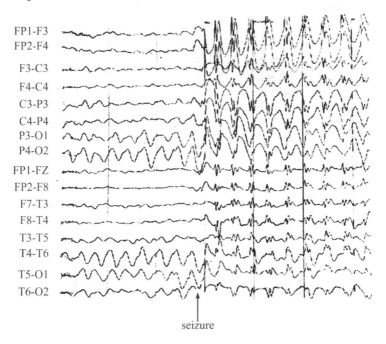

Fig.13.78 : Typical EEG of an epileptic patient showing spikes and waves

Other applications of EEG include, to

- monitor alertness, coma and brain death
- locate areas of damage following head injury, stroke or tumor[1]
- monitor and control anaesthesia 'depth'
- investigate sleep disorders and physiology

---

[1]These are now progressively being diagnosed using CT or MRI scans. However, EEG has some distinct advantages over MRI, such as lower hardware costs and convenience of carrying out the tests, even in mobile conditions.

and a horde of activities related to the brain, mainly by studies of comparison.

Head-ready EEG system(s)

A recent development is the 2-channel wireless EEG recording system in the form of a head band as shown in Fig.13.79. The system is powered by a thermoelectric generator, based on the use of body heat dissipated naturally from the forehead. The entire system, integrated into the head band, is self supporting, the power consumption being only about 0.8 mW.

Fig.13.79 : A head-band form of "EEG" monitor

The potential applications include detection of imbalance between the two halves of the brain, and some kinds of brain trauma, in addition to monitoring brain activity in general.

A low-power digital signal-processing unit encodes the extracted EEG data which are sent to a PC via a wireless link. The device becomes operational in less than a minute after being switched on.

WORKED EXAMPLES

1. An object of mass 4 kg moves round a circle of radius 6 m with a constant speed of 12 m/s. Calculate (a) the angular speed, (b) acceleration, (c) rotational speed in rpm, (d) time of one revolution and (e) the force towards the centre.

Angular velocity, $v = r \times \omega = 12$ m/s

Since $r = 6$ m

$\therefore$ angular speed,

$$\omega = \frac{v}{r} = \frac{12}{6} \quad \text{or} \quad 2 \text{ rad/s}$$

$$\text{Acceleration} \quad = \frac{v^2}{r} = \frac{12 \times 12}{6} \quad \text{or} \quad 24 \text{ m/s}^2$$

$$\text{Revolutions per sec} = \frac{\omega}{2\pi} = \frac{2}{2\pi}$$

$\therefore$ revolutions per minute

$$= \frac{2}{2\pi} \times 60 \quad \text{or} \quad 19 \text{ rpm}$$

Time of one revolution

$$= \frac{2\pi}{\omega} = \frac{2\pi}{2} \quad \text{or} \quad 3.141 \text{ s}$$

The force towards the centre,

$$F = m \times a$$
$$= 4 \times 24 \quad \text{or} \quad 96 \text{ N}$$

2. The resistance of the element of a platinum-resistance thermometer is 2.0 Ω at the ice point and 2.73 Ω at the steam point. What temperature on the platinum-resistance scale would correspond to a resistance value of 8.43 Ω?

The temperature corresponding to a resistance value R_θ is given by

$$\theta_p = \frac{R_\theta - R_0}{R_{100} - R_0} \times 100 \, ^\circ C$$

Substituting,

$$\theta_p = \frac{8.43 - 2.0}{2.73 - 2.0} \times 100$$

$$= \frac{6.43}{0.73} \times 100$$

$$= 881\ ^\circ C$$

3.   In the circuit shown below, the EMF of the standard cell is 1.02 V and this is balanced by the PD across a resistance of 2040 Ω in series with a potentiometer wire AB. The wire AB is 1.0 m long and has a resistance of 4 Ω. If a thermocouple EMF is balanced on the potentiometer such that AC on it is 60 cm, calculate the thermocouple (TC) EMF. Also, if the TC EMF can be expressed as

$$E = a\,\theta + b\,\theta^2$$

where E is the EMF of the TC in μV, θ the temperature (rise) and a and b are constants, equal to 41 and 0.04, respectively (a Cu-constantan TC), calculate the temperature rise measured by the TC.

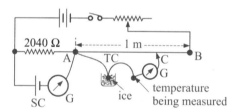

Since 1.02 V is the PD across the resistance 2040 Ω due to the standard cell, and the 4 Ω potentiometer wire AB is in series with 2040,

$$\text{PD across AB} = \left(\frac{4}{2040}\right) \times 1.02$$

$$= 2\ \text{mV, equivalent to 100 cm}$$

With the TC in circuit, the balance is obtained at 60 cm from A

$$\therefore \qquad \text{TC output} = \left(\frac{60}{100}\right) \times 2 \quad \text{or 1.2 mV (or 1200 μV)}$$

∴ from the given expression,

$$1200 = 41\theta + 0.04\theta^2$$

or $0.04\theta^2 + 41\theta - 1200 = 0$

whence

$$\theta = \frac{-41 \pm \sqrt{41^2 - \left[4 \times 0.04 \times (-1200)\right]}}{2 \times 0.04}$$

$$= \frac{-41 + 43.28}{0.08}$$

$$= 28.5 \, ^{\circ}\text{C}$$

4.   In a schematic of fluid flow shown below, the area of cross-section A_1 at X is 4 cm^2 whilst the area A_2 at Y is 1 cm^2. The fluid flows past each section in laminar flow at the rate of 400 cm^3/s What is the pressure difference of the fluid for the flow as above? Assume density of the fluid, $\rho = 1000$ kg/m^3 and g = 9.8 m/s^2.

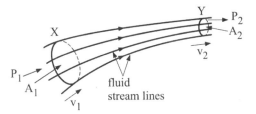

From the given flow rate, the velocity v_1 of fluid at X,

$$v_1 = \frac{400}{4} = 100 \text{ cm/s or 1 m/s.}$$

Similarly, the velocity v_2 at Y is $v_2 = \frac{400}{1} = 400$ cm/s or 4 m/s

Hence, the pressure difference

$$P = \frac{1}{2} \times \rho \times \left(v_2^2 - v_1^2\right)$$

$$= \frac{1}{2} \times 1000 \times \left(4^2 - 1^2\right)$$

$$= 7.5 \times 10^3 \text{ N/m}^2$$

If h is in metres, then from P = h ρ g

$$h = \frac{P}{\rho \, g}$$

$$= \frac{7.5 \times 10^3}{\left(1000 \times 9.8\right)}$$

$$= 0.77 \text{ m (approx) head of the fluid}$$

5.   Water flows steadily along a horizontal pipe of cross section area of 30 cm^2. The static pressure in the pipe is 1.20×10^5 Pa (pascal) and the total pressure is 1.28×10^5 Pa. Calculate the flow velocity and the mass of water per second flowing through the pipe. Take density of water = 1000 kg/m^3.

Total pressure is given by

total pressure = static pressure + "dynamic" pressure

$$= \text{static pressure} + \frac{1}{2}\rho v^2$$

where the velocity of flow is v m/s and ρ the density of fluid in kg/m^3.

$\therefore \qquad 1.28 \times 10^5 = 1.2 \times 10^5 + \frac{1}{2} \times 1000 \times v^2$

whence   v = 4 m/s

Flow of water $= v \times A$

$= 4 \times 30 \times 10^{-4}$ m^3/s

Mass of water flowing per second

$= 4 \times 30 \times 10^{-4} \times \rho$

$= 120 \times 10^{-4} \times 1000$

$= 12$ kg/s

6.   Water flows through a horizontal pipe of cross section 48 cm^2 which has a constriction of cross-section area 12 cm^2 at one place. If the velocity of water at constriction is 4 m/s, calculate the velocity at the wider section. If the pressure in the wider section is 1.0×10^5 Pa, calculate the pressure at the constriction. Density of water = 1000 kg/m^3.

For a given flow, the velocity of fluid is inversely proportional to cross-sectional area

$\therefore$ velocity of water at wider section

$$= \left(\frac{12}{48}\right) \times 4 = 1 \text{ m/s}$$

Also, pressure difference,

$$P = \frac{1}{2}\rho\left(v_2^2 - v_1^2\right)$$

$$= \frac{1}{2} \times 1000 \times \left(4^2 - 1^2\right)$$

$$= 0.75 \times 10^4 \ \text{Pa}$$

∴ pressure at the constriction $= 10 \times 10^4 - 0.75 \times 10^4$

$$= 9.25 \times 10^4 \ \text{Pa}$$

7. If a lamp has an MSCP of 40 cp, calculate the total luminous flux emitted by it in all directions.

Total flux = MSCP × solid angle

$$= 40 \times 4\pi$$

$$= 502 \ \text{lm}$$

8. A lamp of uniform intensity of 100 cp (or candela) is enclosed in a glass globe of 40 cm diameter. If 25% of the light emitted by the bulb is absorbed by the globe, determine the brightness and intensity of light emitted from the globe.

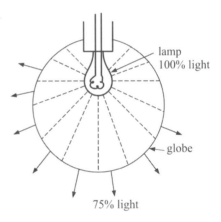

The fixture of the lamp and globe in question is as depicted in the adjoining figure.

Since only 75% light is able to emit from the globe, its intensity of light will be 0.75 × 100 or 75 cp (candela). Now the surface area of the globe is

$$4\pi \times r^2 \text{ or } = 4\pi \times \left(\frac{20}{100}\right)^2 \ \text{m}^2$$

∴ brightness of the *globe* is

$$= \frac{\text{luminous intensity}}{\text{surface area}}$$

$$= \frac{75 \times 10^4}{4\pi \times 400}$$

$$= 149.3 \ \text{cd/m}^2$$

9.   A lamp having an intensity of 1000 candela is hung 6 m over the 'centre' of the floor of a hall. Calculate the illumination due to the lamp directly below it and at each corner of the hall as indicated in the figure. Neglect any absorption or reflection of light owing to the walls.

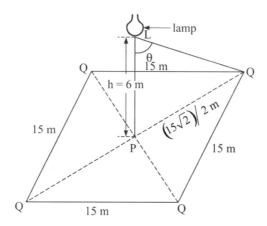

The various distances required for calculations are as marked in the figure. It is desired to obtain illumination at P (directly below the lamp at the 'centre' of the hall) and at corners marked Q.

The illumination at P is simply

$$IL_P = \frac{1000}{(6)^2} = 27.8 \text{ lux } \left(\text{or lm/m}^2\right)$$

The distance from lamp to the corner,

$$LQ = \sqrt{6^2 + \left(15\sqrt{2}/2\right)^2}$$

$$= \sqrt{36 + 225/2}$$

$$= 12.2 \text{ m}$$

Also, the angle between LQ and LP is $\cos^{-1}\left(\frac{LP}{LQ}\right)$ or $\cos\theta = \frac{6}{12.2}$

∴ illumination at any of the corners, Q

$$IL_Q = \frac{cp}{d^2} \times \cos\theta$$

$$= \frac{1000}{(12.2)^2} \times \frac{6}{12.2}$$

$$= 3.3 \text{ lux } \left(\text{or lm/m}^2\right)$$

[Note the drastic reduction of illumination from P to Q].

10. A lamp having a luminous intensity of 500 cd is fitted with a reflector so as to direct 80% of the light from the lamp along a beam having an angle of cone of 20° as shown. Calculate

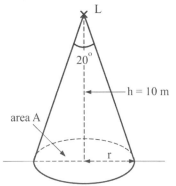

(a) the total flux transmitted along the beam

(b) the average illumination on a surface below the lamp at a distance of 10 m, the surface being normal to the cone of light falling on it.

The total light flux from the lamp

$$= 500 \times 4\pi \ \text{lm}$$

∴ the light flux transmitted along the beam

$$= 0.8 \times 500 \times 4\pi$$

$$= 5024 \ \text{lm}$$

The angle of the cone of light formed by the reflector is 20°

∴ the radius of the circular patch of light on the surface

$$r = h \times \tan \frac{\theta}{2}$$

$$= 10 \times \tan 10°$$

$$= 1.76 \ \text{m}$$

Area of the surface to be illuminated

$$A = \pi \, r^2$$

$$= \pi \times (1.76)^2$$

$$= 9.73 \ \text{m}^2$$

and 'average' illumination on A

$$= \frac{5024}{9.73}$$

$$= 516 \ \text{lux} \ (\text{lm/m}^2)$$

Appendices

Appendix I

Units and Conversions

BASIC AND DERIVED SI[1] UNITS

Basic SI Units

Quantity	Name of the unit	Symbol
Length	metre	m
Mass	kilogram	kg
Time	second	s
Electric current	ampere	A
Temperature	kelvin	K
Luminous intensity	candela	cd
Amount of 'substance'	mole	mol

Derived SI Units (in alphabetical order)

Quantity	Name of the unit	Symbol
Acceleration	metre per second per second	m/s^2
Angular acceleration	radian per second per second	rad/s^2
Angular velocity	radian per second	rad/s
Area	square metre	m^2
Capacitance	farad	F
Density	kilogram per cubic metre	kg/m^3
Dynamic viscosity	newton second per metre per metre	Ns/m^2
Electric charge	coulomb	C
Electric field strength	volt per metre	V/m
Electric intensity	watt per steradian	W/sr
Electric potential	volt	V
Electric resistance	ohm	$V/A(\Omega)$
Energy	joule	J

---

[1]SI is the abbreviation for *"Systeme International d' Units"*. It is the modern form and an extension and refinement of the traditional metric system, and incorporates a number of basic and derived units.

Contd…..

Quantity	Name of the unit	Symbol
Entropy	joule per kelvin	J/K
Force	newton	N
Frequency	hertz	Hz
Illumination	lux	lx
Inductance	henry	H
Luminance	candela per square metre	cd/m^2
Luminous flux	lumen	lm
Luminous intensity	candle power/candela	cp/cd
Magnetic field strength	ampere per metre	A/m
Magnetic flux	weber	Wb
Magnetic flux density	tesla	T
Magnetomotive force	ampere	A
Permeability	henry per metre	H/m (μ)
Permittivity	farad per metre	F/m (ε)
Power	watt	W
Pressure	pascal	Pa
Quantity of heat	joule	J
Specific heat[1]	joule per kilogram kelvin	J/kg K
Stress	pascal	Pa
Thermal conductivity	watt per metre kelvin	W/mK
Velocity	metre per second	m/s
Volume	cubic metre	m^3
Work	joule	J

Multiplying Factors

The multiples of SI units are formed by means of the prefixes as given below.

Factor by which the unit is multiplied	Prefix	Symbol
10^{12}	tera	T
10^9	giga	G
10^6	mega	M
10^3	kilo	k
10^2	hecto	h

[1]Also specified sometimes as *volumetric* specific heat expressed as "joule per cubic metre kelvin", J/m^3K.

Contd…..

Factor by which the unit is multiplied	Prefix	Symbol
10	deca	da
10^{-1}	deci	d
10^{-2}	centi	c
10^{-3}	milli	m
10^{-6}	micro	μ
10^{-9}	nano	n
10^{-12}	pico	p
10^{-15}	femto	f
10^{-18}	atto	a

Definitions of "Basic" SI Units

[See also Chapter I]

1. Unit of length: metre, m

The metre is the length equal to 1650763.73 wavelengths in vacuum of the radiation corresponding to the transition between the levels $2p_{10}$ and 5_{d_5} of the Krypton-86 atom.

2. Unit of mass: kilogram, kg

The kilogram is equal to the mass of the international prototype of the kilogram kept in a vault in Sèvres, near paris[1].

3. Unit of time: second, s

The second is the duration of 9192631770 periods of the radiation corresponding to the transition between two hyper-fine levels of the ground state of the cesium-133 atom.

4. Unit of electric current: ampere, A

The ampere is that constant current which if maintained in two straight parallel conductors of infinite length, of 'negligible' circular cross section and placed one metre apart in vacuum, would produce between these conductors a force equal to 2×10^{-7} newton *per metre length*.

---

[1]It is proposed that the kilogram will be defined by 'fixing' the numerical value of h, the Planck constant, which has units that contain the kilogram.

5. Unit of temperature: kelvin, K

The kelvin is the fraction 1/273.16 of the thermodynamic temperature of the triple point of water (or absolute zero)[1].

6. Unit of luminous intensity: candela, cd

The candela is the luminous intensity, in the perpendicular direction of a surface of 1/600,000 square metre of a black body, at the temperature of freezing platinum under a pressure of 101325 newton per square metre.

7. Unit of "amount of substance": mole, mol

The mole is the amount of substance of a system which contains as many elementary units as there are carbon atoms in 0.012 kilogram of carbon-12.

Note that the definitions of the above basic quantities have undergone several changes and must be considered at their face values only. It is likely that some of these may again be revised in the near future.

CONVERSION OF UNITS

[For normally used *metric* quantities into the SI units]

Quantity	To covert metric unit	into SI units		multiply by
		name	symbol	
Length	angstrom [unit(AU)]	metre	m	10^{-10}
	micron			10^{-6}
Area	are	square metre	m^2	10^2
	hectare			10^4
Volume	litre	cubic metre	m^3	10^{-3}
Angle	degree	radian	rad	$\pi/180$
Velocity				
- linear	km/hr	metre per second	m/s	1/3.6
	knot			0.514 444
- angular	rpm	radian per second	rad/s	0.104 720
Acceleration	'standard' acceleration of free fall	metre per second per second	m/s^2	9.806
Frequency	cycles per second	hertz	Hz	1

---

[1]The international *practical* unit of temperature is degree Celsius (°C) such that 0 °C = 273.15 °K, and 'absolute' temperature = degree Celsius + 273.15

Contd…..

Quantity	To covert metric unit	into SI units		multiply by
		name	symbol	
Mass	gram	kilogram	kg	10^{-3}
	tonne			10^3
Density	gram per cc	kilogram per cubic metre	kg/m^3	10^3
Moment of inertia	kilogram force metre squared	kilogram metre squared	kgm^2	9.806
Force	Dyne	newton	N	10^{-5}
	kg force			9.806
Pressure/stress	dyne per square centimetre	pascal (newton per square metre)	Pa	0.1
	millimeter mercury (torr)			133.322
Torque	dyne centimetre	newton metre	Nm	10^{-7}
	gram centimetre			9.806×10^{-5}
	kilogram force metre			9.806
Energy (also, work and heat)	erg	joule	J	10^{-7}
	kilogram force metre			9.806
	kilowatt hour		kWh	3.6×10^6
	electron volt (ev)			1.602×10^{-19}
	calorie			4.1868
	kilo-calorie			4186.8
Power	erg per second	watt	W	10^{-7}
	metric horse power			735.5
	Kilo-calorie per hour			1.163
Temperature, t	degree Celsius	kelvin	K	$273 + t$ [addition, not multiplication]
Heat flow rate	kilo-calorie per hour	watt or joule per second	W or J/s	1.163
Thermal conductivity	kilo-calorie per second per centimetre per degree Celsius	watt per metre per degree kelvin (or degree Celsius)	W/m/K	418.68
Specific heat capacity	calorie per gram per degree Celsius or kilo-calorie per kilogram per degree Celsius	joule per kilogram per degree Kelvin	J/kg/K	4186.8

Contd.....

Quantity	To covert metric unit	into SI units		multiply by
		name	symbol	
Magnetic flux	maxwell (or lines)	Weber	Wb	10^{-8}
Magnetic flux density	gauss, or lines per square centimetre	Tesla	T	10^{-4}
Magnetic field strength	oersted	ampere per metre	A/m	79.6
Illumination	phot, or lumen per square centimetre	lux lumen per square metre	lx lm/m^2	10^4 10^4
Brightness	lambert	candela per square metre	cd/m^2	3183

[For normally used FPS units into the SI units]

Quantity	To convert the FPS unit	into SI unit		multiply by
		name	symbol	
Mass	ounce pound ton	kilogram	kg	28.35×10^{-3} 0.4536 1016.05
Length/distance	inch foot yard mile nautical mile	metre	m	25.4×10^{-3} 0.3048 0.9144 1609.344 1852
Area	square inch square foot square yard acre square mile	square metre	m^2	0.645×10^{-3} 92.903×10^{-3} 0.836 4.0468×10^3 2.59×10^6
Volume	cubic inch cubic foot gallon (UK) gallon (US)	cubic metre	m^3	16.387×10^{-6} 28.317×10^{-3} 4.546×10^{-3} 3.785×10^{-3}

Contd…..

Quantity	To convert the FPS unit	into SI unit name	symbol	multiply by
Velocity	inch per second	metre per second	m/s	25.4×10^{-3}
	foot per second			0.305
	mile per hour			0.447
Speed	mile per hour	kilometre per hour	km/hr	1.6
	revolution per minute	rad per second	rad/s	0.1047
Acceleration	inch per second per second	metre per second per second	m/s^2	25.4×10^{-3}
	foot per second per second			0.305
Density	pound per cubic inch	kilogram per cubic metre	kg/m^3	27.68×10^3
	pound per cubic foot			16.02
Force	poundal	newton	N	0.1382
	pound force			4.448
	ton force (UK)			9964
Torque	poundal foot	newton metre	Nm	42.14×10^{-3}
	pound force foot			1.356
Pressure/Stress	pound force per square inch	pascal	Pa	6894.76
	pound force per square foot			47.88
	standard atmosphere			101.325×10^3
	inch (mercury)			3.3864×10^3
	inch (water)			2.989×10^3
Energy (work and heat)	foot poundal	joule	J	42.14×10^{-3}
	horse power hour			2.684×10^6
	British thermal unit (BTU)			1055.106

Contd…..

Quantity	To convert the FPS unit	into SI unit		multiply by
		name	symbol	
Power	foot pound force per second	watt	W	1.3558
	horse power			745.7
Temperature, t	degree Fahrenheit	kelvin	K	5/9 × (459.67+t) [not simple multiplication]
Specific heat	BTU per pound per degree F	joule per kilogram per degree kelvin	J/kg/K	4.187×10^3
Specific heat (volume-based)	BTU per cubic foot per degree F	joule per cubic metre per degree kelvin	J/m³/K	67.07×10^3
Thermal conductivity	BTU per foot hour per degree F	watt per metre per kelvin	W/m/K	1.7307
Current density	ampere per square inch	ampere per square metre	A/m²	1550

Appendix II

Magnetic Field Strength due to Forms of Conductors

The basic expression used to derive the magnetic field strength due to various desired forms of conductors is the Ampere's formula (also the Biot and Savart law), given by

$$dH = \frac{I \times dl}{4\pi x^2} \sin \alpha \ \ A/m$$

as depicted graphically in Fig.1, in which dH is the field strength due to an element dl of a conductor carrying a current I, at a point P distant x, inclined to the element at an angle α. All variables are expressed in MKS system of units.

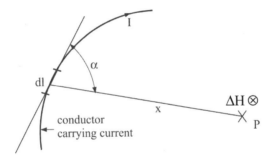

Fig.1: Basics of field strength at point P

1. Field Strength due to a Long Straight Conductor

Fig.2 shows a straight conductor and one of its circular lines of force. At any point P on this circle the magnitude of the field strength due to a current element dx is

$$dH = \frac{I \times dx}{4\pi s^2} \sin \theta \ \ A/m$$

by the Ampere's formula. Now, from the given dimensions,

$$s = r \ \mathrm{cosec} \ \theta \ \text{ and } x = r \cot \theta$$

∴ $$dx = r \ d \ (\cot \theta) = - r \ \mathrm{cosec}^2 \theta \ d\theta$$

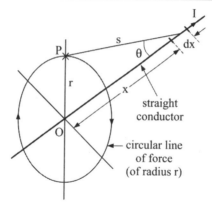

Fig.2 : Field strength due to a long conductor

Hence

$$dH = \left(\frac{I\sin\theta}{4\,\pi}\right) \times \frac{\left(-r\,\mathrm{cosec}^2\theta\;d\theta\right)}{\left(r^2\,\mathrm{cosec}^2\theta\right)}$$

$$= \frac{-I\sin\theta\;d\theta}{4\,\pi\,r}\;\;A/m$$

Infinitely long conductor

For an 'infinitely' long conductor, the limiting values for θ are 0 and π, and so at a radial distance r

$$H = -\frac{I}{4\,\pi\,r}\int_0^\pi \sin\theta\;d\theta$$

or

$$\boxed{H = \frac{I}{2\,\pi\,r}\;\;A/m}$$

and the flux density *in air*

$$B = \mu_0\,H = \frac{\mu_0\,I}{2\,\pi\,r}\;\;T$$

2. Field Strength at the Centre of a Circular coil

The configuration of the coil (of N turns) is shown in Fig.3. Consider an element dl of *one turn* of the coil, carrying a current I. The magnetic field due to this element derived at the centre would act along the axis in the

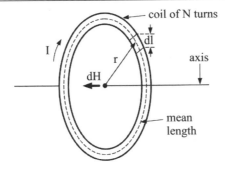

Fig.3 : Field at the centre of a coil

left direction for the assumed direction of the current. The magnitude of the field strength from the Ampere's formula is given by

$$dH = \frac{I\ dl}{4\ \pi\ r^2}$$

with $\sin \theta = 1$ since the angle θ is now equal to $\pi/2$.

Hence for the whole turn

$$H = \frac{I}{4\ \pi\ r^2} \times \sum dl$$

$$= \frac{I}{2\ r}\ A/m \quad \text{since} \sum dl = 2\pi r$$

Given that the coil has N turns and assuming its axial length to be very small, the total field is given by

$$\boxed{H = \frac{N\ I}{2\ r}\ A/m}$$

from which the flux density in air, too, can be deduced.

3. Field Strength at any Point on the Axis of a Circular Coil

The single turn of a coil and its axis is shown in Fig.4. Consider two *diametrically* opposite elements on the coil, of length dl each. The magnetic field at any point on the axis of the coil due to the *lower* element will act in the direction PQ, having the magnitude

$$dH = \frac{I\ dl}{4\ \pi\ s^2}, \quad \theta = \frac{\pi}{2}$$

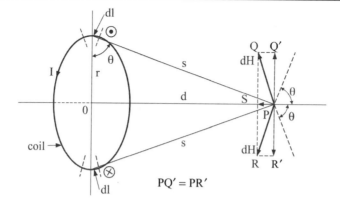

Fig.4 : Field at a point on the axis of a coil

The *upper* element, too, will produce an identical field at the same point directed along PR. It is clear that components of these two fields in the direction perpendicular to the axis at point P will neutralise one another, being in opposite directions. On the other hand, the axial components would add being in the same sense, directed along PS. The axial component due to one element (now regarding dH as an *element* of the resultant field strength) is given by

$$dH = \frac{I\,dl}{4\,\pi\,s^2} \times \cos\theta$$

$$= \frac{I\,dl}{\left\{4\pi\left(d^2 + r^2\right)\right\}} \times \frac{r}{\left(d^2 + r^2\right)^{1/2}}$$

$$= \frac{I\,r\,dl}{4\pi\left(d^2 + r^2\right)^{3/2}}$$

The total field strength at point P due to the whole turn is therefore

$$H = \frac{I\,r}{4\pi\left(d^2 + r^2\right)^{3/2}}\sum dl$$

$$= \frac{I\,r^2}{2\left(d^2 + r^2\right)^{3/2}} \qquad \text{since } \sum dl = 2\pi r$$

If the coil has N turns of axially short length[1], then

$$H = \frac{N\,I\,r^2}{2\left(d^2 + r^2\right)^{3/2}}\ \text{A/m}$$

4. Field Strength inside a Solenoid

A solenoid is a magnetic device, much in use, comprising a tubular former wound on its exterior with a uniform winding of N turns, the turns being placed close to each other. The interior of the former may simply contain air when the device is called an "air-core" solenoid, or contain an iron core. When the winding is excited with a current, an axial magnetic field will be produced, fringing out at the ends.

Fig.5 shows the cross-section of an air-core solenoid having a length l m, assumed to be very much larger compared to its "mean" diameter. The solenoid has N turns, and therefore there are N/l turns per unit length (that is, per metre). Assume the origin as the centre point of the solenoid as shown and consider an element dx, distant x from the origin containing (Ndx/l) turns.

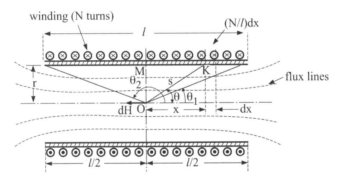

Fig.5 : Field strength inside a solenoid

Regarded as a coil of very short axial length, this element would produce at O a field strength directed along the axis, dH, as derived earlier, given by

$$dH = \frac{N\,dx}{l} \times \frac{I\,r^2}{2\left(x^2 + r^2\right)^{3/2}}$$

Hence for the whole solenoid

---

[1]At the centre of the coil, s = r and the field will simply be (NI/2r) as derived earlier. Also, in air B = μ_0 H, showing that the phenomenon will be magnetically linear.

$$H = \int_{-l/2}^{+l/2} dH = \int_{-l/2}^{+l/2} \frac{NI}{2l} \times \frac{r^2 \, dx}{\left(x^2 + r^2\right)^{3/2}}$$

To integrate this expression, change the variable as follows:

Write $\qquad x = r \cot \theta$

$\therefore \qquad dx = -r \, \mathrm{cosec}^2 \, \theta \, d\theta$

$$= \frac{-r \, d\theta}{\sin^2 \theta}$$

Also, from the triangle OMK

$$s = \frac{r}{\sin \theta} \qquad \text{or} \qquad s^3 = \frac{r^3}{\sin^3 \theta}$$

$\therefore \qquad H = \dfrac{N I}{2 l} \displaystyle\int_{\theta_2}^{\theta_1} \dfrac{\sin^3 \theta}{r} \times \left[\dfrac{(-r \, d\theta)}{\sin^2 \theta} \right]$

$$= 1 - \frac{N I}{2 l} \int_{\theta_2}^{\theta_1} \sin \theta \, d\theta$$

$$= \frac{N I}{2 l} \left[\cos \theta \right]_{\theta_2}^{\theta_1}$$

That is

$$\boxed{H = \frac{NI}{2l} \left[\cos \theta_1 - \cos \theta_2 \right] \text{A/m}}$$

at the 'centre' of the solenoid (or mid-length location).

Field strength at the centre of a 'long' solenoid

In a (very, very) long solenoid, θ_1 would tend to zero and θ_2 to π in the limit and expression for H would reduce to

$$H = \frac{NI}{2l} \left(\cos 0 - \cos \pi \right)$$

or

$$\boxed{H = \frac{NI}{l} \text{ A/m}^{[1]}}$$

---

[1]An air- or iron-core **toroid,** discussed in Chapter XII, is a special application, or example, of a (very) long solenoid having, a large 'mean' diameter compared to the diameter of its cross-section and the magnetic field at *any point along the mean length* of a toroid is given by the above expression.

Again, for an air-core solenoid, $B = \mu_0 H$, and hence at the centre of the solenoid

$$B = \mu_0 \frac{NI}{l} \ T$$

Comment

In practice, a suitably designed long solenoid, that is, the one fabricated choosing appropriate values of r, l, N and I, has great importance in helping calibration of search coils, invariably used in DC and AC magnetic measurements [see Chapter XII]. This derives from the fact that the magnetic field at the centre of the solenoid is

(a)   easily calculated from its physical constants and the given current (to be limited on other considerations such as overheating of the winding); and

(b)   nearly completely uniform as shown in Fig.6, and magnetically linear ($\mu_r = 1$).

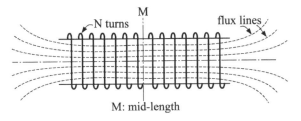

Fig.6 : Magnetic field due to a (long) solenoid

Helmholtz Coils

A case of much practical importance, based on (mutual) electromagnetic induction, used in various magnetic measurements and the basis of fixed coils in all electro-dynamometer instruments, is a device comprising two identical coils placed a short distance apart with their planes parallel to each other as shown schematically in Fig.7. The device, called **Helmholtz** coils, is based on the field of *two* coils, produced axially and simultaneously.

The field along the axis of a *single* coil varies with the distance x from *its* centre as was shown earlier. The two fields combine and result in a uniform field within the space between the two coils as shown.[Fig.7(b)].

Helmholtz used two identical, coaxial, parallel coils of (equal) radius R each and separated centre to centre, also by the distance R[1], each coil having N turns and carrying a current in the *same* direction. The resultant field is

---

[1]A special, specific feature of Helmholtz coils.

then uniform for the distance (R/2), on either side of the point of their axis, mid-way between the coils.

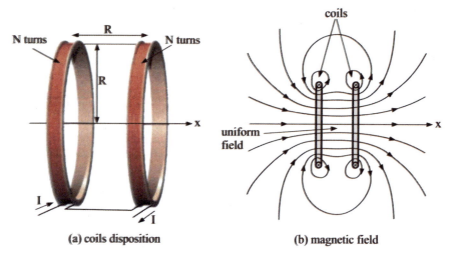

(a) coils disposition   (b) magnetic field

Fig.7 : Helmholtz coils and the magnetic field

Theory

Extending the results for the field due to a single coil at a distance x along its axis, here

$$r = R \quad \text{and} \quad d = \frac{R}{2}$$

Hence, the magnitude of the resultant field due to the two coils

$$H = 2 \times \left\{ \frac{NIR^2}{2} \times \frac{1}{\left[(R/2)^2 + R^2 \right]^{3/2}} \right\}$$

$$= \left(\frac{4}{5} \right)^{3/2} \times \frac{NI}{R}, \qquad \text{after simplification}$$

or

$$\boxed{H = 0.72 \frac{NI}{R} \ \text{A/m}}$$

very approximately, with I in ampere and R in metre.

Usually, the medium between the coil is essentially air and hence the device is magnetically linear.

Therefore, B between the coils is

$$B = \mu_0 H \ T$$

Appendix III

Experiments with the "Special" Transformer

As described in Chapter II, the special feature of this transformer is the "open-ended" core excited by a single 'concentric' coil at the base, wound with large number of turns. The various experiments performed with the help of the device revolve around the *alternating* flux produced by this coil which, accordingly, was named the "mother coil" by its inventer[1]. A sketch of the device and its schematic are shown in Fig.1. [See also Chapter II, Fig.2.31].

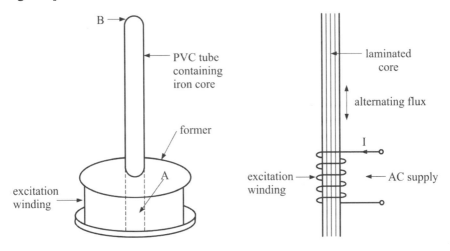

Fig.1 : The "mother coil" and its schematic

For convenience, the excitation winding is of 2300 turns so that the number of turns per volt are 10 when the coil is excited from a 1-phase, 230 V supply. This is helpful in devising the various experiments on the coil.

Magnetic Field of the Coil

When excited, the coil sets up a nearly sinusoidal alternating flux in the core the magnitude of which is maximum at the bottom (location A) and tapers

---

[1](late) A.H. Devadas, Sr.Manager at the Corporate Research and Development Division of Bharat Heavy Electricals Ltd. (BHEL), India.

off to a smaller value at the top (location B) as depicted in Fig.2[1]. The value of flux density B along the height of the core can be actually computed from measured value of the induced EMF at various heights.

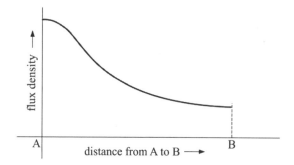

Fig.2 : Flux density along height of the core

Experiments with the coil

A. To Show the Transformer Principle

The simple transformer action due to electromagnetic induction can be demonstrated by fabricating a 'secondary' coil of given number of turns, say 30 so that the induced emf in the coil will be very nearly 3 V *at the bottom of the core* (deduced from 1 V per 10 turns of the mother coil). This can be positioned along the core as shown in Fig.3. After checking the output with the help of a voltmeter, the coil can be moved up to show progressively

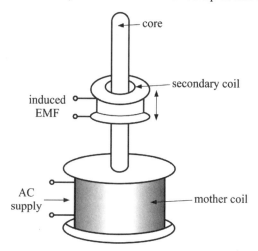

Fig.3 : To show transformer action

---

[1]To avoid much leakage and maintain the sinusoidal variation, care should be taken to avoid saturation of the core by keeping the current low – enough to conduct the various experiments. This is also necessary to avoid overheating of the coil.

reducing electromagnetic coupling, somewhat in accordance with the variation of Fig.2. A visual indication of the effect can be observed by connecting a 3 V lamp across the coil such that the light from the lamp changes from full brightness at the bottom to near zero at the top. The experiment can be repeated by using coils of different number of turns.

Auto-transformer

The principle of operation of an auto-transformer can be demonstrated by making a tubular coil, similar to an air-core solenoid of about 12 to 15 cm length and an inside diameter slightly more than the core of the mother coil and wound with enough number of turns. A part of the winding can be bared of insulation along its length. A lamp, one end of which is connected to one side of the tube and the other sliding on the bare surface would show the auto-transformer action when the lamp glows gradually from the fixed end to the other.

B. To Show Presence of Flux by Production of Sound

By connecting the terminals of a movable coil to a horn or buzzer of appropriate voltage, the presence of alternating flux in the core, of varying strength, can be demonstrated *audibly* in the form of a 50-Hz tone emitted from the horn.

C. Addition and Subtraction of Two EMFs

By fabricating two movable, identical coils of small axial length and placing them near to the bottom of the core so that EMF induced in each is very nearly the same, the addition and subtraction of EMFs can be demonstrated by connecting the two coils first in series and then in series opposition. The resultant EMF in the first case will be 2e and zero in the second case, where e represents the EMF induced in each coil.

D. Magnetic Levitation

The schematic of this experiment is shown in Fig.4. An aluminum ring, about 8 cm OD and 2.5 cm ID, and about 2 mm thick, is inserted in the core, resting on the coil former. When the coil is excited with AC, it is observed that the ring is lifted and levitates itself freely at a certain height. By switching off the excitation, the ring simply drops down or comes to rest. The phenomenon is due to interaction of various currents, magnetic fields and production of forces following the basic laws as shown below:

Let

I_1 be the current in the mother coil at any instant

F_1 the magnetic field due to I_1

I_2 the current in the ring at the same instant due to short circuiting of the EMF induced in the ring

F_2 the magnetic field produced by the current I_2

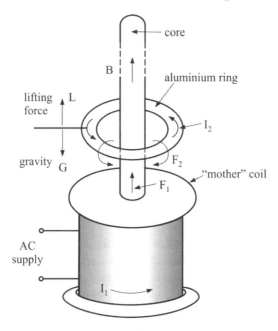

Fig.4 : To demonstrate magnetic levitation

Due to interaction of currents and magnetic fields, the force acting on the ring is such as to lift it vertically. The final floating or levitation position is attained when the upward force, produced electromagnetically, is balanced by the gravitational force acting on the ring in downward direction as depicted in the figure.

E. Heating of Short-circuited 'Secondary'

If the secondary winding of a conventional transformer is short circuited whilst its primary winding is excited at normal voltage, the secondary induced voltage would result in a very-high current flowing through the winding in the absence of a 'load' and consequent 'balancing' MMF. This would lead to I^2R loss, heating and extreme temperature rise, initially in the winding (affecting its insulation) and later the core. The same effect can be demonstrated on the mother coil by performing the above experiment on magnetic levitation, *but by holding the ring firmly at the bottom of the core* and restraining it from levitating. When so held (for example by pressure of fingers), the induced EMF, and hence the currents in the closed ring, result in excessive heating and sharp temperature rise in *a few seconds*.

Induction Heating

In fact, this is a simple demonstration of an important method of electric heating, called "induction heating" employed in steel and other metal industries for melting and/or refining of metals. The basic scheme of induction heating is shown in Fig.5. The core carries a conventional winding, called the primary, to produce the alternating flux which in turn induces EMF in the metal to be melted (called the "charge"), held in a crucible. This forms the short-circuited secondary of the 'transformer' resulting in extremely high induced currents and ultimately melting of the metal due to the heat produced by the induced currents. A number of variations of the process, and devices called "induction furnaces", are available in practice.

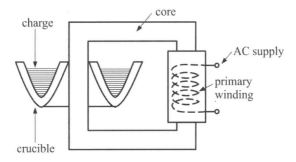

Fig.5 : Induction heating

F. Motor Action

That the induced EMF in a suitably designed 'secondary' winding inserted on to the core can drive a motor is demonstrated by connecting the coil terminals to a bridge rectifier and feed the DC output to a (small) DC motor. The motor speed would be maximum at the bottom of the core, reducing considerably as the coil is moved up.

Many more experiments demonstrating the phenomena of electromagnetic induction can be devised based on the ingenuity of the experimenter.

Appendix IV

DCCT and DCVT

DCCT: Direct Current Current Transductor

DCCTs represent the development over the past few decades for the measurement of large *direct* currents, from several hundred to kiloamperes. An important application of DCCT is in HVDC transmission for measurement and control purposes[1], including control of power flow and protection of the line; for the latter, a very fast response is also necessary. The working principle of a DCCT as a central requirement in a HVDC system is explained by reference to Fig.1.

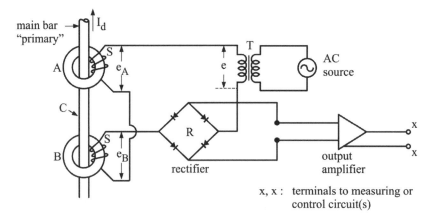

Fig.1 : Schematic of a DCCT

Operation

A and B are two *identical* saturable reactors of toroid construction, placed around conductor C of the line, similar to bar primary of a CT, having equal number of turns, in each of the 'secondary' winding, S, of N_s turns. The two secondary windings are connected in series opposition as shown, across a source of AC excitation via a transformer T and a diode bridge rectifier, R. The other pair of rectifier terminals is fed to an amplifier, the output of

---

[1]In addition, a DCVT – Direct Current Voltage Transductor – as discussed later is also employed for control and regulation of power flow in the (HVDC) transmission systems.

which is connected to the measuring or control circuits. For 'steady' main current, I_d, the 'outputs' of the two secondary windings is balanced so that the net output across xx is zero.

The cores of the reactors consist of a very high permeability material, such as mu-metal, having a sharp saturation point and very low coercivity as shown in Fig.2. The primary current in the conductor exerts an MMF $F_p = N_p I_d = I_d$, since number of 'primary' turns, $N_p = 1$. Even a small MMF, in the absence of a 'secondary' current, drives the cores of *both* the reactors well into saturation, represented by point P on the B/H curve; $H = F_p$, $B = B_{max}$. If the secondaries now carry a direct current I_s, provided by the variable AC source, it would produce an MMF ($= N_s I_s$) in the reactor cores, such that

(a) $F_s = N_s I_s$ is *added* to F_p, say, in reactor A,

(b) $F_s = N_s I_s$ is *subtracted* from F_p in reactor B.

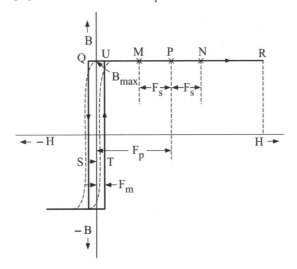

Fig.2 : B/H characteristic of toroid cores

Thus, the total MMFs in the two reactors are

$$F_A = F_p + F_s = I_d + N_s I_s$$

$$F_B = F_p - F_s = I_d - N_s I_s$$

and the points on the B/H curve representing these conditions will shift equidistant horizontally about the 'original' point P as, for example, to M and N. These represent a very small secondary current, or very small change in I_d, such that both reactor cores are still saturated, the two points still lying on Q R, and no output across xx.

However, a larger secondary current drives the points M and N farther apart; in extreme, for example, to R and Q, causing reactor B to start getting unsaturated to a degree, with the net flux density in the cores being proportional to the two MMFs. Thus, for a given secondary current setting, the saturation condition in core A versus core B would depend on the conductor current I_d, governed by the operating condition of the system, that is, whether the line is carrying normal current or overloaded.

On the operating side of the DCCT, if both the reactors are equally saturated, the fluxes in both are equal and there is no output. If one is saturated and the other unsaturated, the fluxes are unequal and a proportional output will appear across terminals xx. This output, which is primarily due to, and proportional to, changes in the main current – a consequential phenomenon in HVDC systems during control of power flow or system faults – can be amplified or treated suitably and passed on to appropriate control circuit or system.

DCVT: Direct Current Voltage Transductor

The schematic of a typical DCVT[1] is shown in Fig.3.

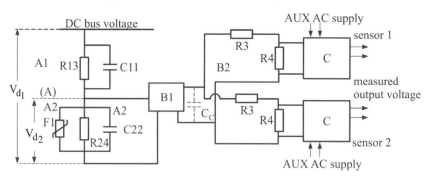

Fig.3 : Schematic of a DCVT

Description

A	:	Voltage divider unit, located in DC yard
B1	:	Auxiliary voltage unit(s), located at the base of the voltage divider unit A
B2	:	Secondary voltage divider, connected to B1
C	:	Isolation amplifiers and distribution or sensor units, located in the control room for external connections.

[1]Again, for example, employed in a HVDC converter station.

The voltage divider unit, A, comprises a high voltage arm A1 and a partial low voltage arm A2, the latter being connected to the isolation amplifiers via auxiliary units B1 and B2 as shown.

The divider has a highly non-inductive, oil-immersed resistance R, divided into R13 and R24 such that

$$\frac{V_{d_2}}{V_{d_1}} = \frac{R24}{R13 + R24}$$

The voltage across R24 is further divided across R3 and R4 in the two parallel-connected circuits, B2.

Finally, the voltage(s) across R4 is measured by electronic voltage measurement techniques; for example, DC instrumentation or operational amplifiers having high input and very low output impedance. Multiple units of B2 may be used depending on the requirement.

Constructional features

The resistor divider A may have inherent inductance owing to large number of turns and distributed capacitance, being connected across several hundred kilovolt (in a HVDC system) and hence unacceptable errors, esp. when measuring high-frequency voltages such as those caused by surges striking the 'pole'. The inductance may particularly be compensated by the use of suitable capacitances such as C11 and C22 [see Fig.3] so that the divider ultimately has practically flat frequency 'response' from DC to a few kHz. C_c represents the measuring cable capacitance, partially compensating the divider inductance.

F1 is a voltage limiting device, similar to a surge arrester, eliminating the effect of the surge into circuits B1.

The resistors used in the divider are of the deposited metal film type with a ceramic base and are thermally stabilised to have an accuracy of 0.5% for an ambient temperature change of up to $50°C$. Also, the divider is housed in an oil-filled insulator with an expansion vessel at the top.

Typical ratings of a DCVT for 400 kV system

Rated direct voltage	:	± 400 kV
Maximum continuous direct voltage	:	± 412 kV
Maximum peak voltage including ripple	:	± 436 kV
Measuring range	:	± 550 kV
Conversion ratio (isolation amplifier output) at nominal bus voltage of 400 kV	:	5 V

Appendix V

Phasor Diagram of Anderson's Bridge at Balance

Referring to the circuit diagram of the bridge shown in Fig.1, the following equations hold at the final balance, in terms of phasor currents and voltages:

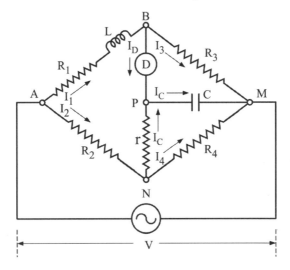

Fig.1 : Anderson's bridge

$$\dot{I}_1 = \dot{I}_3$$

$$\dot{I}_2 = \dot{I}_4 + \dot{I}_C$$

$$\dot{V}_B = \dot{V}_P \qquad [\dot{I}_D = 0]$$

$$\dot{V}_{BM} = \dot{V}_{PM}$$

$$\dot{V}_{AB} + \dot{V}_{BM} = \dot{V}$$

$$\dot{V}_{AN} + \dot{V}_{NM} = \dot{V}$$

$$\dot{V}_{NM} = \dot{V}_{PN} + \dot{V}_{PM}$$

Let $\dot{I}_1$ (or $\dot{I}_3$) be the reference phasor. The phasor diagram at balance is shown in Fig.2.

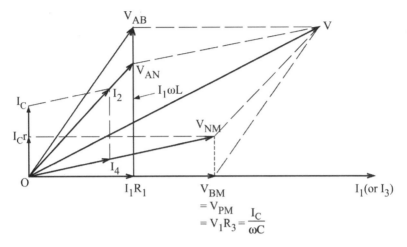

Fig.2 : Phasor diagram of Anderson's bridge

To understand the phasor diagram, note that

- $\dot{I}_1 R_1$ and $\dot{I}_1 R_3$ are in phase with $\dot{I}_1$; also the PD across C, $\dot{I}_C / \omega C$ $[= \dot{V}_{BM}$ or $\dot{V}_{PM}]$

- $\dot{I}_1 \omega L$ is in phase quadrature with $\dot{I}_1$, and the resultant of $\dot{I}_1 R_1$ and $\dot{I}_1 \omega L$ is PD, $\dot{V}_{AB}$

- $\dot{I}_C$ leads $\dot{V}_{PM}$ by 90°; this fixes position of $\dot{I}_C$

- $\dot{I}_C r$ is in phase with $\dot{I}_C$; $\dot{V}_{NM}$ is resultant of $\dot{I}_C r$ and $\dot{V}_{PM}$ (or $\dot{V}_{BM}$)

- $\dot{I}_4$ is in phase with $\dot{V}_{NM}$; $\dot{I}_2$ is resultant of $\dot{I}_C$ and $\dot{I}_4$

- $\dot{V}_{AN}$ is in phase with $\dot{I}_2$; fixes position of $\dot{V}_{AN}$

- $\dot{V}$ is the resultant of $\dot{V}_{AB}$ and $\dot{V}_{NM}$; also that of $\dot{V}_{AN}$ and $\dot{V}_{NM}$

Appendix VI

Alternating Flux and Temperature Rise

One of the Maxwell's electromagnetic field equations to 'define' magnetic flux density, B, is

$$\text{div } \bar{B} = \bar{\nabla} \cdot \bar{B} = 0$$

This relationship implies that, theoretically, $\bar{B}$ can be expressed as curl of another vector, and still satisfy the above identity.

Thus, let $\bar{B}$ be defined as

$$\bar{B} = \text{curl } \bar{A}$$

Then $\bar{A}$ is called the vector potential of the magnetic field $\bar{B}$ and known as *Magnetic Vector Potential* in electromagnetic field problems[1]. Even though a clear physical meaning may not be assigned to $\bar{A}$, similar to flux density $\bar{B}$, it can be shown that contours of $\bar{A}$ in a magnetic region correspond to 'constant' magnetic flux lines. Thus, in practice a magnetic-field problem, or phenomenon, can be expressed and analysed in terms of the magnetic vector potential $\bar{A}$ and then magnetic flux density $\bar{B}$ may be derived from it using the above relationship.

Magnetic Vector Potential in Time-varying Magnetic Field

In a time-varying magnetic field, the two other Maxwell's equations in differential form are

$$\text{curl } \bar{E} = \bar{\nabla} \times \bar{E} = -\frac{\partial \bar{B}}{\partial t}$$

and
$$\text{curl } \bar{H} = \bar{\nabla} \times \bar{H} = \bar{J} + \frac{\partial \bar{D}}{\partial t}$$

where $\bar{E}$ is the electric field intensity

---

[1]However, it is seen that the vector $\bar{A}$ is not *uniquely* defined, since to $\bar{A}$ can be added another vector whose curl is zero and still satisfy the relation div $\bar{B} = 0$, $\bar{B}$ being defined as curl of the new vector.

$\overline{H}$ is the magnetic field intensity

$$\left[= \overline{B}/\mu \right]$$

$\overline{J}$ is the (time-varying) conduction current density

[also represents the *induced* current in metallic media, acted upon by (time-varying) magnetic flux density $\overline{B}$ or magnetic vector potential $\overline{A}$]

$\overline{D}$ is the displacement current density

In cases where displacement currents are non-existent

$$\nabla \times \overline{H} = \overline{J} \text{ (alone)}$$

These set of equations can be shown to lead to the "Diffusion Equation",

$$\nabla^2 \overline{A} = k \frac{\partial \overline{A}}{\partial t}$$

that governs the induction of currents in the metallic media where the solution of the above equation pertaining to each medium would yield the *spatial* distribution of $\overline{A}$ owing to the time-varying ("source") currents, or *vice-versa*. In the equation k is, in general, a constant expressed in terms of angular frequency ω, electrical conductivity σ and permeability μ of the given medium (air, iron etc.); that is, $k = \omega \sigma \mu$. In a two-dimensional analysis generally encountered, where the magnetic vector potential is invariant in the axial or z direction, for example in an axi-symmetric rotating machine, the induced currents in a metallic medium (iron or copper) causing heating and temperature rise, are specifically due to *sinusoidally* time-varying "source" current (for example the excitation). In such cases, the above equation for $\overline{A}$ can be expanded to

$$\frac{1}{\mu}\left\{ \frac{\partial^2 \dot{A}}{\partial x^2} + \frac{\partial^2 \dot{A}}{\partial y^2} \right\} = j\,\omega\,\sigma\,\dot{A} - \dot{J}_s$$

in which

$\dot{J}_s$ is (specifically) the source current (density); steady-state time variation

ω the angular frequency of the current (rad/s)

σ the electrical conductivity (S/m), and

μ the permeability

[both σ and μ assumed constant]

and $\omega\,\sigma\,\dot{A}$ representing the "induced" currents.

Comment

There are many cases where the above equation plays an important role in determining heating due to induced (eddy) currents and consequent temperature rise. For example,

(a) the (useful) case of induction heating (surface hardening of steel rods etc);

(b) the heating and temperature rise in the rotating members of electric machines – for example, turbo generators and induction motors – in the event of their unbalanced-current operation: un-balanced faults in the system to which the turbo-generator is connected, and "single-phasing" of induction motors, being fed from single phase or two-phase supply during distribution-system faults.

Whilst a solution of the above equation using suitable techniques[1], yields the distribution of magnetic vector potential, also representing the flux lines, the *power loss* due to induced currents in the region of interest can be evaluated using the equation

$$P = \frac{1}{2} \int_v \omega^2 \, \sigma \, \dot{A} \dot{A}^* \, dv$$

in a 'small' volume V where $\dot{A}$ is defined as above, and $\dot{A}^*$ is the complex conjugate of $\dot{A}$. With P as power (or heat) input, the (essentially) *transient* heating and temperature rise in the given region is governed by the following Poisson's (or transient heat conduction) equation

$$\nabla(\kappa \, \nabla \, T) + q = C\frac{\partial T}{\partial t}$$

in which

κ is the thermal conductivity

q   the heat input

C   the specific heat, and

T   the resulting temperature rise.

---

[1]Classical methods for 'regular' boundaries and a numerical method such as finite-element technique for irregular geometries or boundaries.

Appendix VII

Search Coils

Basic Principle

The basic principle of a search coil derives from the integral form of the Maxwell's equation

$$\oint_C \bar{E} \cdot d\bar{l} = -\frac{\partial}{\partial t} \int_S \bar{B} \cdot d\bar{s}$$

where $\bar{E}$ is the EMF *induced* in the search coil due to the time-varying magnetic flux density $\bar{B}$; the surface integration is over an "open-ended" surface, S, and the contour of integration is over the boundary C of S as indicated in Fig.1.

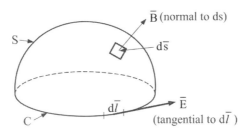

Fig.1 : Basis of EMF in the search coil

Note that the above equation is independent of material properties, being equally valid for empty space and materials. In using a search coil, this fact allows the replacement of the "mathematical" contour C by a metallic contour in the form of a 'thin' wire, the closed loop of the search coil representing $\oint_C \bar{E} \cdot d\bar{l}$.

In practice, to obtain the measure of the induced EMF, the metallic contour has to be 'broken' at some point to connect a recording or monitoring device (for example, a UVR or a CRO) which is usually at some distance from the contour (or the coil). Connecting the points of broken contour (that is, the coil ends) to the terminals of the recorder by a pair of ordinary wires will invariably alter the boundary of the original contour and introduce signals (or EMFs) in the circuit, commonly known as "pick-up, which may frequently be comparable to the signal of interest, that is, the EMF. This can result in serious error of measurement. However, if the leads carrying the coil EMF to the recorder are twisted as shown in Fig.2, the small areas 'enclosed' by the twists (4 shown in the figure), assumed to be in the same plane (at right angles to $\overline{B}$) will reverse the direction of $d\overline{s}$, thus reversing the sign of $\overline{B} \cdot d\overline{s}$ and hence that of the induced EMFs in the adjacent loops of the leads, cancelling the EMFs in pairs. The net effect of this process will be to minimise the error due to induced voltage(s) in the 'contour' formed by the leads. Naturally, the finer the twisting (though rather tedious in practice), the better the reduction, or even complete elimination of the pick-up.

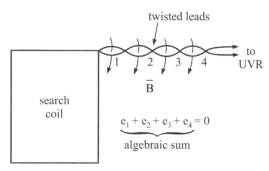

Fig.2 : Search coil with twisted leads

Conductor Size and Single- or Multi-turn Search Coils

A point of practical importance, implicit in the equation of EMF, is the concept of "line" in the (contour) integral. In theory, the line should be of "zero" thickness. This means that when a search coil is constructed in practice, the use of a conductor of finite thickness or size in place of the true mathematical contour would "blur" the periphery of the latter, resulting in a serious error in deriving the flux density value if the conductor size (or the diameter) is comparable to the 'area' enclosed by the coil[1].

Another practical difficulty arises if the coil is wound enclosing a section of the magnetic region; for example, the core of an electric motor when the conductor forming the coil is to be 'threaded' through the holes. Drilling of holes in the magnetic material would result in the "loss of iron" which can

---

[1]For example, if a single-turn search coil is wound around a cross section of 5 mm × 5 mm, using a 40 SWG insulated wire having a diameter of 0.142 mm, the area of the contour could be anywhere from the true 25 mm^2 to 27.93 mm^2, representing an error of approximately 12% in the 'enclosed' area, and a corresponding error in the flux v/s flux density value(s).

be significant if the diameter of the holes is appreciable and the area lost to the holes is comparable to the enclosed area. This means the distortion of local flux distribution with corresponding error of 'measured' flux density. This is apart from the 'difficulty' of drilling a ('small') hole in the magnetic material such as silicon steel.

Single-turn and multi-turn search coils

Ideally, a search coil should be wound with a single turn so as to best correspond to the mathematical contour as discussed. However, when

   (a) the flux density being measured is low or very low;

   (b) the flux-density distribution in the region is non-uniform, necessitating coils of small cross section;

   (c) the measurement of flux is confined to non-magnetic regions, having relative permeability of unity

resulting in low to very low induced EMF, use of multi-turn search coils becomes essential to improve the 'signal level", for the induced EMF is given by

$$e = -N\frac{d(B\,a)}{dt}, \quad \text{or} \quad e \propto N$$

where B is the flux density, a the area of the coil and N the number of turns.

Mult-turn Search Coils

   A multi-turn search coil may generally be arranged in one of the two ways shown in Fig.3.

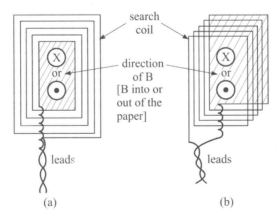

Fig.3 : Two examples of multi-turn search coils in practice

The coil in Fig.(a) illustrates the case of turns 'going round' *in the same plane* and is best suited when the flux to be measured is directed at right angles to the *plane* of the coil and is nearly *uniformly distributed.* Fig.(b) shows a coil, *wound around a cross-section* with the turns placed side-by-side, and is suited to measure the flux *through* the cross section, with little variation along the depth, or 'axis' of the search coil.

In either case, there is considerable blurring of the contour, resulting in pronounced error on account of changed area as discussed earlier. The foregoing factors point to an important practical requirement of a search coil: it should preferably be of a single turn and wound using the finest possible wire. Likewise, when the wire has to be threaded through a hole, the diameter of the hole must be as small as possible, commensurate with the actual coil size[1].

However, whilst a thin, single-turn coil of a small size (or area of cross section) may be very desirable theoretically, it may pose many hazards in practice such as difficulty in handling the thin wire during winding, resulting in snapping of the wire easily, causing an open circuit. Also, the output from a single-turn coil, especially in weak magnetic fields or in non-magnetic regions, is bound to be too small to be measured accurately. Hence, some deviations from the ideal become unavoidable and the design of search coils may, in practice, largely depend on the particular application and nature of investigation. Also, it is necessary to employ a skilled, well-trained operator to perform the task.

One-, two-, and three-dimensional search coils

One- or single-turn search coils are mainly employed to measure flux (or flux density) directed *into* a single plane that is at right angles to the direction of the field. Where the magnetic field is three-dimensional, for example in the end region of an electric machine, search coils can be devised to measure individual as well as resultant flux densities by using an appropriate number of 'small' coils *with their planes fixed in mutually perpendicular directions*[2]. An example of the use of two-dimensional search coils for measuring components of flux, in radial as well as circumferential

---

[1]For coils measuring typically 5 mm x 5 mm, for example, to be used in the core of a machine comprising silicon laminations, the holes of less than 0.5mm diameter may be desirable, to just allow a wire of, say, 44 SWG to pass through the hole(s). In such a case, holes may be best drilled using special techniques; for example, "spark erosion" technology.

[2]Such search coils are now available, prefabricated to suit given requirements.

direction, in the back-of-core of a turboalternator, constructed using a stack of ten laminations, is shown in Fig.4.

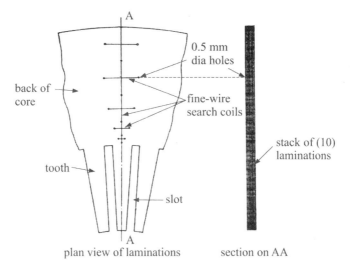

Fig.4 : Search coils in the back-of-core of a turboalternator

As illustrated, the 'grading' of search-coil size is necessary to achieve a reasonable accuracy from near the bottom of the teeth to the outer edge of the core[1]. Note also the size and disposition of holes which were drilled by using smallest possible high-speed drill (of 0.5 mm dia) to allow threading of double-insulated, 44 SWG copper wire.

---

[1] 'Smaller' coils where there is a concentration of flux and relatively 'large' coils where the flux may be sparse, towards the back.

Appendix VIII

Hall Generator and Gauss Meter

The Hall effect was discovered in 1879 at Johns Hopkins University by E. H. Hall and has of late acquired significant importance in a number of applications during the past many decades. The effect manifests as the generation of a voltage across opposite edges of an electrical conductor of finite dimensions, usually rectangular, carrying current and placed in a magnetic field. The basis of this effect is the Lorentz force which depends on the deflection of charged particles moving in a magnetic field – a mutually perpendicular phenomenon.

Hall Effect Equation

The equation governing the Hall effect is

$$\overline{V}_H = w\,R_H\,(\,\overline{j} \times \overline{B}\,)$$

where

      V_H is the Hall output voltage

      R_H is the "Hall coefficient", depending on the material

      w   is the width of the Hall generator

      j   is the current density *through* the Hall generator, and

      B   is the magnetic field strength to which the generator is exposed.

In the above expression, the product of $\overline{j}$ and $\overline{B}$ is written as a *vector cross product* indicating the directional sensitivity of the Hall effect, that is, the mutually-perpendicular-direction property among $\overline{V}_H$, $\overline{j}$ and $\overline{B}$ – a property of importance in practice as brought out later. The equation may also be written as

$$V_H = w\,R_H\,j\,B\,\sin\theta$$

where θ is the angle between the magnetic field direction and the plane of the Hall generator as shown in Fig.1; that is, when the two are not at right angles.

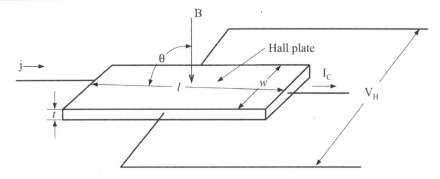

Fig.1 : Configuration of a typical Hall generator

Since current *density* is not a convenient parameter, in practice it is replaced by *total* current through the generator as

$$j = \frac{I_C}{wt}$$

where t is the thickness of the Hall generator (or the crystal/plate) and I_C the total current now being called the "control" current. The equation for V_H then reduces to

$$\overline{V}_H = \frac{R_H}{t}\left(\overline{I}_C \times \overline{B}\right)$$

This shows that the maximum obtainable V_H is essentially independent of the width of the Hall plate, assuming maximum *allowable* current density for all cases.

Materials

The equations for Hall voltage indicate that to obtain a high output voltage, the active Hall element or plate must have a large Hall coefficient, R_H – as high as can be obtained; this means using a particular material. Also, since the output is proportional to the current *density* through the element, its resistance should be as low as practical to prevent excessive (internal) heating. Some of the semiconductor materials commonly used for Hall generators are

- indium antimonide
- indium arsenide, and
- silicon

the latter being although inexpensive has very high resistance and is therefore relatively noisy and inefficient.

For general use, the most commonly employed material is indium arsenide, $I_n A_s$, and has the following properties:

input resistance, R_{in}	:	1.5 Ω
output resistance, R_{out}	:	1.0 Ω
nominal control current, I_C	:	150 mA
magnetic sensitivity	:	150 mV/T
temperature coefficient	:	$- 0.06\%/°C$

Typical Shapes and Sizes

Hall generators are available in a wide variety of shapes and sizes, adaptable to different applications. However, the two basic types are "transverse" and "axial", as illustrated in Fig.2, for measurement of flux at right angles or axially, respectively.

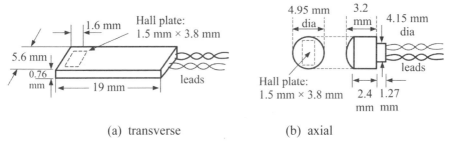

(a) transverse   (b) axial

Fig.2 : Transverse and axial Hall generators[1]

Standard transverse plates as thin as 0.15 mm and axial plates as small as 1.6 mm in diameter are available to suit special requirements.

When measuring flux *density* as accurately as possible, the Hall plate or (active) element area should be ideally smaller than the cross section of the field to be measured so as to obtain the value of B at a 'point'. Also, although the output voltage is proportional to flux density, a Hall plate is not equally sensitive over its entire 'active' area.

Handling Hall Generators

Hall generators or probes[2], especially the actual Hall plate, constitutes a fragile device and must be handled with care. The connections to Hall plate

---

[1]A knowledge of various dimensions, esp. the size and location of 'actual'/active Hall plate, is essential and helpful in 'positioning' of the Hall probe and organise proper and effective measurements.

[2]A Hall "probe" comprises the actual Hall plate (or crystal) and the leads of sufficient length (usually twisted pairs of insulated wires) to allow connections for control current and output.

for supplying the control current and collection of the output voltage are made delicately, using tiniest possible wires. A magnified view of the arrangement of internal connections, along with an actual size of a typical Hall plate, is illustrated in Fig.3.

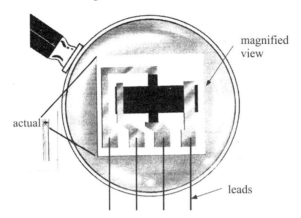

Fig.3: Enlarged view of internal connections of a Hall element and actual size

During operation, care should be taken not to exceed the maximum (control) current rating specified for a given Hall generator to avoid undue heating of the plate as the performance of a hall generator is very sensitive to its temperature, and indeed to the ambient conditions. For making external connections, the following steps are to be taken

(a) the control current should preferably be supplied from a "constant *current*" source and not just a simple stabilised voltage supply[1];

(b) whilst attaching the Hall generator voltage leads to the input terminal of a voltmeter or recorder, its input impedance should be at least ten times greater than the output resistance of the hall generator to avoid 'loading';

(c) the zero balance potentiometer may be adjusted to give minimum reading on the meter when the Hall generator is in minimum magnetic field[2].

TYPICAL APPLICATIONS OF HALL PROBES

Some of the key applications of Hall generators or probes are

1. Magnetic flux density measurement

---

[1]If a current source is not available, a "brute-force" constant-current source may be constructed by connecting a suitable resistor in series with a constant-voltage power supply.

[2]This can be considered as a preliminary calibration of the probe. A proper calibration can be carried out as discussed later.

2. Magnetic field plotting

3. Linear displacement transducer

4. Angular displacement transducer

5. Current sensor

6. Proximity sensor

7. Multiplier

Most of these applications derive from the three-dimensional properties or control action of the Hall generator.

Magnetic Flux Density Measurement

One of the most practical and commonest applications of a (calibrated) Hall probe is in the *direct*, 'absolute' measurement of magnetic flux density, to provide its magnitude, direction and polarity. An usual application is to measure *radial* flux density in the airgap of an electric machine where a thin Hall probe is particularly useful as it is inserted in the airgap and moved along the peripheral space of the gap. Even AC fields may be measured by using a probe still excited by direct current. The output in this case would be an AC voltage at the frequency of the field with its magnitude being proportional to the instantaneous value of the flux density. By applying the output of the Hall probe to an oscilloscope, the true wave shape of the AC field may also be observed.

Magnetic Field Plotting

The magnitude of voltage output from a Hall probe in a given plane is maximum when the magnetic field is at right angles to the plane (or surface) of the Hall plate. This property can be used in obtaining plots of magnetic field variation in a given region: for example, plots (or contours) of constant flux density, varying from the lowest to the highest; the Hall probe being so rotated about its vertical axis so as to indicate the maximum, chosen magnitude at a point. Such plots are particularly useful in the study and analysis of stray or leakage flux in interpolar regions of electric machines or 'window'-spaces of a transformer at a given excitation.

Hall Probe as Linear Displacement Transducer

The basis of this application is that a relative motion between a Hall probe and a magnetic field produces a voltage which is a function of this motion or displacement. A simple example is illustrated in Fig.4. A and B are two *identical* permanent magnets, aligned parallel to each other. Note the relative polarity of the magnets. A "zero" plane, midway between the two N surfaces, will result in zero Hall voltage output when the Hall element is in the plane as shown. Slight displacement of the element to left or right

(X or Y) will result in a proportionate +ve or −ve output that may be related to the movement of a device, to be measured, in either direction.

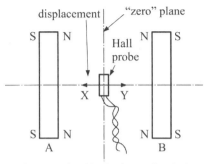

(only one pair of leads shown for clarity)

Fig.4 : A Hall probe as displacement transducer

Angular Displacement Transducer

Since the Hall voltage output is a function of the angle between the plane of the element and the direction of the magnetic field, the angular changes due to the rotation of the field or the Hall probe relative to the former will produce an output voltage which is a sine function of the change. An example of this application is shown in Fig.5 where the probe is made to rotate with the device whose angular motion is to be measured, the probe being held about an axis midway between two permanent magnets which provide the required magnetic field. The output may be connected to, and monitored on, a CRO.

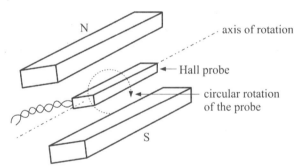

Fig.5 : Hall probe as angular transducer

As a Current Sensor

A simple current sensor using a Hall element is based on the production of a voltage output if the element is positioned as closely as possible to a current-carrying conductor, the Hall voltage then being proportional to the magnitude of the field surrounding the conductor, and thus indirectly to the current through the conductor. The sensor can easily be calibrated by measuring its voltage output for a given current through the conductor, measured by some other means.

A schematic of the simple arrangement is shown in Fig.6. The level of the output voltage can be raised by using a suitable "flux concentrator", located perpendicular to the Hall active area as indicated in Fig.6(b).

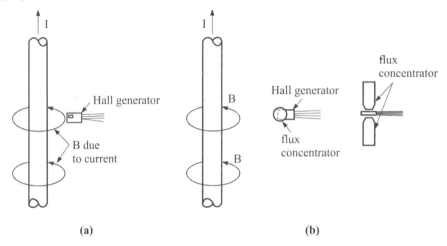

(a)   (b)

Fig.6 : Hall probe as current sensor

Flux Concentrator

A good (flux) concentrator may be fabricated using a one-quarter inch diameter rod of ferrite or high-permeability steel and positioned perpendicularly, above, below or sand-witching, the element as shown. A more sensitive system is shown in Fig.7 that may be used to measure low-level currents as they may result in too low flux densities to be measured by above simple means. The system comprises a core with a Hall device in the core gap. The core encloses the conductor concentrically, the current through which is to be detected, and acts as a flux concentrator. With a proper overall design, current down to near zero can be sensed or 'measured'.

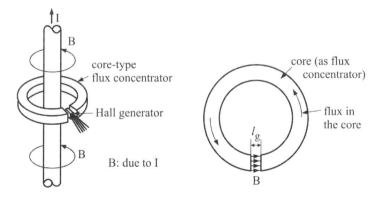

Fig.7 : A current sensor arrangement to measure low currents

Proximity Detector

Since one of the inputs to a Hall probe is a magnetic field, the device is ideal to act as a non-contact proximity 'switch'. The switch may be designed to detect either the presence of a magnetic field (DC or AC) or the disturbance of a magnetic field due to the presence of ferrous materials. An example of this application is shown in Fig.8 where a spatially varying magnetic field is simulated by lateral movement of a permanent magnet relative to a Hall element, and the resulting output voltage. A suitable flux concentrator may be used to increase the sensitivity of the element.

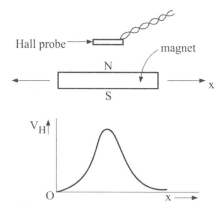

Fig.8 : Hall probe as proximity detector

Hall Multiplier(s)

One of the most useful and practical applications of a Hall generator is its function as a multiplier. The basic principle of operation of a Hall multiplier can be demonstrated as in Fig.9. It consists of a magnetic core and two coils with an airgap in the flux path. A Hall element, positioned in the airgap with appropriate mechanical protection and magnetic shielding, provides the actual multiplication of the two input currents I_f and I_c, with the current I_c 'acting' as control current of the element and I_f producing the magnetic field acting on the element.

The output is a voltage proportional to this product, given by $V_H = K\, I_f\; I_c.$[1] In practice, the accuracy and linearity of the multiplication carried out by the

---

[1]Here, the "cross product" property of the Hall element, that is $\overline{V}_H = K\,(\overline{I}_1 \times \overline{I}_2)$ in which $\overline{I}_1$, $\overline{I}_2$ and $\overline{V}_H$ are all *vector* quantities, is of significance in the multiplier which can account for both magnitude *and* phase displacement of the two current inputs.

element would depend ultimately on the ability to 'deliver' into the multiplier, currents that are proportionately related to the desired input parameters.

An important practical application of a multiplier is in the measurement of electric power, or energy, in conjunction with solid-static electronics, esp. when the quantity to be measured is low in magnitude. [See Chapter VII : Measurement of Power].

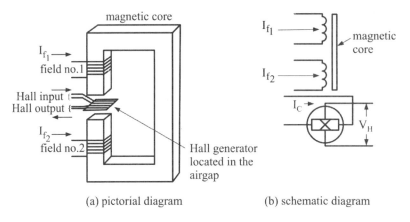

(a) pictorial diagram   (b) schematic diagram

Fig.9 : Schematic of a Hall multiplier

GAUSS METER

The above applications have been discussed with a focus on the use of a Hall generator in various ways, independent of the means of supplying the control current and measuring or monitoring the voltage output. In addition, a compact instrument or apparatus is now commercially available known as "gauss meter", featuring a variety of high-stability, multi-purpose probes designed and manufactured with high precision.

The apparatus incorporates a built-in constant-current supply to provide control current, zero-adjustment control, range-selection device, a sensitive analogue indicating device – or a digital indicator in some modern gauss meters – to give directly the value of flux density to be measured, from a very small to extremely high value. The instrument may be provided with an output socket for external connections for remote control and operation in analogue or digital form, and provision for calibrating the probe in use. A special feature may be an in-built compensation circuit to account for temperature variations, ambient or otherwise. A model of a commercial gauss meter having all the above features is shown in Fig.10.

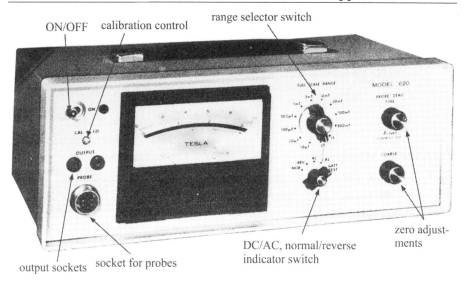

ON/OFF   calibration control   range selector switch

zero adjust-ments

DC/AC, normal/reverse indicator switch

output sockets   socket for probes

Fig.10 : A multi-range gauss meter

The measurement range of the gauss meter may extend from 1mG (10^{-7} T) per scale division (in the 100 mG full-scale range) to 30 kG (3T). Both DC and AC magnetic flux densities can be measured directly on the panel meter[1]. DC polarity can be preserved for direction information when tracing and plotting magnetic fields. The gauss meter is accurate up to ± 0.25% of full scale. A choice of over 10 Hall probes is available to meet a vast number of applications, the accuracy and calibration being independent of the type of probe connected externally. When used with a battery that can be housed inside the instrument, the gauss meter is fully transportable.

Calibration of the Gauss Meter

A Hall probe, whether stand-alone or part of a gauss meter, can be calibrated externally by using specially designed devices.

"Zero-field" chamber

This is a small, closed chamber with a hole (about 8 mm in dia.), made of very-high permeability material, so that the inside of the chamber is completely shielded against *any* external magnetic field. The probe to be tested for "zero" setting is inserted well into the chamber as shown in Fig.11 when the voltage output of the probe should be zero. The device works most satisfactorily for nearly all transverse-flux Hall probes as well as some "axial" probes.

---

[1]AC fields of up to 400 Hz can be measured as *rms* value, for sinusoidally varying fields.

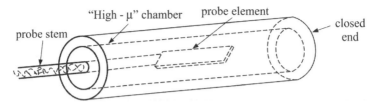

Fig.11 : "Zero-field" chamber for zero calibration of Hall probe

Permanent magnet(s)

For calibration of a Hall probe with reference to a non-zero known flux density, massive permanent magnets of cylindrical shape (usually about 30 cm long and about 15 cm in dia.) are available with a through hole (about 2 cm in dia.) in the middle as shown in Fig.12. The magnet is designed to provide a given, constant flux density (for example 1 T) within a very close tolerance, transverse to the axis of the through hole as shown. The probe to be calibrated is inserted inside the hole and moved in and out laterally, as well as rotated, such that the output of the probe is maximum and should indicate the flux density designated on the magnet.

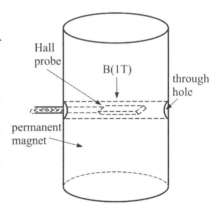

Fig.12 : A permanent magnet for Hall-probe calibration

Long solenoid

Other external means, for example, a well-designed long solenoid of which the field is accurately known at the mid-length location by calculation from its design parameters, may also be used for calibration. Clearly, the accuracy of calibration would depend on the knowledge of dimensions of the solenoid and measured value of excitation.

[See also Chapter XII: Magnetic Measurements].

Appendix IX

Instrumentation

Instrumentation is different from measurement in that it encompasses a number of (additional) aspects, forming a more comprehensive scheme of relating input to output stage in the given process[1]. In the scheme, measurement of an un-known quantity may form one of the many stages. A typical example is the measurement and recording of flux density in an electric machine using a search coil, dealt with in detail later.

General Requirements

A scheme of instrumentation may, in general, consist of

A.   a suitable sensor or transducer

B.   a signal conditioner (device)

C.   measuring, monitoring and/or recording device(s)

as illustrated in Fig.1.

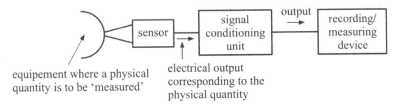

Fig.1: Requirements of a typical instrumentation

Sensors or Transducers

In the scheme of (electrical) instrumentation, a sensor or pick off can be explained as a device that can produce an electrical output *directly* in an equipment; for example, in an electric machine or equipment due to change of electrical, magnetic or electromagnetic parameter such as flux, flux density, current or voltage. Typical examples of sensors being a search coil, a magnetoresistor or a low, standard resistance in series with a load in a circuit.

---

[1]Instrumentation here implies, in general, dealing with *electrical* measurement and monitoring or recording of electrical or non-electrical quantities, from 'start' to 'finish'.

A transducer, in general, refers to a device that produces an electrical 'output' due to changes in *non-electrical* parameters, for example by movement (of core in a solenoid) or similar action as discussed in Chapter XIII. The end result in either case is an electrical signal or output that can be

(i)   extremely low in magnitude, that is, in mV or μV;

(ii)   DC or AC, of power or higher frequency;

(iii)   steady-state or transient in nature;

amongst other features.

Preamplifiers and Amplifiers

In most applications, the electrical output from a sensor or transducer being very small, inadequate to be directly measured on an indicating instrument, to be monitored on a CRO or drive a recorder, it is imperative to use some amplification of the signal. In order to bring down the problem arising from the noise invariably associated with the signal, that is, to deal with poor signal-to-noise ratio, the amplification of the signal is best carried out in *two* stages: pre-amplification followed by amplification.

A pre-amplifier is typically characterised by

(a)   low, very stable gain, of the order of 10 or so

(b)   high input impedance to eliminate 'loading' of the signal

(c)   very low output impedance

(d)   "direct-coupled" configuration, with a flat frequency response over a wide range

(e)   extremely low internal noise

(f)   very-low drift with respect to variation of temperature

(g)   high "common-mode rejection ratio" or CMRR

For the purpose, special high-quality "DC instrumentation" or chopper-stabilised amplifiers are available with best specifications.

The next stage consists of an amplifier, generally of the same properties as above, but with a high gain of the order of 100 to 1000 (or even more to suit some special requirements), to amplify the signal further, to a 'healthy' level for measurement and recording[1].

---

[1]A modern requirement is to digitise the signal for which the signal may have to be, say, at ± 5 to 6 V.

Comment

An improved scheme of amplification, to achieve much higher CMRR is to use an operational amplifier having an "open loop" gain of 10^6 or higher, and with *differential* input. An added advantage of the use of such amplifiers is that their gain can be adjusted to any desired value by proper selection of 'input' and 'output' *resistances*. The other advantage of using an operational amplifier is for integration or differentiation of the input signal before measuring or recording[1]. [See Fig.2].

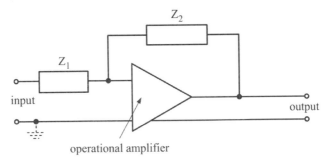

Fig.2 : Connections of a typical operational amplifier

Z_1, Z_2 : resistances for amplification

Z_1 : resistance ⎫
Z_2 : capacitance ⎬ integration (with gain)

Z_1 : capacitance ⎫
Z_2 : resistance ⎬ differentiation

---

[1]An example of signal conditioning

SPECIFICATIONS OF A TYPICAL INSTERUMEMTATION AMPLIFIER

1. Gain

    Range of gain, G   1-1000

    At gain, G = 100

        Gain error   0.07 to 0.25%

        Gain temperature coefficient   25 ppm/$^\circ$C

        Gain non-linearity   0.0006 to 0.01%
                                                    of FS

2. Input offset voltage

    vs. temperature (@ G : 1 - 1000) $1 + 20/G$ μV/$^\circ$C

3. Input bias current vs. temperature	$0.5\,nA/^{\circ}C$
4. Input impedance [differential and common mode]	60 MΩ
5. Input voltage range (common mode range)	± 12 V
CMR (DC – 60 Hz)	
G = 1	86 dB
G = 100	125 dB
6. Input noise (10 Hz to 1kHz)	2 to 1 nV/Hz
7. Output noise voltage (equivalent to 1kHz)	$65\ nV/\sqrt{HZ}$
8. Dynamic response (full-power bandwidth) slew rate (G = 1 to 500)	240 kHz 15 V/μs
9. Power supply	± 15 V
10. Temperature range operation storage	– 40 to +85 °C – 40 to +100 °C

Recorders

These constitute an important requirement of most instrumentation schemes at 'end stage' in which a 'hard', proper record of output of interest is necessary for study and analysis.

Analogue and digital type

In a number of applications, the final output corresponding to the quantity to be measured, or of interest, is to be recorded either in analogue or digital form. Until recently, the *analogue* recorders comprised a set of moving-coil pen-type galvanometers which suffered from limitation of frequency response and not much suited for recording of dynamic or transient signals or inputs. The later models employed sophisticated techniques offering a much better frequency response with high uniform sensitivity and having no moving part(s). The other form of recording device would be a storage oscilloscope, but with limited "window" area/space.

The modern technique of digitisation of signals has revolutionised recording process. With an appropriate sampling rate, matching the anticipated transient form of the input, a very faithful recording of the signal is possible, and for a practically unlimited time interval. Whilst the old recording devices used to be digital tape drives, the modern techniques may comprise CDs or digital storage oscilloscopes. Monitoring in analogue form is possible using DA converters in the end stage.

See Chapter IV for details of various recording devices.

A TYPICAL INSTRUMENTATION EXAMPLE

Search Coil Instrumentation

A search coil forms a very effective and accurate device for measurement of flux or flux density and, as discussed, should preferably be of single turn, constructed using thinnest possible wire, and having a 'small' area, esp. when used in non-uniform magnetic fields for the measurement of flux density. Accordingly, the EMF output from the coil will generally be very small, usually in microvolt to millivolt[1], particularly for flux densities in non-magnetic regions as, for example, the airgap. In most cases, the instrumentation for a search coil (in AC fields) will comprise

- an instrumentation preamplifier, followed by an amplifier, with an overall gain sufficient to raise the final output to a few *volts*, esp. if the signal is to be digitised and recorded digitally
- the amplifiers must
  - also possess high input impedance to avoid 'loading' of the search coil output
  - have high CMR since the output leads of the search coils are 'open' or floating and liable to pick up common mode voltage
  - be with high slew rate if the signals are of transient nature, that is, due to rapidly varying magnetic field.

As explained, the leads from the search coil to the preamplifier input terminals must be suitably twisted to eliminate any (stray) pick-up en-route.

Further, since the EMF output from the search coil is obtained by *differentiation* of the magnetic flux (or flux density), in many measurements the search-coil output must be fed to an integrater to provide *time* integration of the signal before recording, so as to obtain and monitor the un-known flux density *directly*. The integrater, comprising an

---

[1]For example, in a sinusoidally-varying AC magnetic field at 50 Hz, with $B_{rms} = 1$ T, the output of a 1-turn search-coil, measuring 5 mm × 5 mm will be $e_{rms} = 4.44 \times 1 \times 50 \times 25 \times 10^{-6} \times 1 \times \sqrt{2}$ or 7.85 mV; in a non-magnetic region where $B_{rms} = 0.1$ T, the signal would be just 0.785 mV.

operational amplifier, must be of very low noise and drift, to eliminate the errors of signal distortion prior to recording[1]. The schematic of the instrumentation is shown in Fig.3.

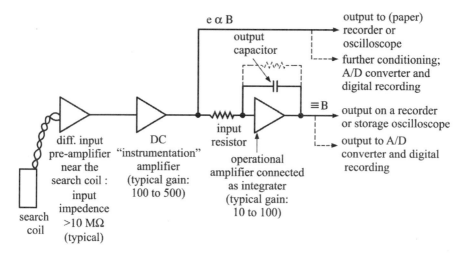

Fig.3 : Schematic of the search-coil instrumentation

Operational Amplifier as an Integrater

An operational amplifier can act as a 'time'-integrater of a signal whilst at the same time amplify it with a known gain depending on the choice of input resistance and output capacitor. The circuit of a general integrater is shown in Fig.4.

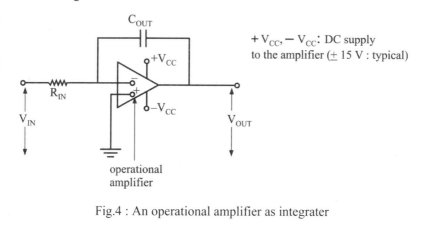

Fig.4 : An operational amplifier as integrater

---

[1]To minimise the error due to drift, a suitable resistor is often connected across the output capacitor; a better solution might be to use a chopper-stabilised operational amplifier.

The output of the integrator, V_{OUT}, is related to input signal, V_{IN}, by the expression

$$V_{OUT} = \int_0^t \frac{V_{IN}}{R_{IN} \, C_{OUT}} \, dt + V_{initial}$$

where $V_{initial}$ is the input voltage at $t = 0$ and can be equal to zero.

If V_{IN} contains some DC voltage, usually called the "offset", the integrated alternating output corresponding to AC part of the input, will result in monotonously increasing drift as shown in Fig.5.

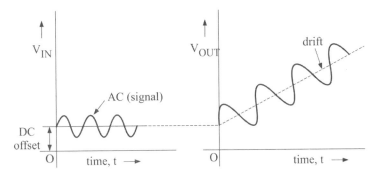

Fig.5 : "Continuous" drift in an integrater

As mentioned earlier, the drift can be controlled to some extent by shunting the (output) capacitor by a resister, of the order of 100 MΩ or so.

Gain

The gain of the integrater is given by

$$G = \frac{1}{R_{IN} \, C_{OUT}}$$

where R_{IN} is in ohm and C_{OUT} in F.

[The product $R_{IN} \, C_{OUT}$ defines the time constant of the circuit and would control the time response of the integrater].

As an example if $R_{IN} = 10 \text{ k}\Omega$ and $C_{OUT} = 0.1 \text{ μF}$

$$G = \frac{1}{10 \times 10^3 \times 0.1 \times 10^{-6}} \qquad \text{or} \quad 1000$$

Output stage

As shown in Fig.3, the output from the amplifier which is an EMF proportional to flux density, or from the integrater which is a 'measure' of actual flux density as picked by the search coil, can be fed to

(a)   an oscilloscope for visual study

(b)   a recorder such as a UVR to have a permanent record

(c)   an A/D converter for digital processing

(d)   for further conditioning of the signal as required.

Other Cases

The above is a typical case of instrumentation describing all necessary stages normally encountered. In many other requirements, once the electrical output is obtained through a transducer corresponding to the variation of the quantity of interest, or the one to be measured, a similar arrangement as in Fig.3 would follow with some variation(s); for example, the integrater may not be required.

Appendix X

Organising Lab Experiments

Experiments in a Measurement Lab

Very often an experiment fails to yield optimum results because it is not properly organised and for want of small measures, even though the instruments and other apparatus have been chosen with due care, with good accuracy and as required.

General Requirements of an Experiment

In general, requirements for a lab experiment consist of some or all of the following:

A. Appropriate power supply

- AC or DC; if AC, 1-phase or 3-phase (four-wire) and if DC at various voltages
- Steady or stabilised, or variable and means to provide the same [3-phase supply must essentially be balanced]

B. Measuring, monitoring or recording instruments of appropriate range and accuracy

- Indicating instruments such as ammeters, voltmeters, wattmeters, p f meters etc, and/or energy meters, UV recorders, CROs

C. Miscellaneous apparatus

- PTs, CTs, variacs, dimmerstats and auto-transformers for supplying variable AC voltage
- Rheostats, decade resistance (or inductance/capacitance) boxes, (4-terminal) standard resistors, fixed or variable inductance and/or capacitance
- Various contact keys, switches etc.
- A variety of wires, leads and cables

D. Machines, equipment and device to be tested or experimented on; for example, in a "machines" lab it could be a motor, dynamo, alternator or a transformer whereas in the "measurements" lab testing is to be carried out on a CT/PT or an energy meter which is to be tested or calibrated.

E. "Load" or loading devices: 1-phase or 3-phase

   - "Lamp" load or resistors

   - Bank of inductors and/or capacitors

   - "Phantom" loads, comprising a phase shifter and special step down transformers

Desk with Panel in the Lab

A desk in the lab is usually provided with a normally "upright" panel at the back on which various power supplies are available, viz. 1-phase and 3-phase (with neutral) AC supply as well as DC supply (at 230 V and 6 or 12 V), in addition to starters and (knife) switches etc.

In a typical "machines" lab, the various terminals pertaining to a particular machine (for example a DC series, shunt or compound motor, or a 3-phase induction/synchronous motor or an alternator etc.) are also brought out on the panel. The height of the desk is generally kept at, say, 75 cm above the floor for ease of making connections and observations whilst *standing*. Necessarily, the readings of instruments should be taken looking directly down the scale, avoiding any error due to parallax. The table/desk where experiments using 3-phase supply at 400 V (L-L) are to be performed, a rubber mat is generally placed in front of the table as a safety measure and, of course, the supply and connections must be handled very carefully.

Organising the Experiment

The following simple steps will be helpful:

A. Making Connections

   (a) Use of wires of proper size

      - For current-carrying circuits, usually the "load" circuits, wires should be chosen to carry the maximum value of current during the experiment, safely and without overheating;

      - For "parallel" connections such as for voltmeters, pressure coils of wattmeters etc. wires of small cross section, even PVC-insulated flexible wires would suffice.

   (b) Colour code

      To avoid any confusion and maintain a 'semblance of order', a "color-code" for wires should be followed for making connections, esp. when dealing with 3-phase supplies and connections.

For example

- for DC supply and connections, use   red wire for positive terminals and green or black for negative;

- for 1-phase AC circuits, use red or yellow colour wire for phase or line terminals and black for the neutral;

- when using 3-phase supply, it is important to maintain identity of the three individual phases and the neutral by essentially using red, yellow and blue wires for the (three) phases and black or brown for the neutral whilst any earth or ground terminals is to be connected (to) using green wires.

(c)   Terminals on the panels/instruments

A common negligence on the part of students is to "hook" the wire *somehow* around the terminal stud in an arbitrary manner which may easily loosen up during the experiment. The correct way to hook the wire is shown in Fig.1 for the generally used terminals with a stud having right-handed threads and a (butterfly) type nut. Thus, when used in the manner shown the rotation of the nut clockwise will ensure good tightening of the wire. Also, it is necessary to avoid more than two connections at a single terminal, as far as possible[1].

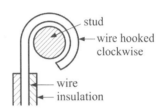

Fig.1: Properly-hooked wire end

B.   Arranging Apparatus

When arranging various apparatus on and around the desk like power-supply control devises – variacs, dimmerstats and regulators – and the instruments before making connections, it is better to

- position the 1-phase and/or 3-phase variacs or the phase-shifter etc. to the *left* of the main desk, using a small side table if required;

- arrange the measuring instruments, such as ammeters, voltmeters and wattmeters on the main desk or table in a proper order;

---

[1]In well-maintained laboratories, leads of different lengths, esp. for "series" or current-circuit connections are made available, fitted with proper lugs at both ends for easy insertion into the terminal studs, to avoid the 'problem' in question.

for example, ammeter(s) to the extreme left, followed by voltmeter(s), then wattmeters, and so on to facilitate reading them in the proper sequence and in a 'logical' manner;

- position the "primary" instruments – ammeters, voltmeters, wattmeters etc. which are generally the indicating type – in the *front*, that is, close to the front edge of the table as far as possible so that these can be read at ease from directly above the instrument, without too much bending over and avoiding parallax error[1];

- arrange the devices to be tested/experimented on, such as the machine or the transformer, or a lamp load or other to the right side of the table – again on a side table, if need be – and make connections *to* these, using leads of appropriate lengths [see Fig.2];

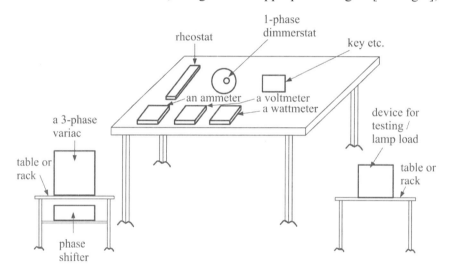

Fig.2 : Arranging apparatus for the experiment

- organise the decade resistance boxes and galvanometer etc on the table to match the 'actual' configuration when performing an experiment such as the Wheatstone bridge to provide clarity of observations as shown in Fig.3.

---

[1]When connecting 3-phase meters, the same should be positioned in the order of their phase sequence: RED followed by YELLOW, followed by BLUE etc. to avoid any confusion during observations.

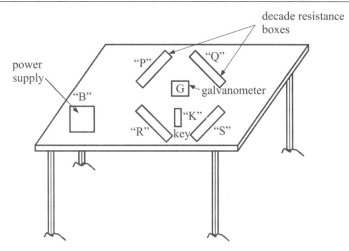

Fig.3 : Arrangement of apparatus for Wheatstone-bridge experiment

C. "Series" and "Parallel" Connections

An often observed tendency of the students is to connect up wires rather haphazardly, without logic of thought, so that it becomes difficult to "check the connections". A process that can help in the matter is to follow a "series and parallel" scheme of carrying out connections as follows:

(a) in experiments where connections involve an ammeter, a voltmeter and a wattmeter – and there are many such experiments – it is bettor to first make the "series" connections, that is, starting from the phase or line terminal of the supply (or positive terminal in the case of DC), connect to the ammeter, followed by current coils of the wattmeter, p f meter and energy meter (if there is one) and the load, if any, and to the other end of the supply (neutral or negative terminal)[1];

(b) the "parallel" connections should be made next which require the line and neutral terminals of the supply (or the positive and negative in the case of DC) rather independently, to be connected to appropriate terminals of instruments like a voltmeter, a frequency meter and pressure coil(s) of wattmeter(s), power factor meter(s), energy meter, all of these connected in *parallel* in the circuit. This is illustrated in Fig.4.

---

[1]It is implied that in the case of experiments involving 3-phase supply and apparatus, there may be three such series circuits, using wire of appropriate colours.

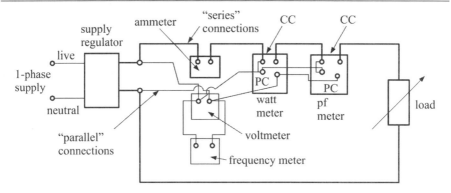

Fig.4 : Typical "series"-"parallel" connections in an experiment
[similar arrangement to follow for 3-phase circuits and connections]

Also, when measuring power using the 2-wattmeter method in a 3-phase circuit/load, correct order of connecting current coils v/s pressure coils must be maintained. [See also Chapter VII: Measurement of Power].

D. Records and Repeatability of Results

In all experiments where measured values and observations are to be recorded, an important requirement is the *repeatability of tests* and experiments if the need arose. For this, a habit which is mostly lacking on the part of student is to invariably make a careful note of *full* specifications of the various instruments, indicating even the serial numbers if possible, used during experimentation. Generally, all instruments have their specifications inscribed on the dial, including the serial number, the type, accuracy and even the "make" (as well as the indication that the instrument should be used in horizontal position). It is a good practice to take down all these details after the connections are made and *before* recording observations, and to employ the *same* set of instruments later when it is required to repeat the experiment.

Bibliography

A. Books

1. Baldwin, C. T.   : Fundamentals of Electrical Measurements
                             George G. Harrap & Co. Ltd., London. 1961

2. Bewley, L. V.   : Two-Dimensional Fields in
                             Electrical Engineering
                             Dover Publications, Inc., New York, 1963

3. Bozorth, Richard M.   : Ferromagnetism
                             Wiley – IEEE Press, 1993
                             [First Published : Bell Lab Series, 1951]

4. Buckingham, H. and   : Principles of Electrical Measurements
    Price, E. M.   English University Press Ltd., London, 1955

5. Cotton, H.   : Applied Electricity
                             Cleaver-Hume Press Ltd., London, 1951

6. Golding,   : Electrical Measurements and
    Edward William   Measuring Instruments
                             Sir Isaac Pitman & Sons, London, 1949

7. Hague, B.   : Alternating Current Bridge Methods
                             Sir Isaac Pitman & Sons, 1971

8. Harris, Forest K.   : Electrical Measurements
                             Wiley, New York, 1952

9. Kimbark, E. W.   : Direct Current Transmission – Vol. I
                             Wiley Inter Science, New York, 1971

10. Laithwaite, Eric R. : Induction Machines for Special Purposes
                             Nrewnes, London, 1966

11. Nelkon, M. and   : Advanced Level Physics
     Parker, P.   Arnold Publishers, 1987

12. Pramanik, Ashutosh   : Electromagnetism : Theory and Applications
                             Prentice-Hall of India, New Delhi, 2003

13. Stevenson (Jr.), W. D. : Elements of Power System Analysis
                             McGraw Hill Book Co., Inc., New York, 1962

14. Sunil S. Rao   : EHV-AC and HVDC Transmission &
                             Distribution
                             Khanna Publishers, New Delhi, 1984

15. Taylor, E. O.   : Utilisation of Electric Energy
                              Orient Longman Pvt. Ltd., 1971

16. Weber, Ernst   : Electromagnetic Theory
                              Dover Publications Inc., New York, 1965

B. Publications/Technical Papers

  I By the Author

    1. The Salient-pole Hysteresis Coupling
        IEEE Transactions on Magnetics, Vol. MAG-11, No. 5, 1975, pp
        1461-63
    2. The Use of Search Coils for Magnetic Measurements
        Int. J. Elect. Eng. Edn., Vol. 19, 1982
    3. Energy Flow in a 2-pole Hysteresis Coupling by Poynting Theorem
        IEE Proceedings, Vol.130, Pt.A, No. 6, 1983, pp 301-05
    4. Influence of Unbalanced Currents in Turbogenerators and the
        Analysis of the Associated Problems by Finite Element Method
        B H E L Journal, Vol. 4, 1983
    5. Negative-Sequence Currents, Losses and Temperature Rise in the
        Rotor of a Turbogenerator During Transient Unbalanced
        Operation
        Electric Machines and Power Systems, Vol. 8, 1983
    6. Measurement of Residual Flux in Permanent Magnets
        Int. J. Elect. Eng. Edn., Vol. 20, 1983
    7. Mapping of Magnetic Field due to DC-Excited Unsaturated
        Salient-Pole Field System
        J. Inst. Engrs.(India), Vol.64, Pt.EL-4, 1984
    8. Magnetic Suspension Bearings for AC Energy Meters
        B H E L Journal, Vol.6, 1985
    9. A New Technique for Measurement of B/H Characteristics of
        Annular Permanent Magnets
        Int. J. Elect. Eng. Edn., Vol.22, 1985

  II. Others

    1. Booth, J. : A Short History of Blood Pressure Measurement
        Proceedings of the Royal Society of Medicine, Vol. 70, 1977
    2. Fortescue, C. L. : Method of Symmetrical Coordinates Applied to
        the Solution of Poly-phase Networks
        Trans. AIEE, Vol.37, 1918, pp 1027-1140

S C BHARGAVA

Non-synchronous Operation of a Hysteresis Machine
Ph. D. Thesis, University of Aston in Birmingham, U K, 1972

Index

A

AC bridge(s)
- Anderson's 583-585, 807, 834
- De Sauty's 136, 561, 591, 615
- Hay's 19, 580
- Heaviside-Campbell's 586
- Maxwell's 577, 579, 613, 669
- Owen's 585, 613, 615
- Schering 595, 597, 599
- Wien's 593, 598, 615

A/D converter(s) 395, 453, 454, 627, 783,796, 862

ALNICO 45, 427, 446, 678, 689, 690

alpha waves 793

aluminium 378, 572, 573
- cover 712
- cup 713
- diaphragm 716, 720, 748-750
- disc 426, 427, 437, 429-31, 433, 439
- former 216, 220
- pointer 221
- resistivity of 218
- vane 214

amber 42

ammeter-voltmeter method 575, 590

Ampere
- 's formula 59, 817

amplifier(s)
- bandwidth 187
- chopper-stabilised 856
- differential-input 188
- horizontal 179, 187, 673, 674
- instrumentation
  - DC 833
- operational 188, 857, 860
- vertical 179, 186, 673, 674

Anderson
- 's bridge 583, 807, 834

aneroid gauge 779

anisotropic 654

antimony 730

anti-resonance 139

Argand diagram 119, 490

Aristotle 41

arsenide
- indium 845, 846

astatic 541

atomic clock 23

attenuation 299

ausculatory
- method 779

Ayrton-Perry 571
- inductometer 571
- winding 26, 571

B

B-H
- characteristics 680, 831
- curve 620, 655, 679, 682
- loops 652, 658-660, 673, 678, 682
  - recoil 678

Bakelite 614

Barlow, H.M. 391

bearings
- bush 428
- jewel 204, 249, 429
- magnetic(suspension) 439
- pivot 424, 425

Bell, Alexander Grahm 753

belows 719

Bequerel, Edmund 770

Bernoulli 737
- 's principle 737

beta waves 793

bifilar 25, 26
- coil 81
- winding 568

Biot and Savart 59
- law 59, 60, 817

Notes

Notes

Notes